TAURINE IN NUTRITION AND NEUROLOGY

ADVANCES IN EXPERIMENTAL MEDICINE AND BIOLOGY

Recent Volumes in this Series

Volume 134
HAMSTER IMMUNE RESPONSES IN INFECTIOUS AND ONCOLOGIC DISEASES
Edited by J. Wayne Streilein, David A. Hart, Joan Stein-Streilein, William R. Duncan, and Rupert E. Billingham

Volume 135
DIET AND RESISTANCE TO DISEASE
Edited by Marshall Phillips and Albert Baetz

Volume 136
BIOLOGICAL REACTIVE INTERMEDIATES II: Chemical Mechanisms and Biological Effects
Edited by Robert Snyder, Dennis V. Parke, James J. Kocsis, David J. Jollow, G. Gordon Gibson, and Charlotte M. Witmer

Volume 137
THE RUMINANT IMMUNE SYSTEM
Edited by John E. Butler

Volume 138
HORMONES AND CANCER
Edited by Wendell W. Leavitt

Volume 139
TAURINE IN NUTRITION AND NEUROLOGY
Edited by Ryan Huxtable and Herminia Pasantes-Morales

Volume 140
COMPOSITION AND FUNCTION OF CELL MEMBRANES: Application to the Pathophysiology of Muscle Diseases
Edited by Stewart Wolf and Allen K. Murray

Volume 141
BIOCHEMISTRY AND FUNCTION OF PHAGOCYTES
Edited by F. Rossi and P. Patriarca

Volume 142
BIOCHEMISTRY AND BIOLOGY OF CORONAVIRUSES
Edited by V. ter Meulen, S. Siddell, and H. Wege

Volume 143
RELAXIN
Edited by Ralph R. Anderson

TAURINE IN NUTRITION AND NEUROLOGY

Edited by

Ryan J. Huxtable

The University of Arizona
Tucson, Arizona

and

Herminia Pasantes-Morales

Universidad Nacional Autónoma de México
Mexico City, Mexico

PLENUM PRESS • NEW YORK AND LONDON

Library of Congress Cataloging in Publication Data

Main entry under title:

Taurine in nutrition and neurology.

(Advances in experimental medicine and biology ; v. 139)
Proceedings of a symposium held Nov. 17-19, 1980, at the Universidad Nacional Autónoma de México, Mexico City.
Includes bibliographies and index.
1. Taurine—Physiological effect—Congresses. 2. Taurine—Metabolism—Congresses. 3. Neurochemistry—Congresses. I. Huxtable, Ryan. II. Pasantes-Morales, Herminia. III. Series. [DNLM: 1. Taurine—Congresses W1 AD 559 v. 139 / QU 60 S994t 1980]
QP801.T3T383 612'.01575 81-15699
ISBN 978-1-4757-0404-4 ISBN 978-1-4757-0402-0 (eBook) AACR2
DOI 10.1007/978-1-4757-0402-0

Proceedings of a symposium on Taurine: Questions and Answers, held November 17–19, 1980, at the Universidad Nacional Autónoma de México, Mexico City, Mexico

Softcover reprint of the hardcover 1st edition 1982
A Division of Plenum Publishing Corporation
233 Spring Street, New York, N.Y. 10013

PREFACE

There is increasing recognition of the biological importance of taurine. Disturbances in taurine metabolism are seen in dysfunctional states as diverse as epilepsy, cardiomyopathy, Friedreich's ataxia and congestive heart failure, which combined with the antiepileptic and cardiotonic actions of taurine underscore the emerging clinical importance of this fascinating substance.

This is the third book to appear on taurine since 1976. The last book (Taurine and Neurological Disorders, Raven Press, 1978) emphasized the involvement of taurine in central nervous system phenomena. These actions are further explored in this volume. In addition, nutritional aspects of taurine, particularly as these relate to development, have been increasingly recognized to be important parts of the taurine story. The marked emphasis placed on nutrition by a number of contributors to this volume is recognition of the meteoritic growth of this area.

Are the multitudinous actions of taurine differential expressions of a universal mechanism? There is surprising unity of opinion expressed by the scientists contributing to this volume, despite their differences in viewpoint and background, as the reader, tantalized we hope by this remark, will discover.

This volume is the outcome of a Symposium, "Taurine: Questions and Answers" held at the Universidad Nacional Autónoma de México in November, 1980. An unusual aspect of the meeting was the devotion of half the time to open, relatively unstructured, discussions of various problems in taurine research. Transcripts of these discussions are included in this volume, and the editors hope that workers both well aquainted and peripheral to the area will find them of value.

Ryan J. Huxtable

Herminia Pasantes-Morales

ACKNOWLEDGMENTS

This symposium on which this book is based was made possible by the backing of the Centro de Investigaciones en Fisiología Celular, the Secretaria Ejecutiva del Consejo de Estudios de Posgrado and the Dirección General de Asuntos del Personal Académico, all of the Universidad Nacional Autónoma de México. The support and commitment of these units permitted the gathering of a distinguished group of scientists in the superb setting of the University Botanical Gardens.

We also thank the University of Arizona for generous support, which enabled the editing of this volume. We thank Dr. Kassander, Vice President of Research, for a generous gift, and the Administration of the College of Medicine for help with the multifarious expenses of typing and mailing.

The format of the meeting posed special problems with the recording and transcriptions of discussion. Furthermore, no one, without attempting to do it, can appreciate the difficulties of coordinating communication with thirty busy scientists strung around the world and combating the varying efficiencies of the postal and telephone services needed to communicate with these scientists. For all of this, and the retyping of the whole book, we are indebted to Rita Wedell. Without her help, there would be no book. We appreciate her dedication to every aspect of this meeting, including passing six days in Mexico City without the opportunity of sightseeing.

For the Editors, working together has been enriching, both scientifically and culturally. The international cooperation involved in an undertaking of this nature has been a pleasant, rewarding and educational experience. The United States of Mexico and America have much to offer each other, and we hope that in organizing this symposium we have taken one small step towards increasing scientific interchange with our colleagues across the Rio Grande.

Ryan J. Huxtable

Herminia Pasantes-Morales

CONTENTS

PART I

PHYSICOCHEMICAL PROPERTIES OF TAURINE

PART II

TAURINE IN NUTRITION AND DEVELOPMENT

PART III

TRANSPORT AND METABOLISM OF TAURINE

PART IV

TAURINE AND THE HEART

PART V

THE NEUROCHEMISTRY OF TAURINE

PART VI

THE NEUROPHARMACOLOGY OF TAURINE

PART I

PHYSICOCHEMICAL PROPERTIES OF TAURINE

Ryan J. Huxtable

Department of Pharmacology
Health Sciences Center
University of Arizona
Tucson, Arizona

The biological actions of a compound are an ineluctable consequence of its physical properties. However, basic physicochemical data on compounds of biological interest are often unavailable. Those interested in interpreting biological actions in terms of physical properties frequently do not have the background to make the latter measurements themselves, and those capable of making the measurements are frequently unaware of the importance of these data to biomedical science.

Taurine, $C_2H_7NO_3S$, is a simple substance of molecular weight 125.1. Some of its physical constants are shown in table 1, and compared to those of other neuro-active amino acids. The most noteworthy difference is the high acidity of the sulfonic acid function compared to the carboxylic acid function in, for example, β-alanine. The amino group of taurine is also more acidic (i.e. less basic) than those of related compounds.

Many of the biological effects of taurine, as is clear from this book, are mediated via an action on calcium. Direct complex formation between taurine and metal ions such as calcium were early proposed to be important in the biological actions of taurine. Although evidence against this has been available for some time, it has been slow to make itself felt, and theories proposing direct interactions of taurine and calcium are regularly put forward in the literature and at meetings. Dolara and his coworkers (2,3), using natural abundance ^{13}C NMR, were the first to attempt a quantitative estimate of calcium-taurine complexation under biological conditions. They found the stability constant to be low, but as a result of the high taurine to calcium ratio in the heart, they estimated that approxi-

TABLE 1

PHYSICAL CONSTANTS OF NEUROACTIVE AMINO ACIDS

	Solubility[a] (g/100 ml in water)		Ionization Constants[b]			Isoelectric Point
	25°	100°	pK_1	pK_2	pK_3	
TAU	10.48		1.5	8.82		5.16
GLY	24.99	67.17	2.34	9.60		5.97
ASP	0.5	6.9	1.88	3.65	9.60	2.77
GLU	0.86	14.0	2.16	4.32	9.96	3.24
β-ALA	freely sol.		3.55	10.24		6.89
GABA	freely sol.		4.03	10.56		7.30
ASN	2.98	55.1	2.02	8.80		5.41
GLN	3.6		2.17	9.13		5.65

[a]Data from Refs. 5,6, and 13.

[b]Data from Refs. 1,7, and 13.

mately 8% of the free calcium in the heart would be bound to taurine. In the following article, Irving and his coworkers (4) have refined this approach by using ^{13}C enriched taurine, and have calculated association constants even lower than those of Dolara et al. As a result, they estimate that slightly over 1% of the free calcium in the heart is associated with taurine. This eliminates a direct interaction as the mechanism of the marked calcium modulatory actions of taurine.

Another metal ion whose biological actions appear to be intertwined with those of taurine is zinc. To the best of my knowledge, quantitative studies of zinc-taurine complexes, with accurate measurements of stability constants, have not been done. Metal stability constants with amino acids and related compounds form an important data base for many aspects of the biological sciences. Extension of this data base is sorely needed.

Subtle structural features apparently determine other biological effects of taurine and related compounds apart from interactions with metal ions. Why do taurine, hypotaurine and β-alanine have affinity for the taurine transport site, whereas the closely related compounds thiotaurine, aminoethylphosphonate and 1,2-dimethyltaurine have no affinity? Why, on the other hand, is nipecotic acid an effective inhibitor of both taurine (12) and GABA transport (11)? To approach an answer to questions like this, detailed structural information is needed such as that available from an x-ray crystallographic determination. X-Ray crystallography has been performed for taurine (10), β-alanine (8) and 2-aminoethylphosphonate (9). However, there are numerous other compounds of biological interest related to these for which crystallographic structures have yet to be done.

Metal stability constants and bond lengths and bond angles - these are but a few of the physicochemical parameters needed in biology. Cooperation between biomedical reseachers and scientists capable of performing physicochemical measurements will be of benefit to both groups and will pave the way for the more rapid solution of some of the problems posed by small molecular weight compounds such as taurine.

REFERENCES

1. Bender, D.A., Amino Acid Metabolism, John Wiley (1975).
2. Dolara, P., Franconi, F., Giotti, A., Basosi, R., and Valensin, G., 1978, Taurine-calcium interaction measured by means of ^{13}C nuclear magnetic resonance, Biochem. Pharmacol., 27:803-804.
3. Dolara, P., Ledda, F., Mugelli, A., Mantelli, L., Zilletti, L., Franconi, F., and Giotti, A., Effect of taurine on calcium, inotropism and electrical activity of the heart, in: "Taurine and Neurological Disorders," A. Barbeau and R. J. Huxtable, eds., Raven Press, New York (1978), pp151-159.
4. Irving, C.S., Hammer, B.E., Danyluk, S.S., and Klein, P.D., Co-ordination and binding of taurine as determined by nuclear magnetic resonance measurements on ^{13}C-labeled taurine, this volume.
5. Jacobsen, J.J., and Smith, L.H., 1968, Biochemistry and physiology of taurine and taurine derivatives, Physiol. Reviews, 48:424-511.
6. Jakubke, H.D., and Jeschkeit, H., Amino Acids, Peptides and Proteins, John Wiley (1977).
7. Johnston, G.A.R., Biochemistry of glycine, taurine, glutamate and aspartate, in: "Handbook of Psychopharmacology Vol. 4. Amino Acid Neurotransmitters," L.L. Iversen, S.D. Iversen and S.H. Snyder, Plenum Press (1975), pp59-82.
8. Jose, P. and Pant, L.M., 1965, The crystal and molecular structure of β-alanine, Acta Cryst., 18:806-810.
9. Okaya, Y., 1966, Crystal structure of the stable modification of 2-aminoethyl phosphonic acid, β-ciliatine, Acta Cryst., 20:712-715.
10. Okaya, Y., 1966, Refinement of the crystal structure of taurine, 2-aminoethyl phosphonic acid. An example of computer-controlled experimentation, Acta Cryst., 21:726-735.
11. Schousboe, A., Effects of GABA-analogs on the high affinity

uptake of GABA in astrocytes in primary cultures, *in*: "Advances in Exp. Med. and Biol. Vol. 123, GABA-Biochemistry and CNS functions," P. Mandel and F.V. DeFeudis, eds., Plenum Press (1979), pp219-237.
12. Sebring, L., and Huxtable, R.J., unpublished data.
13. Windholz, M., Budavari, S., Stroumtsos, L.Y., and Fertig, M.N., The Merck Index: An Encyclopedia of Chemicals and Drugs, 9th Edition, Merck (1976).

COORDINATION AND BINDING OF TAURINE AS DETERMINED BY NUCLEAR MAGNETIC RESONANCE MEASUREMENTS ON ^{13}C-LABELED TAURINE

Charles S. Irving[1], Bruce E. Hammer[2], Steven S. Danyluk[2] and Peter D. Klein[1]

[1]Children's Nutrition Research Center
Department of Pediatrics
Baylor College of Medicine
Houston, Texas
[2]Division of Biological and Medical Research
Argonne National Laboratory
Argonne, Illinois

A number of pharmacological actions of taurine (5,8,9,16,17,19, 21,30,31) on central nervous and peripheral tissues have suggested that taurine might modulate cation flux (15) by either direct (10, 13) or indirect interactions (4) with calcium. The ability of taurine to influence calcium flux by direct coordination with calcium depends on the fraction of free intracellular calcium which can be trapped by taurine. The fraction of calcium that exists as a taurine complex can be estimated readily from intracellular taurine, calcium and hydrogen ion levels, and the formation constants of taurine-calcium complexes. The latter formation constants can be determined *in vitro* and applied to *in vivo* conditions. Dolara and coworkers (11,12) using calcium-taurine formation constants estimated from natural abundance ^{13}C chemical shift titration curves, have calculated that approximately 8% of the total calcium in mammalian myocardium exists in taurine complexes. The inherent difficulties associated with making very precise measurements on noisy, low sensitivity natural abundance ^{13}C NMR spectra suggested that the determination of calcium-taurine formation constants be repeated using ^{13}C enriched taurine. We have determined ^{13}C chemical shift titration curves of the calcium-taurine system using ^{13}C-taurine and have obtained new formation constants for calcium-taurine complexes by rigorous complex equilibrium analysis of the titration curves. These formation constants predict that only neglible amounts of calcium are bound directly to taurine at the taurine and calcium levels found in myocardial cells. Additional ^{13}C NMR measurements of taurine in solution have revealed some unexpected electronic properties of the molecule, which may be important to its biological function.

MATERIALS AND METHODS

^{13}C-Taurine

$^{13}C_1$-Taurine (93% atom %^{13}C), consisting of an equimolar mixture of taurine-[1 ^{13}C] and taurine-[2 ^{13}C], was obtained from the Stable Isotope Resource Laboratory at Los Alamos, New Mexico. The ^{13}C-taurine sample moved as a single spot (ninhydrin spray) on cellulose thin-layer plates developed with n-butanol-acetic acid-water (60:15:25) and was used without further purification.

Sample Preparation

For titration measurements above pH 2, 2 ml aqueous solutions (triple distilled water) of 0.75 M $^{13}C_1$-taurine and either 0.75 M calcium chloride or 0.75 M sodium chloride (analytical grades) were prepared and their pH sequentially adjusted by addition of concentrated HCl or NaOH at 40°C. For titration measurements below pH 2, 0.75 mmole $^{13}C_1$-taurine and 0.75 mmole of either calcium chloride or sodium chloride were dissolved in 2 ml of 0.1, 0.5, 1.0, 2.0, 4.0, 6.0, 8.0 and 10.0 N HCl. The titration solutions were not degassed. All glassware and NMR tubes were soaked in concentrated ethylenediaminetetraacetic acid and rinsed with triple distilled water before use.

NMR Measurements

^{13}C NMR measurements were made at 15.089 MHz on a Nicolet-TT14 multinuclear Fourier transform spectrometer equipped with a MONA probe. Samples were contained in a 10 mm tube inserted concentrically into an 18 mm tube containing 8% dioxane in D_2O and maintained at 40° ± 1°C by a stream of nitrogen. In titration measurements, 4 free induction decays following 37° pulses were collected in 16K data points over 8.19 sec and Fourier transformed to yield proton noise-decoupled (10 W) spectra with a 1000 Hz sweep width and 0.122 Hz (0.0081 ppm) digital resolution. The dioxane resonance (67.53 ppm downfield from TMS) provided an external reference. For each sample, chemical shifts were determined repeatedly until reproducible to within digital resolution, which occurred when the sample was in thermal equilibrium. No corrections were made for possible susceptibility changes as a function of pH or Ca^{++} concentration.

RESULTS AND DISCUSSION

The pH Titration Curve and Coordination

The coordination of a protic ligand to a non-paramagnetic metal ion often can be detected and measured by changes introduced into the pH chemical shift titration curve of the ligand (29). Deprotonation of a carboxylic acid group, a sulfonic acid group, or an amino group releases electron density into the orbitals of the ligand, which affects the chemical shifts of the ligand nuclei. The changes in chemical shift resulting from changes in pH comprise the chemical shift titration curve. Complexation of a positively charge metal ion to the deprotonated group leads to the partial return of electron density to the deprotonated group. This results in a chemical shift intermediate between that of the protonated and deprotonated forms of the ligand. The formation constants of the ligand-metal complex can be obtained directly from the titration curves in an empirical fashion when the titration curves of ligand in the presence and absence of metal ion are very different and the point of crossover of the two curves can be measured accurately, as shown in Fig. 1. Rabenstein (29) has shown that for simple 1:1 metal-ligand complexes the fraction of ligand bound to metal ion, P_{ML}, is given by:

$$P_{ML} = \frac{\nu_{obs} - \nu_L}{\nu_{ML} - \nu_L} \qquad (1)$$

where ν_{obs} and ν_L are the chemical shifts of the ligand-metal ion system and the ligand at a particular pH, and ν_{ML} is the chemical shift of ligand-metal ion complex. The value of ν_{ML} corresponds to the chemical shift at which the two titration curves cross over each other. The concentration of the metal ligand complex is given by:

$$[ML] = P_{ML}\,[L]_{total} \qquad (2)$$

Taurine ^{13}C pH Titration Curves

The proton decoupled ^{13}C NMR spectrum of ^{13}C-taurine (0.75M)

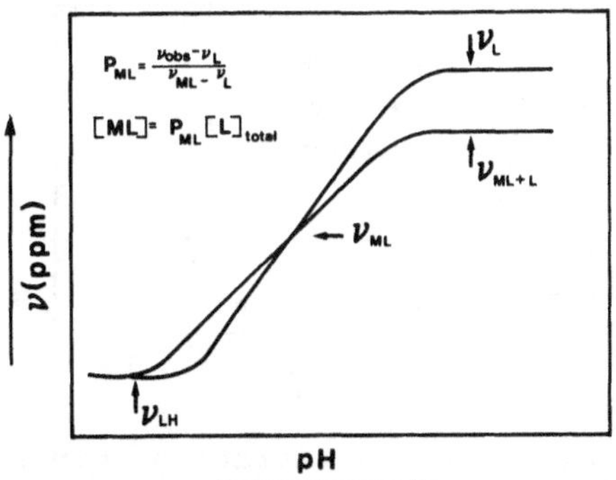

Fig. 1. Schematic representation of a method (29) for determining metal ion ligand formation constants from titration data. See text for explanation of symbols.

at pH 6.6 displays two resonances at 48.59 and 36.73 ppm originating from taurine-[1-^{13}C] and taurine-[2-^{13}C] which are in good agreement with those obtained in natural abundance ^{13}C NMR measurements by Norton (25), who assigned them to the S-methylene (C_1) and N-methylene (C_2) carbons, respectively. This assignment is consistent with recently reported ^{13}C chemical shifts of sulfonic acids (23). The 36.73 ppm resonance displays a broader linewidth which is consistent with unresolved ^{13}C-^{14}N scalar interactions.

The pH titration curves of the S-methylene and N-methylene carbons of ^{13}C-taurine (0.75M, 40°C, 0.75M NaCl) are shown in Fig. 2. The titration was carried out in the presence of 0.75M NaCl to facilitate comparison with the taurine-calcium chloride system. The pH titration curves of $^{13}C_1$-taurine (0.75M, 40°C) in the presence of 0.75M $CaCl_2$ are shown in Fig. 3, along with the regression line of the free taurine curves. The presence of calcium produced a downfield shift of the N-methylene and S-methylene resonances of zwitterionic taurine and a shift of the apparent pK_b of taurine toward lower pH.

The effect of calcium on the taurine titration curve apparently resulted from calcium-taurine complexation rather than from secondary effects. Dissociation of possible taurine zwitterionic dimers by calcium would be expected to produce shifts in the opposite direction

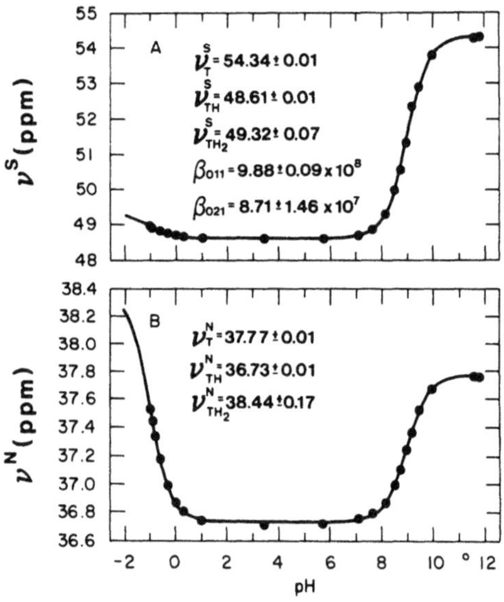

Fig. 2. The ^{13}C chemical shift pH titration curves of the S-methylene carbon (A) and the N-methylene carbon (B) of $^{13}C_1$-taurine (0.75M, 40°C, 0.75 NaCl). The solid line represents the titration curve calculated using the dissociation constants and chemical shifts shown. All symbols are the same as those given in the text, except T=Tau.

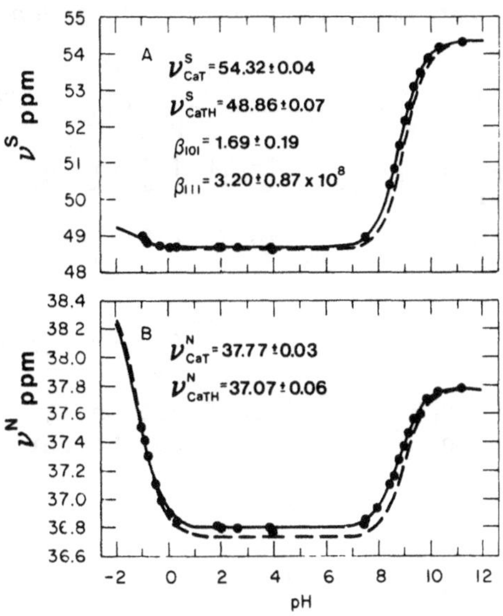

Fig. 3. The ^{13}C chemical shift pH titration curves of the S-methylene (A) and N-methylene (B) carbons of $^{13}C_1$-taurine (0.75M, 40°C) in the presence of $CaCl_2$ (0.75M). The solid line represents the titration curve calculated using the dissociation constants, stability constants, and chemical shifts shown. All symbols are defined in the text, except T=Tau.

from those observed. The complex combination of media effects on chemical shifts pK_a and pK_b values required to explain the effect of calcium on the taurine titration curve makes this an unlikely explanation.

Although the calcium-taurine titration curves appeared to cross-over their free taurine counterparts in the low pH region, the exact point of crossover could not be estimated accurately by visual inspection. No apparent crossover point was observed in the high pH region, where stronger complexation was expected to occur. This, together with the possible presence of more than a single type of calcium-taurine complex at each pH, precluded the use of empirical methods (Fig. 1) for estimating the formation constants of metal-ligand complexes.

Complex Equilibria Analysis of Titration Curves

The object of complex equilibrium analysis is to derive the formation constants (β) of calcium-taurine complexes from observed chemical shift data (ν^N and ν^S) at various hydrogen ion concentrations [H] given by the pH, and the total concentrations of calcium and taurine, $[Ca]_{total}$ and $[Tau]_{total}$. This requires a simple nomenclature (3,26) for describing the species present in the taurine and taurine : calcium system, which are given by equation 3 and 4.

$$pCa^{+} + qH^{+} + rTau^{-} \rightleftharpoons Ca_pH_qTau_r^{(2p+q-yr)+} \qquad (3)$$

$$\beta_{pqr} = [Ca_p\ H_q\ Tau_r] \ /\ [Ca]^p\ [H]^q\ [Tau]^r \qquad (4)$$

The observed chemical shift, ν^i_{obs} (i= N or S), at a particular pH can now be described by equation 5

$$\nu^i_{obs} = \sum_p \sum_q \sum_r \nu^i_{pqr}[Ca_pH_qTau_r] \ /\ [Tau]_{total} \qquad (5)$$

In order to calculate $[Ca_pH_qTau_r]$ from Equation 4, [Tau] and [M] at equilibrium must be known. These values can be derived from the mass balance equations and are

$$[M] = \frac{[M]_{total}}{1+\Sigma\ \beta_{1q1}\ [H]^q\ [L]} \qquad (6)$$

$$[L] = \frac{-[B + (E-A)(C+D)]+\sqrt{[B + (E-A)(C+D)]^2 + 4AB(C+D)}}{2\ B\ (C+D)} \qquad (7)$$

where $A = [Tau]_{total}$, $B = 1+\beta_{011}\ [H] + \beta_{021}\ [H]^2$, $C = \beta_{101}$, $D = \beta_{111}\ [H]$, and $E = [Ca]_{total}$. Non-linear least squares regression analysis routines, such as the SAS NLIN routine (33) can be used to obtain estimates of β_{pqr} and ν^i_{pqr} by fitting the calculated values of ν^i obtained from Equations 4,5,6, and 7 to ν^i_{obs} at various pH for fixed values of $[Ca]_{total}$ and $[Tau]_{total}$.

Taurine Dissociation Constants

Accurate estimates (Table 1) of taurine zwitterion and cation formation constants (β_{011} and β_{021}, respectively) and ^{13}C chemical shifts of the taurine anion, zwitterion, and cation (ν_{001}, ν_{011}, and ν_{021}, respectively) were obtained by simultaneous least squares regression analysis (33) of the S-methylene and N-methylene ^{13}C chemical shift titration curves of $^{13}C_1$-taurine. Good convergence was obtained below pH 1.0 when [H] was obtained directly from HCl concentrations, but not when Hammet-Deyrup H_0 acidity values for HCl were used. Remarkably small standard deviations (Table 1) were obtained for the chemical shifts and formation constant of the taurine cation, even though the titration curve could not be followed beyond the mid-point of protonation of the sulfonic acid group. The estimated taurine pK_a ($-\log(\beta_{011}/\beta_{021})$) of -1.05 is considerably lower than the reported values of 1.5 and -0.3, (1,7,14) but is consistent with the reported failure to observe taurine protonation down to pH 0 (25). The pK_b ($-\log(\beta_{011})$) of 8.99 is in good agreement with reported values (1,6,7,14,18).

Taurine-Calcium Complex Formation Constants

TABLE 1. Stability Constants and ^{13}C Chemical Shifts for the Calcium-Taurine System

p	q	r	log (β_{pqr})	ν^S	ν^N
0	0	1	———	54.34 ± 0.02	37.77 ± 0.02
0	1	1	8.995 ± 0.008[a]	48.61 ± 0.02	36.73 ± 0.01
0	2	1	7.915 ± 0.150	49.32 ± 0.15	38.43 ± 0.30
1	0	1	0.217 ± 0.099	54.32 ± 0.07	37.77 ± 0.06
1	1	1	8.426 ± 0.268	48.86 ± 0.13	37.08 ± 0.15

[a] All errors represent the asymptotic 95% confidence interval.

Having the values of β_{011}, β_{021}, ν^i_{001}, ν^i_{011}, ν^i_{021} of free taurine, it was possible to use simultaneous non-linear least squares regression analysis of the N-methylene and S-methylene ^{13}C chemical shift titration curves of taurine (0.75M) in the presence of Ca^{2+} (0.75M) to obtain the formation constants and ^{13}C chemical shifts of taurine-calcium complexes. Convergence was not obtained in the regression analysis when only one type of calcium-taurine complex (CaTau or CaTauH) was accounted for. Convergence was obtained by taking into account both CaTau and CaTauH type complexes, but did not require the inclusion of $Ca(Tau)_2$ type complexes. This is in agreement with solubility studies (14,29), that showed that amino acids formed only 1:1 complexes with Mn, Mg, Ca, and Ba.

The estimated values of β_{101}, ν^i_{101}, β_{111}, ν^i_{111} for the two 1:1 calcium taurine complexes are given in Table 1. The formation constants ($K_{CaTau} = \beta_{011} = [CaTau] / [Ca] [Tau]$ and $K_{CaTauH} = \beta_{111}/\beta_{011} = [CaTauH]/[Ca][TauH]$) of 1.62 and 0.32 obtained for the CaTau and CaTauH complexes, respectively, are significantly smaller than the previously reported values of 2.23 and 5.02 (13).

The literature (14) contains much indirect evidence pointing to the high degree of instability of calcium-taurine complexes. Albert (1) showed that β-alanine transition metal-ion complexes were considerably more unstable than α-amino acid complexes and that taurine complexes were so unstable that they could not be measured accurately. Compared to other metal ions, calcium-amino acid complexes are among the least stable (14).

Estimation of Cellular Levels of Taurine Bound Calcium

The formation constants, β_{pqr}, given in Table 1, and equations

4,6, and 7 can be used to determine how the fraction of calcium bound to taurine varies as a function of pH and taurine and calcium concentrations. The change in the distribution of calcium complexes ([complex] / $[Ca]_{total}$) as a function pH is shown in Fig. 4 for the levels of taurine and calcium in the myocardial cell. Below pH 7.0 the CaTauH complex is the dominant species. Between pH 0 and 7 the fraction of bound calcium is constant. Above pH 7.0 a small increase in pH produces a marked increase in the fraction of bound calcium.

In order to determine the actual fraction of calcium bound to taurine in the myocardial cell, it is necessary to consider the amounts of total intracellular taurine and calcium available for complexation. Since calcium forms only 1:1 complexes, the fraction of bound calcium will be independent of $[Ca]_{total}$ and proportional to $[Tau]_{total}$ below 100mM Tau. This makes it possible to estimate the fraction of free calcium that could be coordinated to taurine, without knowing how much of the total cellular Ca exists in a non-bound state. The fraction of calcium (not bound to other cellular constituents) that exists as CaTau and CaTauH is approximately equal to 0.3369 $[Tau]_{total}$. In the myocardial cell, where $[Tau]_{total}$ = 0.039M (13), the fraction of calcium coordinated to taurine is only 0.014. Taurine binding to other cellular constituents would lower the amount of taurine available for calcium binding and further reduce the fraction of taurine-bound calcium. We conclude that direct taurine-calcium complexation is not sufficient to explain adequately the effect of calcium on taurine flux. The absence of significant amounts of 1:2 calcium taurine complexes in the titration mixtures appears to rule out the formation of $Ca(Tau)_2$ complexes that could serve as neutral calcium carriers. Direct taurine-calcium-macromolecule complexes also would be very stable.

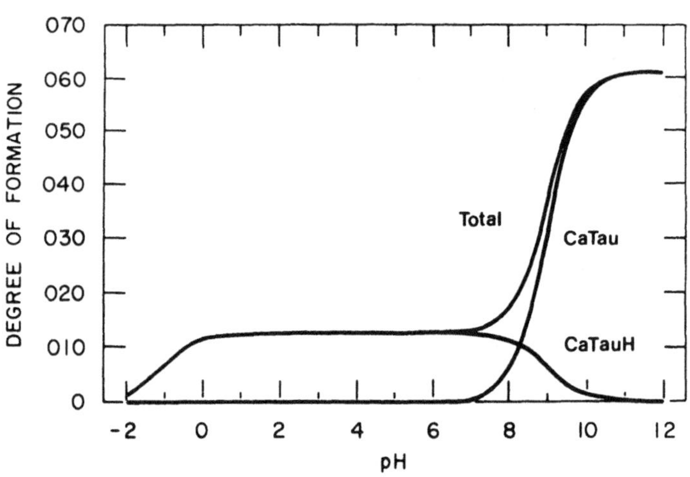

Fig. 4. The calcium species distribution curve for 39mM taurine at 40°C.

Electronic Properties of Taurine and Calcium-Taurine Complexes

Insights into the electronic properties of taurine can be obtained from ^{13}C titration shifts since ^{13}C chemical shifts are determined by charge density, bond order, the energies of low lying electronic excited states, (22) and electric field effects (2). Comparison of the ^{13}C titration shifts of taurine (Fig. 5a) with those of ethanesulfonic acid, β-alanine, propionic acid, and ethylamine (Figs. 5b,c,d, and e, respectively) reveals some unusual electronic properties of taurine. Deprotonation of the taurine sulfonic acid group shifts both the N-methylene and S-methylene ^{13}C resonances *upfield*, not *downfield* as is the case with carboxylic acids, α-amino acids, and β-alanine. The upfield shift of both taurine resonances requires the presence of the amino group, since it is not found in alkyl sulfonic acids (23), in which only the C_1 methylene carbon shifts upfield while the other methylene groups shifts downfield. The taurine sulfonic acid titration shifts cannot be explained by electric field effects (2), since the shifts in β-alanine occur in the opposite direction (28). The unique sulfonic acid titration shift of taurine most likely is related to the electronic properties of the sulfonic acid group which, compared to the carboxylic acid group, is larger, more symmetric, and less affected by deprotonation (28). The conclusion, therefore, with respect to interactions involving the acidic groups, is that taurine and β-alanine cannot be considered analogues. The ^{13}C titration shifts associated with deprotonation of the amino group are comparable for taurine, β-alanine, and ethylamine (Fig. 5). Interactions involving the amino group of taurine should not differ significantly from those of other amino groups.

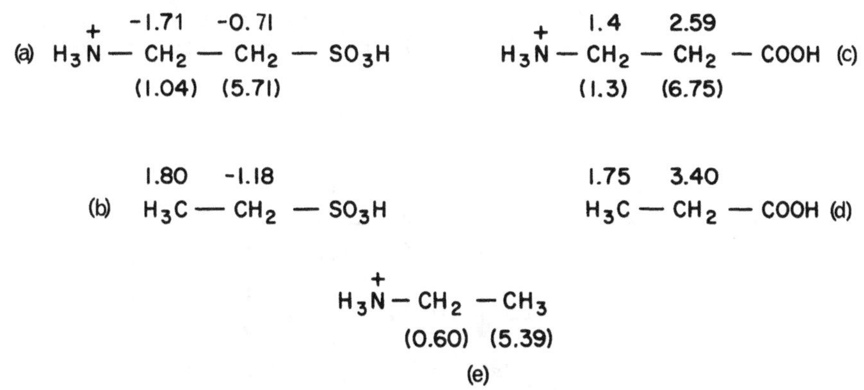

Fig. 5. Titration shifts of compounds related to taurine. The values with and without parentheses refer to downfield chemical shifts upon deprotonation of the amino and acid groups, respectively, that were obtained from this study and Refs. 23 and 28.

Some interesting insights into the structure of the calcium-taurine complex are provided by their estimated ^{13}C chemical shifts $(\nu_1^i p_1)$. As seen in Fig. 6, formation of the CaTauH complex from TauH produces shifts which qualitatively are similar but less intense than the addition of a proton. This is consistent with the direct interaction of the Ca^{2+} ion with the sulfonic acid group. As seen in Fig. 6 formation of CaTau from Tau has remarkably little affect on taurine ^{13}C resonances. If Ca interacted with the deprotonated amino group in addition to the acid group, as is the case with transition metal ion-amino acid complexes, then the taurine ^{13}C resonance would be expected to shift in the direction of those of TauH. Since this was not observed, it was concluded that Ca remains exclusively associated with the sulfonic acid group at high pH and that the formation of calcium-taurine complexes is more like salt formation than coordination.

Biological Implications

Whatever the biological functions of taurine may be, they must be based on the physico-chemical properties of taurine and most probably arise from those properties of taurine that distinguish it from its carboxylic acid analogues. Proposed functions for taurine must be congruent with the properties of taurine and the way it interacts with other molecules. With respect to the suggestion that Ca flux is modulated by the formation of direct calcium-taurine complexes both the studies reported here and the chemical literature demonstrate that calcium and taurine are uniquely unsuited for complexation with one another. Proposed formation of calcium-taurine or direct ternary taurine-calcium-macromolecule complexes are not consistent with the know physicochemical properties of taurine. This leaves allosteric or electrostatic effects generated by taurine binding at one site in a macromolecular matrix on the calcium binding or transport at another site as the most likely explanation of the effect of taurine on Ca^{2+} and K^+ flux in neurologically active tissues.

SUMMARY

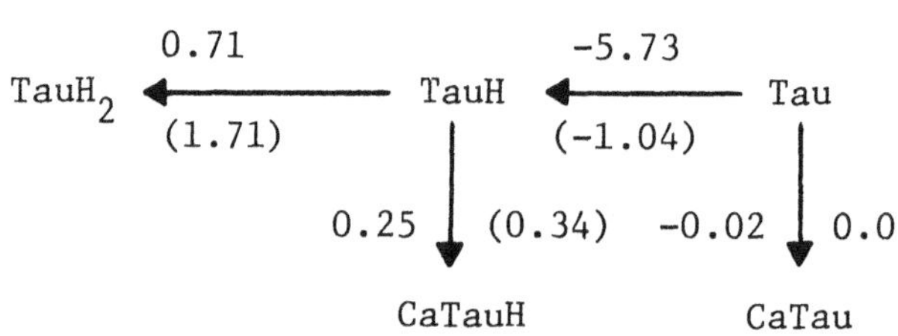

Fig. 6. Comparison of changes in ^{13}C chemical shifts of taurine resulting from complexation with calcium and protonation. The values with and without parentheses refer to downfield chemical shifts of the N-methylene and S-methylene carbons, respectively.

The proposal that taurine modulates calcium flux in neurologically active tissue by formation of calcium-taurine complexes was evaluated in vitro by obtaining the formation constants of 1:1 calcium:taurine complexes from complex equilibrium analysis of ^{13}C nuclear magnetic resonance chemical shift titration curves of ^{13}C-taurine in the presence and absence of calcium. The formation constants of 1.62 and 0.32 for $(H_2NCH_2CH_2SO_3Ca)^+$ and $(H_3NCH_2CH_2SO_3\text{-}Ca)^{2+}$, respectively, indicated that the fraction of calcium bound to taurine is 0.3369 times the concentration of taurine available for binding. In the rat myocardial cell, in which the taurine level is 39mM, only 1.4% of calcium available for complexation will be bound to taurine. These data suggest that the formation of direct calcium-taurine or ternary taurine-calcium-macromolecule complexes are not likely explanations for the effect of taurine on calcium flux.

ACKNOWLEDGMENTS

We would like to thank Professor Gerald E. Gaull for bringing to our attention some of the unsolved problems of the function of taurine and for his constant encouragement during the course of this study.

This work was partially supported by the United States Department of Energy under contract No. W-31-109-ENG-38. ^{13}C-Taurine was made available from the Stable Isotope Resource, Los Alamos Scientific Laboratory through National Institutes of Health grant RR 00962. This work is a publication of the USDA/SEA, Children's Nutrition Research Center, Department of Pediatrics, Baylor College of Medicine and Texas Children's Hospital.

REFERENCES

1. Albert, A., 1950, Quantitative studies of the avidity of naturally occurring substances for trace metals 1. Amino acids having only two ionizing groups, Biochem. J. 47, 531-538.
2. Batchelor, J.G., Feeney, J. and Roberts, G.C.K., 1975, C-13 NMR protonation shifts on amines, carboxylic-acids and amino-acids, J. Magn. Res. 20, 19-38.
3. Beattie, J.K., Fensom, D.J. and Freeman, H.C., 1976, A reinterpretation of paramagnetic line broadening in the nuclear magnetic resonance spectra of amino acids and peptides. I. The copper (II)-glycine system, J. AM. Chem. Soc. 98, 500-507.
4. Chovan, J.P., Kulakowski, E.C., Benson, B.W. and Schaffer, S.W., 1979, Taurine enhancement of calcium binding to rat heart sarcolemma, Biochem. Biophys. Acta 551, 129-136.
5. Chubb, J. and Huxtable, R.J., 1978, Transport and biosynthesis of taurine in stressed heart, in Taurine and Neurological Disorders (A. Barbeau and R.J. Huxtable, Eds.), Raven Press, New York, pp161-178.
6. Datta, S.P. and Grzybowski, A.K., 1959, The stability constants

of the silver complexes of some aliphatic amines and amino acids, J. Chem. Soc. 1091-1095.

7. Dawson, R.M.C., Elliott, D.C., Elliott, W.H. and Jones, K.M., 1969, Data for Biochemical Research, Oxford University Press, New York, pp56-57.
8. Dietrich J. and Diacona, J., 1971, Comparison between ouabain and taurine effects on isolated rat and guinea pig hearts in low calcium media, Life Sci. 10, 499-507.
9. Dolara, P., Agresti, A., Giotti, A. and Pasquini, G., 1973, Effect of taurine on calcium kinetics of guinea pig heart, Eur. J. Pharmacol. 24, 352-358.
10. Dolara, P., Agresti, A., Giotti, A. and Sorace, E., 1976, The effect of taurine on calcium exchange of sarcoplasmic reticulum of guinea pig heart studied by means of dialysis kinetics, Can. J. Physiol. Pharmacol. 54, 529-533.
11. Dolara, P., Franconi, F., Giotti, A., Basosi, R. and Valensin, G., 1978, Taurine-calcium interaction measured by means of ^{13}C nuclear magnetic resonance, Biochem. Pharmacol. 27, 803-804.
12. Dolara, P., Ledda, F., Mugelli, A., Mantelli, L., Zilletti, L., Franconi, F. and Giotti, A., 1978, Effect of taurine on calcium, inotropism, and electrical activity of the heart, In: Taurine and Neurological Disorders, (A. Barbeau and R.J. Huxtable, Eds.), Raven Press, New York, pp151-159.
13. Dolara, P., Marino, P. and Buffoni, F., 1973, Effect of 2-aminoethanesulfonic acid (taurine) and 2-hydroxyethane sulfonic acid (isethionic acid) on calcium transport by rat liver mitochondria, Biochem. Pharmacol. 22, 2085-2094.
14. Greenstein, J.P. and Winitz, M., 1961, Chemistry of the amino acids, John Wiley & Sons, Inc., New York, p487.
15. Grosso, D.S. and Bressler, R., 1976, Taurine and Cardiac physiology, Biochem. Pharmacol. 25, 2227-2232.
16. Guidotti, A., Bandiani, G. and Giotti, A., 1971, Potentiation by taurine of inotropic effect of strophanthin K on guinea pig isolated auricles, Pharmacol. Res. Comm. 3, 29-38.
17. Guidotti, A. and Giotti, A., 1970, Taurina e Sistema Cardio Vascolare, Rec. Prog. Med. (Roma) 49, 61-71.
18. Hargreaves, M.K., Stevinson, E.A. and Evans, J., 1965, The apparent dissociation constants of various weak acids in mixed aqueous solvent, J. Chem. Soc., 4582-4583.
19. Huxtable, R., 1976, Metabolism and function of taurine in the heart, In: Taurine, (R. Huxtable and A. Barbeau, Eds.), Raven Press, New York, pp99-119.
20. Igisu, H., Izumi, K., Goto, I. and Kina, K., 1976, Effects of taurine on the ATPase activity in the human erythrocyte membrane, Pharmacology 14, 362-366.
21. Jacobsen, J.G. and Smith, L.H., Jr., 1968, Biochemistry and physiology of taurine derivatives, Physiology 48, 424-511.
22. Karplus, M. and Pople, J.A., 1963, Theory of carbon NMR chemical shifts in conjugated molecules, J. Chem. Phys. 38, 2803-2807
23. Kosugi, Y. and Takeuchi, T., 1979, The ^{13}C NMR titration shifts

of sulphonic acids, Org. Magn, Reson. 12, 435-437.
24. Monk, C.B., 1951, Electrolytes in solutions of amino acids. Part V.--The solubilities of calcium, barium and lanthanum iodates in glycine, alanine, and glycylglycine, Trans. Faraday Soc. 47, 1233-1240.
25. Norton, R.S., 1979, Identification of mollusc metabolites by natural abundance ^{13}C NMR studies of whole tissue and tissue homogenates, Comp. Biochem. Physiol. 63B, 67-72.
26. Österberg, R. and Sjöberg, B., 1968, The metal complexes of peptides and related compounds. III. Copper (II) complexes of glycylglycylglycine in 3.0 M $Na(ClO_4)$ medium, J. Biol. Chem. 243, 3038-3050.
27. Paul, M.A. and Long, F.A., 1956, H_0 and related indictor acidity functions, Chem. Rev. 57, 1-45.
28. Rabenstein, D.L., 1972, Nuclear magnetic resonance studies of the solution chemistry of metal complexes III. Acetylglycine complexes of cadmium, zinc, and lead, Can. J. Chem. 50, 1036-1043.
29. Rabenstein, D.L. and Sayer, T.L., 1976, Carbon-13 chemical shifts parameters for amines, carboxylic acids, and amino acids, J. Magn. Reson. 24, 27-39.
30. Read, W.O. and Welty, J.D., 1963, Effect of taurine on epinephrine- and digoxin-induced irregularities of dog heart, J. Pharmacol. Exp. Ther. 139, 283-289.
31. Schaffer, S.W., Chovan, J.P. and Werkman, R.F., 1978, Dissociation of cAMP changes and myocardial contractility in taurine perfused heart, Biochem. Biophys. Res. Commun. 81, 248-253.
32. Winegrand, S. and Shanes, A.M., 1962, Calcium flux and contractility of guinea pig atria, J. Gen. Physiol. 45, 371-394.
33. SAS User's Guide 1979 Edition, SAS Institute Inc., P.O.B. 10066, Raleigh, NC 27605, 1979: 317-330.

PART II

TAURINE IN NUTRITION AND DEVELOPMENT

Ryan J. Huxtable

Department of Pharmacology
Health Sciences Center
University of Arizona
Tucson, Arizona

Interest in nutritional aspects of taurine has grown rapidly, and much recent research has been in this area. The high level of interest is made abundantly clear in the discussions recorded in this book.

This recent attention to nutritional aspects of taurine is fueled by the light these studies may throw on the biological functioning of taurine, and the direct relevance of this work to humans. For example, taurine concentrations are extraordinarily high in developing brain, and fall off sharply thereafter. Other neuroactive amino acids show an opposite developmental pattern being higher in adult brain than in developing brain. This demand for taurine is conjoined in humans with an apparent inability to synthesize it. The major, if not the sole, source of taurine in infants is milk. At the beginning of lactation, taurine is the most abundant amino acid in milk, comprising 20% or more of the total free amino acids (4). The concentration falls over the first few days, but remains high throughout nursing. Babies that are raised on milk formulae, however, are deprived of this source of taurine. This deprivation results in a number of biochemical abnormalities, but no obvious clinical ones. Given the association of taurine with neuronal development, however, it is of concern that deprivation at an early stage may lead to subtle alterations in development, that may ultimately result in learning, behavioral or neurological defects.

It is clear that different mammals have different taurine economies. Dietary intake and biosynthetic abilities, as judged by cysteine sulfinic acid decarboxylase and other enzymatic activities, vary enormously. At one extreme, a herbivore such as the guinea pig normally receives no taurine in its diet, but maintains high tissue concentrations, and excretes taurine routinely in its urine. On the other hand, a pure carnivore like the cat receives an abundant dietary intake of taurine, and has little or no biosynthetic capacity. Few quantitative studies on taurine balance have been done, however. In the following article, Huxtable and Lippincott report on a series of experiments designed to give information on the handling of taurine in the newborn and adult rat, and on the comparative importance of biosynthesis versus diet as sources of taurine. Chesney and coworkers examine developmental changes in taurine resorption in the kidney. Their work, showing a difference between immature and mature rats, give some insight into the developmental changes affecting taurine regulation. Their work also suggests that examination should be made for similar changes in the brain, as a possible explanation for the peculiar biphasic variation in brain taurine levels during development. Maturational changes in taurine transport may also be related to the defects in transport associated with disorders such as Friedreich's ataxia (3), epilepsy (1) and cardiomyopathy (2).

The best established pathological consequence of taurine depletion is the retinal degeneration that occurs in cats maintained on taurine-deficient diets. Sturman and coworkers report on an extensive and detailed investigation of tapetal changes that occur on taurine deficiency. Their work also reinforces the connection between zinc and taurine that was discussed at the first meeting on taurine. The biochemical basis for this association is unclear, but that taurine affects zinc balance and excretion in the body is undoubted. The investigations of Sturman and coworkers may serve as a paradigm for other phenomena associated with taurine, in that work in this area has progressed from the establishment of a phenomenon through to its morphological and biochemical basis.

Hayes and Sturman write insightfully on the comparative aspects of growth in relation to taurine, pointing out the enormous variation in growth rates between species. The importance of bile salt conjugation as a determinate in taurine turnover is often overlooked. Hayes and Sturman make it clear that where there is a obligate requirement for taurocholic acid, as occurs in felines with their high protein intake, this is quantitatively the most important route of taurine elimination. Species, such as man, in which the taurine: glycine ratio in bile salts can alter are protected to a certain extent against the consequences of dietary depletion of taurine, albeit at the expense of other aspects of nutrition. For example taurocholic acid is a superior surfactant compared to glycocholic acid, and a disturbance in the normal ratio of these two substances may lead to defects in digestion and absorption.

The final chapter in this section comes from the laboratory most responsible for molecular knowledge of taurine in development. Gaull reviews comparative aspects of taurine in milk, pointing out the enormous variations in concentrations between species. He summarizes the evidence that for humans there is a dietary requirement for taurine.

REFERENCES

1. Airaksinen, E.M., 1979, Uptake of taurine, GABA, 5-HT and dopamine by blood platelets in progressive myoclonus epilepsy, Epilepsia, 20:503-510.
2. Darsee, J.R., and Heymsfield, S.B., 1981, Decreased myocardial taurine levels and hypertaurinurea in a kindred with mitral-value prolapse and congestive cardiomyopathy, New Eng. J. Med., 304:129-135.
3. Filla, A., Butterworth, R.F., and Barbeau, A., 1979, Pilot studies on membranes and some transport mechanisms in Friedreich's ataxia, Can. J. Neurol. Sci., 6:285-289.
4. Rassin, D.K., Sturman, J.A., and Gaull, G.E., 1978, Taurine and other free amino acids in milk of man and other mammals, Early Human Devel., 2:1-13.

SOURCES AND TURNOVER RATES OF TAURINE IN NEWBORN, WEANLING, AND MATURE RATS

Ryan J. Huxtable and Shirley E. Lippincott

Department of Pharmacology
Health Sciences Center
University of Arizona
Tucson, Arizona

The pharmacology of taurine has received close study, and taurine has been shown to be beneficial in conditions as diverse as genetic cardiomyopathy (1), cardiac arrhythmias (2), and epilepsy (5). Deficiencies in taurine have also been associated with the development of such pathological processes as retinal degeneration in cats (9) and cardiomyopathy in humans (4). These investigations have linked taurine with various phenomena without yielding insight into the mechanism by which taurine is acting, or its normal functional involvement in cellular processes.

There is a major need to place these processes on a mechanistic basis, and future advances in our understanding of taurine will be dependent on this. There is considerable evidence that taurine modifies the biological actions of calcium, for example. These modifications may be important physiologically and pharmacologically. As yet, however, in trying to explain in molecular terms the interactions of taurine and calcium we can offer nothing but speculation.

One of the fundamental gaps in our knowledge concerns the handling of taurine by mammals. Where does it come from? How important is the diet as a source? How do animals respond when deprived of taurine during development?

That the problem of the quantitative source of taurine has remained unexamined for so long is puzzling. The biochemistry of taurine has received intense scrutiny. Numerous investigations have been performed on the putative mechanisms whereby it is biosynthesized. Although there are laboratories carrying out investigations on the cysteine sulfinic acid decarboxylase pathway, the cysteamine dioxygenase pathway, or the PAPS transferase pathway, no one has approached the problem of taurine biosynthesis from the quantitative viewpoint. Enzyme studies alone are insufficient to answer the question as to what proportion of taurine is biosynthesized. First, as has been amply discussed before, all the proposed routes of taurine biosynthesis contain missing enzymatic steps, and the relative importance of each route has not been determined. The demonstration, for example, that cat liver contains no cysteine sulfinate decarboxylase activity is insufficient in and of itself to point the conclusion that the cat is unable to synthesize taurine. Secondly, even where a complete sequence of enzymes can be demonstrated, in the absence of other information this only demonstrates the potential to synthesize taurine. No deductions may be drawn about the quantitative importance of the pathway. For example, guinea pig liver contains enzymes capable of synthesizing taurocholate, and cat liver contains enzymes capable of synthesizing glycocholate, yet in practice the guinea pig makes only glycocholate (except under unusual conditions of dietary loading (7)) and the cat makes only taurocholate.

We have tried to determine the quantitative importance of endogenous biosynthesis and diet as sources of taurine. The experimental approaches are relatively straightforward, designed to answer a series of simple questions concerning the handling of taurine by juvenile and mature rats.

The experimental approach we have used is described in more detail below, but in principle consists of maintaining animals on diets containing substrates for taurine at a constant specific radioactivity. The contribution of that substrate to the pool of taurine within any tissue can be simply determined at any time by measurement of the specific activity of taurine. Quantitative information may thus be obtained without any assumption as to the route by which the taurine has been made.

SOURCES OF TAURINE IN THE ADULT RAT

Principles

A typical experiment is run between time t1 (initiation) and time t2 (termination). At time t2, there are four possible sources for the taurine within the rat: (i) it could have been biosynthesized from dietary substrates between time t1 and t2, either by the animal or by digestive flora; (ii) it could have been obtained from the diet between time t1 and t2; (iii) the taurine could have been already present in the rat at time t1; or (iv) it could have been biosynthesized between t1 and t2 from substrates already present in the animal at time t1.

For the contribution of dietary taurine to total body taurine, rats are maintained chronically on diets containing ^{3}H-taurine of a known (and constant) specific activity. At time t2, the dietary contribution to any tissue can be simply determined by comparing the specific activity of the taurine within that tissue to the specific activity of the taurine in the diet.

The amount contributed by biosynthesis is determined, in principle, by a similar experiment. The rat is maintained on a diet deficient in sulfur amino acids. ^{35}S-Methionine, of known specific activity, and in an amount commensurate with normal dietary levels (8), is added to the diet. At time t2, the specific activity of the taurine in any tissue will yield the proportion derived by biosynthesis. It is important to note that this experiment does not show where the taurine was biosynthesized, but only that it was biosynthesized somewhere in the animal.

The proportion of taurine at time t2 derived from the taurine already present in the animal at time t1, or formed from substrates present at time t1, may be estimated by subtraction of the amounts biosynthesized and obtained from the diet from the total taurine content of the tissue.

Experiment 1. Dietary Taurine (Protein Diet)

In this experiment, male rats were given free access to a taurine-free diet and to drinking water containing 5 mM ^{3}H taurine. The daily intake of taurine averages between 0.05 to 0.1% by weight of total food consumption, an amount that approximates the taurine intake of animals in our facility maintained on standard laboratory chows. We did this experiment to provide a comparison for subsequent experiments with nonprotein containing, amino acid diets. The tissue concentration of taurine derived from this diet at various time intervals is shown in Table 1, as is the percentage of the total taurine in the tissue this represents. Equilibrium with the diet in tissues outside of the central nervous system was rapidly achieved. The kidney and intestines, organs of elimination and absorption respectively, have equilibrated with the diet within 4 days. The lung and spleen equilibrate within a week. In general, for the peripheral tissues, the diet has contributed an average of 34% of the total taurine by day 4, and this average does not increase thereafter. Brain areas are slower to equilibrate, and deviate more. However, little difference is observed between 30 and 60 days, the diet contributing 50 ± 20% of taurine in the various areas of brain. On average, for all the tissues analyzed, the diet has contributed 42% of the total taurine.

TABLE 1

TAURINE DERIVED FROM THE DIET

μmole/g tissue (% of total taurine)

DAYS	4		7		10	
Heart	4.4 ±0.4	(25)	8.4 ±0.5	(35)	8.6 ±0.5	(28)
Lung	6.0 ±0.5	(37)	6.2 ±0.5	(44)	4.8 ±0.5	(36)
Liver	2.8 ±0.4	(34)	2.4 ±1.0	(34)	5.0 ±0.6	(35)
Muscle	2.7 ±0.6	(19)	4.2 ±0.1	(29)	4.2 ±0.6	(30)
Kidney	3.6 ±0.5	(36)	3.9 ±0.1	(35)	4.9 ±0.6	(33)
Spleen	3.7 ±0.4	(43)	4.7 ±0.1	(45)	4.0 ±0.0	(41)
Intestine	5.2 ±0.7	(41)	4.7 ±0.2	(38)	3.9 ±0.2	(35)
Cerebellum	0.99±0.04	(17)	1.5 ±0.1	(28)	1.9 ±0.1	(38)
Pons Medulla	0.76±0.01	(16)	1.0 ±0.1	(26)	0.88±0.17	(25)
Midbrain	0.81±0.07	(40)	0.70±0.55	(36)	1.2 ±0.2	(60)
CH*	0.73±0.30	(9)	1.4 ±0.3	(20)	2.3 ±0.1	(30)

DAYS	15		30		60	
Heart	10.3 ±0.1	(37)	11.0 ±0.3	(41)	10.6 ±0.5	(38)
Lung	5.6 ±0.1	(37)	4.4 ±0.5	(36)	5.3 ±0.3	(35)
Liver	3.0 ±0.8	(30)	1.5 ±0.3	(35)	4.1 ±0.1	(30)
Muscle	6.1 ±0.1	(38)	6.3 ±0.7	(39)	7.7 ±0.5	(38)
Kidney	3.8 ±0.1	(33)	4.0 ±0.1	(34)	4.5 ±0.3	(36)
Spleen	4.7 ±0.1	(40)	4.8 ±0.3	(36)	5.3 ±0.3	(39)
Intestine	3.9 ±0.0	(35)	3.9 ±0.7	(32)	3.4 ±0.1	(25)
Cerebellum	1.9 ±0.1	(39)	1.8 ±0.1	(62)	1.8 ±0.4	(70)
Pons Medulla	1.2 ±0.1	(38)	1.3 ±0.3	(23)	1.2 ±0.1	(26)
Midbrain	1.4 ±0.1	(74)	1.6 ±0.1	(83)	1.7 ±0.4	(60)
CH*	2.3 ±0.3	(36)	2.3 ±0.1	(45)	2.6 ±0.1	(45)

Rats (initial weight 100g) were maintained on a taurine-free diet (Biomix 900, Bioserv, Inc., Frenchtown, NJ) and drinking water containing 5 mM ^{3}H-taurine (24,600 dpm/μmole) for the number of days indicated.

Data (means ± SEM for 2 animals per time) are μmole/g ^{3}H-taurine.

The percentage of the total taurine in the tissue this represents is shown in parentheses.

* Cerebral Hemispheres

Experiment 2. Contribution of Biosynthesis

Whatever the route of biosynthesis, the sulfur atom in taurine, though not the carbon skeleton, must ultimately derive from methionine. In order to control the specific activity of methionine, it was necessary to use a nonprotein-containing, amino acid diet. That is to say, the diet must contain the individual free amino acids that are normally components of protein. This relatively short term experiment (15 days) was designed to test the practicality of using such a diet. The basal diet (see legend to Table 2) contained no taurine, cysteine or methionine. The animals were given free access to this diet and to drinking water containing 5 mM unlabeled taurine and 67 mM ^{35}S methionine. The daily intake of methionine ranged between 0.7 to 1.3 % of total food intake. We chose to use methionine in place of cysteine as the source of sulfur in order to include putative pathways of taurine biosynthesis not involving cysteine. Furthermore, the rat is able to convert methionine to cysteine, and if enough of the former is available cysteine ceases to be essential.

The results of this experiment are shown on Table 2. Despite the major difference in diet compared to the experiment shown in Table 1, the data are consistent with those derived from the first experiment. For example, the peripheral tissues examined in the first experiment derived 36% of the taurine from the diet by day 15. In the experiment illustrated here, by day 15 the same seven tissues have derived 48% of their taurine from biosynthesis. Essentially, therefore, approximately 84% of the taurine in these tissues have been accounted for.

Contribution of biosynthesis to the pool of taurine in the brain proceeds at a slower rate. Nevertheless, by day 15, 41% of the taurine in the brain has been derived from biosynthesis, with little variation from area to area.

Experiment 3. Simultaneous Quantitation of Biosynthesis and Diet

The concordance between the two experiments described above encouraged us to attempt a more ambitious one: simultaneous quantitation of diet and biosynthesis as sources of taurine, and a study of the effect of dietary deprivation of taurine on the contribution of biosynthesis to the total body pool.

^{35}S Methionine and ^{3}H taurine (where present) were mixed in with an amino acid chow, which was supplied to the animals as an agar gel. Distilled water with no additives was used as drinking water. Methionine constituted 0.58% by weight of the diet, and taurine, where present, 0.4%. The taurine level was increased above that used in the previous experiment in order to compare the effect on biosynthetic contribution of the extremes of a high and a zero intake

TABLE 2
TAURINE DERIVED FROM BIOSYNTHESIS FROM METHIONINE
μmole/g tissue (% of total tissue)

DAYS	4		7		11		15	
Heart	3.3 ±0.7	(12)	4.7 ±1.1	(18)	9.0 ±0.3	(28)	11.5 ±0.3	(36)
Lung	3.1 ±0.7	(21)	4.3 ±0.5	(35)	6.3 ±0.2	(46)	6.5 ±0.1	(53)
Liver	1.6 ±1.1	(26)	0.9 ±0.3	(37)	5.5 ±0.9	(57)	5.7 ±0.7	(71)
Muscle	2.0 ±0.1	(11)	2.5 ±0.5	(14)	4.6 ±0.7	(25)	6.0 ±1.2	(31)
Kidney	3.4 ±0.6	(28)	3.8 ±0.5	(31)	7.2 ±0.3	(49)	7.7 ±1.0	(51)
Spleen	3.5 ±0.1	(27)	4.4 ±0.7	(32)	6.9 ±0.4	(46)	6.8 ±1.1	(48)
Intestine	2.8 ±1.0	(22)	3.5 ±0.5	(31)	6.3 ±0.5	(45)	6.2 ±0.5	(47)
Stomach	1.7 ±0.1	(22)	2.4 ±0.3	(32)	4.1 ±0.3	(45)	4.2 ±0.5	(51)
Thymus	2.9 ±0.3	(27)	4.0 ±0.3	(32)	6.3 ±1.0	(56)	6.4 ±0.3	(63)
Pancreas	0.59±0.03	(29)	0.78±0.11	(33)	1.3 ±0.1	(50)	1.4 ±0.1	(58)
Testes	0.42±0.01	(21)	0.53±0.05	(27)	0.83±0.07	(39)	0.91±0.05	(43)
Eye	1.2 ±0.2	(12)	1.7 ±0.4	(17)	3.1 ±0.3	(27)	3.5 ±0.2	(33)
Cerebellum	0.74±0.05	(14)	0.95±0.14	(23)	1.6 ±0.1	(38)	1.8 ±0.2	(45)
Pons Medulla	0.33±0.12	(16)	0.42±0.19	(24)	0.77±0.24	(45)	0.94±0.13	(47)
Midbrain	0.66±0.07	(10)	1.0 ±0.1	(16)	1.7 ±0.4	(30)	2.1 ±0.2	(37)
Cerebral Hemispheres	0.62±0.03	(12)	0.91±0.17	(19)	1.6 ±0.2	(31)	1.8 ±0.2	(39)
Frontal Cortex	0.88±0.08	(11)	1.3 ±0.1	(19)	1.9 ±0.1	(26)	2.4 ±0.4	(35)
Inferior Colliculus	0.20±0.21	(8)	0.56±0.20	(17)	1.0 ±0.4	(28)	1.4 ±0.1	(38)
Hypothalamus	0.58±0.10	(20)	0.55±0.10	(18)	0.93±0.53	(27)	1.6 ±0.1	(40)
Spinal Cord	0.23±0.08	(18)	0.43±0.05	(28)	0.64±0.06	(38)	0.75±0.11	(47)
Serum	0.17±0.02	(26)	0.22±0.04	(29)	0.42±0.06	(47)	0.44±0.10	(48)

Data (Means ± SD for 3 animals) are μmole/g taurine derived by biosynthesis from ^{35}S-methionine since day 0. The percentage of the total taurine this represents is shown in parentheses. Diet (%): Isoleucine 0.5; leucine 0.8; lysine 1.12; phenylalanine 0.8; threonine 0.5; tryptophan 0.15; valine 0.7; histidine 0.4; arginine 1.34; tyrosine 0.8; alanine 0.53; aspartic acid 1.23; glycine 0.89; proline 1.96; serine 1.04; glutamic acid 5.67; vitamin E acetate 0.005; vitamin mixture (CDC 742) 2.2; Fiber 5; corn starch 6; dextrose 15; salt mix (HMW-RM) 5; sodium bicarbonate 2; corn oil 10; sucrose 36.365. Animals received drinking water containing 67 mM ^{35}S-methionine (19,200 dpm/μmole) and 5 mM taurine.

of taurine. Groups of rats were maintained on these diets for 63 and 87 days before sacrifice.

Data obtained on the "carcasses" are shown in Table 3. For our purpose, we use the word "carcass" for what is left of the animal after the individual organs have been taken. It thus consists mainly of muscle, skin and bone, and contains the bulk of the taurine present in the animal. Animals on the taurine-enriched diet have obtained 18% of the taurine in the carcass from biosynthesis by 63 days, and 51% from the diet, leaving 32% still to be accounted for. This unaccounted for taurine we call depot taurine; it represents either taurine present in the animal at the beginning of the experiment, or taurine biosynthesized from other substances present at the beginning of the experiment. The dietary contribution observed here is rather close to that found in experiment 1, even though there is more taurine in the diet. However, the biosynthetic contribution is considerably lower than that observed in experiment 2. This may also be related to the higher level of taurine in the diet, as animals deprived of taurine for 63 days derive 41% of carcass taurine from biosynthesis. This is in good agreement with the data of Table 2.

A similar pattern is seen at 87 days, with progression in the amounts. Of the taurine in the deprived animals, 46% is depot taurine, tenaciously held for three months, as opposed to only 13% in the animals receiving dietary taurine. However, it seems that the deprived animals are having problems maintaining their tissue taurine pools, their carcasses containing an average of 250 μmole taurine less than the enriched animals (Table 3). The general decrease in taurine is also evidenced by the lower tissue concentrations in the deprived animals as compared to the enriched. Among the more affected organs, liver levels are 53% lower, small intestines 40%, lung 31% and spleen 28%.

Data on individual tissues for the 63 day time period are shown in Table 4. The data for the 87 day period, which are not reported, showed continuations of the same trends.

At true equilibrium, which may not be attained within the life of the animal, all of the taurine present should be accounted for in terms of diet or biosynthesis, and in addition the pool of sulfur-containing amino acids in body protein should be uniformly labeled to the same specific activity as that of the methionine being fed. In this experiment, how close are we to an equilibrium state? After 87 days, even in the taurine-enriched animals, 13% of the taurine cannot be accounted for. This suggests, *a priori*, that complete equilibrium has not been obtained with dietary taurine. After about 50 days, on the other hand, all the taurine in the urine can be accounted for in animals on the taurine supplemented diet (Fig. 1). The diet contributes roughly two thirds of the urinary taurine, and

TABLE 3

QUANTITATION OF SOURCES OF TAURINE IN RAT CARCASS

	63 Days		87 Days	
Taurine in Diet?	Yes	No	Yes	No
	μmole taurine			
Total Taurine	1925 ± 225	1857 ± 88	1817 ± 314	1564 ± 214
Biosynthesis	352 ± 146	761 ± 354	526 ± 160	689 ± 301
Diet	995 ± 379	0 ± 0	1060 ± 324	0 ± 0
Depot	592 ± 377	1095 ± 316	231 ± 374	874 ± 330
	% of total taurine			
% Biosynthesis	18.4 ± 6.1	40.7 ± 17.9	28.6 ± 7.0	53.6 ± 20.2
% Diet	51.0 ± 15.3	0 ± 0	58.1 ± 14.6	0 ± 0
% Depot	31.6 ± 19.7	59.3 ± 17.9	13.2 ± 21.1	46.4 ± 20.2

Data are means ± SD for 5 animals (initial weight 240 ± 11g) per group and represent total taurine in carcass (i.e. the residue after the tissues listed in table 4 were removed), and amounts and percentages derived from the various sources over a 63 day and 87 day period.

Diet was as for table 2, with the following changes; arginine was added to 0.6%, and 0.58% ^{35}S-methionine (19,000 dpm/μmole) and, when present, 0.4% ^{3}H-taurine (122,000 dpm/μmole) mixed in with the diet. The diet was mixed 1:1 with a 4.5% agar solution, and after gelling was cut into cubes for feeding. Deionized water was supplied as drinking water.

TABLE 4

SOURCES OF TAURINE IN RAT TISSUES AFTER 63 DAYS
μmole/g tissue (% total taurine)

	Taurine Containing Diet			Taurine Free Diet	
	Total Taurine	Diet	Biosynthesis	Total Taurine	Biosynthesis
Heart	25.5±4.3	11.1±1.4 (44)	4.2 ±1.1 (17)	30.0±1.1	21.6±1.7 (72)
Lung	17.7±2.7	7.9±2.6 (45)	3.3 ±0.8 (18)	12.0±1.0	8.4±0.7 (74)
Liver	6.7±3.2	2.8±1.3 (42)	2.4 ±2.0 (36)	2.6±1.2	2.3±1.0 (87)
Muscle	24.3±2.5	13.7±3.5 (56)	5.0 ±1.3 (21)	16.6±3.3	11.0±2.7 (67)
Kidney	15.4±3.1	6.4±1.0 (42)	3.1 ±1.0 (21)	12.0±0.4	9.7±0.7 (81)
Spleen	18.9±9.5	9.4±5.5 (50)	3.8 ±2.2 (20)	16.5±0.5	12.3±0.6 (75)
Small Intestine	24.2±2.0	12.4±2.7 (51)	4.7 ±0.7 (20)	20.4±2.2	15.5±1.6 (76)
Large Intestine	22.5±6.9	10.5±2.9 (47)	4.7 ±1.9 (21)	17.4±4.2	14.0±3.5 (81)
Stomach	15.1±1.5	8.4±1.2 (56)	3.4 ±0.5 (23)	11.9±1.1	8.2±0.8 (69)
Thymus	26.3±3.6	14.3±1.8 (54)	6.6 ±1.0 (25)	16.0±2.2	9.0±0.8 (56)
Pancreas	8.4±1.3	3.7±0.9 (44)	1.5 ±0.2 (18)	6.9±0.5	3.3±1.3 (48)
Testes	3.3±0.8	2.7±0.3 (82)	0.8 ±0.1 (24)	1.9±0.2	1.8±0.8 (93)
Eye	19.8±7.5	10.9±5.3 (55)	4.4 ±2.1 (22)	11.7±2.1	7.1±4.6 (61)
Carcass	9.5±0.2	4.9±1.5 (51)	1.7 ±0.6 (18)	10.0±0.5	4.1±1.7 (41)
Cerebellum	2.7±0.8	1.6±0.4 (61)	0.35±0.09 (13)	3.2±0.5	4.0±1.1 (124)
Pons Medulla	4.5±0.4	1.9±0.4 (42)	0.50±0.26 (11)	4.3±1.3	2.8±0.3 (64)
Midbrain	4.6±0.9	1.8±0.3 (40)	0.38±0.11 (8)	3.4±1.1	2.4±1.2 (68)
Cerebral Hemispheres	2.7±0.7	1.6±0.2 (57)	0.41±0.09 (15)	4.6±0.9	5.5±1.8 (121)
Frontal Cortex	5.0±0.6	2.2±0.4 (44)	0.69±0.19 (14)	5.5±0.8	3.7±0.7 (67)
Inferior Colliculus	4.6±1.2	2.6±0.3 (56)	0.20±0.04 (4)	3.7±1.4	3.2±0.7 (88)
Hypothalamus	4.0±0.8	2.2±0.5 (56)	0.49±0.25 (12)	4.3±2.1	2.1±0.8 (49)
Spinal Cord	4.1±0.5	2.2±0.5 (54)	0.44±0.31 (11)	4.2±1.7	3.0±2.6 (71)

Data are means ± SD for 5 animals per group. Conditions as for table 3. Data for each diet consist of the taurine concentration in each tissue, and the amount and percentage derived from biosynthesis and (where appropriate) diet over the previous 63 days.

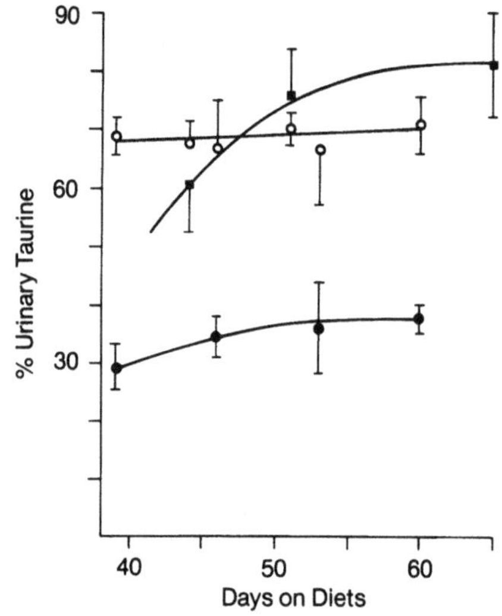

Fig. 1. Derivation of urinary taurine. Conditions are described in legend to Table 3.
A. Animals on a taurine-free diet: ■ Percentage of taurine biosynthesized from methionine over indicated period.
B. Animals on a taurine-containing diet: ● Percentage of taurine biosynthesized. O Percentage of taurine obtained from diet.
Data are means ± SD of 10 animals.

endogenous biosynthesis one third. About 80% of the urinary taurine in the taurine-deprived animals originates from biosynthesis (Fig. 1).

Have other sulfur compounds equilibrated with ^{35}S methionine? If the heart and muscle are taken as typical tissues, the concentration of radioactive sulfur in these tissues, other than ^{35}S-taurine, is greater than the combined concentrations of cysteine, methionine and glutathione (data not shown). This finding is consistent with the pool of soluble sulfur compounds being in a state of pseudoequilibrium with the dietary methionine (there is presumably some input of unlabeled amino acids from protein breakdown).

In the absence of taurine in the diet, it is clear that a greater proportion of the body taurine pool is derived from biosynthesis (Tables 3,4). Does this mean that the rate of biosynthesis is stimulated in response to the lack of taurine in the diet, or does it mean that the rate of excretion of taurine is decreased, and that the half life of elimination of biosynthesized taurine is longer? A direct answer to this question could be obtained by adding up all the (^{35}S)taurine in the animal and all the (^{35}S)taurine eliminated over the course of the experiment. Most taurine is eliminated in the urine, and the largest pool of endogenous taurine is in the carcass. These pools are shown in Table 5. Although the contribution of biosynthesis to body taurine content is doubled in animals deprived of taurine, it is clear that there has been no increase in synthetic rate. The rat has responded to the lack of dietary taurine by making the same amount, but excreting less. Although the total excretion of taurine in deprived animals is decreased markedly, the decrease in excretion of taurine derived from biosynthesis is surprisingly slight. Excretion of (^{35}S)taurine in deprived animals is 76% of that in the enriched animals.

TAURINE IN NURSING AND WEANED RAT PUPS

Principles

This experiment was designed to answer the following: In the developing rat, how much taurine is obtained from the mother, either *in utero* or via the milk, and how much does the animal make itself? After being weaned, at what rate is the taurine obtained from the mother lost, and how does the presence of taurine in a diet onto which the animal is weaned affect this rate of loss?

Female Sprague-Dawley rats were maintained on a taurine-free diet and drinking water containing 5 mM ^{3}H taurine. They were kept under these conditions for one week, by which time, as shown in experiment 1, taurine pools outside the central nervous system are uniformly labeled. The taurine solution was then removed while the animals were mated. Subsequently, they were then given 5 mM ^{3}H taurine of the same specific activity, and maintained on it throughout pregnancy and lactation. The degree of labeling of body taurine pools in these animals at weaning is shown in Table 6. This is in close agreement with the findings of Table 1.

Radioactivity in the pups is a measure of the transfer of taurine from the mothers. Twenty-one days after birth, the pups were weaned to either a taurine-free diet or a diet containing 0.4% taurine. Animals were sacrificed at intervals, and the total taurine and total ^{3}H taurine content of various tissues determined. From the measurements of radioactivity, the turnover of the taurine that was present at weaning could be calculated.

Taurine Levels in Rat Neonates

Total taurine content of various tissues at birth, midnursing (11 days), and immediately prior to weaning (21 days) are shown in Table 7. All organs show a continuous increase in total taurine content, in keeping with the rapid increase in body weight following birth. During nursing, the whole body taurine content of pups

TABLE 5

EFFECT OF DIET ON TAURINE BIOSYNTHESIS AND URINARY EXCRETION

DIET	TAURINE-CONTAINING	TAURINE-FREE
Biosynthesized Taurine		
In Carcass	352 ± 146	808 ± 358[a]
In Urine	2743 ± 438	2092 ± 597
TOTAL	3094 ± 458	2901 ± 862

[a] $p<.05$ compared to other diet. All values are μmole ± SD for 5 animals. These data were derived on animals sacrificed after 63 days on the diets. Urine values are the total excretion of (^{35}S) methionine over this period.

increases from 40 μmoles at birth to 300 μmoles at weaning (Fig. 2). These figures are in good agreement with those of Sturman (11). In the three week period following weaning, there is a marked increase in total body taurine in both the animals receiving taurine in the diet and those on a taurine-deficient diet (Fig. 2).

Sources of Taurine During Nursing

The proportion of the taurine in the pups obtained from the mothers can be readily calculated from a comparison of the specific activity (dpm/μmole) of the taurine in the pups and the mothers. The percentages of taurine derived from the mother are shown on Table 7. A surprising observation is that at birth only 66% of body taurine has been obtained from the mother, indicating that the rat *in utero* is already synthesizing a considerable proportion of its taurine. By the time the animal is weaned, more than half of its total taurine has been endogenously synthesized. All tissues in the pups at birth have lower specific activities than the mothers, indicating biosynthesis *in utero*. However, although these calculations indicate where biosynthesized taurine is *found*, they give no indication where it is *made*. During nursing, a considerable biosynthesis of taurine is occurring. At weaning, the rat has biosynthesized 177 μmole of taurine, or 60% of its total. The findings are in disagreement with the conclusions of others. The major route of taurine biosynthesis in the rat has been presumed to be via the enzyme cysteine sulfinic acid decarboxylase. Kaczmarek (6) has shown that little activity of this enzyme can be detected at birth, and that activity is not appreciable until 30 days of age. Others

TABLE 6

PERCENTAGE OF TAURINE IN MOTHERS OBTAINED FROM THE DIET

Eye	44 ± 10
Intestine	51 ± 14
Kidney	34 ± 0
Spleen	46 ± 15
Stomach	43 ± 10
Liver	43 ± 14
Lung	46 ± 10
Heart	41 ± 9
Muscle	43 ± 9
Frontal Cortex	58 ± 9
Brain	51 ± 27
Whole Body	42 ± 9

Data are means ± SD. Mothers were maintained on a taurine-free diet, and drinking water containing 5mM (^{3}H)taurine (124,000 dpm/μmole) from 10 days before conception until pups were weaned at 21 days. The mothers were then sacrificed.

have concluded that de novo synthesis in a rat is low up to 18 days following birth (12).

Taurine Turnover after Weaning

Figures 3a to 3t illustrate the increase in (^{3}H)taurine content of various tissues during nursing, and the rate of loss of (^{3}H)-taurine when animals are weaned onto taurine-enriched or taurine-deficient diets. For most tissues, the loss of radioactivity is well fitted by a semi-logarithmic decay curve, as shown by the correlation coefficients in Table 8. Correlations are poor for data derived from animals on the taurine-deficient diet for the liver, uterus, frontal cortex and cerebellum. For all other tissues, coefficients are in excess of 0.83. Half lives of taurine present at weaning calculated from the regression curves are given in Table 8. For rats receiving a taurine-enriched diet, half lives for visceral organs average 4.8 days with a standard deviation of 1.1 days (±23% of mean). Thus, the half lives cluster with a narrow deviation. The half life of elimination from the brain is not significantly different from this, the different areas averaging 5.5 ± 1.1 days. In rats on a taurine-deficient diet, however, the visceral half lives increase to 8.7 ± 2.0 days, with a standard deviation remaining at 23% of the mean. In these animals, the half life of elimination from the brain has increased by only 13%, making

TABLE 7

THE SOURCE AND CONTENT OF TAURINE IN RAT NEONATES

μmole/organ (% from mother)

AGE	BIRTH	11 DAY	21 DAY
Eye	0.23 ± 0.05 (69)	0.40 ± 0.22 (79)	1.5 ± 0.2 (37)
Intestine	0.36 ± 0.14 (80)	1.6 ± 0.5 (41)	4.9 ± 0.7 (40)
Kidney	0.63 ± 0.14 (64)	1.9 ± 0.5 (37)	6.6 ± 1.6 (42)
Spleen	0.13 ± 0.06 (74)	0.69 ± 0.23 (41)	2.4 ± 0.7 (53)
Stomach	0.26 ± 0.08 (75)	0.78 ± 0.23 (33)	3.2 ± 1.2 (46)
Liver	2.1 ± 0.4 (59)	3.9 ± 1.8 (34)	25.2 ± 6.1 (35)
Lung	0.91 ± 0.19 (59)	3.3 ± 1.9 (32)	7.5 ± 1.6 (35)
Heart	0.58 ± 0.11 (59)	1.9 ± 0.7 (33)	7.7 ± 1.5 (28)
Thymus	0.51 ± 0.11 (69)	0.64 ± 0.18 (32)	2.6 ± 0.7 (46)
Pancreas	0.028 ± 0.014 (101)	0.10 ± 0.05 (62)	0.24 ± 0.10 (53)
Whole Brain	2.2 ± 0.4 (96)	11.5 ± 2.5 (37)	13.5 ± 1.5 (34)
Muscle[a]	6.3 ± 1.2 (50)	8.8 ± 3.8 (30)	9.3 ± 2.3 (40)
Carcass	32 ± 7 (65)	87 ±19 (50)	222 ±43 (55)
Whole Body	41 ± 8 (66)	114 ±26 (46)	297 ±50 (40)

Values are means ± SD of 5 organs.

[a] μmole taurine/g muscle.

Figures in parentheses are the percentage of the taurine in the tissue obtained from the mother *in utero* (birth) or *in utero* and via the milk (11 and 21 days old). This was derived from a comparison of the specific activity in the pups and the mothers.

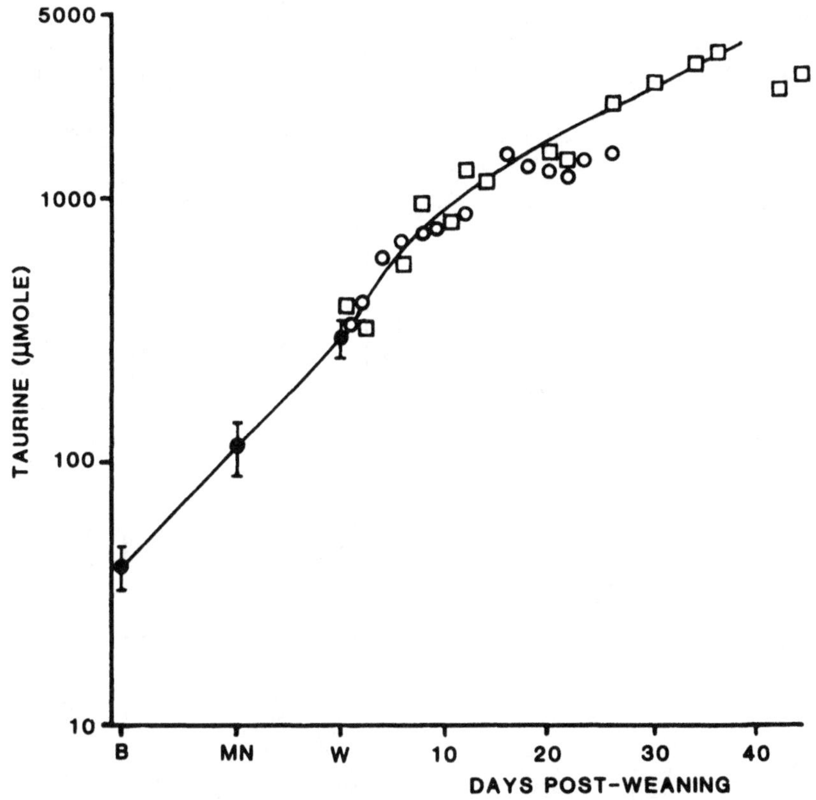

Fig. 2. Total body taurine content of pups as a function of time. Data for birth (B), midnursing (MN, 11 days) and weaning (W, 21 days) are shown as means ± SD for 5 rats. For days post-weaning, values for individual animals are shown. Circles represent rats weaned onto a taurine-deficient diet (Biomix 900), and squares rats weaned onto a taurine-enriched diet (Biomix 900 + 0.4% taurine).

it 2½ days less than for the other organs. Although the total taurine content of these animals is increasing rapidly (Fig. 2), the whole body half life of taurine only changes from 11.4 to 15.0 days when the taurine-containing diet is compared to the deficient one.

The similarity of half-lives in the supplemented rats indicates that exchange between organs is faster than elimination from the body. Thus, visceral organs are in mutual equilibrium. The findings for the brain contrast with other claims that in the neonatal rat taurine turns over very slowly. In one study, the half life of taurine in the brain of newborn rats was measured at 44 days, and in the mothers at 22 days (13). Others have measured half lives in the brain ranging from 0.4 days to 10 days (3,10). From our data,

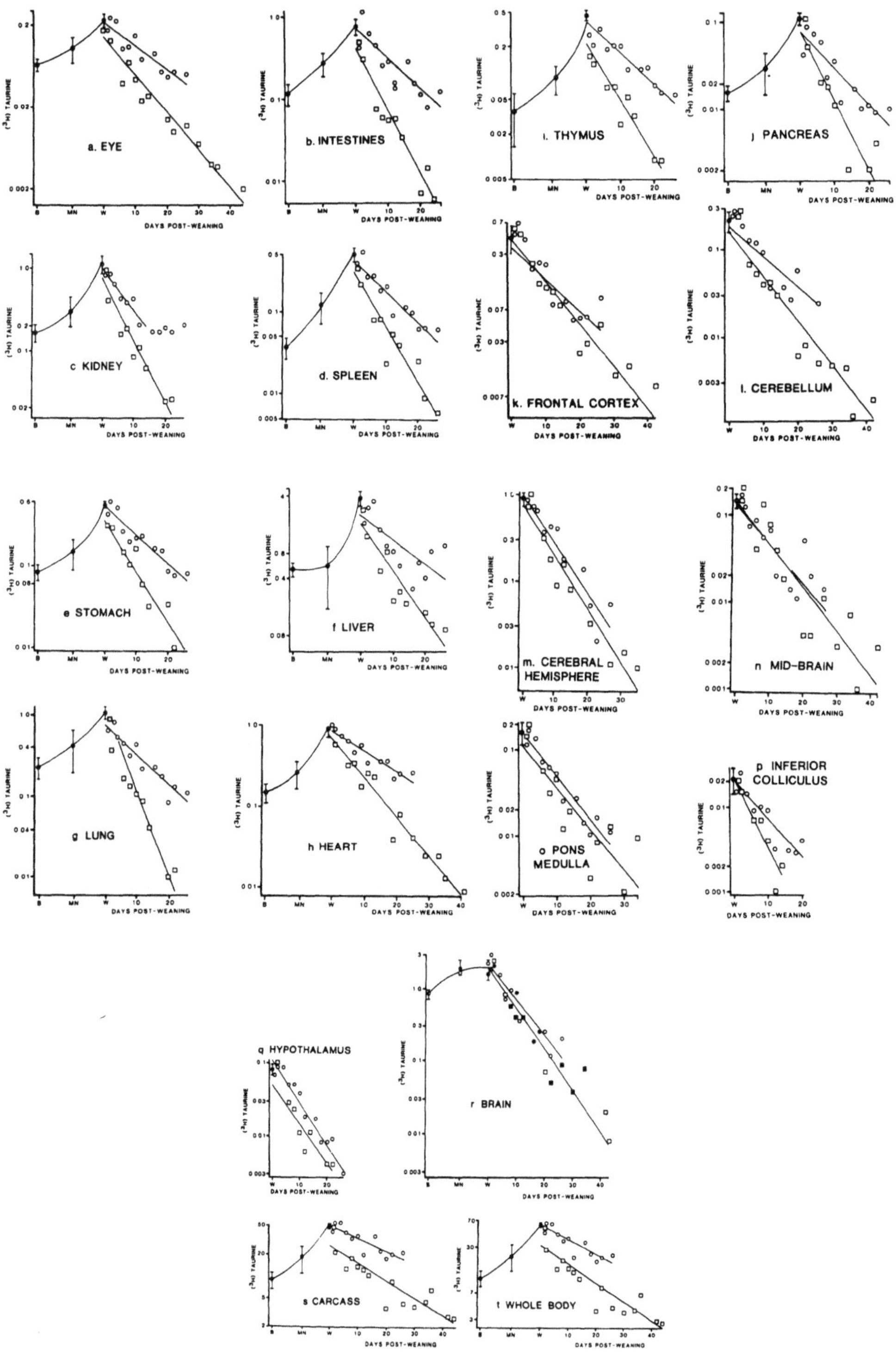
a. EYE
b. INTESTINES
i. THYMUS
j PANCREAS
c KIDNEY
d. SPLEEN
k. FRONTAL CORTEX
l. CEREBELLUM
e STOMACH
f LIVER
m. CEREBRAL HEMISPHERE
n MID-BRAIN
g LUNG
h HEART
o PONS MEDULLA
p INFERIOR COLLICULUS
q HYPOTHALAMUS
r BRAIN
s CARCASS
t WHOLE BODY
(^{3}H) TAURINE
DAYS POST-WEANING

←

Fig. 3. Organ content of (^{3}H)taurine during nursing and after weaning. Symbols and axes as for Fig. 2. Taurine values are shown arbitrarily as μmole ^{3}H-taurine at the specific activity of that in the diet the mothers received (i.e. divide by 0.423 for absolute μmole taurine). Fig. 3r. Filled symbols are males, open symbols are females. Certain brain areas at weaning showed slight but significant differences between the sexes in levels of ^{3}H-taurine accumulation.

it appears that the rapid increase in brain taurine content following birth is not associated with a slower turnover rate compared to the rest of the body.

Rats receiving the taurine-deficient diet show a somewhat shorter half life for taurine in the brain as compared to viscera. It therefore appears that the half life of exchange of taurine from the brain in weanling rats is unaffected by dietary deprivation of taurine. Under these conditions, the viscera are still exchanging taurine at a faster rate than taurine is eliminated from the body. No significant differences occurred in tissue taurine concentrations (i.e. μmole/g) between the supplemented and deprived animals, although the former tended to have slightly higher levels (data not shown). Thus, rats are not responding to dietary deprivation by becoming avid conservers. The similarity of levels also indicates that the concentration of taurine is closely regulated, and that dietary intake, biosynthetic rate and elimination mutually interact to achieve this regulation.

Sturman (10) has reported finding both "fast" and "slow" exchanging pools of taurine in each organ, and that in rats not receiving taurine in the diet, the slow half lives are increased much more than the fast. As the half life data in this experiment are calculated over a three week period, we are probably measuring "fast" half life, as defined by Sturman. However, radioactivity in rats receiving dietary taurine falls to zero over a six week period with no indication of biphasic kinetics of elimination (Fig. 3).

TAURINE IN NEWBORN RATS

Principles

This experiment was designed to answer the following questions: How much taurine does the fetus receive from the mother *in utero*, and how much does the animal synthesize? In addition, how rapidly is the taurine obtained *in utero* lost after birth?

Female rats were labeled as described in Table 7 with either (^{35}S)taurine or (^{3}H)taurine. They were mated, and at birth litters

TABLE 8

HALF-LIFE OF ELIMINATION OF ^{3}H TAURINE PRESENT AT WEANING
(Calculated from ^{3}H-Taurine/organ)

	Taurine-Enriched Diet			Taurine-Deficient Diet		
	-R	Intcpt.	$t_{1/2}$	-R	Intcpt.	$t_{1/2}$
Eye	.97	-0.87	6.9	.93	-0.66	10.4
Intestine	.96	-0.40	4.0	.88	-0.16	8.2
Kidney	.97	-0.14	4.1	.97	0.02	5.9
Spleen	.95	-0.51	4.5	.93	-0.37	8.0
Stomach	.95	-0.45	4.9	.95	-0.28	8.7
Liver	.98	-0.27	5.2	.74	0.37	10.1
Lung	.90	0.16	2.9	.74	-0.10	8.1
Heart	.98	-0.11	6.2	.95	-0.05	12.7
Thymus	.95	-0.66	4.6	.94	-0.47	9.5
Pancreas	.93	-1.11	3.9	.91	-1.11	7.2
Uterus[a]	.98	-1.19	4.8	.29	-1.43	26.9
Testes	.95	-0.66	5.8	.86	-0.53	6.4
Frontal Cortex	.96	-0.35	6.1	.66	-0.45	8.2
Cerebellum	.96	-0.78	5.9	.70	-0.73	8.7
Cerebral Hemispheres	.97	-0.12	4.8	.93	0.00	5.1
Midbrain	.91	-0.84	6.0	.85	-0.86	7.2
Pons Medulla	.87	-0.94	6.1	.96	-0.80	5.9
Inferior Colliculus	.95	-1.63	3.5	.92	-1.69	6.6
Hypothalamus	.83	-1.32	5.7	.97	-0.96	5.2
Whole Brain	.94	0.24	5.5	.94	0.31	6.2
Carcass	.90	1.42	12.5	.86	1.71	16.5
Whole Body	.91	1.51	11.4	.88	1.78	15.0

[a] Uterus, Fallopian tubes and ovaries. These data were obtained from the plots in figure 3, over the periods indicated by the solid lines. R is the correlation coefficient and $t_{1/2}$ the half-life in days (Calculated from the slope of the regression curve: $t_{1/2} = -\frac{\log 2}{\text{slope}}$).

were crossfostered such that animals born to (^{35}S)-labeled mothers were nursed by (^{3}H)-labeled animals, and vice versa. By such means, the amounts of exogenous taurine received before and after birth could be calculated.

Sources of Taurine

Figures 4a - 41 show the contribution of the three sources of taurine (mother *in utero*, milk, endogenous synthesis) to the total taurine pool in various organs as a function of time after birth. Except for the brain, taurine obtained *in utero* is eliminated with first order kinetics; i.e. a semilogarithmic plot of taurine remaining versus time yields a straight line (Fig. 4). For the brain, the taurine obtained *in utero* actually increases after birth, reaching a maximum on day 8, an increase of 38% over that present at birth. For all tissues except the stomach, the taurine obtained from endogenous synthesis follows a growth curve that corresponds to mixed order kinetics, falling between zero and first order. That is to say, a straight line is not obtained either from a linear or semilogarithmic data plot. The same holds true for taurine derived from the milk for liver, lung, brain, thymus, carcass and the whole body. The amount of taurine derived from the milk at a given time is a function of the taurine concentration in the milk, which falls off sharply during lactation, the volume of milk consumed, which increases during nursing, and the rate of elimination of taurine from the body. It is to be expected, therefore, that the amount derived from the milk is a complex function of age.

These data reinforce the conclusion from the previous experiment concerning the taurine-synthesizing ability of the newborn rat. At birth, rats have 42 μmole of taurine of which 68% was obtained from the mother (Table 9). These findings are identical to those of the previous experiment (Table 7). The increase in total taurine content during nursing is also indistinguishable from that in the previous experiment. However, the proportion of taurine synthesized during nursing is greater in this experiment. By 20 days, 83% as opposed to 60% of the total body taurine has been synthesized. Why these values differ in unknown. It may be of no significance, or it may be caused by some stress induced by crossfostering, or the increased handling of the pups.

The half-lives of turnover of taurine obtained *in utero* are shown in Table 10. These may be compared to the half-lives for taurine turnover following weaning, shown in Table 8. For most tissues, taurine turnover is slower following birth then following weaning. The carcass, however, containing the bulk of the taurine in the body, has a half-life of elimination following birth midway between the half-lives of animals on taurine-deficient and enriched diets following weaning.

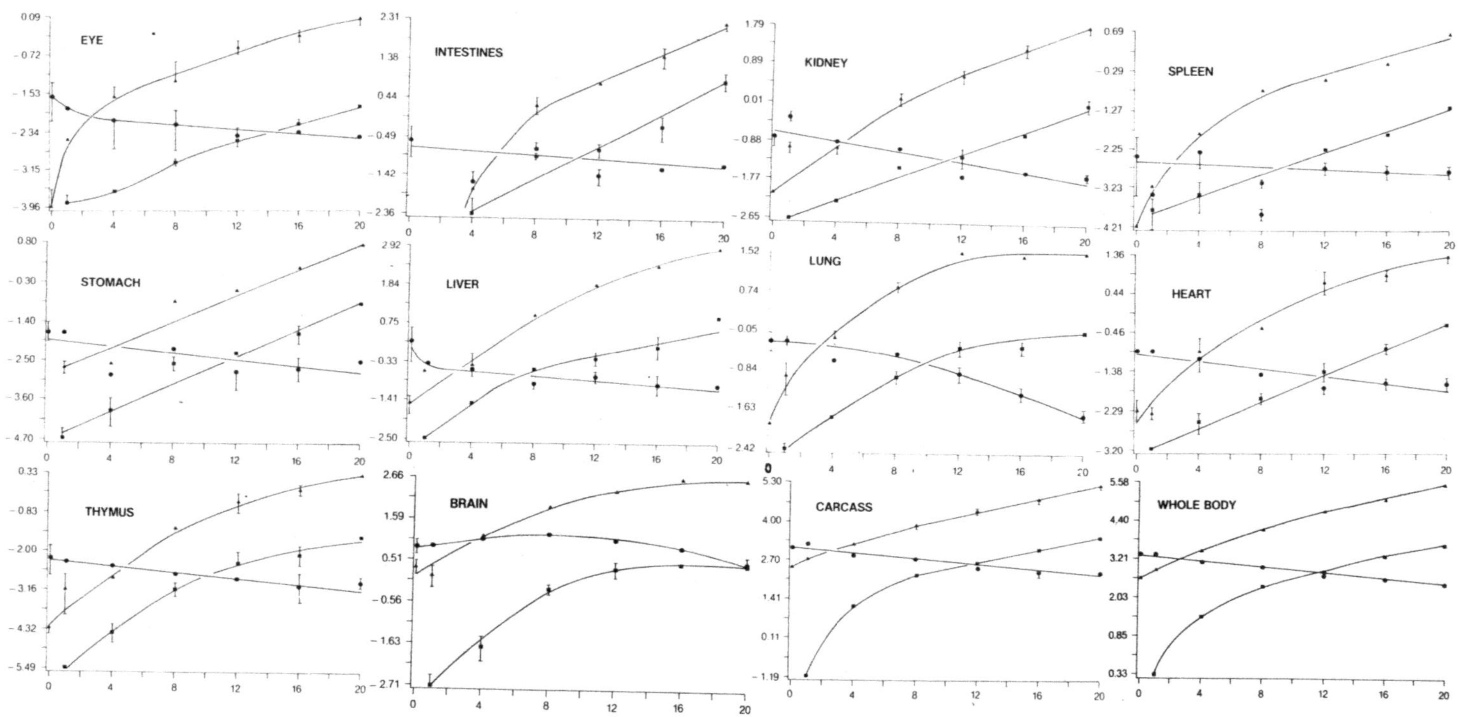

Fig. 4. Sources of taurine during nursing. Conditions as described in legend to Table 9. These figures show for the indicated tissues the total amount of taurine in those tissues derived from the mother *in utero* (●), the nurse via the milk (■), and from endogenous synthesis within the pup (▲).
Ordinates are loge (μmole taurine) and abscissa are days following birth.
Data points are means ± SD. (Deviations too narrow for clarity are not shown)

TABLE 9

TAURINE AND ITS SOURCES IN RAT NEONATES

	AGE (DAYS)						
	0	1	4	8	12	16	20
Total Taurine[a]	42	47	55	95	149	210	318
% *in utero*[b]	68	61	38	22	13	7	4
% Milk[c]	0	2	7	12	12	13	13
% Synthesized[d]	32	38	54	67	78	81	83

Values are for whole animals.
[a] μmole. [b] Percent obtained from mother *in utero*. [c] Percent obtained from nurse via milk. [d] Percent synthesized within pup.

The average doubling times for taurine obtained from milk and endogenous synthesis are also shown on Table 9. Although correlation coefficients are high, inspection of Fig. 4 shows clearly that in many cases the lines are not well described by logarithmic growth. The doubling times as calculated must be used with circumspection, therefore.

SOME CONCLUSIONS

1. Even fetal rats have high biosynthetic rates for taurine.

2. The biological half-life of taurine during nursing is 15 days.

3. The biological half-life of taurine after weaning is 15 days for rats placed on a taurine-deficient diet, and 11 days for rats placed on a taurine-enriched diet.

4. By weaning, rats have obtained more taurine from endogenous synthesis than from their mothers.

5. In rats of all ages, most organs interchange taurine faster than taurine is eliminated from the body.

6. Adult rats on a high taurine diet can obtain up to 58% of their taurine from the diet.

7. When taurine is absent from the diet, rats lower the rate of excretion of taurine, but do not increase the rate of synthesis.

TABLE 10

SOURCES OF TAURINE IN THE NURSING RAT

	Uterine		Milk		Synthesis	
	-R	$t_{1/2}$	R	2t	R	2t
Eye	.91	18.7	1.00	6.1	.94	4.1
Intestine	.35	37.1	.97	3.8	.96	3.0
Kidney	.90	11.7	1.00	5.1	.98	3.9
Spleen	.01	-	.99	5.0	.95	3.2
Stomach	.57	18.8	.98	3.6	.97	3.6
Liver	.85	15.7	.99	4.2	.98	3.3
Lung	.97	10.4	.96	5.6	.91	4.2
Heart	.91	16.4	1.00	4.6	.97	3.6
Thymus	.97	18.3	.97	3.4	.96	3.1
Pancreas	.68	7.1	.55	7.1	.83	5.0
Whole Brain	.45	50.1	.91	4.2	.94	5.1
Carcass	.96	14.6	.91	3.3	.99	4.9
Whole Body	.98	16.3	.94	3.6	.99	4.6

Correlation coefficients (R); half-lives ($t_{1/2}$) and doubling times (2t) are calculated from the data of figure 4, with the assumption of exponential changes with time.

Female rats were maintained on a taurine-deficient diet and on drinking water containing 5mM taurine (either ^{3}H taurine 167,000 dpm/µmole, or ^{35}S taurine 135,000 dpm/µmole). After 2 weeks, they were mated, but kept on radioactive taurine throughout pregnancy and lactation. At birth, pups born to ^{35}S-labeled mothers were fostered by ^{3}H-labeled mothers, and vice-versa.

ACKNOWLEDGMENT

This work was supported by USPHS grant HL 19394.

REFERENCES

1. Azari, J., Brumbaugh, P., and Huxtable, R., 1980, Prophylaxis by taurine in the hearts of cardiomyopathic hamsters, J. Mol. Cell. Cardiol., 12:1353-1366.
2. Chazov, E., Mal'Chikova, L., Lipina, N., Reshetnaya, N., Asafoy, G., and Smirnov, V., 1974, Taurine effect on the electrical activity of the heart, Kardiologiya 14 (7):18-26.
3. Collins, G.G.S., 1974, The rates of synthesis, uptake, and disappearance of (^{14}C) taurine in eight areas of the rat central nervous system, Brain Res., 76:447-458.
4. Darsee, J.R., and Heymsfield, S.B., 1981, Decreased myocardial taurine levels and hypertaurinuria in a kindred with mitral valve prolapse and congestive cardiomyopathy, New Eng. J. Med, 304;129-135.
5. Huxtable, R., and Laird, H., 1978, The prolonged anticonvulsant action of taurine on genetically determined seizure susceptibility, Can. J. Neurol. Sci., 5:215-221.
6. Kaczmarek, L.K., A comparison of the evidence for taurine and GABA as neurotransmitters, in: "Taurine," R. Huxtable and A. Barbeau, eds., Raven Press, New York (1976), pp.283-292.
7. Kibe, C., Wake, C., Kuramoto, T., and Hoshita, T., 1980, Effect of dietary taurine on bile acid metabolism in guinea pigs, Lipids, 15:224-229.
8. National Academy of Sciences: Nutrient Requirements of Domestic Animals, Washington, D.C., (1978).
9. Schmidt, S.Y., Berson, E.L., Hayes, K.C., 1976, Retinal degeneration in cats fed casein 1. Taurine deficiency, Invest. Ophthalmol., 15:47-52.
10. Sturman, J.A., 1973, Taurine pool sizes in the rat: Effects of vitamin B-6 deficiency and high taurine diet, J. Nutrition, 103 (11):1566-1580.
11. Sturman, J.A., 1979, Taurine in developing rat brain: Changes in blood-brain barrier, J. Neurochem., 32:811-816.
12. Sturman, J.A., Rassin, D.K., and Gaull, G.E., 1977, Taurine in developing rat brain: Transfer of (^{35}S) taurine to pups via the milk, Pediat. Res., 11:28-33.
13. Sturman, J.A., Rassin, D.K., and Gaull, G.E., 1977b, Taurine in developing rat brain: Maternal-fetal transfer of (^{35}S) taurine and its fate in the neonate, J. Neurochem., 28:31-39.

STUDIES ON THE RENAL HANDLING OF TAURINE: CHANGES DURING MATURATION AND AFTER ALTERED DIETARY INTAKE

Russell W. Chesney, Aaron L. Friedman, Patti W. Albright, Diane K. Jax, Robben Gingery, and Naomi Gusowski

Pediatric Renal Disease Laboratory
Department of Pediatrics
University of Wisconsin Center for the Health Sciences
600 N. Highland Avenue
Madison, Wisconsin

The main route for the removal of taurine from the body is by urinary excretion. As a consequence, the renal tubular transport and handling of taurine is important in regulating the taurine content in the remainder of the body. In our previous report, we have described the use of stop-flow, free-flow micropuncture, continuous microperfusion and endogenous clearance as techniques to examine the renal handling of taurine (6). This chapter also covered our studies on the accumulation of taurine by thin cortex slices of mouse (4) and rat (5) kidney. These studies indicate that under normal conditions taurine excretion is 4 to 10% of the filtered load and that uptake by slices occurs by two processes: a low-Km, high-affinity uptake system and a high-Km, low-affinity uptake system. The kinetics of taurine accumulation are determined by Lineweaver-Burk analysis and the Eadie-Augustinson transformation. Other characteristics of taurine accumulation include transport by the β-amino acid transport system (such that β-amino acids but not α-amino acids will block uptake), sodium dependency of uptake, dependency on oxidative metabolism for active accumulation and enhanced efflux of taurine when incubated in a solution containing other β-amino acids.

More recently, our laboratory has examined the transport of taurine in slices from immature animals, since young mammals have a physiologic aminoaciduria which disappears with maturity (2). Again, from the clearance *in vivo* and uptake by slices *in vitro*, it was evident that Sprague-Dawley rat pups had reduced net tubular reclamation of taurine and that reabsorptive capacity rose between the 2nd and 4th weeks of life. The excreted taurine fell despite an increase

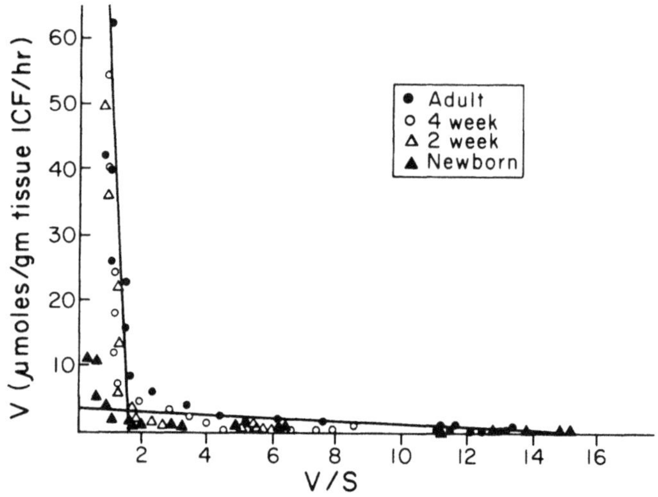

Fig. 1. Concentration-dependent kinetics for taurine uptake by rat cortex slices examined over a 4000-fold variation in substrate concentration. The Eadie-Augustinson transformation reveals at least two modes of transport at each age, indicating that both systems are present at birth. (ICF is intracellular fluid) (Reprinted from ref. 8 with permission of Pediatric Research)

in glomerular filtration rate with maturity, strongly suggesting that transepithelial transport increases with maturation (8). The uptake of taurine again occurs by two transport processes (Fig. 1) so that maturational events in taurine reabsorption do not include the emergence of a previously absent transport site (9). The ontogeny of renal taurine transport in cortex slices appears to involve an increase in the initial rate of accumulation with age, an increase in efflux and a greater dependence on aerobic metabolism.

Younger animals, as has been indicated, have excessive taurinuria and two-week-old rat pups excrete 46% of the filtered taurine load (8). Transport studies were aimed at finding differences between immature and mature animals. Three other phenomena deserve comment. First, preincubation of cortex slices in a taurine-free medium prior to uptake studies led to enhanced accumulation only in immature tissue (9). This finding may be explained by the observation of Blazer-Yost et al (1) indicating that efflux of amino acids from the intracellular pool is much slower in cortex slices taken from neonatal rat kidney. The maintenance of a "normal" amino acid pool may account for this enhanced accumulation either by maintaining the energy required for uptake processes or by the synthesis of new transport proteins. Secondly, preincubation of tissue in a medium containing cyclic nucleotides increases accumulation by 50% in mature cortex slices, but not in immature cortex (Fig. 2a,2b) (9). Thirdly,

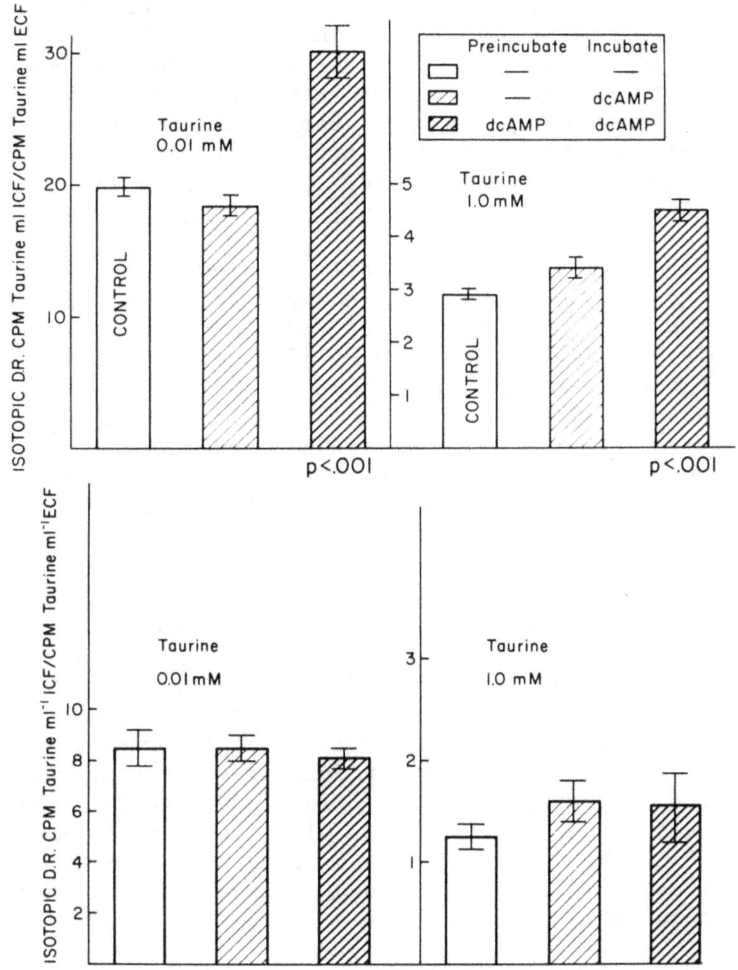

Fig. 2. The influence of dbcAMP on taurine accumulation by cortex slices from adult (A) or newborn (B) animals. Slices were pre-incubated in taurine-free medium for 120 minutes and then incubated in taurine-containing medium for 60 minutes at 37°. The dibutyryl cAMP (dcAMP) level was 1.0 mM. Taurine uptake in adult tissue at 0.01 and 1.0 mM is greater ($p < .0001$) when tissue is both pre-incubated and co-incubated with dibutyryl cAMP. No influence of cyclic nucleotides was found in newborn tissue. (ECF is extracellular fluid; DR is the distribution ratio - see text.) (Reprinted from ref. 9 with permission of Pediatric Research)

the glutathione-oxidizing agent, diamide, enhances taurine efflux from pre-loaded cortex slices in adult animals (5), but does not alter the rate of exit in slices from immature cortex.

Since our studies in slices had detected differences in the initial rate of accumulation, it was important to measure initial rate with accuracy. There are problems with cortex slices, even when cut thin, when measuring initial rate kinetics; thus, it became necessary to move to another system. Because of slice thickness, diffusional constraints, and the real possibility that medium would be unevenly distributed within the slices, we chose to examine taurine accumulation in collagenase-isolated tubule fragments (13).

Other techniques could potentially be used to measure initial rate kinetics in young animals, including continuous microperfusion, as studied by Dantzler and Silbernagl (12), and isolated brush border membrane vesicles (16,20). Hammerman and Sacktor (16), using renal tubular brush border membranes, found two transport systems for the uptake of β-alanine across luminal membranes from adult animals. Rozen et al. (20) report only one taurine uptake system in luminal membranes from adult rats. Yet, the utilization of these techniques in rats of age 2 weeks or less is fraught with immense difficulties. For example, large numbers of rats must be used to obtain only a few mg of brush border membranes.

Isolated tubule preparations are heterogeneous, but brush border membrane preparations also come from numerous regions of the nephron. In addition, brush border membranes do not permit analysis of intracellular events which may effect luminal transport. We have modified the collagenase method of Berg and Orloff (3) to examine taurine uptake and efflux in tubules from animals of various ages and to study other maturational transport events. In a separate but related series of studies, we examined the effect of alteration in dietary taurine and methionine intake on renal transport, both _in vivo_ and _in vitro_. The remainder of this paper will discuss the characteristics of uptake in isolated tubules in detail.

Sprague-Dawley rats less than 36 hours old, two weeks old, four weeks old, eight weeks old and 10 weeks old were used in all studies; the latter two groups were felt to represent adult animals. Timed urine samples were obtained in metabolic cages. Blood samples were taken from the bifurcation of the iliac arteries and carefully centrifuged to obtain plasma free from contamination by platelets. Plasma samples without a visible buffy coat were rejected. The concentration of taurine in protein-free filtrates of plasma and urine was measured in a Beckman-120 automated amino acid analyzer.

The levels of taurine in plasma and urine from animals of different ages are shown in Table I. From these values, it is clear that neonatal and 2-week-old animals have taurinuria far exceeding that found in older animals. Since plasma levels are comparable at all ages, this taurinuria is not of the overflow type but represents a renal tubular taurine leak. The taurinuria of immaturity disappears between two and four weeks of age.

Table 1. Concentrations of taurine in plasma and urine and fractional reabsorption of taurine

	Neonate (≤ 36 hr)	2-Week-Old	4-Week-Old	8-Week-Old	10-Week-Old
Plasma: μmole/L plasma H_2O	(6)246 ± 43	(7)236 ± 56	(11)204 ± 28	(8)332 ± 27	(8)424 ± 89
Urine: μmole/mg creatinine	(6)235 ± 127*	(7)190 ± 36*	(6)6.1 ± 0.4	(9)6.84 ± 0.93	(10)26.1 ± 8
Fractional reabsorption: $\frac{U/P_{taurine}}{U/P_{creatinine}}$	----	(7)0.536 ± 0.179*	(6)0.928 ± 0.003*	(8)0.904 ± 0.013	(8)0.961 ± 0.009

* Different from values in adult animals by $p < .001$. Data are means ± SEM for number of samples shown in parentheses.

U/P: urine:plasma ratio

Tubules were prepared from finely minced renal cortex incubated in medium containing 0.375 to 0.75% collagenase (Worthington II) as described elsewhere (13,3). Studies of taurine uptake were performed at 37° under continuous 95% O_2:5% CO_2 gassing, and incubations were performed using unlabeled taurine plus 1 μCi (0.1 μmole) of C^{14}-labeled taurine. The measurement of intracellular taurine concentration was ascertained in isolated tubules which were placed in tared centrifuge tubes and centrifuged for 10 min at 16,000 rpm at 4°C. Tissue and medium samples were taken and placed in counting solution (Aquasol), and from these the distribution ratio (DR) of uptake could be determined. In effect, the DR is equivalent to the DPM x ml^{-1} ICF divided by DPM x ml^{-1} ECF (medium). Using radio-labeled polyethyleneglycol and drying at 110° for 48 hr, the appropriate percent of the tubule pellet that represents ECF and ICF was estimated (7).

The accumulation of taurine by tubules is concentrative (Fig. 3). The greatest accumulation occurs in Krebs-Ringer bicarbonate medium. No concentrative uptake occurs in the Krebs-Ringer phosphate medium. Using Tris-Ringer medium, uptake was only 40% of that found using the bicarbonate medium. The uptake of taurine at 10 μM reached a steady state at 20 to 30 min in tubules from adult rats, whereas the uptake of taurine reached a steady state only after 90 min (not shown). When uptake over the initial 10 min is compared to uptake at steady state (Fig. 4), it can be appreciated that uptake is more rapid in adult cortex (14).

Initial rate kinetics were studied over 10 min and are shown in Fig. 5. Concentrative uptake over the range 5 μM to 20 μM reveals that both limbs are present in adult and newborn tissue and that Km values are comparable on both limbs. The Vmax at both sites is significantly higher in adult tissue, indicating the same delay in initial rate of uptake as found in slice studies.

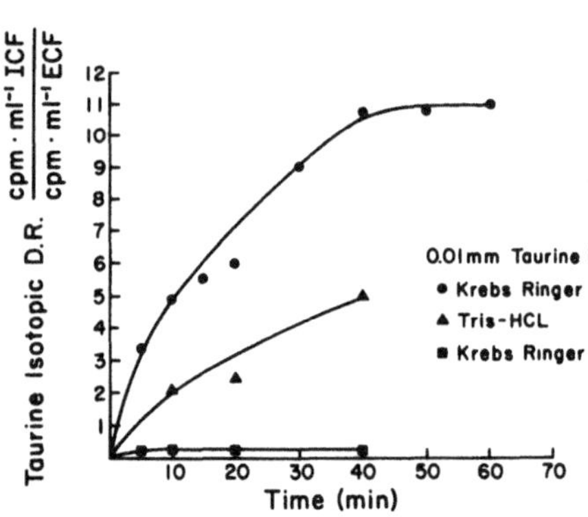

Fig. 3. The time course of taurine accumulation by isolated cortical tubule segments in the three buffers shown. Shown are the mean ± SE of at least four determinations each performed in triplicate. (Reprinted from ref. 7 with permission of Renal Physiology)

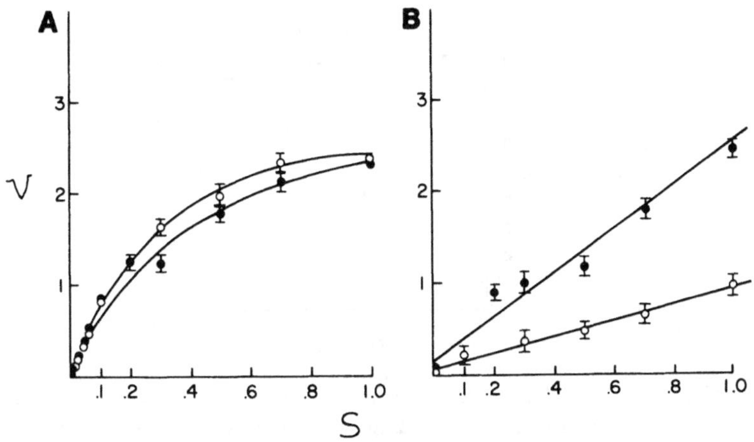

Fig. 4. Velocity of uptake (V) of taurine over concentration (S) range 0.01 to 1.0 mM. Velocity of steady state is measured in adult as 10^{-3} moles x ml^{-1}ICF x 30 min, and in newborn as 10^{-3} moles x ml^{-1}ICF x 90 min. Each point is the mean ± SE of at least four determinations. (Reprinted from ref. 4 with permission of Pediatric Research)

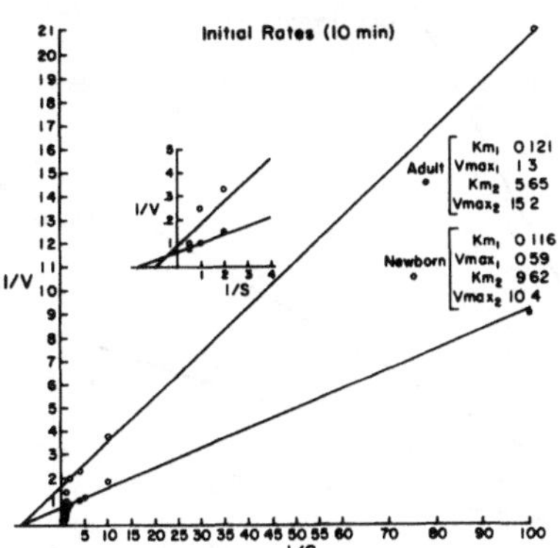

Fig. 5. Lineweaver-Burk double reciprocal analysis of taurine uptake. Concentration range was 0.005 mM to 20 mM taurine and incubation period was 10 min. Insert represents enlargement of analysis at 1/s values less than 1.5. Each point is derived from at least four determinations. V represents velocity and S represents substrate concentrations. (Reprinted from ref. 4 with permission from Pediatric Research)

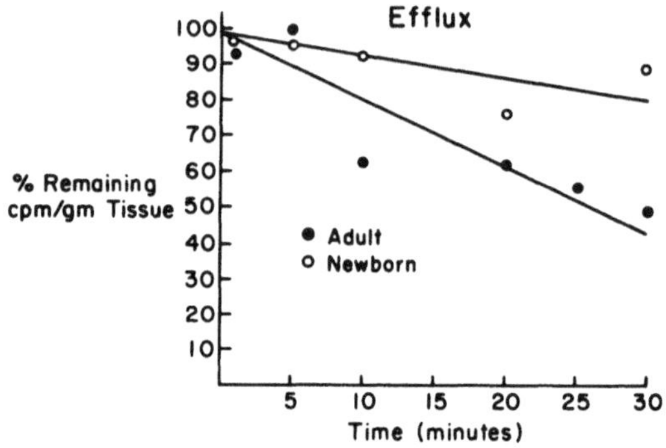

Fig. 6. Efflux of taurine from cortical tubule segments. Ordinate represents percent taurine remaining in tissue as explained in the text. (Reprinted from ref. 14 with permission from Pediatric Research)

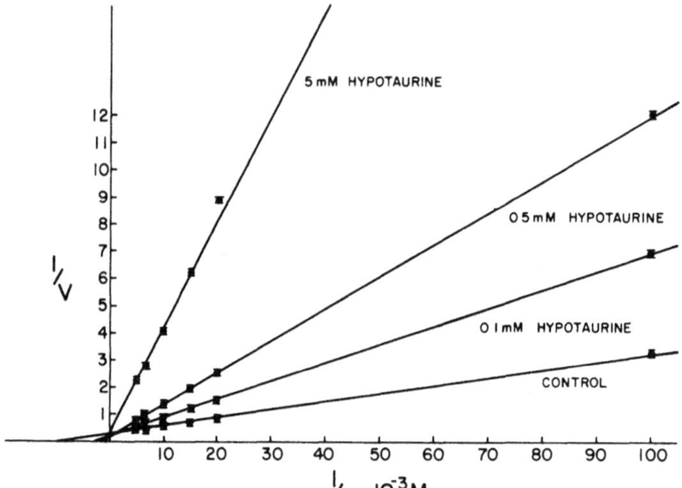

Fig. 7. A Lineweaver-Burk plot of hypotaurine inhibition of taurine accumulation by isolated cortex tubules over the concentration range 0.01 to 20 mM. V represents velocity of transport (micromoles per ml ICF per 30 min). (Reprinted from ref. 13 with permission of Proceedings of the Society for Experimental Medicine and Biology)

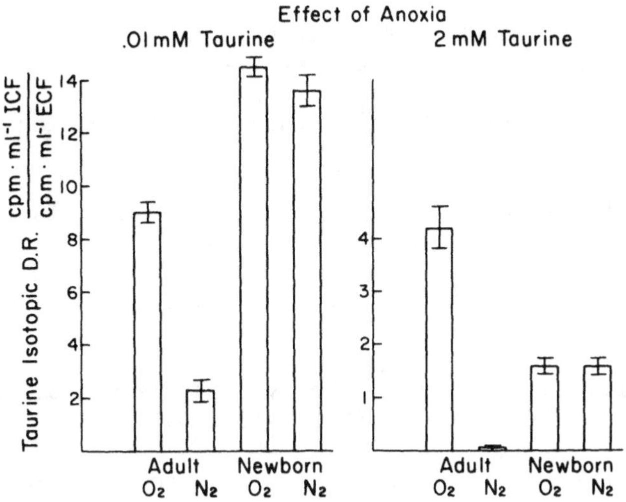

Fig. 8. Effect of incubation under 95% N_2:5% CO_2 atmosphere. Bars represent taurine accumulation in cortical tubules after 30 min incubation in adult and 90 min incubation in newborn. Taurine concentration in the external medium is noted. Data represents mean $\pm$ SE of at least 4 determinations. (Reprinted from ref. 14 with permission of Pediatric Research)

The efflux of taurine from preloaded tubules is slower in neonatal tissue as shown in Fig. 6. Again, efflux characteristics are comparable in slices *in vivo* (8) and *in vitro* (9).

The transport of taurine by tubules is inhibited only by β-amino acids and not by α-amino acids. The influence of hypotaurine on taurine uptake was examined and was found to inhibit uptake in a competitive fashion (Fig. 7). Both uptake limbs at all ages are inhibited by hypotaurine and β-alanine, although the affinity of the β-amino acid transport site is greater for the former compound (15).

Tubules from neonates are relatively resistant to ablation of concentrative uptake by hypoxia or anoxia (Fig. 8). Prolonged survival under hypoxic conditions is a characteristic of very young animals and is perhaps a holdover from the situation *in utero*. Also implicit is that oxidative metabolism may not be as important in taurine accumulation in neonatal cortex as in mature cortex.

Certain definite maturational changes in the renal handling of taurine are apparent from these studies (8,9,11,14). Many of these findings are consistent in both slice studies and isolated tubules. Table 2 is a partial list of differences between accumulation in mature and immature cortex. What is not clear is whether all or only one of these differences can account for physiologic neonatal taurinuria.

Table 2. Difference in characteristics of taurine transport in kidney cortex from young and older animals

1. Renal tubular taurine "leak" persists until weaning at age 4 weeks.

2. Initial rate of accumulation (over 5 to 10 min) is lower in immature cortex slices and isolated tubules.

3. Decreased efflux or rate of exit in immature cortex slices and tubules.

4. Almost full capacity for uptake by immature slices and tubules under conditions of anoxia (95% N_2:5% CO_2).

5. Greater uptake by neonatal cortex in presence of NaCN or iodoacetate.

6. Preincubation in dibutyryl cyclic AMP stimulates increased uptake only in mature cortex.

7. Glutathione oxidation by diamide results in increased efflux from slices only in adult tissue.

8. Preincubation of slices in taurine-free medium enhances uptake only by immature cortex slices at low-Km, high-affinity site.

9. Adaptation to low-taurine diet is evident in tubules from mature, but not immature animals.

Recently, there has been interest in studies of the remal adaptation to change in diet. When man or other mammals are placed on a diet low in sodium or phosphate, they excrete almost no sodium or phosphate in their urine after a period of 2 to 3 days. The reduction in dietary intake serves as a tocsin to the reabsorptive mechanism in the kidney, and these ions are tightly conserved. Dietary phosphate deprivation results in hypophosphaturia, increased uptake of phosphate anion by tubules and slices, greater uptake by brush border membrane vesicles and a resistance to the phosphaturic effect of exogenously-administered parathyroid hormone (17,21,24). Thyroparathyroidectomy, followed by phosphate depletion, causes this same adaptive response. Even the inbred mouse, Hyp/Y, which has X-linked hypophosphatemic rickets and a renal tubular phosphate leak, demonstrates renal adaptation to phosphate deprivation (23).

Renal adaptation to alterations in dietary amino acid intake has not been examined and, indeed, such a study would be difficult to do for a number of reasons. One cannot deprive animals of essential amino acids without greatly altering growth and development. Non-essential amino acids are produced endogenously by an array of intermediary metabolic steps. Frank starvation leads to catabolism of protein, loss of lean body mass and release of amino acids from the intracellular pool.

Two recent studies had examined the effect of alteration in diet on tissue, plasma and urinary taurine concentrations in rats. Sturman (22) found that a vitamin B_6-deficient diet led to a marked reduction in urinary and plasma taurine concentrations, but not in tissues. Supplementation with 0.1% taurine restored urinary taurine excretion and plasma taurine concentration to normal levels. Medina and Lombardini (18) found that a soy protein-based diet fed to rats led to a fall in plasma taurine levels and in many tissues as well. Methionine supplementation (0.41%) and taurine (0.34%) restored plasma and tissue levels to normal. Urinary taurine levels were not measured in this study. We chose to study the influence of change in dietary taurine on the renal transport, of urinary excretion and plasma levels of taurine. Taurine is particularly suitable for such a study, since the availability can be lowered by decreasing the cysteine and methionine content to sub-optimal levels, thereby reducing taurine synthesis. Dietary intake can be increased by direct addition of taurine. Other non-essential amino acids do not permit this type of manipulation, as they can be biosynthesized from several metabolic sources. Since taurine is not incorporated into proteins, catabolism of proteins does not release taurine.

Four experimental diets were used (Table 3). Soy protein was utilized since it is devoid of taurine and deficient in methionine and cysteine (18). Supplementation of the normal taurine diet (NTD) with 0.5% methionine provided levels greater than the minimal daily requirement, since soy protein contains 0.2%, thus providing a total methionine level of 0.7%. The control diet contains a mixture of proteins from many sources and a total protein content of 24%. Results obtained using the control diet (CD) and the NTD were not significantly different. Pregnant rats were placed on a given diet and maintained on this diet during pregnancy and lactation. Weanlings were fed the same diet as their mothers.

The plasma and urinary taurine concentrations are given in Table 4. Rats fed the low-taurine diet (LTD) had a reduction in plasma and urinary concentrations and a significantly greater fractional reabsorption of taurine. The HTD-fed rats had higher plasma taurine levels, a 25-fold increase in urine taurine concentration and a significant reduction in the fractional reabsorption of taurine (Fig. 9). When the urinary excretion pattern is examined in 4-week-old animals, the same changes are detected. Also evident from comparison of the table and the figure is that urinary taurine excretion is comparable in rats fed either the CD or NTD.

Table 3. Diets fed to rats for studies of renal adaptation to altered taurine load

COMPONENT	CONTROL(a) (CD)	LOW(b) (LTD)	NORMAL(b) (NTD)	HIGH(b) (HTD)
Cornstarch	66 %	67 %	66.5 %	63.5 %
Oil -- corn	--	5	5	5
-- soy	5	--	--	--
Salt mix Vitamin mix	8	8	8	8
Protein	24(c)	20(d)	20(d)	20(d)
Methionine supplement	--	--	0.5	0.5
Taurine supplement	--	--	--	3

(a) Lab Blox Chow, Wayne Foodstuffs, Waynesville, Illinois

(b) ICN, Cleveland, Ohio

(c) Protein = fish meal, soy, wheat, casein

(d) Protein = soy

The renal adaptation to dietary alteration was examined in renal tubule segments prepared from adult animals on the different diets. As shown in Fig. 10, the uptake distribution ratio (DR) is highest in tubules from animals fed the LTD and lowest in animals fed the HTD for their entire lives. Furthermore, uptake by tubules from animals fed either the CD or NTD fall between the values found in tubules from animals fed the other diets. From 5 min up to 45 min, the uptake DR is significantly greater in the LTD tubules, and after 10 min is significantly lower in the HTD-fed tubules.

In an attempt to analyze the alterations in transport kinetics after different diets were used, we examined uptake over a variety of

Table 4. Concentration of taurine in plasma and urine of rats on different diets.

	Normal Diet	Low-Taurine Diet	High-Taurine Diet
Plasma Concentration (percent of normal)	(8) 100.0± 8.1	(21) 53.0±12.5*	(10) 344.9± 7.2*
Urine Concentration (percent of normal)	(9) 100.0±13.5	(17) 8.2± 4.2*	(10) 2573 ±13*

The plasma concentration in animals on the normal diet was 332 ± 27 µmoles/liter plasma H_2O, and the urine concentration was 6.84 ± 0.93 µmoles/mg urine creatinine. The symbol (*) indicates different from animals on the normal diet at $p < .001$. Values are means ± SEM for numbers shown in parentheses.

concentrations (0.005 to 20 µM). Using Lineweaver-Burk analysis of the uptake over the first 5 min, a two-limbed system of uptake was found (Table 4). As indicated, dietary alteration did not change the "apparent Km" of uptake on either limb, but the $Vmax_1$ increased significantly in the LTD-fed animals. Although the $Vmax_2$ was increased, this change was not statistically significant.

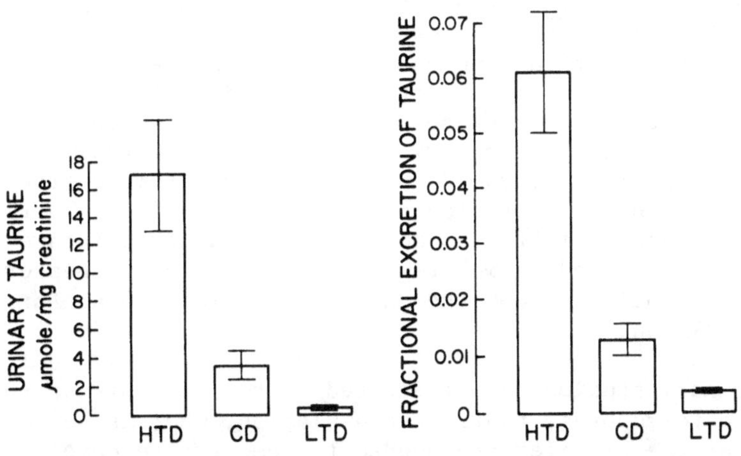

Fig. 9. (a) Urinary taurine excretion by adult rats on the three diets expressed as µmoles/mg creatinine. (b) Fractional excretion of taurine from animals on three diets. The significance of the differences between these values is given in the text.

TAURINE ISOTOPIC DISTRIBUTION RATIO

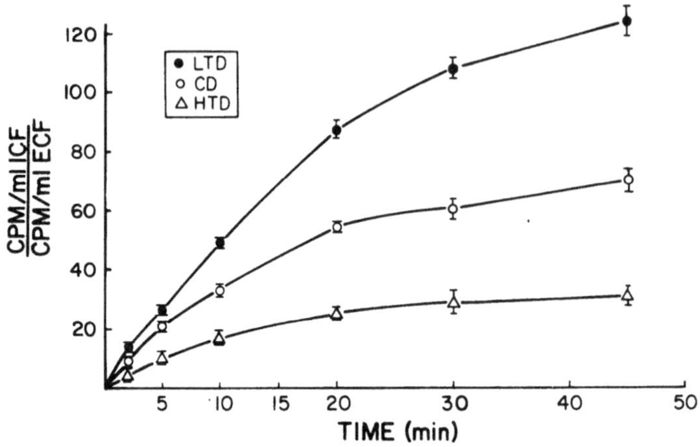

Fig. 10. The time course of taurine accumulation at 10 μmoles/liter in tubules from adult animals fed the three diets. Each point represents the mean ± SE of 12 determinations.

This study shows that there is renal adaptation of taurine transport to changes in dietary taurine and methionine intake. As shown in Table 5, the animals receiving the LTD for their entire lives weighed less than animals fed either the LD or HTD. Animals were not pair-fed, since the LTD-fed animals actually consumed more food. If the CD or HTD-fed animals were pair-fed, their weights would actually be higher.

Both in vivo and in vitro data are concordant in that animals fed the LTD have lower plasma levels, lower urinary excretion, enhanced reabsorption of taurine and greatly augmented uptake. This enhanced uptake involves mainly the low-Km, high-affinity system, since uptake at 2 μM demonstrates much smaller differences when tubules from animals fed the various diets are compared. The enhanced accumulation appears to involve the acquisition of more transport sites, rather than an alteration of the affinity of the carrier to taurine. The mechanism of this adaptation to altered intake may be likened to the "up" and "down" regulation of hormone receptors by substrate availability. Nonetheless, the precise signal for these changes is unknown.

Preliminary studies have indicated that three days of the HTD will suffice to suppress taurine accumulation, whereas between 3 and 6 days of the LTD diet are needed to demonstrate complete adaptation of tubule uptake. Again, a kinetic analysis indicates that the $Vmax_1$ is altered by these dietary changes of only a few days duration.

Table 5. Kinetic parameters of taurine accumulation by collagenase isolated tubules in 4-week-old rats fed different diets

Diet	Weight* (gm)	Km_1 (μM)	$Vmax_1$**	Km_2 (μM)	$Vmax_2$**
CD	185 ± 8.5	0.059	0.79	2.46	9.82
HTD	180 ± 5.6	0.074	0.672	3.17	8.55
LTD	133 ± 4.0^{+}	0.050	1.39^{+}	2.41	12.03^{++}

* Mean ± SE for 12 samples per group
** Values given in μmole/ml intracellular fluid/5 min
+ $p < .005$
++ $p < .1$ and $>.05$

An obvious question is whether immature rats will have this renal adaptation to decreased taurine intake. Preliminary studies do not show the increase in uptake when tubules from 2-week-old rats are used. Nonetheless, the breast milk levels of taurine in nursing mothers fed the various diets are not known, and caution is indicated before interpreting any data from pups who are still nursing.

The studies reported in this paper have indicated that the kidney is important in regulating the total body level of taurine. Urinary excretion patterns reflect the ability of isolated tubules to accumulate taurine. When the reabsorption of taurine from the lumen of the renal proximal tubule is reduced, such as in immature animals or following administration of increased dietary taurine, the uptake by tubules is also decreased. Conservation of taurine is seen with maturity and following the switch to a diet that restricts the amount of taurine synthesized. Again, studies in tubules indicate greater uptake. Future studies using isolated luminal membrane vesicles will indicate whether these maturational and adaptive changes identified are taking place at a membrane level.

ACKNOWLEDGMENTS

This research was supported by funds from the Medical School and Graduate School of the University of Wisconsin in Madison and from NIH grants AM-19489 and HD-05484. Dr. Chesney is the recipient of a Research Career Development Award K04 AM00421.

REFERENCES

1. Blazer-Yost, B., Reynolds, R., and Segal, S., 1979, Amino acid content of rat renal cortex and the response to in vitro incubation, Am. J. Physiol., 236:F398-F404.
2. Brodehl, J., and Gellisen, K., 1968, Endogenous renal transport of free amino acids in infancy and childhood, Pediatrics, 42: 395-404.
3. Burg, M.B., and Orloff, J., 1962, Oxygen consumption and active transport in separated renal tubules, Am. J. Physiol., 203:327-330.
4. Chesney, R.W., Scriver, C.R., and Mohyuddin, F., 1976, Localization of the membrane defect in transport of taurine by parallel studies in vivo and in vitro in hypertaurinuric mice, J. Clin. Invest., 57:183-193.
5. Chesney, R.W., and Jax, D.K., 1977, Heterogeneity of the β-amino preferring transport system in rat kidney cortex; differential influence of glutathione oxidation, Biochim. and Biophys. Acta., 466:84-96.
6. Chesney, R.W., Jax, D.K., Scriver, C.R., and Mohyuddin, F., Taurine transport in mammalian kidney, in: "Taurine in Neurological Disorders," R. Huxtable, and A. Barbeau, eds., Raven Press, New York (1978), pp.73-94.
7. Chesney, R.W., Jax, D.K., Mohyuddin, F., and Scriver, C.R., 1978, In vitro use of polyethylene - (1,2^{14}C)-glycol to measure extracellular space in renal cortex slices from neonatal, immature and adult animals, Renal Physiol., 1:166-170.
8. Chesney, R.W., and Jax, D.K., 1979, Developmental aspects of renal β-amino acid transport; I. Ontogeny of taurine reabsorption and accumulation in renal rat cortex, Ped. Res., 13:854-860.
9. Chesney, R.W., and Jax, D.K., 1979, Developmental aspects of renal β-amino acid transport; II. Ontogeny of uptake and efflux processes and effect of anoxia, Ped. Res., 13:861-867.
10. Chesney, R.W. and Jax, D.K., 1979, The influence of glutathione oxidation on renal cortex taurine transport, Life Sci., 25:1497-1506.
11. Chesney, R.W., Friedman, A.L., Albright, P.W., and Gingery, R.R., The use of taurine as a model of developing amino acid transport processes, in: "Developmental Nephrology," A. Spitzer, ed., in press.
12. Dantzler, W.H., and Silbernagl, S., 1976, Renal tubular reabsorption of taurine, γ-aminobutyric acid (GABA) and β-alanine studied by continuous microperfusion, Pfluegers Arch., 367:123-128.
13. Friedman, A.L., Jax, D.K., and Chesney, R.W., 1980, β-amino acid transport in isolated tubule segments, Renal Physiol., 2:21-28.
14. Friedman, A.L., Jax, D.K., and Chesney, R.W., Developmental aspects of renal β-amino acid transport; III. Ontogeny of transport in isolated renal tubule segments, Ped. Research, in press.
15. Gingery, R., and Chesney, R.W., 1980, The influence of hypotaurine on taurine transport in isolated renal cortex tubules, Proc. of Soc. for Exp. Biol. and Med., 164:18-22.

16. Hammerman, M., and Sacktor, B., 1978, Transport of β-alanine in renal brush border membrane vesicles, Biochim. Biophys. Acta., 509:338-347.
17. Kempson, S.A., and Dousa, T.P., 1979, Phosphate transport across renal cortical brush border membrane vesicles from rats stabilized on a normal, high or low phosphate diet, Life Sci., 24: 881-888.
18. Lombardini, J.B., and Medina, E.V., 1978, Effects of dietary inorganic sulfate, taurine and methionine on tissue levels of taurine in the growing rat, J. of Nutrition, 108:428-433.
19. Rogers, Q.R., and Harper, A.E., 1965, Amino acid diets and maximal growth in the rat, J. of Nutrition, 87:267-273.
20. Rozen, R., Tenenhouse, H.S., and Scriver, C.S., 1979, Taurine transport in renal brush border membrane vesicles, Biochem. J., 180:245-248.
21. Steele, T.H., Underwood, J.L., Stronberg, B.A., and Larmore, C.A., 1976, Renal resistance to parathyroid hormone during phosphorus deprivation, J. Clin. Invest. 58:1461-1464.
22. Sturman, J.A., 1973, Taurine pool sizes in the rat: Effects of vitamin B-6 deficiency and high-taurine diet, J. of Nutrition, 103:1566-1580.
23. Tenenhouse, H.S., and Scriver, C.R., 1979, Renal adaptation to phosphate deprivation in Hyp mouse with X-linked hypophosphatemia, Canad. J. Biochem., 57:938-944.
24. Trohler, U., Bonjour, J.P., and Fleisch, H., 1976, Inorganic phosphate homeostasis; renal adaptation to the dietary intake of phosphate in intact and thyroparathyroidectomized rats, J. Clin. Invest., 54:264-273.

TAURINE AND TAPETUM STRUCTURE

J. A. Sturman[1], G. Y. Wen[1], H. M. Wisniewski[1],
W. H. Niemann[2] and K. C. Hayes[3]

[1]Department of Pathological Neurobiology
Institute for Basic Research in Mental Retardation
Staten Island, New York
[2]Department of Experimental Pathology and Toxicology
Hoffmann-La Roche Inc.
Nutley, New Jersey
[3]Department of Nutrition
Harvard School of Public Health
Boston, Massachusetts

The tapetum lucidum is a multi-layer reflecting surface situated in the eyes of a number of vertebrates (Fig. 1). It is located behind the photoreceptors and gives them a second opportunity to absorb light not initially absorbed (see 15 for review). Our interest in the physiological role of taurine and recent discovery that dietary taurine deprivation causes disorganization and degeneration of the tapetum lucidum of the cat (21) lead to the present investigations of the chemistry and structure of the tapetum lucidum. We have further examined the tapetum in normal and taurine-depleted cats and in a naturally occurring tapetal abnormality in the beagle dog.

The tapetum lucidum of the cat's eye consists of layers of specialized cells lying between the light-sensitive cells of the retina and the vascular layer of the choroid which form the highly reflecting surface sometimes called "eyeshine". The cat's eye reflects approximately 130 times more light than does the human eye, for example (16). The feline tapetum consists of modified choroidal cells, which are rectangular in cross-section and hexagonal when sectioned tangentially (see Fig. 2). The individual cells are packed with long slender regularly-spaced rods 0.1 μm by 4 to 5 μm (3) with their long axes roughly parallel to the plane of the surface of the retina.

The existence of an inherited tapetal abnormality in the beagle dog was first described by Bellhorn et al. (2). This condition appears to be inherited as an autosomal recessive trait, a conclusion that is supported by the additional information presented here. Beagles with this condition do not have an ophthalmoscopically visible tapetum but rather have a uniform brownish-red fundus reflex. The original study clearly demonstrated that such dogs did have a histologically verifiable tapetum lucidum and electron microscopic examination showed that the tapetal cells were normal in out-line (2). However, the tapetal cells lacked or contained only few electron-dense tapetal rods which are present in the normal beagle tapetum. Instead, the abnormal tapetal cells appeared to contain inclusions surrounded by a membrane, and some electron-dense inclusion bodies which appear to be spherical or ellipsoidal in shape. These last structures resemble some features described in the degenerating tapetal cells of the taurine-depleted cat (21).

The composition of the tapetal rods of the cat has been identified as free riboflavin (7) which contrasts with that of the dog which is zinc cysteinate (19,20). In each species, the compound in question is present in greater amounts than found in other tissues. In addition, we examined the tapetum lucidum of the cow, which is composed of extracellular collagen fibrils (15). In all three species the regular arrays of tapetal rods allow constructive interference.

We have measured the taurine concentration in the tapetal area of the choroid and in the non-tapetal area of the choroid of eyes from these three species, including nutritionally taurine-depleted cats (Table 1). The concentration of taurine in the tapetal area of the cat choroid is approximately twice that of the choroid itself (Table I). Both tapetum and choroid taurine concentrations were decreased to a similar extent in cats maintained for 9 months on a taurine-free diet. Further confirmation of the presence of taurine in the tapetum is provided by the autoradio-graphic localization of [^{35}S]taurine in the tapetal region (Fig. 3). The heavily labelled layers of retina are the regions occupied by the photo-receptors and the tapetum cells.

Electron microscopic examination of the tapetum lucidum of taurine-depleted cats reveal cells in various degrees of disorganization (Fig. 2), as previously reported in a

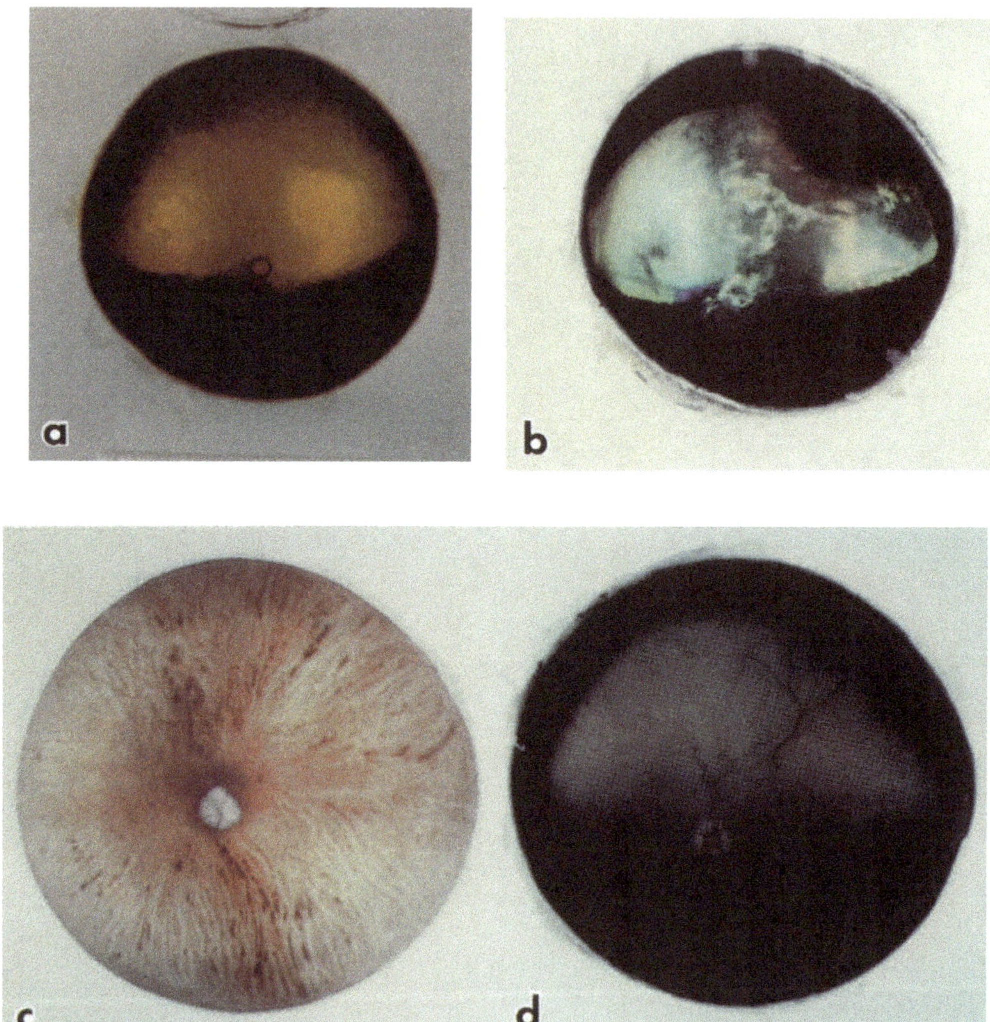
a
b
c
d

Fig. 1. Tapetum of (a) normal cat, (b) cow, (c) normal beagle, and (d) beagle with tapetal abnormality. The photographs of the cat and cow are taken from freshly removed, hemisected eyes whereas those of the beagles are taken after perfusion with glutaraldehyde saturated with H_2S. This procedure does not alter the appearance of the tapetum of beagles with the tapetal abnormality, but darkens the color and reflection of the tapetum of normal beagles.

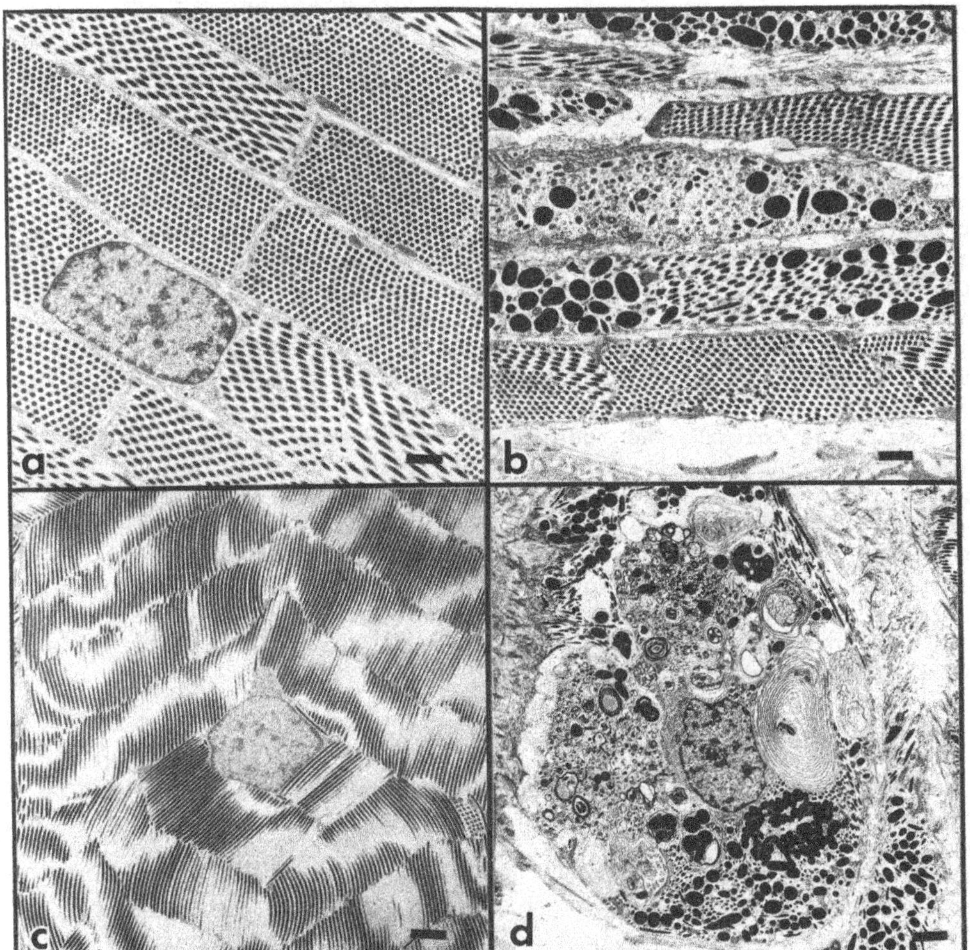

Fig. 2. Electron micrographs of cross sections of center of tapetum from (a) cat fed synthetic diet supplemented with taurine, and (b) cat fed synthetic diet alone (Scale bars 1 μm). Electron micrographs of tangential sections of center of tapetum from (c) cat fed synthetic diet supplemented with taurine, and (d) cat fed synthetic diet alone (Scale bars 2 μm). The tapetal cell illustrated in (d) is an advanced state of degeneration.

preliminary communication (21). Examination of tapetal rods at high magnification suggests that the electron-dense core is surrounded by a membrane (Fig. 4). Of especial note is the observation that tapetal rods of tapetal cells from taurine-depleted cats which look apparently normal in the low power electron micrographs (i.e. the cells have the normal shape, the nucleus has the usual sharp angles, the regular spacing and shape of the tapetal rods appears normal) can be seen to have some changes when examined at high magnification (Fig. 4). The membrane surrounding the tapetal rods appears fragmented and disorganized.

Further examination of the tapetal rods using an electron microscope equipped with a stage tilting device and electron probe (TEMSCAN JEM-100 CX) revealed that the membrane does indeed completely surround the tapetal rod. The electron probe analysis indicated that sulfur is present in this membrane, and that the amount of sulfur is decreased in taurine-depleted cats. No sulfur could be detected in the membranes surrounding the tapetal cells, nor in the intercellular space. This observation rules out the possibility of contamination with sulfur from the embedding materials. In addition, the electron probe analysis indicated that large amounts of zinc were associated with the membrane surrounding the tapetal rods, but not with the core of the rods themselves. This observation was confirmed by the results of perfusion with H_2S-saturated glutaraldehyde for fixation of the zinc, and detection as silver grains. The electron micrograph (Fig. 5), clearly shows that the zinc is on the outside of the tapetal rods, possibly associated with the membrane, and that no zinc occurs within the core of the rod. It is of note that the amount of zinc detectable by this method is substantially reduced in the tapetum of a taurine-depleted cat (Fig. 5).

Our conclusion from the high magnification electron micrographs and the results of examination of tilted sections is that each tapetal rod is completely surrounded by a membrane. Furthermore, since this membrane contains sulfur, as shown by electron probe analysis, which is decreased in amount in taurine-depleted cats, we conclude that taurine is structurally related to this membrane. Thus, when the taurine is depleted from the tapetum, as it clearly is in cats fed taurine-free diets (Table 1), the membrane disintegrates (Fig. 4). We suggest that this is the first stage of the degenerative process, which is followed by loss of shape of the tapetal rods and formation of the electron-dense droplets (Fig. 2).

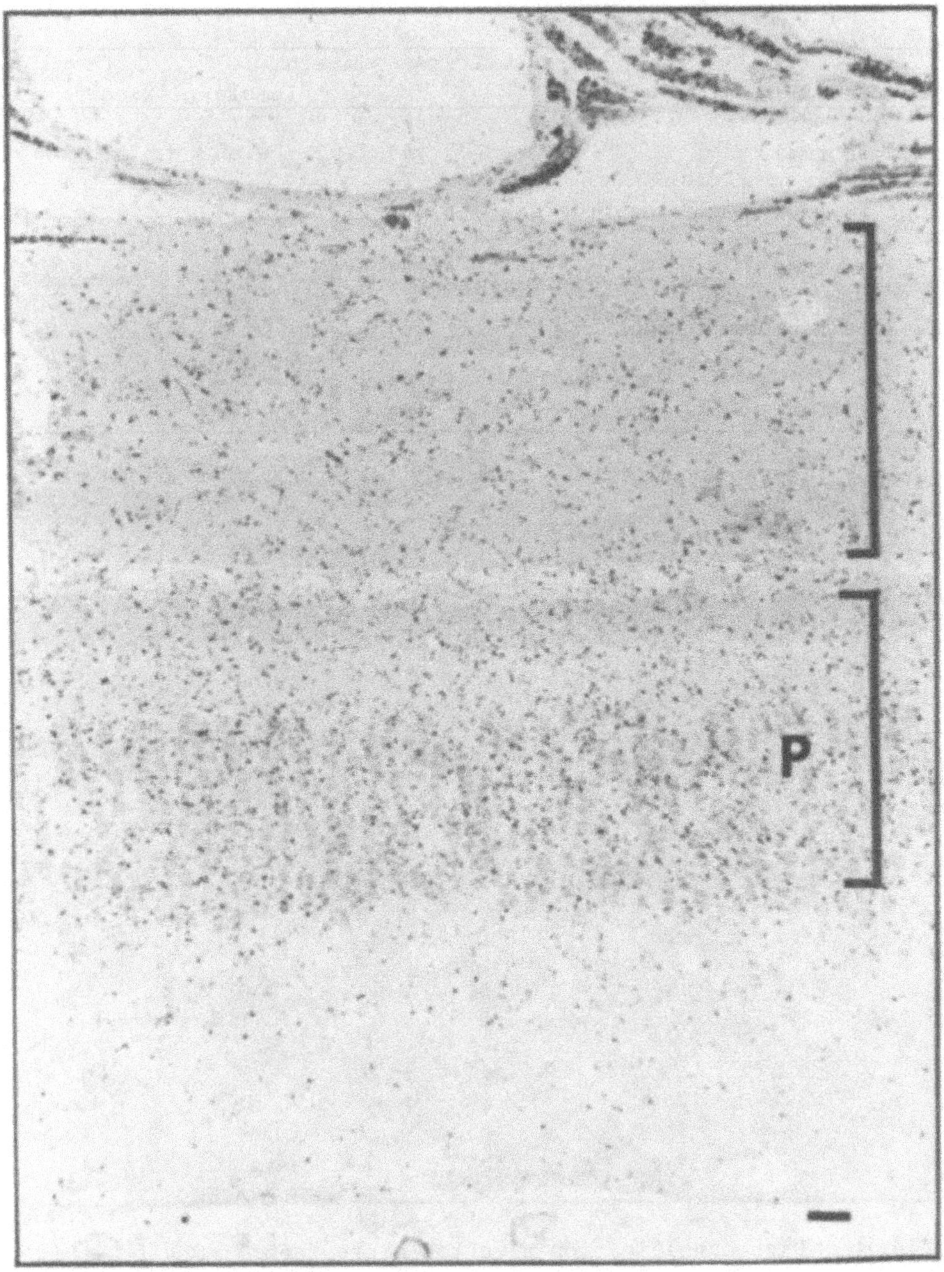

Fig. 3. Autoradiograph of cat retina 24 hr after intravenous injection of 5 mCi (^{35}S)taurine. Heavy labelling is evident in the region of the tapetal cells (T) and in the region of the photoreceptor (P) cells (Scale bar 10 μm).

Table 1. Taurine concentrations in tapetum and choroid

Species		Tapetum	Choroid
		(μmole/g tissue)	
Cat (Normal)	(4)	7.92 ± 0.64	3.57 ± 0.59
Cat (9 months taurine-free diet)	(3)	2.31 ± 0.77	1.92 ± 0.43
Beagle (Normal)	(8)	5.48 ± 0.43	3.69 ± 0.10
Beagle (abnormal tapetum)	(6)	5.94 ± 0.55	3.94 ± 0.23
Cow	(7)	2.53 ± 0.21	3.73 ± 0.45

Each value represents the mean ± SEM from the number of animals in parentheses.

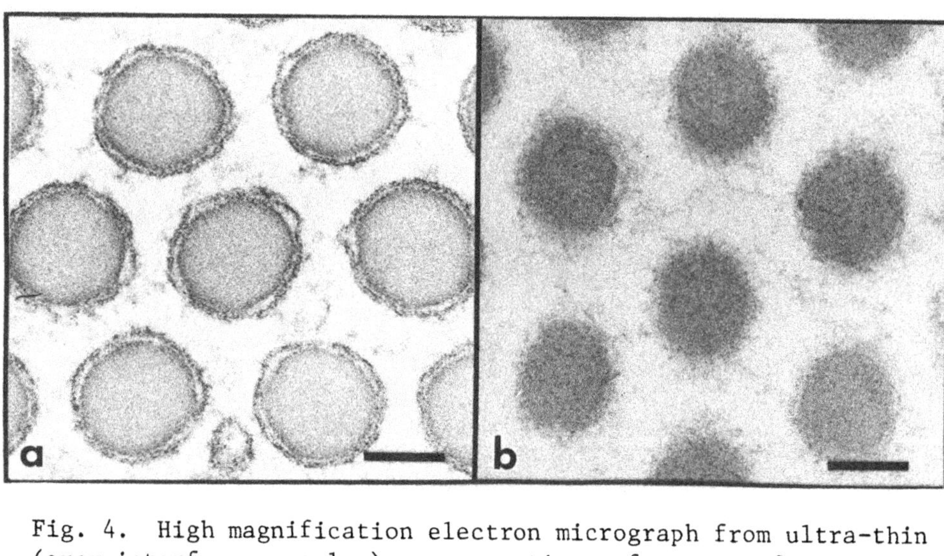

Fig. 4. High magnification electron micrograph from ultra-thin (gray interference color) cross sections of center of tapetum from (a) cat fed synthetic diet supplemented with taurine, and (b) cat fed synthetic diet alone (Scale bars 0.1 μm).

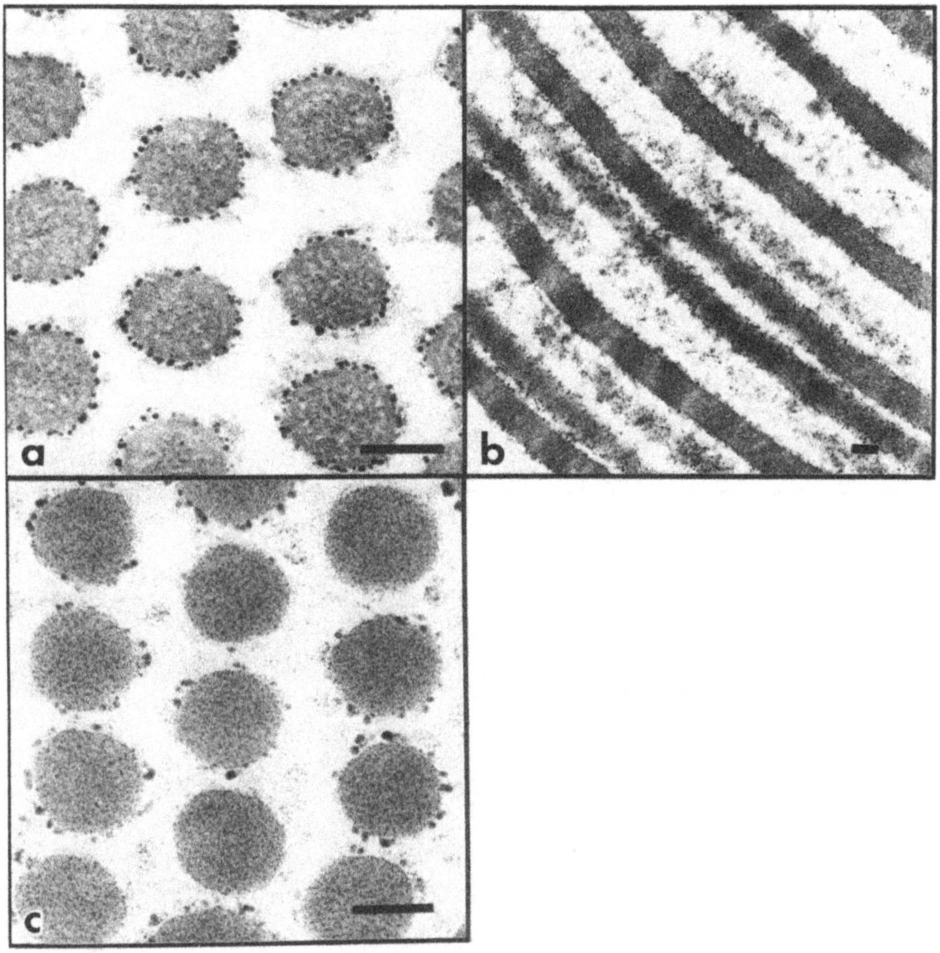

Fig. 5. Ultrastructural localization of zinc in the cat tapetum by the Timm's sulfide-silver method. (a) Cross section and (b) longitudinal section of center of tapetum from cat fed synthetic diet supplemented with taurine (c) cross section of center of tapetum from cat fed synthetic diet alone (Scale bars 0.1 μm).

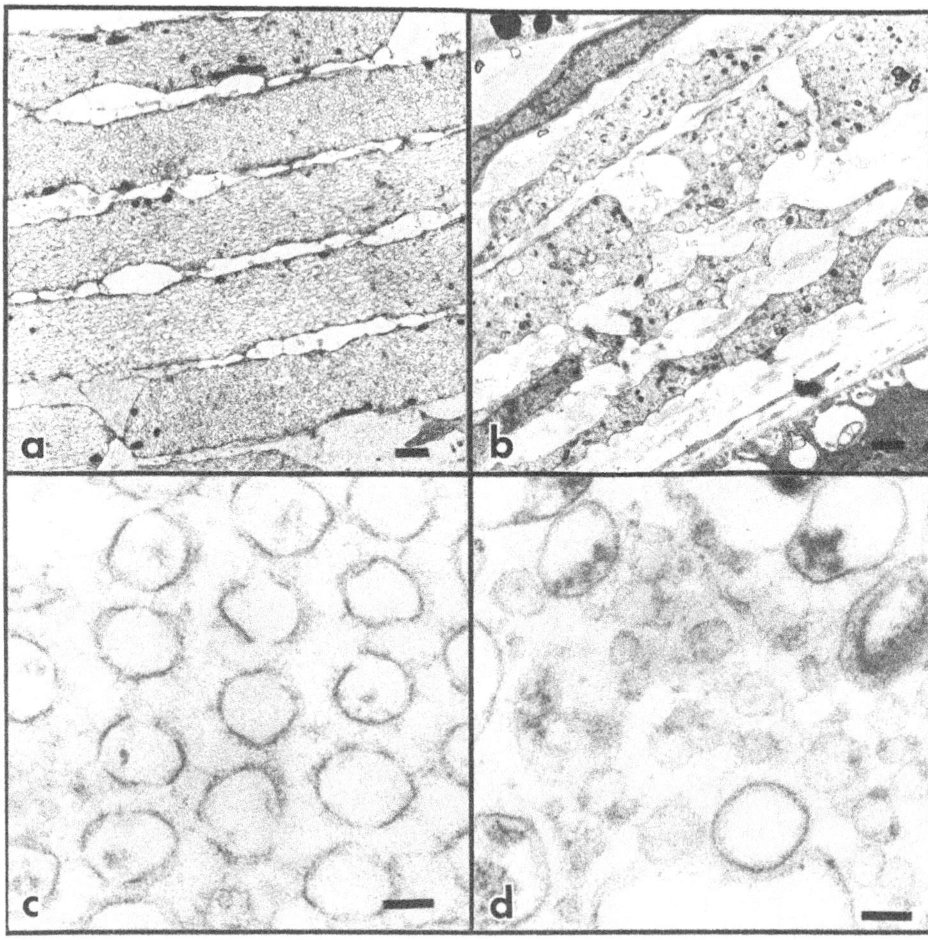

Fig. 6. Electron micrographs of cross sections of center of tapetum from (a), (c) normal beagle, and (b), (d) beagle with inherited tapetum abnormality. Scale bars, (a) and (b), 1 μm; scale bars (c) and (d), 0.1 μm.

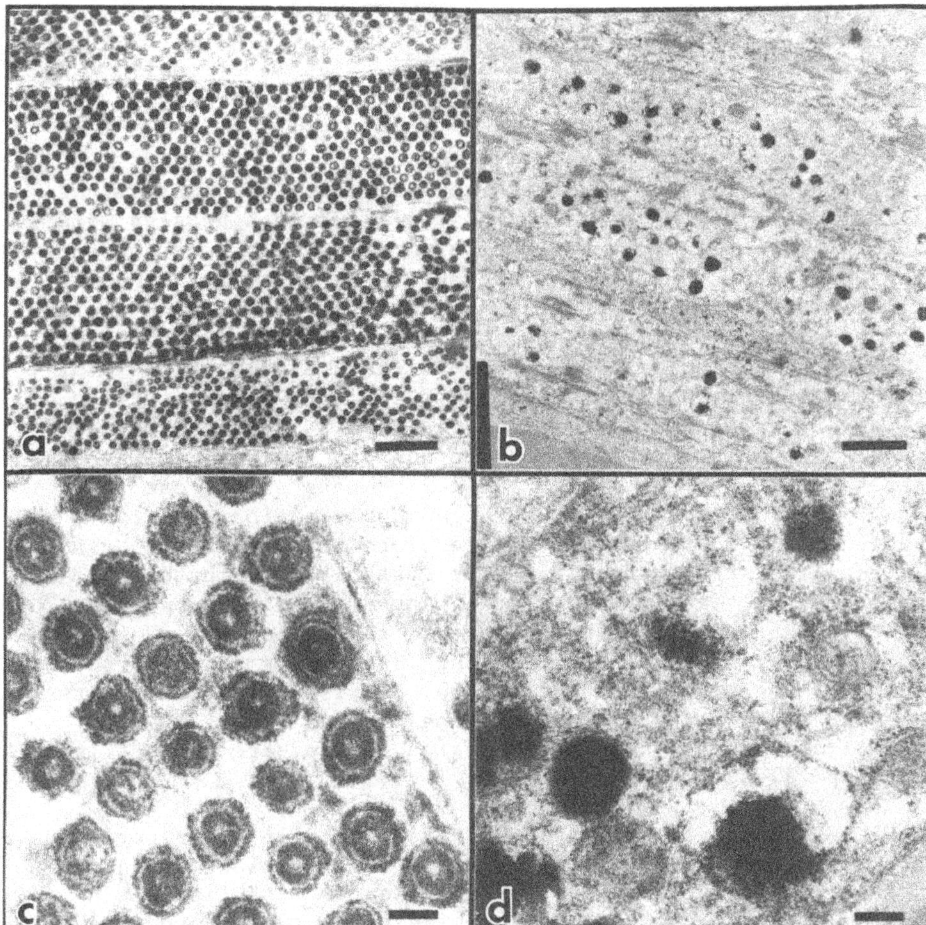

Fig. 7. Ultrastructural localization of zinc in the beagle tapetum by the Timm's sulfide-silver method. Cross sections of center of tapetum from (a), (c) normal beagle, and (b), (d) beagle with inherited tapetum abnormality. Scale bars, (a) and (b), 1 μm; scale bars, (c) and (d), 0.1 μm.

Our results confirm the presence of large amounts of zinc in the feline tapetum, first reported in 1954 by chemical analysis (20), and demonstrate for the first time that the zinc surrounds the tapetal rods, and is not part of the core of the tapetal rods. Whether or not the zinc is involved in the reflecting phenomenon is not known. It seems more likely that it is involved with the membrane structure, especially because of the association of zinc and taurine in other tissues (1,10,11), and the likelihood that the tapetum of all species reflects light by interference (6).

The results of the studies with the beagle dog confirm the original observations of Bellhorn et al. (2) that beagles with no ophthalmoscopically visible tapetum lucidum do have such a structure, albeit altered considerably from that of the normal beagle. Electron microscopic examination of the tapetum lucidum from normal beagles shows the highly organized layers of cells filled with regular arrays of tapetal rods (Fig. 6). It is of note that the tapetal rods are not as electron dense as those of the cat. The last finding may be the result of the different compositions of the tapetal rods in the two species, in the dog consisting of zinc cysteinate (19,20) whereas in the cat they consist of riboflavin (7). The high magnification electron micrographs indicate that each tapetal rod is surrounded by a membrane-like structure, similar to that observed in the cat. Ultrastructural examination of the tapetum from the abnormal beagles confirmed the observation that the cells have a relatively normal shape and that the basic abnormality is the absence of the tapetal rods (Fig. 6). It is of speical interest that many of the tapetal rod membranes appear intact. Other abnormal structures, such as inclusion bodies, are present in these tapetal cells also.

Ultrastructural examination of tapetal cells after perfusion with glutaraldehyde saturated with H_2S for fixation of zinc and later detection as silver grains clearly establishes the presence of large amounts of zinc within the tapetal rods of normal beagles, and its virtual absence from the tapetal rods of the abnormal beagles (Fig. 7). This observation was confirmed by the absence of any significant amounts of zinc detectable by atomic absorption spectrometry (approximately 100 ppm in the abnormal beagle tapetum compared with approximately 25,000 ppm in the normal beagle tapetum). The virtual absence of zinc in the tapetal rods of the abnormal beagles is accompanied by the absence of detectable cysteine (Table 2), which is normally present in very large amounts. The taurine concentration in the tapetum of the abnormal beagles is unchanged, however (Table 1).

Table 2. Cysteine concentrations in the beagle tapetum and choroid

	Tapetum (μmole/g)	Choroid
Normal	(8) 241 ± 18	0
Homozygote	(6) 0	0
Heterozygote	(4) 248 ± 21	0

Each value represents the mean ± SEM from the number of animals in parentheses.

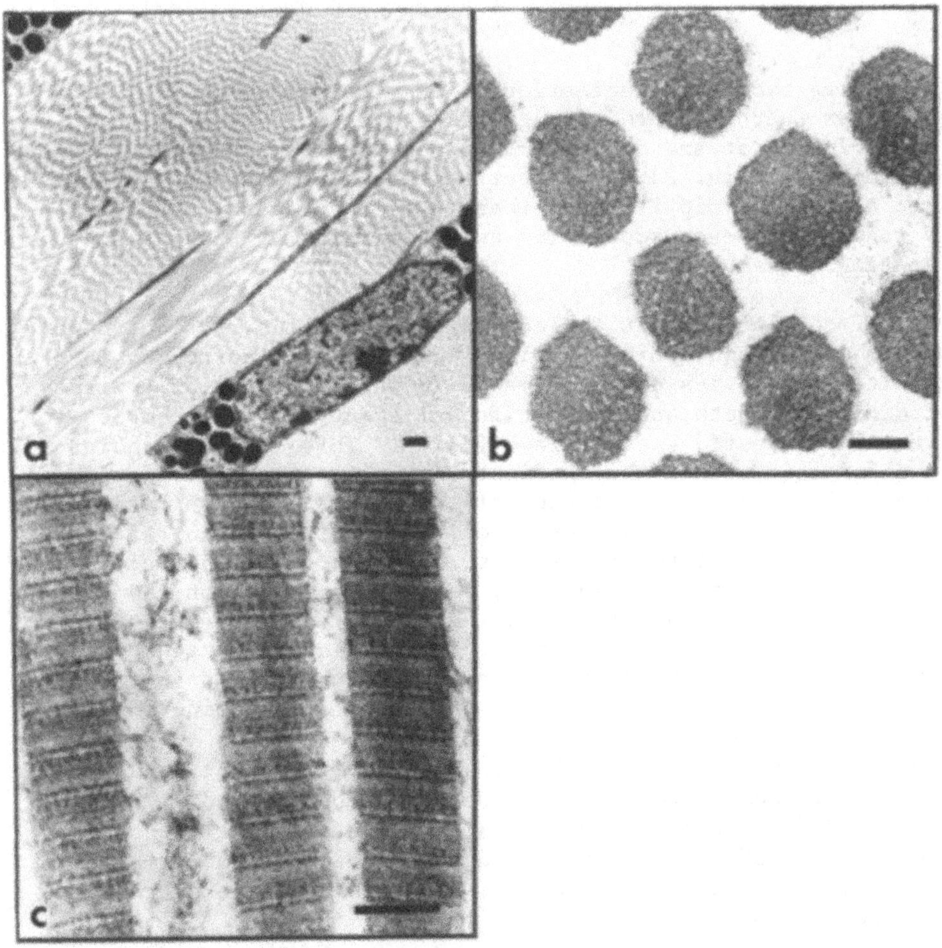

Fig. 8. Electron micrographs of (a), (c) cross sections and (b) longitudinal section of center of tapetum of cow. Scale bar (a) 1 μm; scale bars (b) and (c) 0.1 μm.

We consider that these results support our conclusion that taurine is associated with the membranes surrounding tapetal rods. Our results also establish that the defect in the tapetum of the abnormal beagle is the absence of the zinc-cysteine complex which constitutes the normal electron dense core, which is different from its location in the tapetum of the cat, where it is present _outside_ the tapetal rods, probably as part of the tapetal rod membrane. These different localizations are in accord with the original observations of Weitzel et al. (20) that the zinc content of dog tapetum is an order of magnitude higher than that in cat tapetum. All of the observations of ultrastructure and chemistry of the tapetum from beagles known to be heterozygotes for this condition were indistinguishable from those on normal beagles, supporting the original suggestion that this condition is inherited as an autosomal recessive trait (2).

Ultrastructural examination of the tapetum lucidum of the cow shows that it contains regular arrays of rods, similar to those of the cat and dog, but does not have a membrane surrounding the rods (Fig. 8). The taurine concentration of the tapetal area of the choroid is _lower_ than in the nontapetal area, which we take as evidence supporting an association of taurine with tapetal rod membranes.

Our conclusion that the cat tapetum contains rods surrounded by a membrane containing taurine and zinc is supported by the results of a series of experiments in which the ocular toxicity of zinc pyridinethione, an antifungal agent used in shampoos for treating dandruff, was examined (5,17). This compound chelates zinc and this action would effectively alter the normal structural role of zinc in the tapetal rod membrane of the cat and result in the tapetal degeneration which was noted in this species. The same result was also observed in the normal beagle dog, but this was accompanied by retinal degeneration and blindness. Although the chemical action of zinc pyridinethione remains the same in this species, i.e. chelation of zinc, the consequences are more severe because the tapetal rod in the dog consists of zinc cysteinate, and chelation of the zinc would free large amounts of cysteine, a potent neurotoxic agent, especially to the retina (12,13,14). The same result was observed in beagle dogs treated with ethambutol (4,8,9,18), a drug used for treating tuberculosis, which also chelates zinc. The dramatic effects of these drugs are absent in abnormal beagles, which is to be expected, since there is no cysteine to be released.

ACKNOWLEDGMENTS

This research was supported by the Office of Mental Retardation and Developmental Disabilities of the State of New York, and by NIH Public Health Service Grant HD-11129. We thank Mr. Richard Weed for expert photography.

REFERENCES

1. Barbau, A. and Donaldson, J., 1974, Zinc, taurine, and epilepsy, Arch. Neurol., 30:52-58.
2. Bellhorn, R.W., Bellhorn, M.B., Swarm, R.L., and Impellizzeri, C.W., 1975, Hereditary tapetal abnormality in the beagle, Ophthalmic Res., 7:250-260.
3. Bernstein, M.H., and Pease, D.C., 1959, Electron microscopy of the tapetum lucidum of the cat, J. Biophys. Biochem. Cytol., 5:35-40.
4. Capiello, V.P., and Layton, W.M., 1965, A one year study of the toxicity of ethambutol in dogs: Results of gross and histopathologic examination, Toxicol. Appl. Pharmacol., 7:844-849.
5. Cloyd, G.G., Wyamn, M., Shadduck, J.A., Winrow, M.J. and Johnson, G.R., 1978, Ocular toxicity studies with zinc pyridinethione, Toxicol. Appl. Pharmacol., 45:771-778.
6. Coles, J.A., 1971, Some reflective properties of the tapetum lucidum of the cat's eye, J. Physiol., 212:393-409.
7. Elliott, J.H., and Futterman, S., 1963, Fluorescence in the tapetum of the cat's eye, Arch. Ophthalmol., 70:137-140.
8. Figueroa, R., Weiss, H., Smith, J.C., Hackley, B.M., McBean, L.D., Swassing, C.R., and Holstad, J.D., 1971, Effect of ethambutol on the ocular zinc concentration in dogs, Amer. Rev. Resp. Disp., 104:592-594.
9. Kaiser, J.A., 1964, A one-year study of the toxicity of ethambutol in dogs: Results during life, Toxicol. Appl. Pharmacol., 6:557.
10. McLardy, T., 1962, Zinc enzymes and the hippocampal mossy fibre system, Nature, 194:300-302.
11. McLardy, T., 1974, Prolonged blockage by taurine of dentatohippocampus synapse in rats, IRCS 2:1696.
12. Olney, J.W., and Ho, O.L., 1970, Brain damage in infant mice following oral intake of glutamate, aspartate or cysteine, Nature, 227:609-610.
13. Olney, J.W., Ho, O.L., Rhee, V., and Schainker, B., 1972, Cysteine-induced brain damage in infant and fetal rodents, Brain Res., 45:309-313.

14. Pedersen, O.O., and Karlsen, R.L., 1980, The toxic effect of L-cysteine on the rat retina: A morphological and biochemical study, Invest. Ophthalmol. Vis. Sci., 19:886-892.
15. Pirie, A., The chemistry and structure of the tapetum lucidum in animals, in: "Aspects of Comparative Ophthalmology," O Graham-Jones, ed., New York, Pergamon Press, (1966) pp. 57-68.
16. Rodieck, R.W., The Vertebrate Retina: Principles of Structure and Function, San Francisco, W.H. Freeman and Comapny (1973).
17. Snyder, F.H., Buehler, E.V. and Winek, C.L., 1965, Safety evaluation of zinc 2-pyridinethiol 1-oxide in a shampoo formulation, Toxicol. Appl. Pharmacol., 7:425-437.
18. Voge, A.W., and Kaiser,·J.A., 1963, Ethambutol-induced transient change and reconstitution (in vivo) of the tapetum lucidum color in the dog, Exp. Mol. Pathol. Suppl., 2:136-149.
19. Weitzel, G., Buddecke, E., Fretzdorff, A.M., Strecker, F.J., and Roester, U., 1955, Structur der im Tapetum lucidum von Hund und Fuchs enthaltenen Zinkverbindung, Hoppe Seyler Z Physiol. Chem., 299:193-213.
20. Weitzel, G., Strecker, F.J., Roester, U., Buddecke, E., and Fretzdorff, A.M., 1954, Zinc in Tapetum lucidum, Hoppe Seyler Z Physiol. Chem., 296:19-30.
21. Wen, G.Y., Sturman, J.A., Wisniewski, H.M., Lidsky, A.A., Cornwell, A.C., and Hayes, K.C., 1979, Tapetum disorganization in taurine-depleted cats, Invest. Ophthalmol. Vis. Sci., 18:1201-1206.

TAURINE DEFICIENCY: A RATIONALE FOR TAURINE DEPLETION

K. C. Hayes[1] and John A. Sturman[2]

[1]Department of Nutrition
Harvard School of Public Health and Nutrition Division
Harvard Medical School
New England Regional Primate Research Center
Southborough, Massachusetts
[2]Department of Pathological Neurobiology
Institute for Basic Research in Mental Retardation
Staten Island, New York

Investigators concerned with the metabolism of taurine are confronted by two distinct aspects of the problem. The first concerns the function of taurine at the molecular level and the second concerns those factors that set the limits of taurine supply and demand as they affect taurine depletion and deficiency. This review discusses the latter in terms of the comparative aspects of taurine metabolism that facilitate our understanding of taurine deficiency.

Taurine Depletion and Deficiency

Evidence has been provided for both cats (1,3) and monkeys (4) that insufficient tissue stores of taurine are associated with retinal degeneration and blindness or growth depression in the two species, respectively. The ultimate question is whether these or other afflictions of taurine deficiency can be expected to affect humans. The answer requires an appreciation of those factors influencing taurine balance, i.e., input versus demand, in the various species, including man. Factors contributing to taurine are endogenous synthesis and diet, whereas demand is reflected in the specific requirements of the various tissues coupled with conservation or loss of taurine via excretion. A schematic relationship between these factors is depicted in Figure 1.

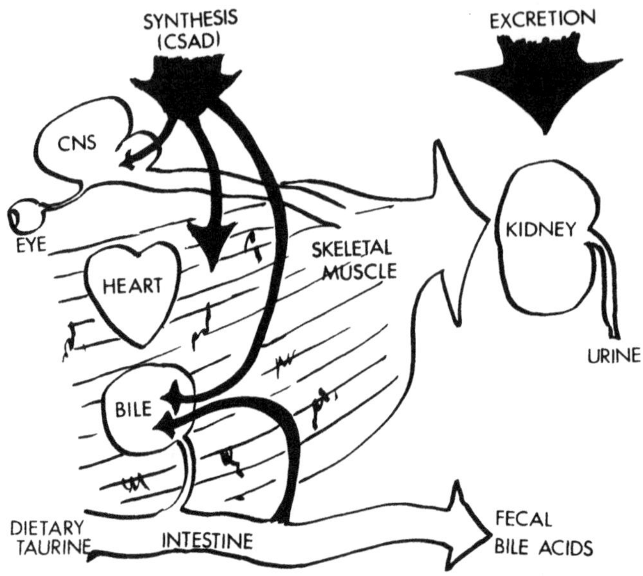

Fig. 1. This scheme depicts the relationship between taurine input (diet and synthesis) and the requirement of major tissues (central nervous system, retina, bile acids, skeletal muscle and heart). Taurine excretion via urine and fecal bile acids fluctuates with the supply and demand imposed by various metabolic circumstances.

Endogenous Synthesis

Figure 2 outlines the currently accepted pathway for biosynthesis of taurine (15). Examination of this scheme indicates that synthesis is dependent upon normal sulphur amino acid (SAA) metabolism and conversion of methionine to taurine via cysteine.

The actual rate of taurine biosynthesis in tissues of various species depends on the activity of the enzyme, cysteinesulfinic acid decarboxylase (CSAD), which converts cysteinesulfinic acid to hypotaurine. As seen in Figure 3, CSAD activity is greatest in omnivorous rats and dogs, with lesser synthesis in herbivores, such as the rabbit and guinea pig, and the least activity represented by the carnivorous cat and omnivorous primates. Age is also an important factor affecting the synthetic capacity, since immature animals, including human infants, have a limited enzymatic ability to convert methionine to cysteine and possibly cysteine to taurine (15).

Secondly, the biosynthetic pathway for taurine depends heavily on the coenzyme function of vitamin B6 as pyridoxal-5'-pyrophosphate (14). Thus, restriction of this vitamin has the potential of limiting the taurine supply. This possibility has been examined in rats (13) and mentioned in monkeys (7). Vitamin B6 deficiency in rats results in reduced taurine biosynthesis, accompanied by substantial reductions

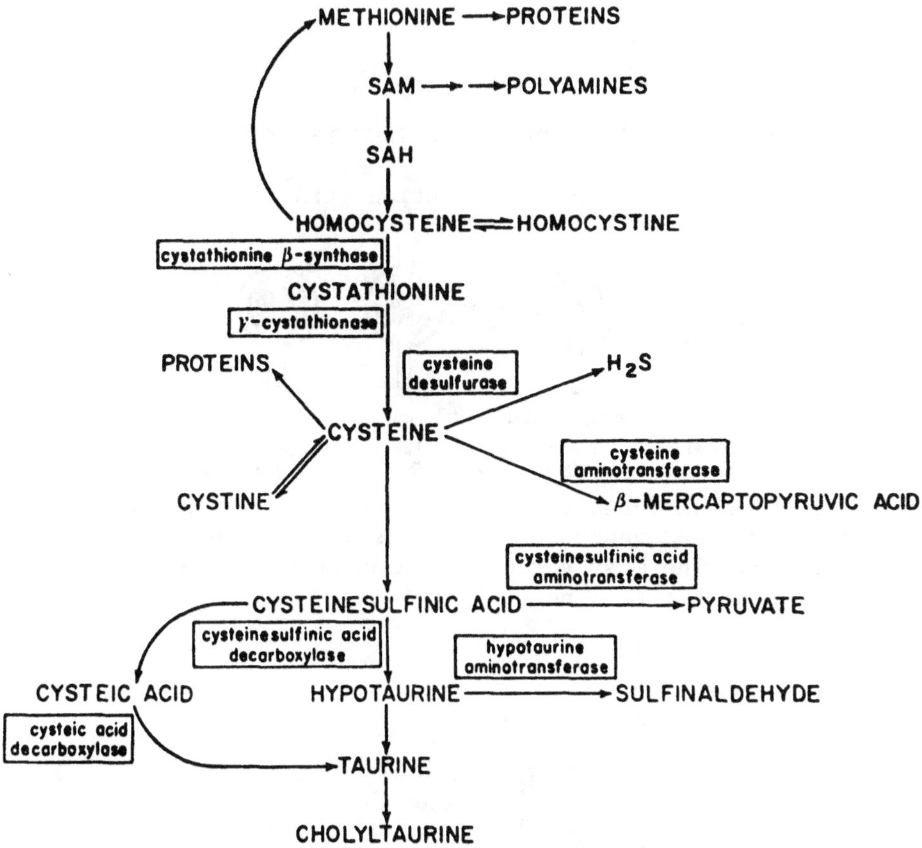

Fig. 2. The metabolic pathway for sulfur amino acids includes the biosynthesis of taurine. The major pathway involves hypotaurine synthesis by decarboxylation of cysteinesulfinic acid. Enzymes requiring vitamin B_6 are noted in the pathway.

in the turnover and urinary excretion of taurine (13), whereas this deficiency in cebus monkeys decreased the taurine conjugated bile acids while decreasing the chenodeoxycholic acid pool (7).

A third influence on the endogenous synthesis of taurine is the level of dietary SAA, and factors such as choline, which spare the methyl-donating requirement of methionine and thus influence the availability of methionine for further metabolism to taurine (12). For example, in adult cats a higher level of SAA (1.5% vs 0.78%) is associated with a greater plasma taurine concentration (6), and methionine supplementation (0.8%) of a low SAA diet (0.6%) increased the pool of taurine available for bile acid conjugation (9). This effect was not demonstrated in growing kittens (9), but cysteine or methionine supplementation of the same low-SAA diet tended to

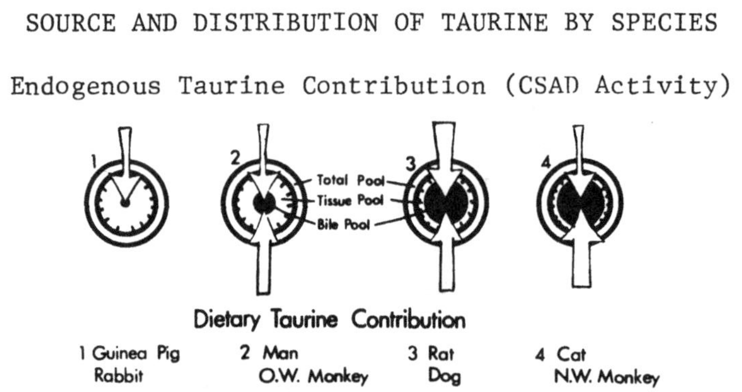

Fig. 3. Species differences are depicted in the supply of taurine originating from endogenous biosynthesis (top arrow) and diet (bottom arrow). The most variable demand derives from the bile acid requirement for taurine conjugation.

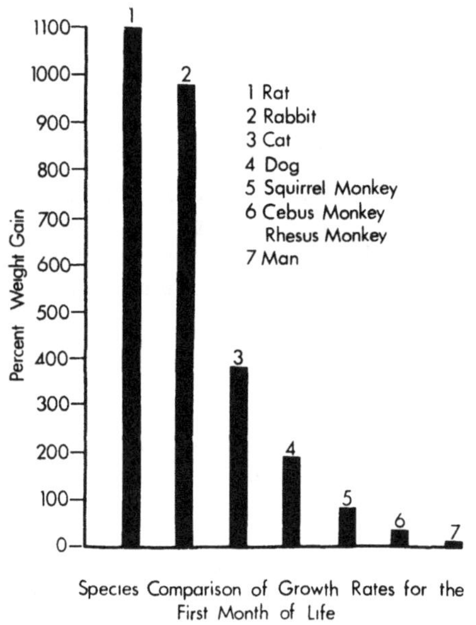

Fig. 4. Differences in neonatal growth rates between species are depicted.

diminish the severity of retinal deterioration in kittens, even though the need for dietary taurine persisted (1). In rat studies, a diet low in SAA has an adverse effect on the hyperlipemia of cholesterol feeding, which appears to be independent of taurine status (12). In contrast, the similarly challenged cebus monkey reduces its hypercholesterolemia when supplemented with taurine (8). In addition, supplementation by sulphate of the diet of growing rats can reduce SAA oxidation to sulphate by the gut microflora; the reduced oxidation spares taurine in the process (18).

Dietary Input

Since the major source of dietary taurine is meat (11), the diet of the carnivore contains a substantial amount of taurine. The essential absence of taurine from the plant kingdom (5) precludes taurine consumption by herbivores. Omnivores, including man, obviously fall somewhere between these extremes in terms of the potential dietary contribution of taurine to their total body pool.

Dietary-physiological Interrelationships Related to Taurine

When species differences between endogenous synthesis and the dietary requirement for taurine are considered, it is relevant to review other aspects of metabolism and digestive physiology, particularly that related to lipid metabolism and the conjugation of bile acids with either taurine or glycine.

One of the major distinctions between species conjugating bile acids with glycine, as opposed to those conjugating with taurine, is their susceptibility to atherosclerosis (16). Thus, the glycine-conjugating rabbit is most susceptible to cholesterol and saturated fat-induced hyperlipemia, whereas the taurine-conjugating cat, dog, and rat are extremely resistant to this diet-induced hyperlipemia. Ambivalent (both taurine and glycine conjugating-) man falls between these extremes. The key to this association is an inherent distinction between species in their capacity to metabolize cholesterol. Carnivores with their naturally high-fat, cholesterol-containing diet have adapted metabolically to the load of dietary cholesterol encountered, whereas herbivorous animals have not. The implication from this species comparison is that if saturated fat and cholesterol are to be consumed and metabolized normally by lipoproteins, conjugation of bile acids with taurine offers an advantage. The mechanism underlying the beneficial role of taurine in this situation has not been examined, but taurine-conjugated bile acids are thought to have superior emulsifying capabilities because they are less easily deconjugated in the intestine than their glycine counterparts.

Figure 3 represents an attempt at comparing specific aspects of the taurine requirement in terms of the variables discussed, i.e., endogenous synthesis, dietary intake, and the tissue demand. No one variable is totally independent of another, but the most crucial determinant of the total body requirement appears linked to the taurine demand imposed by the bile acid pool. For example, the variables are interdependent such that a species not requiring taurine for bile acid conjugation (rabbit) can exist without appreciable synthesis or a dietary supply of taurine, whereas a species with a considerable bile demand for taurine is protected against deficiency if hepatic synthesis is appreciable (the rat or dog). The predicament arises when extreme demand, i.e., an obligate taurine requirement for bile acids, is coupled with poor synthetic capability (cat and New World monkey) and a limited dietary source of taurine.

Within the category of demand, a second major species difference that correlates with taurine deficiency is the rate of growth, which presumably reflects the requirement attributable to the expanding muscle mass. Comparison of the growth rate of species (Figure 4) demonstrates the extraordinary need for taurine implicit in the growth rate of rats. This graph also predicts potential differences between other species. Thus, even though monkeys and cats appear equally vulnerable on the basis of limited synthesis and a large requirement for bile acid taurine conjugation, the greater growth rate in cats elicits the larger requirement, which is expressed by the retinal and tapetal degenerations that occur in that species.

In light of these relationships, including the generally rapid growth of neonates, it is noteworthy that the milk of early lactation is rich in taurine. In fact, taurine usually represents the first or second-most abundant amino acid in such milk (10). This would seem to assure a source of dietary taurine for the rapidly growing neonate while its own synthetic capability matures. Even in the slow-growing human infant, the absence of taurine in the diet significantly depressed plasma, urinary, and bile acid taurine concentrations (12); yet this restriction in dietary taurine availability, with its associated increase in glycine-conjugated bile acids, has no apparent effect on the bile acid pool size or the dynamics of bile metabolism (17).

In summary, susceptibility to dietary taurine deficiency reflects the balance between endogenous synthesis and the total body demand for taurine. Demand, in turn, reflects two chief components, that contributed by bile acid conjugation and that established by growth rate and expansion of muscle mass. Based on these assumptions, it appears that the human infant is unlikely to have a dietary requirement for taurine or develop taurine deficiency. For even though taurine input may be minimal, two factors are in the infant's favor, the first being the reduced demand attributable to an ability to conjugate bile acids with glycine and the second being a relatively slow growth rate.

TABLE 1

TAURINE CONCENTRATIONS IN TISSUES OF ADULT MAMMALS

Species	Plasma μmole/dl	TISSUE Liver	Heart	Brain	Muscle
			μmole/g		
Human	8	2	6	2	5
Monkey					
Rhesus	10	3	11	2	4
Cynomolgus (6 mo)	19	5	14	8	14
Cebus (6 mo)	17	16	17	7	20
Cat	11	10	14	2	8
Rat	45	3	27	3	7
Dog	--	--	10	2	--
Rabbit	--	--	16	2	5
Mouse	--	--	43	9	--

ACKNOWLEDGMENTS

Dr. Hayes's laboratory is supported by NIH grant RR00168 from the Division of Research Resources and the Fund for Research and Teaching, Department of Nutrition, Harvard School of Public Health. Dr. Sturman's laboratory is supported by the Office of Mental Retardation and Developmental Disabilities of the State of New York, and by NIH Public Health Service grant HD-11129.

REFERENCES

1. Berson, E.L., Hayes, K.C., Rabin, A.R , Schmidt, S Y., and Watson, G., 1976, Retinal degeneration in cats fed casein. II. Supplementation with methionine, cysteine, or taurine, Invest. Ophthalmol., 15:52-58.
2. Gaull, G.E., Rassin, D.K., Raiha, N.C., and Heinonen, K., 1977, Milk protein quantity and quality in low-birth-weight infants. III. Effects on sulfur amino acids in plasma and urine, J. Pediatr., 90:348-355.
3. Hayes, K.C., Carey, R.E., and Schmidt, S.Y., 1975, Retinal degeneration associated with taurine deficiency in the cat, Science, 188:949-951.
4. Hayes, K.C., Stephan, Z.F., Sturman, J.A., 1980, Growth depression in taurine-depleted infant monkeys, J. Nutr., 110: 119-125.
5. Jacobsen, J.G., and Smith, L.H., 1968, Biochemistry and physiology of taurine and taurine derivatives, Physiol. Revs., 48:424-511.
6. O'Donnell, J.A., III, Rogers, W.R., and Morris, J.G., 1980, Effect of diet on plasma taurine in the cat, J. Nutr., 110: (in press).
7. Portman, O.W., 1962, Importance of diet, species, and intestinal flora in bile acid metabolism, Fed. Proc., 21:896-902.
8. Portman, O.W., and Mann, G.V., 1955, The disposition of taurine-S^{35} and taurocholate-S^{35} in the rat: dietary influences, J. Biol. Chem., 213:733-747.
9. Rabin, B., Nicolosi, R.J., and Hayes, K.C., 1976, Dietary influence on bile acid conjugation in the cat, J. Nutr., 106: 1241-1246.
10. Rassin, D.K., Sturman, J.A., and Gaull, G.E., 1978, Taurine and other free amino acids in milk of man and other mammals. Early Hum. Dev., 2:1-13.
11. Roe, D.A., and Weston, M.D., 1965, Potential significance of free taurine in the diet, Nature, 205:287-288.
12. Seidel, J.C., Nath, N., and Harper, A.E., 1960, Diet and cholesterolemia: V. Effects of sulfur-containing amino acids and protein, J. Lipid Res., 1:474-481.
13. Sturman, J.A., 1973, Taurine pool sizes in the rat: effects of vitamin B_6 deficiency and high-taurine diet, J. Nutr., 103:

1566-1580.
14. Sturman, J.A., Vitamin B_6 and sulfur amino acid metabolism, in: "Vitamin B_6 Analytical Methodology and Criteria for Assessing Nutritional Status," R.D. Reynolds and J.E. Laklem, eds., Plenum Press, New York (in press) 1981.
15. Sturman, J.A., and Hayes, K.C., 1980, The biology of taurine in nutrition and development, Adv. Nutr., 3:231-299.
16. Truswell, A.S., McVeigh, S., Mitchell, W.D., and Bronte-Stewart, B., 1965, Effects in man of feeding taurine on bile acid conjugation and serum cholesterol levels, J. Athero. Res., 5:526-532.
17. Watkins, J.B., Jarvenpaa, A.L., Raiha, N., Szczepanik, V.P., Klein, P.D., Rassin, D.K., and Gaull, G.E., 1979, Regulation of bile acid pool size: Role of taurine conjugates, Ped. Res., 13:410 (Abstract).
18. Whittle, B., and Smith, J.T., 1974, Effect of dietary sulfur on taurine excretion by the rat, J. Nutr., 104:666-670.

TAURINE NUTRITION IN MAN

Gerald E. Gaull

Department of Human Development and Nutrition
New York State Institute for Basic Research in Mental Retardation
Staten Island, New York
Department of Pediatrics
Mount Sinai School of Medicine
City University of New York
New York, New York

The high brain taurine concentration during fetal development of the rat, monkey, rabbit and man at a time when the presumed synthetic pathway, via cysteinesulfinic acid decarboxylase, has little measurable activity suggested a dietary source of taurine to be necessary (3,19,20). In studies investigating the effects on preterm infants of feeding formulas containing different protein quantity and quality and comparing them with infants fed pooled human milk, an interesting pattern of plasma and urine amino acid concentrations was observed. The concentrations of most plasma and urine amino acids were either increased or unchanged when compared with those of infants fed pooled human milk (10,11). Taurine was the striking exception (4). The concentration of taurine in the urine of infants fed casein-predominant formulas was lower from the first week of study than was that of infants fed pooled human milk. The plasma taurine decreased steadily and by the fourth week of study was significantly lower in the plasma of infants fed such formulas than it was in infants fed the pooled human milk. Plasma and urine concentrations of preterm infants fed whey-predominant formulas did not have such consistent decreases, and whether or not there was a decrease was a function of the amount of taurine provided by the formula (4). This effect of commercial formulas on plasma and urine taurine concentrations has been shown to occur in infants on total parenteral nutrition, which is devoid of taurine (14). The small concentrations of taurine in plasma of human preterm infants fed artificial formulas have been documented by others, but not commented on (2).

The finding of a dietary requirement for taurine in the human infant is consistent with the negligible activity of cysteinesulfinic acid decarboxylase present both in fetal and in mature human liver (4).

This evidence for the relative inability of man to synthesize taurine and the evidence of taurine deficiency in plasma and urine led us to an examination of the milk of various species, as well as of some currently available commercial formulas. Taurine is a major constituent of the free amino acid pool of milk in a number of species (12). Human milk contains a considerable amount of taurine, whereas mature bovine milk, from which most of the formulas are prepared, contains only minimal amounts of taurine (4). The lack of taurine in the casein-predominant formulas in this study resulted from the lack of taurine in the dairy cow milk from which the formulas were prepared (Table 1). Four major commercial infant formulas available in the United States contain only small amounts of taurine (Table 1), and soy-based formulas contained none at all. For most species studied, taurine is a major constituent of the free amino acid pool of milk, and it has a greater concentration early in lactation (Table 1). It is noteworthy that even though the dairy cow provides man with a milk containing only 1 μmole/dl, the lactating cow provides its own calf with 30-fold higher concentrations of taurine.

Rassin and I, in collaboration with our colleagues in Helsinki (13), then turned our attention to the systematic study of the effect of casein- and whey-predominant formulas on plasma and urine taurine concentrations in the term infant. Again, a 1.5 g % whey-predominant protein formula as well as a 1.5 g % casein-predominant protein formula failed to provide sufficient taurine to maintain plasma and urine taurine concentrations equal to those found in breast-fed infants. Greater excretion of taurine by full-term human infants fed human milk than by those fed bovine milk also had been documented, but not commented on (2,8,16). A finding of interest is that in term infants breast fed ad libitum the taurine concentrations (approximately 10 μmole/dl in plasma and 40 μmole/dl in urine) are considerably higher than those of preterm infants fed 170 ml human milk/kg body weight/day (approximately 6 μmole/dl in plasma and 20 μmole/dl in urine). It is interesting as well that in adult man fed defined liquid formulas devoid of taurine over a long period that virtually no taurine is excreted in the urine (E. Ahrens, personal communication). In adult man only 1% of an oral load of L-cysteine was recovered as increased urinary excretion of taurine, giving further evidence that mature human beings also have a relatively limited ability to synthesize taurine and may be largely dependent on dietary taurine (22). The rat, in striking contrast, has considerable ability to convert dietary cysteine to taurine (17).

The effects of reduced taurine intake on plasma taurine concentrations are seen much later (4-5 weeks) than those on urine concentrations (1-2 weeks). These results remind me of experiments we did some years ago using vitamin B_6 deficiency in the rat as a model for the study of disturbed sulfur metabolism (18). In confirmation of earlier workers we found that a striking decrease in taurine excretion was one of the earliest effects of B_6 deficiency. Furthermore, the concentration of taurine in various organs was carefully maintained. What we found that was new was that the plasma concentration of taurine was also maintained, resulting in a striking decrease in taurine clearance by the kidney. Thus, even in the rat, an animal with considerable capacity to synthesize taurine, its body pools were maintained by a global mechanism like renal clearance. One wonders whether or not nature is telling us something about the importance of taurine.

We have recently completed a study of the effects of taurine supplementation of formulas. When a whey-predominant formula is supplemented with the amount of taurine found in pooled human milk (30 μmole/dl), the taurine concentration of plasma and urine of preterm infants is the same as that found in such infants fed pooled human milk. Even more interesting is the fact that the infants fed pooled human milk and taurine-supplemented formula remain predominantly taurine conjugators of bile acids, whereas those fed the taurine-deficient formula converted to predominantly glycine conjugation of bile acids (1,23). Furthermore, Wharton (personal communication) has now shown that cystine-supplementation will not substitute for taurine-supplementation of formulas in the maintenance of taurine-predominant bile-acid conjugation.

The finding that formula-fed infants tend to become predominantly glycine conjugators rather than taurine conjugators of bile acids suggests that the dietary requirement for taurine may be significant. Taurine conjugates of bile salts may be better detergents for fat absorption in the gut because they are not precipitated at the acidic pH frequently present in the proximal small intestine (6). Since taurine conjugates are fully ionized at body pH, passive absorption does not occur. A higher intraluminal concentration of bile acids can be maintained, thereby facilitating the solubilizing activity of these conjugated bile acids. If taurine conjugation decreases passive absorption in the gut, bile acids conjugated with taurine may be more efficient than those conjugated with glycine (15). This complicated problem requires further investigation.

Because of the alternative means of bile acid conjugation, the human infant may be able better to conserve taurine than the kitten who can conjugate bile acids with taurine only and who goes blind when taurine deficient. The ability to conjugate bile acids both with taurine and with glycine may thus serve to ameliorate the effects of taurine deficiency in man. In addition, even those human

Table 1

Taurine Content of Milk from Various Species

Species	More than 5 days after birth	Less than 5 days after birth
Gerbil	595	
Cat	287	
Beagle Dog	191	264
Mouse (C57 Bl/6J)	75	
Rhesus Monkey	56	61
Baboon	38	
Man	34	41
Chimpanzee	26	71
Water Buffalo	19	
Guinea-pig	17	
Rat	15	63
Java Monkey	14	
Rabbit	14	
Sheep	14	68
Horse	3	
Cow	1	31
Enfamil[R]	2.8	
Similac[R]	1.8	
SMA[R]	2.8	
PM 60/40[R]	1.6	

Data are μmole/dl of milk. Samples were not available for all species before 5 days after birth. (Adapted from Rassin et al., 1978 and unpublished data from the author's laboratory.)

infants fed the currently available taurine-deficient formulas usually have supplementary foods added to their diet within a few weeks or months of birth. Even in the cat, after 10 weeks of a taurine-deficient diet, the bile taurine pool is minimally affected, albeit at the expense of the liver pool of taurine (21). In the beagle puppy fed by total parenteral nutrition using an amino acid solution devoid of taurine, the liver pool of taurine falls rapidly while the bile taurocholate pool is much slower to reflect the taurine deficiency (9). Indeed, in man in relatively short-term experiments (up to 9 days) the pool of taurocholate reflected the pool of taurine available in the liver and the diet, rather than that in plasma or muscle, indicating that the bile taurine pool in man cannot be predicted from plasma taurine measurements (5).

In summary, it seems likely that man's ability to synthesize taurine is limited and that he has a dietary requirement for it. His ability to conjugate bile acids either with glycine or with taurine may well be important in allowing the feeding of low-taurine formula to infants without devastating short-term effects. Nevertheless, the long-term effects of the resulting low concentrations of plasma and urine taurine and of the conjugation of bile salts predominantly with glycine deserves thorough investigation before being dismissed as without nutritional consequence.

ACKNOWLEDGMENTS

Original work in the author's laboratory supported in part by the New York State Office of Mental Retardation and Developmental Disabilities and by grants-in-aid from Wyeth Laboratories, Ross Laboratories and Mead-Johnson Laboratories.

The collaboration of Dr. David Rassin, and of Dr. Niels Räihä and his clinical colleagues in Helsinki, Finland, is acknowledged with gratitude.

REFERENCES

1. Brueton, M.J., Berger, H.M., Brown, G.A., Ablitt, L., Lyangkarer, N., and Wharton, B.A., 1978, Duodenal bile acid conjugation patterns and dietary sulphur amino acids in the newborn, Gut 19:95-98.
2. Dickinson, J.C., Rosenblum, H., and Hamilton, P.B., 1970, Ion exchange chromatography of the free amino acids in the plasma of infants under 2,500 gm at birth, Pediatrics, 45:606-613.
3. Gaull, G.E., and Rassin, D.K., 1979, Taurine and brain development: Human and animal correlates, in: "Neural Growth and Differentiation," E. Meisami and M.A.B. Brazier, eds., Raven Press, New York, pp. 461-477.
4. Gaull, G.E., Rassin, D.K., Räihä, N.C.R., and Heinonen, K., 1977, Milk protein quantity and quality in low-birth-weight

infants. III. Effects on sulfur amino acids in plasma and urine, J. Pediatr., 90:348-355.

5. Hardison, W.G.M., 1978, Hepatic taurine concentration and dietary taurine as regulators of bile acid conjugation with taurine, Gastroenterology, 75:71-75.
6. Hofman, A.F., and Small, D.M., 1967, Detergent properties of bile salts: Correlation with physiological function, Ann. Rev. Med., 18:333-376.
7. Jagenburg, O.R., 1959, The urinary excretion of free amino acids and other compounds by the human, Scand. J. Clin. Lab. Invest., 11:3-183.
8. Jonxis, J.H.P., 1951, The influences of differences in food on the amino acid excretion, Arch. Dis. Child., 26:272.
9. Malloy, M.H., Rassin, D.K., Gaull, G.E., and Heird, W.C., 1980, Development of taurine metabolism in beagle pups: Effects of taurine-free total parenteral nutrition, Biol. Neonate, in press.
10. Rassin, D.K., Gaull, G.E., Heinonen, K., and Räihä, N.C.R., 1977, Milk protein quantity and quality in low-birth-weight infants. II. Effects on selected aliphatic amino acids in plasma and urine, Pediatrics 59:407-422.
11. Rassin, D.K., Gaull, G.E., Räihä, N.C.R., and Heinonen, K., 1977, Milk protein quantity and quality in low-birth-weight infants. IV. Effects on tyrosine and phenylalanine in plasma and urine, J. Pediatr. 90:356-360.
12. Rassin, D.K., Sturman, J.A., and Gaull, G.E., 1978, Taurine and other free amino acids in milk of man and other mammals, Early Hum. Develop., 2:1-13.
13. Rassin, D.K., Järvenpää, A.-L., Räihä, N.C.R., and Gaull, G.E., 1979, Breast feeding versus formula feeding in full-term infants: Effects on taurine and cholesterol, Pediatr. Res., 13:406.
14. Rigo, J., and Senterre, J., 1977, Is taurine essential for the neonate? Biol. Neonate, 32:73-76.
15. Schersten, T., 1979, Bile acid conjugation, in: "Metabolic Conjugation and Metabolic Hydrolysis," W.H. Fishman, ed., Academic Press, New York.
16. Souchon, F., 1952, Papierchromatographische Untersuchungen der freien amino sauren im Sauglingsham, 2. Ges. Exp. Med., 118: 219-229.
17. Sturman, J.A., and Cohen, P.A., 1971, Cystine metabolism in vitamin B_6 deficiency, evidence of multiple taurine pools, Biochem. Med., 5:245-268.
18. Sturman, J.A., Cohen, P.A., and Gaull, G.E., 1969, Effects of deficiency of vitamin B_6 on transsulfuration, Biochem. Med., 3:244-251.
19. Sturman, J.A., Rassin, D.K., and Gaull, G.E., 1977, Taurine in development, Life Sci., 21:1-22.
20. Sturman, J.A., Rassin, D.K., and Gaull, G.E., 1978, Taurine in the development of the central nervous system, in: "Taurine and Neurological Disorders," A. Barbeau and R.J. Huxtable, eds., Raven Press, New York, pp. 49-71.

21. Sturman, J.A., Rassin, D.K., Hayes, K.C., and Gaull, G.E., 1973, Taurine deficiency in the kitten: Exchange and turnover of (^{35}S) taurine in brain, retina, and other tissues, J. Nutr., 108: 1462-1476.
22. Swan, P., Wentworth, J., and Linksweiler, H., 1964, Vitamin B_6 depletion in man: Urinary taurine and sulfate excretion and nitrogen balance, J. Nutr., 84:220-228.
23. Watkins, J.B., Järvenpää, A.-L., Räihä, N., Szczepanik Van-Leweeen, P., Klein, P.D., Rassin, D.K., and Gaull, G., 1979, Regulation of bile acid pool size: Role of taurine conjugates, Pediat. Res., 13:410.

PART III

TRANSPORT AND METABOLISM OF TAURINE

Ryan J. Huxtable

Department of Pharmacology
Health Sciences Center
University of Arizona
Tucson, Arizona

Active transport is an important aspect of taurine biochemistry. Certain organs, such as the heart, rely on transport to maintain their taurine content, and in species that are unable to synthesize taurine, such as the cat, the transport and distribution of taurine is all important in taurine regulation. It is becoming apparent that defects in taurine transport are of clinical importance. Such defects, for example, have now been found in epilepsy, Friedreich's ataxia, and genetic cardiomyopathy. These defects are associated with an impaired ability to retain taurine in the body. Furthermore, it has been shown that the accumulation of taurine that occurs in hearts in congestive failure is probably due to a stimulation of taurine transport.

A new and exciting stage in the study of taurine transport has been set by the findings reported in this section on a reconstituted system. Taurine binding proteins isolated from sarcolemma have been incorporated into proteoliposomes, which are then able, under appropriate conditions, to transport taurine. The characteristics of transport are remarkably similar to those seen in the intact heart. Binding affinities of taurine and its analogues, and the structure-activity requirements of the site precisely mimic the physiological system. It is to be hoped that this system will allow the detailed study of conditions or agents, such as adrenergic agonists, that modulate the transport of taurine. Furthermore, the reconstituted system may be of value in the study of calcium-taurine interactions.

The unanswered questions of taurine metabolism have been discussed many times. One of the major problems concerns the oxidation of hypotaurine. This is an interesting substance in its own right, and is apparently a key substance in taurine metabolism. This rather reactive compound is clearly well converted to taurine in intact animals. Concentrations to be found in tissues are low, except in male reproductive structures. Despite the evidence for its conversion, and despite intensive study, convincing demonstration and isolation of a hypotaurine oxidase activity have yet to be achieved. The puzzling characteristics of hypotaurine oxidation in the brain are discussed in one chapter in this section. Another chapter explores alternative routes of hypotaurine metabolism not proceeding to taurine. Fellman reports that hypotaurine can undergo transamination in the brain and other organs. The transaminase activity in the brain differs from that in peripheral tissues in its co-substrate requirements. This pathway represents another way of generating inorganic sulfate from sulfur amino acids, and a further catabolic route for cysteine sulfinic acid.

There is abundant discussion elsewhere in this volume that point to the most marked effects of taurine being on disturbed systems. In view of the effects of taurine in such systems on calcium-dependent phenomena, membrane excitability, and amino acid balance, it is logical to examine for the effects of taurine in cells whose regulation has been disturbed to the extent that they have become cancerous. Very little work has been done in this area. The chapter by Baskin et al., represents a pioneering effort in this difficult but important area. There are clear alterations in hypotaurine and taurine biochemistry in rapidly regenerating systems, such as the liver after partial hepatectomy, and the other strand of methionine metabolism proceeding through to the polyamines is intimately associated with cell cycling and cell division. It is possible, therefore, that this could develop into one of the important future areas of taurine research.

HYPOTAURINE AMINOTRANSFERASE

J. H. Fellman and E. S. Roth

University of Oregon Health Sciences Center
School of Medicine
Department of Biochemistry
Portland, Oregon

Taurine is a substance widely distributed in the animal world (1). Certain mammalian tissues contain remarkably high concentrations of this sulfur amino acid suggesting an important biological role and yet the detailed account of its metabolic origin remains incomplete. An understanding of the metabolic turnover of taurine necessarily would extend insight into the control mechanisms for regulation of biosynthesis and transport.

It has been accepted for some time that the biosynthesis of taurine proceeds from cysteine via its immediate precursor hypotaurine (Fig. 1). Recently, Sturman and we carried out experiments which directly confirmed this metabolite as an intermediate *in vivo*. Rats were injected with tracer ^{35}S-cystine and hourly for five hours, injected with 50 μmole of unlabeled hypotaurine as a "cold" trap. The liver of the animals sacrificed four hours after the ^{35}S cystine injection was shown to contain ^{35}S hypotaurine, while those not receiving unlabeled hypotaurine did not contain detectable amounts of labeled hypotaurine in their livers (2). Furthermore, the conversion of hypotaurine to taurine was clearly endogenous and not from microbial origin since we observed that germ-free mice were able to carry out this metabolic step as briskly as conventional mice (3).

The overall steady state concentration of each metabolic is determined by the activities of the enzymes which govern their rates of synthesis and catabolism. It was from this conceptual platform that we undertook a study of enzymes which may participate in determining the metabolic fate of hypotaurine and thus influence the rates of competing pathways including taurine biosynthesis (Fig. 1). We were interested in the possibility of transamination of hypotaurine

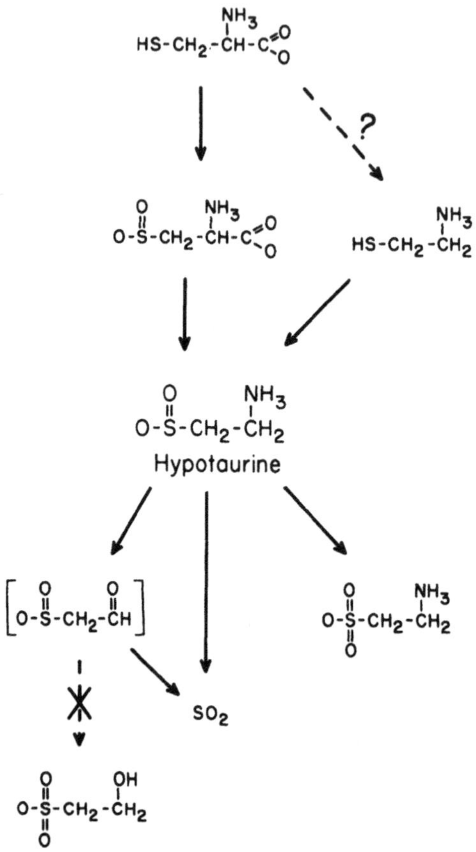

Fig. 1. Taurine biosynthesis from cysteine. The question mark indicates uncertainty as to the magnitude of the contribution of this pathway to taurine biosynthesis. The x indicates that this step cannot be detected and probably does not occur.

for this central reason; also because we wished to examine an hypothesis that the transamination may lead to an intermediate sulfinoaldehyde which by rearrangement might undergo an hypothetical internal oxidation-reduction leading to isethionate. While we were able to observe hypotaurine transaminase activity, associated with cell particles, we did not observe any isethionate product (3).

This communication primarily concerns the properties of the hypotaurine aminotransferase activity found in the particulate fraction of mammalian tissue.

We examined and contrasted this enzyme activity with γ-aminobutyrate transaminase activity since the latter enzyme also accepts β-alanine, the carbon analogue of hypotaurine, as substrate.

MATERIALS AND METHODS

2-^{3}H-Hypotaurine was synthesized by the tritiation of sulfoacetonitrile to form 2-^{3}H-taurine (4) followed by conversion of the latter to 2-^{3}H-taurylchloride (2-aminoethanesulfonylchloride) and reduction with metallic zinc to 2-^{3}H-hypotaurine (5).

Hypotaurine was purchased from Calbiochem-Behring Corporation. Taurine, sodium isethionate, α-ketoglutarate and pyridoxal-5'-phosphate were purchased from Sigma Chemical Co. ^{35}S-Taurine and 1,2-^{14}C-taurine were purchased from Amersham Corporation and were converted to hypotaurine as described previously (5). Two ion exchange resins used throughout were Bio-Rad AG-50W-X8 200-400 mesh, hydrogen form and Bio-Rad Bio-Rex 5 100-200 mesh, chloride form.

Samples of 4-amino-5-fluoropentanoic acid and 4-amino-5-chloropentanoic acid were the generous gift of Dr. R.B. Silverman (6). A sample of 4-aminohex-5-enoic acid (vinyl GABA) was generously supplied by Merrill National Laboratories, Cincinnati, Ohio.

Selenohypotaurine was synthesized from selenocysteamine by oxidation with bromine according to the method of Pichat (7) and selenotaurine was prepared from selenocysteamine by oxidation with hydrogen peroxide according to the procedure of Rinaldi et al. (8).

Assays for hypotaurine aminotransferase activity were carried out in Thunberg tubes. Routinely mixtures contained enzyme sources, β-mercaptoethanol 1mM, α-ketoglutarate 0.5 mM, pyridoxal-5' phosphate 0.01 mM, hypotaurine 5 mM, 2-^{3}H-hypotaurine 0.5 μCi and 0.05 M imidazole-HCl (pH 8.2) in 1.0 ml volume. After incubation with shaking at 37°, usually for one hr, reactions were stopped by addition of 0.2 ml 20% sulfosalicyclic acid. Tubes were evacuated and ^3H_2O formed was distilled at 60° and collected in ice. One-half ml aliquots were counted.

Assays for γ-aminobutyrate/β-alanine aminotransferase were carried out in a similar manner in Thunberg tubes. Routine assay mixtures contained enzyme sources, dithiothreitol 1 mM, α-ketoglutarate 1 mM, pyridoxal-5'-phosphate 0.01 mM, β-alanine 5 mM, 3-^{3}H β-alanine (obtained from Amersham Corporation) and 0.05 M imidazole buffer (pH 8.2) in 1.0 ml final volume. Incubation and aliquot counting was as above.

Determination of alanine and glutamate was carried out with a Beckman Model 116 Amino Acid Analyzer. ^{35}S-Sulfate was determined using a dual ion exchange resin procedure previously described (9).

RESULTS

Substrate and Products

When we launched these investigations of hypotaurine metabolism, we started by devising a radiochemical synthesis of hypotaurine. This, we reasoned, would enable us to conveniently assay for transaminase activity in complex mixtures of tissue enzymes. Employing the appropriate isotopically labeled form of hypotaurine we could measure the rate of appearance of $^{3}H_2O$, $^{35}SO_3{=}^{35}SO_4{=}$ or ^{14}C-acetate or acetaldehyde. Our initial efforts were rewarded; we observed $^{3}H_2O$ released from $2^{3}H$-hypotaurine and $^{35}SO_4{=}$ from ^{35}S-hypotaurine with the appearance of alanine when pyruvate was employed as co-substrate (Fig. 2). We also observed that the enzyme activity was associated with the heavy mitochondria fraction when tissue preparations were separated on a sucrose gradient (10). This hypotaurine aminotransferase activity was found to be greatest in brain followed by liver, testes and kidney (Table 1). The liver particulate fraction contains hypotaurine transaminase activity which accepted α-ketoglutarate and also pyruvate as co-substrate. A striking difference was exhibited by brain particulate preparations in that α-ketoglutarate was an obligate co-substrate; pyruvate did not participate with the brain enzyme. The activity of hypotaurine aminotransferase is dependent on the presence of a keto acid co-substrate. When α-ketoglutarate was used as co-substrate in hypotaurine aminotransferase incubation mixtures the product observed was glutamate (Table 2) and when pyruvate was present as co-substrate alanine was the observed product (3). We also discovered a $^{3}H_2O$ "releasing" activity in the cytosol fraction of liver with only traces of activity in the other tissue. We conducted an intense study of this enzyme activity which we found contrasted sharply with the properties of mitochondrial hypotaurine aminotransferase activity. The cytosol enzyme appeared only to catalyse the release of the

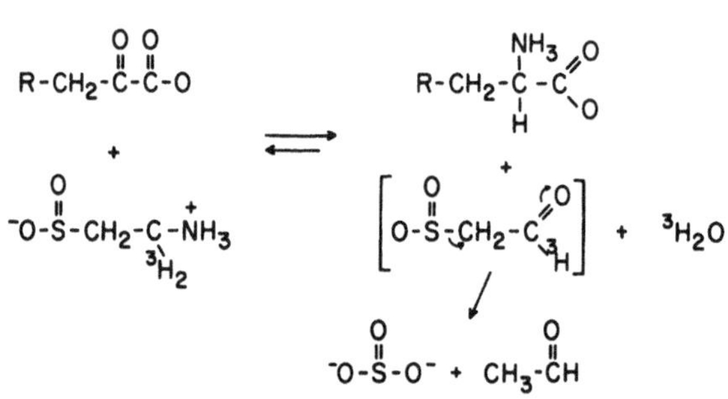

Fig. 2. Biochemical substrates and products encountered in the transamination of hypotaurine.

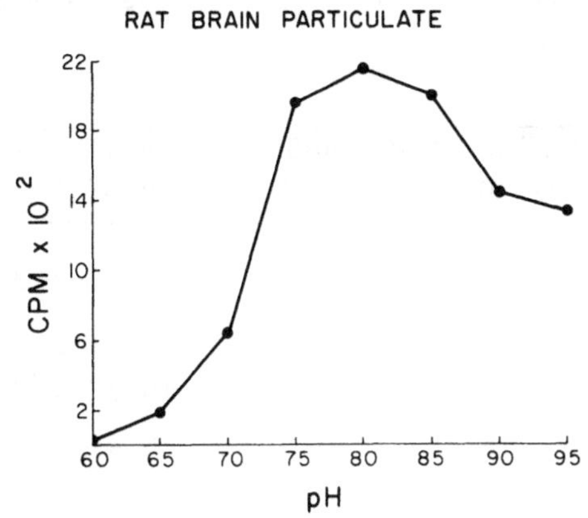

Figure 3. A pH vs activity curve for hypotaurine aminotransferase activity. Washed rat brain particulate fraction was incubated 1 hr with 0.01 mM pyridoxal 5'-phosphate, 1 mM β-mercaptoethanol, 0.5 mM α-ketoglutaric acid, 5 mM hypotaurine, 0.5 μCi/ml 2-^{3}H-hypotaurine, and 50 mM imidazole-HCl at the pH indicated. An aliquot of the distillate, containing ^3H_2O formed, was counted. Activity is expressed as cpm x 10^{-2}.

proton. No net transamination, measured by the stoichiometric accumulation of co-product amino acid could be detected (see Discussion below).

The body of this report concerns studies which employed brain and liver washed particulate fractions and thus focuses on the hypotaurine aminotransferase activity. The hypotaurine aminotransferase activity was optimal at approximately pH 8.2 (Fig. 3) and was stimulated by the addition of pyridoxal-5'-phosphate. Addition of 10^{-5} M of this assumed co-enzyme caused a 20% increase in enzymatic activity. Consistant with this assumption was the observation that aldehyde reagents known to inhibit pyridoxal phosphate-requiring enzymes also inhibit hypotaurine aminotransferase activity (see Table 3). Aminooxyacetic acid profoundly inhibited the enzyme activity while hydroxylamine and semicarbazide exhibited less potency. The inhibition could be reversed by the addition of pyridoxal-5'-phosphate when the hydroxylamine-treated enzyme was dialyzed against a buffer.

Inhibitor Studies

Concentrations of γ-aminobutyrate and β-alanine, but not taurine, equimolar to the substrate hypotaurine inhibited the activity of the enzyme. A reciprocal plot of the effect of γ-aminobutyrate on the velocity of the reaction with respect to substrate concentrate provided evidence that the inhibition was competitive (Fig. 4).

Table 1. Distribution of hypotaurine aminotransferase activity in rat tissues (particulate fraction). The values are means ± SD from three male rats.

HYPOTAURINE AMINOTRANSFERASE ACTIVITY nmole hypotaurine/mg protein/hr	
Liver	13.67 ± 3.49
Kidney	4.70 ± 0.89
Brain	18.53 ± 3.86
Pancreas	0.41 ± 0.25
Heart	0.17 ± 0.17
Testes	7.36 ± 2.64
Seminal Vesicles, Prostate & Epididymis	1.37 ± 0.28

Table 2. Products of hypotaurine aminotransferase in washed rat liver particulate fraction. Values are the average of two determinations ± SD. ^3H_2O and glutamate were determined from the same tubes, sulfate from a second assay series using ^{35}S-hypotaurine.

PRODUCTS OF HYPOTAURINE AMINOTRANSFERASE STOICHIOMETRY

Substrate		μmole/hr/g liver	
	^3H_2O	^{35}S-Sulfate	Glutamate
2-^{3}H-hypotaurine & α-ketoglutarate	3.22 ± 0.02	----	2.49 ± 0.40
^{35}S-hypotaurine & α-ketoglutarate	----	2.57 ± 0.15	----

Table 3. Effect of inhibitors on hypotaurine aminotransferase in washed brain particulate fraction.

INHIBITORS OF HYPOTAURINE AMINOTRANSFERASE IN BRAIN PARTICULATE

		μmole/hr/g liver	10% Inhibition
No inhibitor		2.0	----
Aminooxyacetic acid	10^{-5}M	0	100%
Hydroxylamine	10^{-5}M	0.04	98%
Semicarbazide	10^{-3}M	0.83	58%
	10^{-4}M	1.68	16%
β-aminoisobutyric acid	5×10^{-3}M	1.15	42%
	10^{-3}M	1.69	15%
GABA	5×10^{-3}M	0.14	93%
	10^{-3}M	0.53	73%
β-alanine	5×10^{-3}M	0.67	66%
	10^{-3}M	1.29	35%
Taurine	5×10^{-3}M	1.77	11%

Furthermore, in a similar study of the effect of 4-amino-5-fluoropentanoic acid on the activity of particulate hypotaurine aminotransferase, we observed that this halomethyl-GABA acted competitively when added simultaneously with the substrate but behaved as a mixed competitive-non-competitive inhibitor when pre-incubated for 10 min with the enzyme assay mixture before substrate was added (Fig. 5). The results of similar studies using the 5-chloro-4-aminopentanoic acid revealed only non-competitive type inhibition when the inhibitor was added simultaneously or pre-incubated for 30 min with the enzyme (Fig. 6). Finally, experiments were carried out in a similar manner using the inhibitor 4-amino-hex-5-enoic acid ("vinyl-GABA"). Non-competitive type inhibition was observed and the amount of inhibition increased when the vinyl-GABA was pre-incubated with the enzyme (Fig. 7).

Both selenotaurine and selenohypotaurine were found to be inhibitors of hypotaurine aminotransferase at concentrations as low as 10^{-4} M (Table 4). A reciprocal plot of the effects of selenohypotaurine on the velocity of the hypotaurine aminotransferase activity was carried out. When the inhibitor was added simultaneously with the substrate non-competitive inhibition was observed. When the

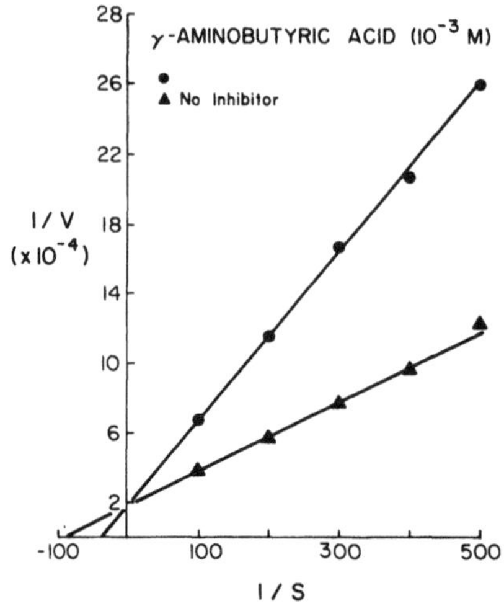

Fig. 4. A reciprocal plot of the effect of γ-aminobutyrate on the velocity of hypotaurine aminotransferase. Mouse liver homogenate dialyzed against 50 mM imidazole HCl (pH 8.2) was the enzyme source. Assay components were 0.01 mM pyridoxal 5' phosphate, 1 mM dithiothreital, 1 mM sodium pyruvate, 5 mM hypotaurine, 0.5 μCi/ml 2-^{3}H-hypotaurine and 50 mM imidazole-HCl (pH 8.2). Incubation time @ 37°C was 30 min.

inhibitor was preincubated without substrate for 30 min with the enzyme assay mixture and then the velocity determined at the substrate concentrations indicated, profound non-competitive inhibition was observed. These inhibitors were then tested for their effects on γ-aminobutyrate/β-alanine aminotransferase activity. Marked inhibition of γ-aminobutyrate/β-alanine aminotransferase activity by these inhibitors was observed (Table 4.) Furthermore, we observed that hypotaurine at 5 mM (substrate levels) did not inhibit GABA/β-alanine transaminase (Table 4).

DISCUSSION

The presence of hypotaurine aminotransferase has been demonstrated in microorganisms (13). Purified and crystallized taurine:α-ketoglutarate aminotransferase from *Achromobacter superficialis* was reported to transaminate hypotaurine at six times the rate at which taurine was transaminated. These investigators found this process to be irreversible, probably due to the instability of the sulfinoacetaldehyde. The facile rupture of the S-C bond probably occurs rapidly and spontaneously due to the electron pull of the neighboring carbonyl carbon and the fact that SO_2 is an excellent leaving group. Thus these workers observed only sulfite and acetaldehyde to be products of hypotaurine transamination, in addition to the glutamate formed as a co-product of this enzymatic process. It should be noted and underscored that the microbial hypotaurine aminotransferase differed from the mammalian enzyme described here, at least with respect to substrate specificity, in that taurine was <u>not</u> a substrate for the mammalian enzyme.

Narrow substrate specificity of microbial enzymes was observed recently in a report on γ-aminobutyrate transaminase from *Pseudomonas* sp F-126. The enzyme purified to homogeneity did not accept β-alanine nor taurine as substrate, but would accept hypotaurine (14).

In a recent study of the transamination of selenohypotaurine by rat liver and kidney mitochondria, the investigators reported that hypotaurine was not a substrate (15). They employed phosphate buffers pH 7.4 and used α-ketoglutarate as co-substrate since pyruvate and oxalacetate could not act as co-substrates. We found maximal activity at pH 8.2 using an imidazole-HCl buffer and found that pyruvate, oxalacetate and α-ketoglutarate served as amino acceptors when liver particulate was the source of the enzyme, while only α-ketoglutarate would function as co-substrate when brain particulate was the enzyme source. Also, we employed dithiothreitol or mercaptoethanol in incubation mixtures and dialysis solutions, which protected and increased the activity of the enzyme. When pyruvate was employed as co-substrate the product observed was alanine, when α-ketoglutarate was used glutamate accumulated as product. Close to stoichiometric amounts of the latter as well as sulfate were observed. Since mitochondria contain generous sulfite oxidase activity we would expect that the sulfite formed in the transaminase reaction was rapidly oxidized to sulfate. When 1,2-^{14}C-hypotaurine was used, the principle radiolabelled product was acetaldehyde (3). The hypothetical internal oxidation/reduction of the sulfinoacetaldehyde to isethionate did not take place. Rather the formation of the observed products is initiated by the loss of the SO_2, an event which could occur at the active site of the enzyme as the pyridoxal-5'-phosphate aldimine intermediate (13).

The work reported in this communication taken together with an earlier preliminary report (3) sketches out some of the properties of a mammalian enzyme capacity which to our knowledge was previously undescribed. The hypotaurine aminotransferase activity appears to be restricted to the "heavy" mitochondrial fraction from primarily brain and liver tissue homogenates. It is of interest that testes and ancillary male sexual organs also posses hypotaurine transaminase activity since hypotaurine has been reported to occur in uniquely high concentration in male sexual tissue (16). Whether this aminotransferase plays a regulatory role in setting the steady state of hypotaurine in this tissue can only be conjectured. However, from previous work (2) it is doubtful that hypotaurine aminotransferase

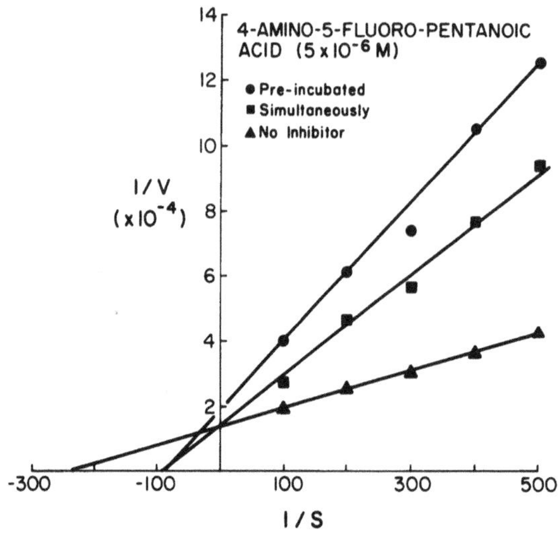

Fig. 5. A reciprocal plot of the effects of 4-amino-5-fluoro-pentanoic acid added simultaneously or preincubated for 10 min on the velocity of hypotaurine aminotransferase activity. The assay system contained washed rat brain particulate fraction, 0.01 mM pyridoxal 5' phosphate, 1 mM β-mercaptoethonal, 0.5 mM α-ketoglutaric acid, 5 mM hypotaurine, 0.5 μCi/ml 2-^{3}H-hypotaurine and 50 mM imidazole-HCl (pH 8.2) and inhibitor as indicated. Incubation time was 75 min.

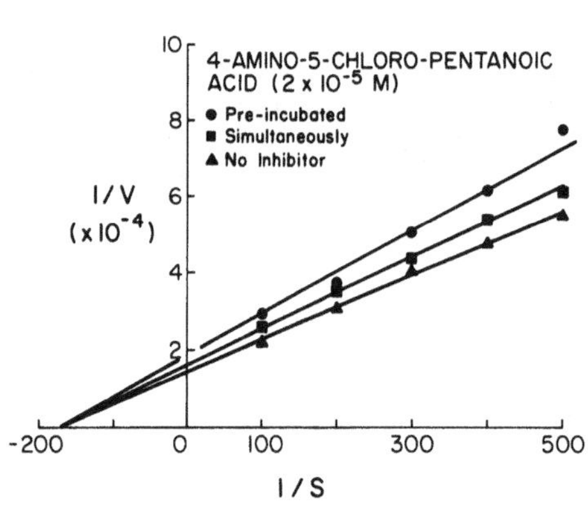

Fig. 6. A reciprocal plot of the effect of 4-amino-5-chloro-pentanoic acid added simultaneously or preincubated for 30 min on the velocity of hypotaurine aminotransferase activity. Assay system was as in Fig. 5. Incubation time was 60 min.

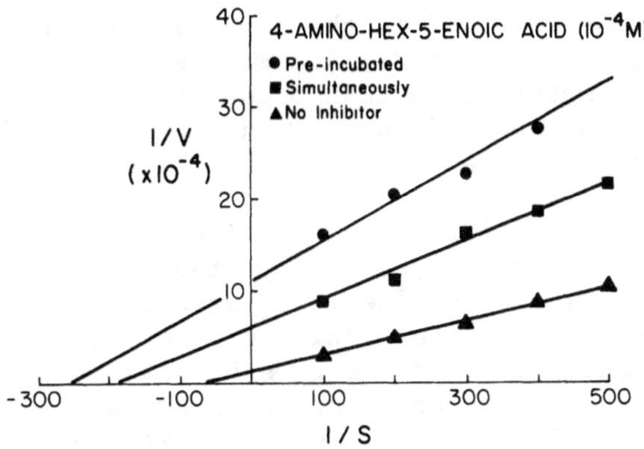

Fig. 7. A reciprocal plot of the effect of 4-amino-hex-5-enoic acid. Conditions as for Fig. 6.

Table 4. Comparison of inhibition of hypotaurine aminotransferase and β-alanine aminotransferase by seleno-compounds and substrate analogs. Washed particulate fraction from rat brain was used.

Inhibitor		% Inhibition Hypotaurine Aminotransferase	β-alanine Aminotransferase
Selenohypotaurine	10^{-3}M	98%	96%
	10^{-4}M	8%	15%
Selenotaurine	10^{-3}M	93%	83%
	10^{-4}M	56%	37%
Hypotaurine	$5x10^{-3}$M	--	5%
β-alanine	$5x10^{-3}$M	66%	--
GABA	$5x10^{-3}$M	93%	76%

occupies such a role in liver. In experiments which utilized rats subjected to partial hepatectomy to study sulfur amino acid metabolism, hypotaurine was observed in the remaining regenerating liver 4 hr but not 24 hr after hepatectomy. Studies of the activity of hypotaurine aminotransferase in such preparations revealed no change of enzyme activity in 4 hr and thus could not account for the transient accumulation of hypotaurine (2).

The data reported in this communication make clear that pyridoxal-5'-phosphate is a co-enzyme for hypotaurine aminotransferase as with other transaminases. Aldehyde reagents such as amino-oxyacetic acid, hydroxylamine, etc. inhibited the enzyme activity at a concentration far below that of the co-substrate, the only other likely carbonyl substance which might be targeted in the system. When the liver particulate fraction was dialyzed overnight after treatment with hydroxylamine (10^{-3} M) for 10 min, no detectable activity could be observed. However, addition of pyridoxal-5'-phosphate to the assay mixtures produced a greater than 50% return of enzyme activity. These observations taken together make reasonable the conclusion that pyridoxal-5'-phosphate is the co-enzyme for hypotaurine transaminase.

The products of the enzyme are $SO_3^{=}/SO_4^{=}$, acetaldehyde, and alanine or glutamate depending upon the ketoacid co-substrate employed. An additional product is a proton from the amino carbon of hypotaurine and the product we routinely measured in assaying the enzyme. It was as a result of this convenient assay procedure that we uncovered another enzymatic process which released a proton from hypotaurine but did not function as a transaminase. The activity was observed associated primarily with the liver cytosol, while trace amounts of activity was detected in the kidney, brain, and male sex organ cytosols (at levels 35 to 70 times less than liver). The enzyme activity failed in the absence of a thiol activator such as mercaptoethanol or dithiothreitol but would proceed if the activity was measured in the absence of oxygen. The enzyme activity was not inhibited by GABA, vinyl-GABA, β-aminoisobutyrate or β-alanine tested at substrate levels but was profoundly inhibited by aldehyde reagents such as amino oxyacetic acid, and hydroxylamine. The nature of the products of this liver cytosol enzyme on hypotaurine is not fully defined. However, sulfite release was also observed while a keto acceptor was not necessary for enzyme activity. These studies are currently in progress and a clear definition of this enzyme awaits further investigation. However, these observations underscore the importance of defining all the products in an enzymatic process.

The enzyme-activated ("suicide") inhibitors described in this report have all been reported to inhibit γ-aminobutyrate aminotransferase by forming a site for nucleophilic attack as a result of enzymatically converting the inhibitor to a carbonyl adjacent to a double bond as in the case of vinyl-GABA or to a carbonyl adjacent to a halo-methyl group as in the cases of 5-fluoro- and 5-chloro-4-amino-pentanoate (7). In all instances the inhibitor has been activated to a reactive alkylating agent by the action of the enzyme (6, 17-19). Thus, an active site-directed ("suicide") inhibitor of γ-aminobutyrate transaminase also inhibited hypotaurine transaminase activity. Furthermore, it has been reported that the chronic administration of vinyl-GABA to rats caused a significant elevation of the hypotaurine content in the brain tissue (20). These observations taken together form an argument that hypotaurine aminotransferase activity and 4-aminobutyrate aminotransferase activity are capacities of the same enzyme. However, studies on purified pig brain 4-aminobutyrate aminotransferase have revealed that when the substrate specificity was investigated, hypotaurine was <u>not</u> a substrate for the purified enzyme (12). Added to this is our observation that liver hypotaurine aminotransferase proceeds with pyruvate, oxalacetate or α-ketoglutarate as a co-substrate while brain hypotaurine transaminase and γ-aminobutyrate transaminase accepts only α-ketoglutarate. From the evidence above we tentatively conclude that there are at least two enzymes which carry out transamination on these substrates. One, found in brain, requires α-ketoglutarate as co-substrate and will accept hypotaurine, γ-aminobutyrate, β-alanine and β-aminoisobutyrate as substrate. It is unlikely to be identical to the purified mouse GABA-T (11) or pig GABA-T I but may be identical to GABA-T II (12). The other form of hypotaurine transaminase found in liver particulate accepts pyruvate, oxalacetate and α-ketoglutarate as co-substrate as well as β-alanine and GABA and may be identical to the GABA-T enzyme activity reported in non-neural tissue (21). This enzyme, however, appeared to be identical by immunodifusion studies with the brain GABA-T specific antibody. The resolution of the identity or non-identity of the two enzymes awaits purification of hypotaurine aminotransferase from both tissues, an undertaking which is now in progress.

ACKNOWLEDGMENT

These investigations were supported in part with a grant from the National Science Foundation 1179 and from a grant from the National Institutes of Health NS 1572.

REFERENCES

1. Jacobsen, J.G., and Smith, L.H., 1968, Biochemistry and physiology of taurine and taurine derivatives, <u>Physiol. Res.</u>, 48:424-511.
2. Sturman, J.A., and Fellman, J.H., Taurine metabolism in the rat: effect of partial hepatectomy, manuscript in preparation.
3. Fellman, J.H., Roth, E.S., Avedovech, N.A., and McCarthy, K.D., 1980, Mammalian hypotaurine aminotransferase: isethionate is not a product, <u>Life Sci.</u>, 27:1999-2004.

4. Fellman, J.H., Roth, E.S., and Fujita, T.S., 1978, Taurine is not metabolized to isethionate in mammalian tissue, in: "Taurine and Neurological Disorders," A. Barbeau and R. Huxtable, eds., Raven Press, New York.
5. Fellman, J.H., The synthesis of 2^3H-taurine and 2-^{3}H-hypotaurine, J. Lab. Comp. Radio. Pharm., in press.
6. Silverman, R.B., and Levy, M.A., 1980, Irreversible inactivation of pig brain γ aminobutyric acid α-ketoglutarate transaminase by 4-amino-5-halopentanoic acid, Biochem. Biophys. Res. Commun., 95:250-254.
7. Pichat, L., Herbert, M., and Thiers, M., 1961, Etudes sur les methodes de preparation de la selenocystamine, selenohypotaurine et selonotaurine, Tetrahedron, 12:1-6.
8. Rinaldi, A., Fadda, M.B., Dernini, S., Cossu, P., and DeMarco, C., 1975, Copper catalysed alkaline autoxidation of selenocystamine, It. J. Biochem., 1:219-230.
9. Fellman, J.H., Roth, E.S., Avedovech, N.A., and McCarthy, K.D., 1980, The metabolism of taurine to isethionate, Arch. Biochem. Biophys., 204:560-565.
10. Rowan, G.L., 1974, Isolation of lysosomes, Methods in Molecular Biology, 5:89-101.
11. Schousboe, A., Wu, J.-Y., and Roberts, E., 1973, Purification and characterization of the 4-aminobutyrate-2-ketoglutarate transaminase from mouse brain, Biochemistry, 12:2868-2873.
12. Block-Tardy, M., Rolland, B., and Gonnard, P., 1974, Pig brain 4-aminobutyrate-2-ketoglutarate transaminase. Purification, kinetics and physical properties, Biochimie, 56:823-832.
13. Tanaka, H., Toyama, S., Tsukahara, H., and Soda, K., 1974, Transamination of hypotaurine by taurine: α-ketoglutarate aminotransferase, FEBS Letters, 45:111-113.
14. Yonaha, K., and Toyama, S., 1980, γ-aminobutyrate: α-ketoglutarate aminotransferase from Pseudomonas species F-126: purification, crystallization, and enzymologic properties, Arch. Biochem. Biophys., 200:156-164.
15. Rinaldi, A., Floris, P., Cossu, P., and DeMarco, C., 1978, Transamination of selenohypotaurine, Bull. Mol. Biol. Med., 3: 234-242.
16. Kochakian, C.D., 1975, Free amino acids of sex organs of the mouse: regulation by androgen, Am. J. Physiol., 228:1232-1235.
17. Abeles, R.H., and Maycock, A.L., 1976, Suicide enzyme inactivators, Acc. Chem. Res., 9:313-319.
18. Jung, M.J., Lippert, B., Metcalf, B.W., Schechter, P.J., Bohlen, P., and Sjoerdsma, A., 1977, The effect of 4-amino hex-5-ynoic acid (γ-acetylenic GABA, γ-ethynyl GABA) a catalytic inhibitor of GABA transaminase, on brain GABA metabolism in vivo, J. Neurochem., 28:717-723.
19. Jung, M.J., Lippert, B., Metcalf, B.W., Bohlen, P., and Chechter, P. J., 1977, γ-vinyl GABA (4-amino-hex-5-enoic acid) a new selective irreversible inhibitor of GABA-T: effects on brain GABA metabolism, J. Neurochem., 29:797-802.

20. Perry, T.L., Kish, S.J., and Hansen, S., 1979, γ-vinyl GABA: effects of chronic administration on the metabolism of GABA and other amino compounds in rat brain., J. Neurochem., 32:1641-1645.
21 Wu, J.-Y., Moss, L.G., and Chude, O., 1978, Distribution and tissue specificity of 4-aminobutyrate-2-oxoglutarate aminotransferase, Neurochem. Res., 3:207-219.

HYPOTAURINE UPTAKE IN MOUSE BRAIN SLICES

S.S. Oja and Pirjo Kontro

Department of Biomedical Sciences
University of Tampere
Box 607
SF-33101 Tampere 10
Finland

Hypotaurine, 2-aminoethanesulfinic acid, an immediate precursor in taurine biosynthesis (3), is converted to taurine also in brain tissue (10). Hypotaurine concentration in the rat brain is 0.07 mmole/kg fresh weight, i.e. higher than that of the biogenic amine precursors, phenylalanine and tryptophan (21). Taurine transport has been studied in different nervous tissue preparations (16,19) and its penetration into brain tissue has been found to be very slow both in vivo (18) and in vitro (8) when compared to the putative amino acid transmitters. Now we have characterized the properties of hypotaurine by mouse brain slices in vitro. The uptake was found to be strikingly fast and efficient, generating exceptionally high concentration gradients and consisting of two sodium-dependent high-affinity and low-affinity transport systems.

MATERIALS AND METHODS

(^{35}S)Hypotaurine was synthesized from (^{35}S)cystamine (1.15 - 4.37 TBq/mole, the Radiochemical Centre, Amersham) (10), thiotaurine from hypotaurine (2) and β-guanidinoethylsulfonic acid (GES) from taurine (5). The purity of these preparation was checked by thin-layer chromatography (10,17). Mouse brain slices 0.4 mm thick were cut in the frontal plane from the whole cerebrae with a non-wetted tissue chopper of the McIlwain type. From 50 to 100 mg slices were incubated in Krebs-Ringer-HEPES-glucose medium (pH 7.4) at 37°C under O_2 for 15 min (17). The results were calculated and the transport kinetics analyzed as in Kontro and Oja (8) and Oja and Kontro (17).

RESULTS AND DISCUSSION

Fig. 1 shows the time course of hypotaurine uptake. The uptake was linear for about 40 min and an approximate equilibrium between influx and efflux was attained only after 3 hours of incubation. The slices effectively concentrated hypotaurine from the medium. The tissue/medium concentration ratios obtained were exceptionally high for an amino acid, comparable only to those for GABA in cerebral cortex slices (1,7). The corresponding ratios for taurine have been considerably lower (7,12,15).

Hypotaurine uptake was very temperature-sensitive, being almost abolished at 5°C (Fig. 2). The optimum temperature was about 37°C, since the uptake was reduced by about 33% at 45°C. The activation energy needed for hypotaurine uptake (92.7 ± 11.5 kJ/mole, mean ± SEM) was calculated from the integrated Arrhenius plot (Fig. 2). The Q_{10} value of the uptake was 1.8 for the interval of 27 to 37°C.

The energy dependence of the uptake was further confirmed when metabolic poisons were added to incubation medium containing two different hypotaurine concentrations, 0.01 and 1.0 mM (Table 1). Iodoacetate was effectual only at the hypotaurine concentration of 1.0 mM, whereas ouabain and cyanide significantly reduced uptake at both hypotaurine concentrations, but again more so at the concentration of 1.0 mM. Furthermore, the presence of 1.0 mM 2,4-dinitrophenol and the anaerobic incubation conditions both separately inhibited hypotaurine uptake by about 80% (hypotaurine concentration: 0.1 mM) (17).

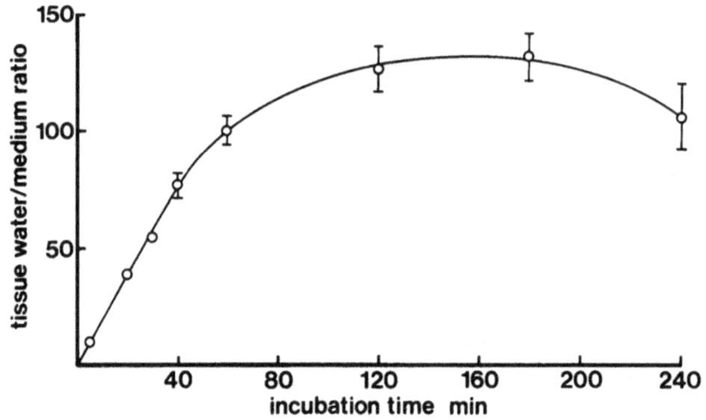

Fig. 1. Time course of hypotaurine uptake by mouse brain slices. The slices were first preincubated for 30 min at 37°C in Krebs-Ringer-HEPES-glucose medium (pH 7.4) under O_2 and then incubated with an initial (^{35}S)hypotaurine concentration of 0.1 mM for the periods indicated. Each point is a mean ± SEM of three determinations.

Table 1

Effects of Metabolic Poisons on Hypotaurine Uptake

Inhibitor	Uptake in percent of controls at hypotaurine concentrations	
(Concentrations in mM)	0.01 mM	1.0 mM
None (control)	100.0 ± 3.2 (8)	100.0 ± 4.2 (4)
NaCN (1.0)	$66.4 \pm 7.8^{+}$ (8)	$24.5 \pm 0.9^{\dagger}$ (4)
Sodium iodoacetate (1.0)	98.6 ± 2.4 (5)	$78.2 \pm 2.0^{\dagger}$ (4)
Ouabain (0.01)	$53.3 \pm 1.5^{\dagger}$ (5)	$30.0 \pm 0.1^{\dagger}$ (4)

Mouse brain slices were incubated as in Fig. 2. The results are means ± SEM. Number of experiments in parentheses. Significance of differences from the corresponding controls: $^{\dagger}p<0.01$.

The uptake of hypotaurine was saturable, consisting of two transport components, the kinetic parameters of which are given in Table 2. The transport constant for the high-affinity uptake was of the same order of magnitude as the constants generally reported for amino acid transmitter candidates (4,19,20). The transport capacities of both high-affinity and low-affinity hypotaurine uptake systems were remarkably large when compared to taurine uptake (8,10).

The uptake of hypotaurine was influenced by the concentrations of the extracellular cations. In the absence of sodium ions only 3.5 ± 0.4% (mean ± SEM, n=6) of the uptake persisted (hypotaurine concentration: 10 μM), indicating an absolute sodium requirement for transport. In this respect hypotaurine uptake mimics the uptake of neurotransmitter amino acids (4,19). On the other hand, certain structural analogues of GABA, i.e., L-2,4-diaminobutyric acid (L-DABA) (22), β-alanine (20) and nipecotic acid (6), have also been found in brain slices to be similar in this respect. The omission of potassium and calcium ions inhibited the uptake 63.5 ± 1.4 and 39.8 ± 5.7% (means ± SEM, n=6), respectively. The optimal potassium ion concentration for the uptake was 5 mM, as depolarizing potassium ion concentrations again reduced uptake (Fig. 3). The dependence of the uptake on the sodium and potassium ions shows that the ion gradients generated by an active (Na^+, K^+)ATPase are essential to sustain hypotaurine transport. This inference is also supported by the ouabain sensitivity of the uptake (Table 1).

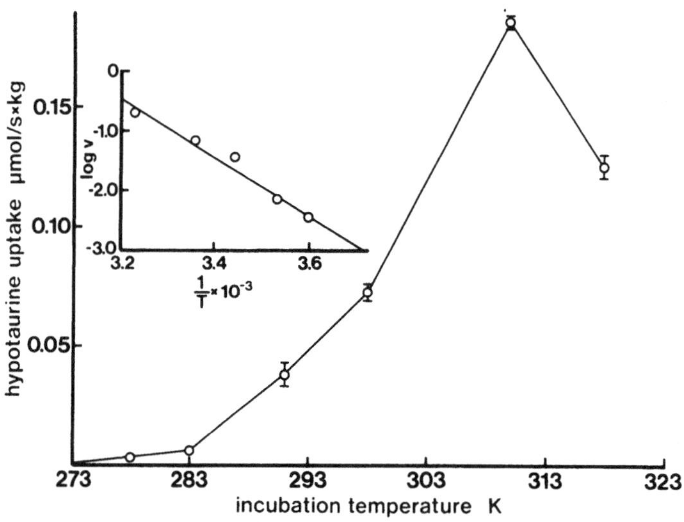

Fig. 2. Temperature dependence of hypotaurine uptake by brain slices. The slices were incubated as in Fig. 1 but with 10 µM of (^{35}S)hypotaurine for 15 min. Each point is a mean ± SEM of three or six determinations calculated per wet weight of the incubated slices. The insert shows the estimation of activation energy by the least-squares straight line in an Arrhenius plot. Transport units are µmole/sec/kg.

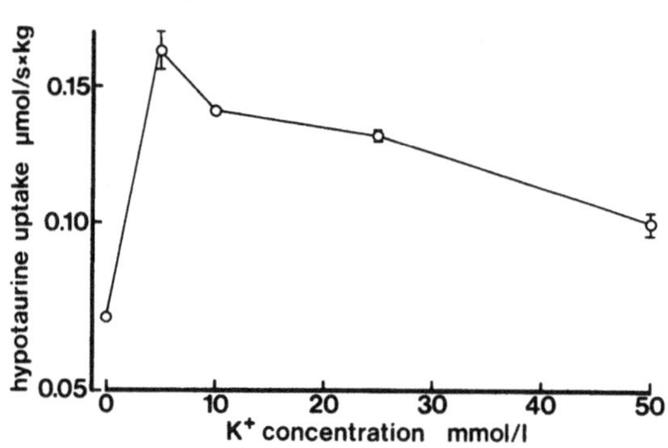

Fig. 3. Potassium dependence of hypotaurine uptake by brain slices incubated as in Fig. 2. Potassium ions in medium were replaced by equimolar concentrations of choline chloride. At elevated potassium ion concentrations the results were corrected for hyperosmolarity (11). Each point is a mean ± SEM of four or eight determinations.

The interactions of sodium ions with hypotaurine transport were further analyzed in a low-sodium (60 mM) medium, where the uptake was still saturable and consisted of the two transport components (Table 2). The transport constants for both components were now more than three times higher than in the normal-sodium (control) medium, whereas the maximal velocities were not markedly altered. This suggests that sodium ions are needed mainly for the attachment of hypotaurine molecules to their postulated carrier sites in neural membranes, but it is less likely that they also increase the rates of translation of either unloaded or hypotaurine-loaded carriers across the cell membranes. The optimal sodium concentration for hypotaurine uptake was well above the normal concentration range of the extracellular sodium ions in vivo (Fig. 4). The sodium dependence curve of hypotaurine uptake was also similar to those obtained for taurine and GABA (9,13,14,23). Sodium ions mimic allosteric effectors of hypotaurine transport, showing positive cooperativity. The slope of the straight line obtained in a Hill plot was 2.4 ± 0.2 (Fig. 4). This may indicate that hypotaurine carriers in the cell membrane contain several (more than two) binding sites for sodium. The corresponding Hill coefficients were 3 for taurine uptake by synaptosomes (9) and brain slices (P. Kontro, unpublished results).

Hypotaurine uptake was inhibited by several structural analogues (Table 3). The strongest inhibitors were GABA, L-DABA, cysteic acid and β-alanine. Glycine, cysteine, cystamine and cysteine sulfinic acid were only moderately potent. Taurine had only a slight effect

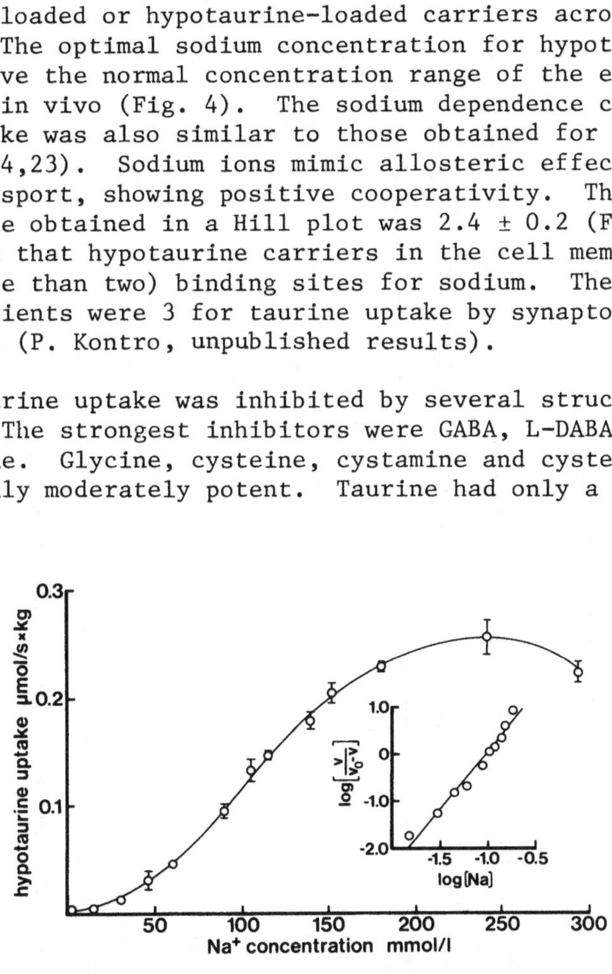

Fig. 4. Sodium dependence of hypotaurine uptake by brain slices incubated as in Fig. 2. Sodium ions in medium were replaced by equimolar concentrations of choline chloride. At elevated sodium ion concentrations the results were corrected for hyperosmolarity (11). Each point is a mean ± SEM of six determinations. The insert shows the Hill plot of the same data. v_0 Is the apparent maximal velocity of uptake at 240 mM Na^+ and v denotes the uptake velocities at other Na^+ concentrations.

Table 2

Kinetic Constants of Hypotaurine Uptake

Incubation conditions	Low-affinity uptake		High-affinity uptake	
(Concentrations in mM)	K_m (mM)	V (μmole/sec/kg)	K_m (μM)	V (nmole/sec/kg)
Control	2.9 ± 0.2	34.1 ± 0.9	24 ± 3	314 ± 19
Low-sodium (60)	11.2 ± 1.5†	22.1 ± 1.6†	109 ± 17†	394 ± 46
GABA (0.5)	2.8 ± 0.3	21.7 ± 0.7†	-	-
β-Alanine (1.0)	2.2 ± 0.1†	21.7 ± 0.4†	-	-
Cysteic acid (1.0)	1.9 ± 0.1†	23.9 ± 0.7†	25 ± 6	271 ± 44

Significance of differences from the corresponding controls: $^\dagger p<0.01$.

Table 3

Effects of Structural Analogues on Hypotaurine Uptake

Inhibitor	Hypotaurine uptake (percent of control)	
None (control)	100.0 ± 4.5	(21)
Taurine	79.2 ± 3.8*	(6)
β-Alanine	61.4 ± 3.7†	(10)
Glycine	66.1 ± 5.1†	(3)
GABA	40.1 ± 2.1†	(6)
L-DABA	47.7 ± 2.3†	(7)
Cysteine	67.8 ± 5.3†	(6)
Cysteinesulfinic acid	73.4 ± 2.3*	(3)
Cysteic acid	51.9 ± 1.2†	(3)
Cystamine	68.2 ± 4.1†	(6)
N-Methyltaurine	102.1 ±13.0	(3)
Isethionic acid	92.1 ± 6.3	(6)
Amidosulfonic acid	99.1 ± 3.0	(3)
Methionine sulphoxide	92.9 ± 2.3	(3)
Thiotaurine	99.3 ± 4.8	(3)
GES	77.6 ± 1.2*	(4)
Methionine	90.5 ± 3.9	(4)
Ethanolamine Phosphate	101.6 ± 3.3	(4)

The slices were incubated as in Fig. 2. The concentration of hypotaurine was 0.1 mM and that of the structural analogues 1.0 mM. The results are means ± SEM. Number of experiments in parentheses. Significance of differences from the control: *$p<0.05$, †$p<0.01$. Abbreviations: L-DABA = L-2,4-diaminobutyric acid, GES = β-guanidinoethylsulfonic acid.

at this hypotaurine concentration (0.1 mM). The effects of GABA, taurine, β-alanine and cysteic acid were subjected to further kinetic analyses. These inhibitors reduced hypotaurine uptake equally when they were added only to the final incubation (15 min) medium or were already present at the beginning of the 30-min preincubation. Hypotaurine uptake was thus obviously (at least mainly) affected only

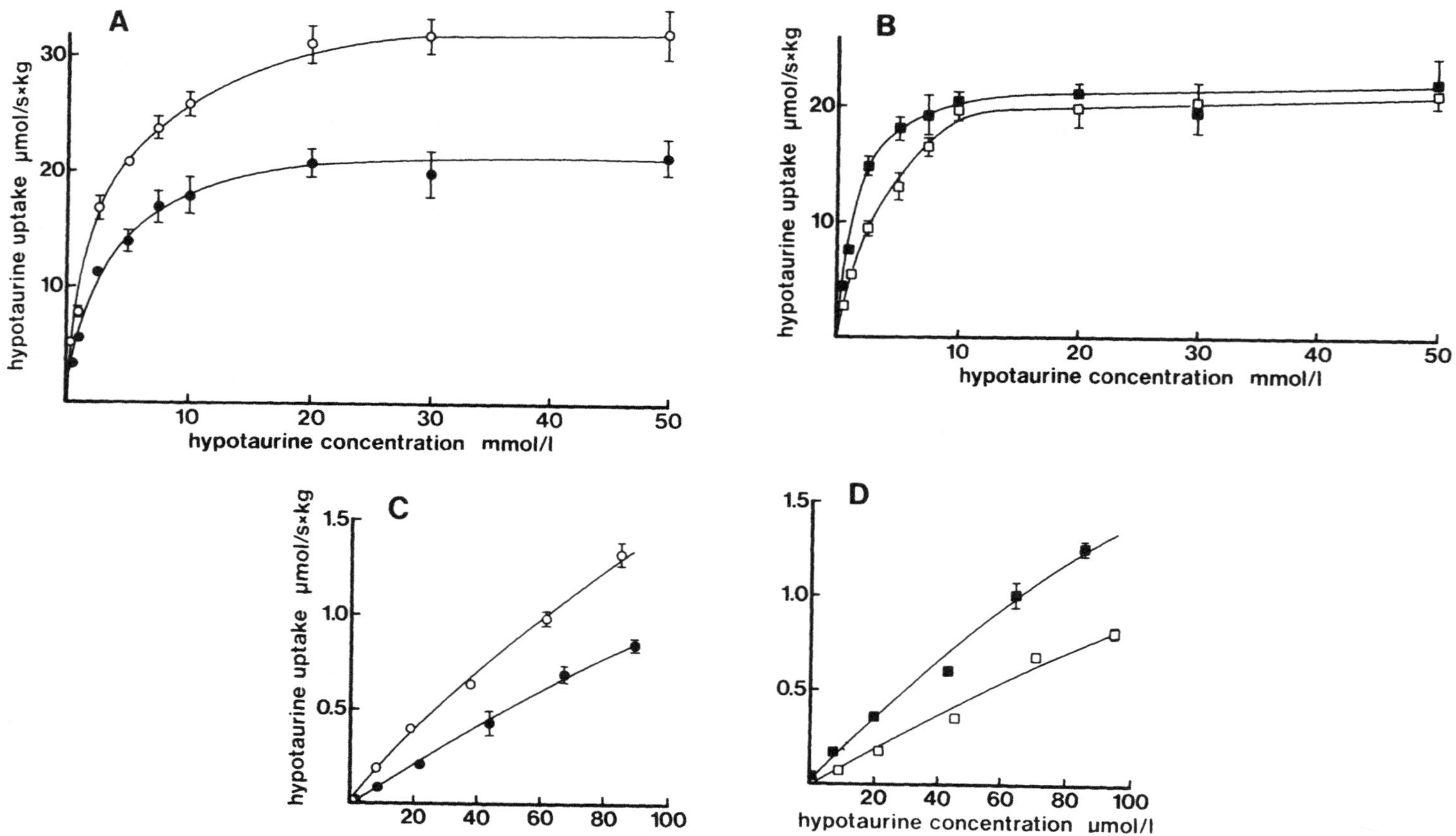

Fig. 5. Effects of some structural analogues on hypotaurine uptake in brain slices incubated as in Fig. 2. Hypotaurine uptake within the millimolar concentration range: (A) in the normal unsupplemented medium (○——○); and in the presence of 1.0 mM β-alanine (●——●); (B) in the presence of 1.0 mM cysteic acid (■–■) or 0.5 mM GABA (□–□). (C–D) Hypotaurine uptake within the micromolar concentration range in the presence of the aforementioned structural analogues (Symbols as for 5A,B). Each point is a mean ± SEM of three or six determinations.

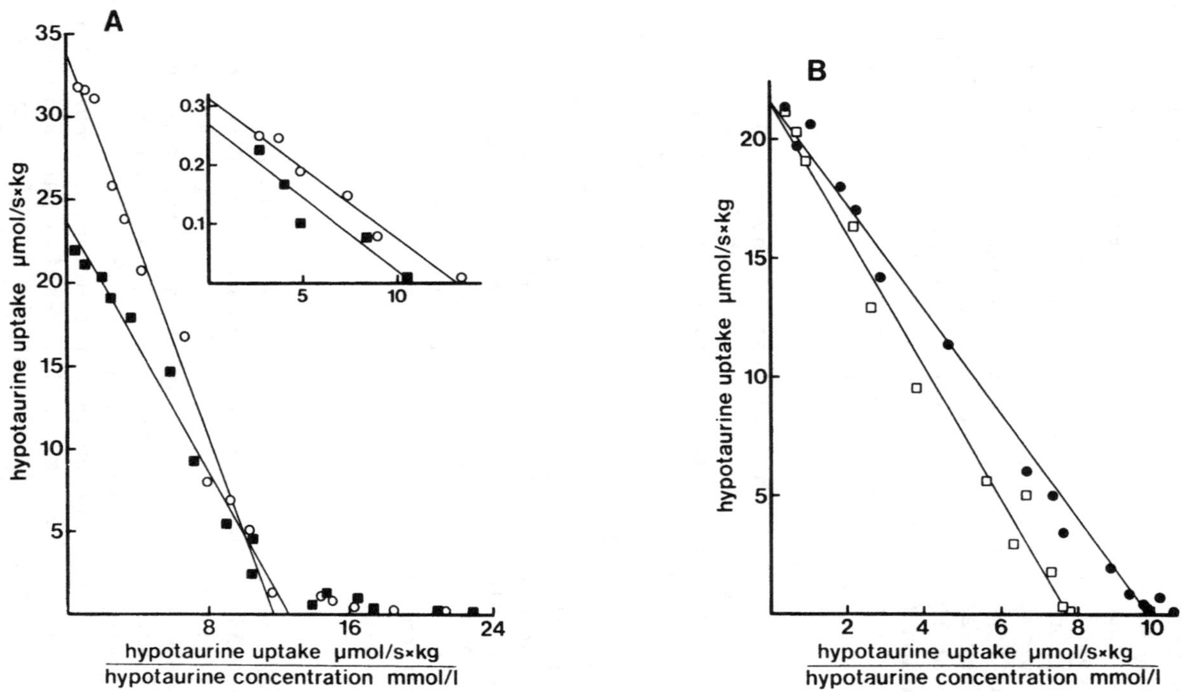

Fig. 6. Hypotaurine uptake (A) in the normal unsupplemented (○-○) medium, in the presence of cysteic acid (■-■), (B) in the presence of β-alanine (●-●) or GABA (□-□) depicted in v vs. v/s plots. The plots are recalculated from the data presented in Fig. 5. In the presence of β-alanine or GABA the experimental points could be fitted to one straight line. The pattern conforms to a linear transform of the simple Michaelis equation (only the low-affinity uptake operating). In the normal unsupplemented medium or in the presence of cysteic acid two saturable uptake systems (the low-affinity and high-affinity) must be operating simultaneously. The insert in (A) shows the fitting of the high-affinity uptake component after subtraction of the share of the low-affinity transport (8). The estimated kinetic parameters are compiled in Table 3.

by extracellular inhibitors in the medium and not by the inhibitor molecules gradually accumulating intracellularly in the incubated slices. It also became obvious that the two components of hypotaurine uptake were differently affected by these structural analogues. In the presence of GABA (0.5 mM) or β-alanine (1.0 mM) the high-affinity uptake of hypotaurine was completely blocked, the low-affinity uptake being inhibited considerably less and the type of inhibition being at least in this case apparently not competitive (Figs. 5 and 6, Table 2). We also tested 0.1 mM GABA, and even this eliminated the high-affinity uptake. On the other hand, cysteic acid (1.0 mM) inhibited only the low-affinity transport component, without significantly affecting the high-affinity uptake. A high concentration (30 mM) of taurine increased the transport constant and diminished the maximal velocity of the low-affinity hypotaurine uptake (Table 4). A low concentration (0.5 mM) of taurine - which does not detectably affect the low-affinity uptake of hypotaurine - did not significantly alter the kinetic constants of the high-affinity uptake component either.

The above inhibition studies reveal that the hypotaurine uptake systems may not be very substrate (hypotaurine)-specific. In particular, the properties of the transport mechanisms of hypotaurine on the one hand, and those of GABA and taurine on the other, must be still more closely comparatively characterized. The present results demonstrate the potential of brain cells to take up extracellular hypotaurine. The physiological significance of this capacity remains to be settled, however, since it is not known whether or not there is any circulating hypotaurine available for brain tissue in vivo. In any case the extracellular hypotaurine levels must be very low, less than 10^{-6} molar when extrapolated from the tissue water/medium concentration ratios now obtained and from the reported cerebral hypotaurine concentration (21).

CONCLUSIONS

Hypotaurine uptake by mouse brain slices is highly concentrative, temperature-, energy- and sodium-dependent, consisting of two transport components, low- and high-affinity, both of them possessing relatively large transport capacities. Hypotaurine transport thus exhibits features characteristic for transport of a neurotransmitter candidate. In particular, the uptake of hypotaurine is considerably faster and more efficient than the uptake of taurine. However, the specificity of the transport processes still remains partially unsettled. Taurine (and cysteic acid) affects the low-affinity transport component without significantly affecting the high-affinity uptake. On the other hand, GABA (and β-alanine) effectively interferes with the high-affinity hypotaurine uptake.

Table 4

Kinetic Constants of Hypotaurine Uptake in the Presence of Taurine

	Low-affinity uptake		High-affinity uptake	
	K_m mM	V µmole/sec x kg	K_m µM	V nmole/sec x kg
Control	2.5 ± 0.1	28.1 ± 0.5	26 ± 6	206 ± 23
Taurine (30 mM)	5.9 ± 0.3†	18.0 ± 0.4†		
Taurine (0.5 mM)			35 ± 4	230 ± 16

Significance of differences for the corresponding controls: $^{\dagger}p<0.01$.

ACKNOWLEDGMENTS

The authors acknowledge the financial support of the Medical Research Council of the Academy of Finland. The study was also supported by grants from the Magnus Ehrnrooth Foundation and the Emil Aaltonen Foundation to P. Kontro. Mrs. Pirkko Erkkilä, Miss Päivi Kyrölä and Mr. Ari-Pekka Viirtelä offered their expert technical assistance.

REFERENCES

1. Bond, P.A., 1973, The uptake of gamma-(^{3}H)-aminobutyric acid by slices from various regions of rat brain and the effect of lithium, J. Neurochem., 20:511-517.
2. Cavallini, D., De Marco, C. and Mondovi, B., 1958, Sur la synthèse de la thiotaurine, Bull, Soc. Chim. Biol., 40:1711-1715.
3. Cavallini, D., Scandurra, R., Duprè, S., Santoro, L., and Barra, D., 1976, A new pathway of taurine biosynthesis, Physiol. Chem. Physics, 8:157-160.
4. Davidson, N., Neurotransmitter Amino Acids, Academic Press, London (1976).
5. Huxtable, R.J., Laird, H.E., II, and Lippincott, S.E., 1979, The transport of taurine in the heart and the rapid depletion of tissue taurine content by guanidinoethyl sulfonate, J. Pharmacol. Exp. Ther., 211:465-471.
6. Johnston, G.A.R., Stephanson, A.L., and Twitchin, B., 1976, Uptake and release of nipecotic acid by rat brain slices, J. Neurochem., 26:83-87.

7. Kontro, P., Comparison of the uptake processes of taurine, hypotaurine and GABA, in: "Amino Acid Transmitters," P. Mandel and F.V. De Feudis, eds., Raven Press, New York, in press.
8. Kontro, P., and Oja, S.S., 1978, Taurine uptake by rat brain synaptosomes, J. Neurochem., 30:1297-1304.
9. Kontro, P., and Oja, S.S., 1978, Sodium dependence of taurine uptake in rat brain synaptosomes, Neuroscience, 3:761-765.
10. Kontro, P., and Oja, S.S., 1980, Hypotaurine oxidation by mouse tissues, in: "Natural Sulfur Compounds: Novel Biochemical and Structural Aspects," D. Cavallini, V. Zappia and G.E. Gaull, eds., Plenum Press, New York (1980), pp. 201-212.
11. Kontro, P., and Oja, S.S., Properties of hypotaurine uptake in mouse brain slices, in: "The Action of Taurine on Excitable Cells," S.I. Baskin, J.J. Kocsis and S.W. Schaffer, eds., Spectrum Publications, New York, in press.
12. Levi, G., Regional differences in cerebral amino acid transport, in: "Progr. Brain Research, Vol. 29," A. Lajtha and D.H. Ford, eds., Elsevier, Amsterdam (1968), pp. 219-228.
13. Martin, D.L., and Shain, W., 1979, High affinity transport of taurine and β-alanine and low affinity transport of γ-aminobutyric acid by a single transport system in cultured glioma cells, J. Biol. Chem., 254:7076-7084.
14. Martin, D.L., and Smith, A.A., III, 1972, Ions and the transport of gamma-aminobutyric acid by synaptosomes, J. Neurochem., 19: 841-855.
15. Oja, S.S., 1971, Exchange of taurine in brain slices of adult and 7-day-old rats, J. Neurochem., 18:1847-1852.
16. Oja, S.S., and Kontro, P., Neurotransmitter actions of taurine in the central nervous system, in: "Taurine and Neurological Disorders," A. Barbeau and R.J. Huxtable, eds., Raven Press, New York (1978), pp. 181-200.
17. Oja, S.S., and Kontro, P., 1980, Uptake of hypotaurine in brain slices from adult and 8-day-old mice, J. Neurochem., 35:1303-1308.
18. Oja, S.S., Lehtinen, I., and Lähdesmäki, P., 1976, Taurine transport rates between plasma and tissues in adult and 7-day-old mice, Quart, J. exp. Physiol., 61:133-143.
19. Oja, S.S., Kontro, P., and Lähdesmäki, P., 1977, Amino acids as inhibitory neurotransmitters, Progr. Pharmacol., 1/3 1-119.
20. Okamoto, K., and Namima, M., 1978, Uptake, release and homo- and hetero-exchange diffusions of inhibitory amino acids in guinea-pig cerebellar slices, J. Neurochem., 31:1393-1402.
21. Perry, T.L., and Hansen, S., 1973, Quantification of free amino compounds of rat brain: identification of hypotaurine, J. Neurochem., 21:1009-1011.
22. Weitsch-Dick, F., Jessel, T.M., and Kelly, J.S., 1978, The selective neuronal uptake and release of (^{3}H)-DL-2,4-diaminobutyric acid by rat cerebral cortex, J. Neurochem., 30:799-806.
23. Wheeler, D.D., and Hollingsworth, R.G., 1979, A model of GABA transport by cortical synaptosomes from the Long-Evans rat, J. Neurosci. Res., 4:265-289.

THE SULFUR-CONTAINING AMINO ACID PATHWAY IN NORMAL AND MALIGNANT CELL GROWTH

S. I. Baskin[1], K. Wakayama[2], T. Knight[1], J. H. Jepson[3] and E. C. Besa[1]

[1]Departments of Pharmacology and Medicine
The Medical College of Pennsylvania
3300 Henry Avenue
Philadelphia, Pennsylvania
[2]Showa University School of Medicine
Yokohama, Japan
[3]City of Hope Medical Center
Duarte, California

Taurine is a highly acidic sulfur-containing amino acid which is not incorporated into protein, but appears to be an end product of methionine metabolism. Its high acidity (pKa 1.5 or less) and its oxidized nature suggest that by normal mechanisms it might be a readily excreted molecule and, thus. not to found in wide abundance. However, taurine is found to be widely distributed in animal tissues, especially during periods of cellular growth (15).

Piez and Eagle (28) were the first to find large differences in the tissue/medium ratio of taurine in cultured Hela cells. The high tissue/medium ratio does not necessarily reflect a high intracellular concentration in comparison to normal cells. but may reflect a very effective accumulation mechanism by cells which are incubated in media not containing taurine. Related studies have found that other cancer cells such as ascites carcinoma cells (9), neoplastic murine mast cells (12), and cultured glial tumor cells (30) similarly accumulated taurine selectively to a high degree.

In addition, cysteine sulfinic acid decarboxylase, an important enzyme in the biosynthesis of taurine, has been shown to be progressively less active over time in Guerin epithelioma cells and in sarcoma tissue derived from stomach grown *in vitro* (8).

The taurine biopathway has been shown to play an important role in the growth of benign and malignant mammalian cells. Methionine and cysteine which are precursor amino acids of taurine were found to be an essential requirement for the growth of human malignant cells (10,21,31). Methionine is essential in the initiation of polypeptide chains and is the chief methyl donor for methylation of DNA, RNA, protein and other biological substrates needed for cellular proliferation. In a recent survey of both benign and malignant mammalian cells growing in tissue cultures, Kries (20) found adenocarcinoma of the breast, two types of lymphoblastic leukemias and neuroblastoma to have an absolute requirement for methionine. Enzymatic depletion of this amino acid by L-methionase resulted in complete inhibition of cell growth in both normal fibroblast and the acute lymphoblastic leukemic cells *in vitro*. Rescue with a homocysteine analog was partly successful in normal but not in malignant cells.

It has been shown that the *in vitro* and *in vivo* uptake of ^{35}S L-cystine is greater in leukemic leukocytes than in normal leukocytes (34,36). L-cystine is the oxidized form of L-cysteine and these two compounds occur in normal human plasma in a respective ratio of 2-5:1 (6). The increased uptake of ^{35}S L-cystine is believed to be due to low intracellular L-cysteine levels caused by the inability of leukemic leukocytes to metabolize L-cystathionine, the immediate precursor of L-cysteine in the biosynthetic pathway. In keeping with this, α-cystathionase has been observed in abnormally low levels in leukemic leukocytes (11,14,22,25).

The different growth requirements in leukemic cells caused by this metabolic blockage have been utilized for many therapeutic approaches. L-cysteine depleted diets and the administration of L-cysteine analogues have produced considerable reductions in leukopoiesis *in vitro* and *in vivo* (32,37). The detrimental effects of L-cysteine depleted diets and the toxic effects of the analogues, however, have minimized their therapeutic usefulness.

Studies of amino acid requirements in human leukemia demonstrated that cells from a patient with acute lymphoblastic and myelocytic leukemia required L-cysteine in the presence of adequate levels of methionine and serine *in vitro* (29). Normal cells can grow if cystathionine is substituted for cysteine but leukemic cells cannot. Leukemic cells were found to have one tenth the level of α-cystathionase of normal cells (38).

Similar observations were found in other malignant cells (27). Livingston et al. (26) measured a five to ten-fold decrease of α-cystathionase activity in thymomas in AKR mice occurring spontaneously or induced by Maloney virus (25). Since taurine is an end product of methionine metabolism, it is not certain whether taurine itself or any of the precursor amino acids are essential in leukemic or malignant cell growth. Our studies are designed to determine whether these cell proliferation changes in the leukemic cells are unique for taurine or whether it is a reflection of an abnormality in a precursor amino acid for taurine.

METHODS

Cell Preparation

Leukemic Lymphocytes

L1210 cells were obtained from A.D. Little Company (Cambridge, Mass.) and 1 x 10^5 cells were innoculated into the peritoneal cavity of recipient (DBA/2J) mice. Five mice were sacrificed every 24 hr through 168 hr after injection of L1210 cells and cells were collected by washing the peritoneal cavity with 9-15 ml of Hank's balanced salt solution without Ca^{+} or Mg^{++}(HBSS). This cell suspension was centrifuged at 100 xg for 15 min. The pellet was resuspended in HBSS, and layered on 3 ml of Ficoll-Hypaque (Pharmacia Fine Chemicals, d=1.077). The cell suspension was centrifuged at 400 xg for 30 min, then the buffy coat was collected. These cells which layered in the density for lymphocytes were designated as L-cells. Additionally, the L1210 cells that sedimented at the bottom of the tube were collected and referred to as Bo-cells. Both groups of cells were washed with HBSS at 100 xg for 15 min and resuspended in HBSS and counted using a hemocytometer. An aliquot of each L1210 cell suspension was used for cell volume analysis and a smear was prepared from each aliquot. The remaining cells were stored at -20°C until use.

Normal Lymphocytes

Control lymphocytes were obtained from 10 lymph nodes, the spleen and the thymus of a normal DBA/2J female mouse. These tissues were minced with scissors in a plate containing a small amount of HBSS, and the cells were dispersed further by aspirating them through an 18 gauge needle. Then the cell suspension was passed through a 1-2 cm column which was lightly packed with glass wool, and the filtrate was then layered on 3 ml of Ficoll-Hypaque. Cells were similarly separated into L and Bo-cell fractions using the Ficoll-Hypaque cell separation method.

Cell Volume Analysis

Separated L1210 cell suspension fluid (L and Bo-cell suspensions) was diluted with phosphate buffered saline (PBS). Cell volume distribution and mean cellular volumes (MCV) were measured using a Coulter Channel analyzer CHL2 and Coulter counter Model ZBI. The aperture diameter was 100 μm and calibration was performed by standard polystyrene beads with a diameter of 9.69 μm. MCV was given by the following (26):

$$MCV = \{(\text{mean channel No} \times WW/100) + BCT)\} \times TF$$

WW: Window width (100)

BCT: Base channel threshold (5)

TF: Threshold factor (10.34 was provided by calibration)

Protein Analysis

Protein concentrations of cells were measured by the Bio-Rad method as described by Bradford (5).

DNA Analysis

DNA content of cells was determined using a fluorometric assay modified from Kissane and Robins (17).

Taurine Assay

Colorimetric assay by 4-Dimethyl-aminoazobenzene 4' - sulfonyl chloride (DABS-Cl) using high pressure liquid chromatography (HPLC, Waters Co.) was initially employed (24). However, DABS-Cl behaved as an oxidizing agent and conversion from hypotaurine to taurine and CSA to CA were recognized. Conversion was reduced when the amount of DABS-Cl was reduced.

We changed the above method to a fluorometric assay using O-phthaldialdehyde. This method was more sensitive and efficient and was faster than the first technique. In this method, modified from Stuart et al. (3), oxidation of hypotaurine to taurine did not occur and less cells were required than the DABS-Cl method for the assay. The methods for taurine assay are as follows: 100 μl of cell suspension fluids (0.5-10 x 10^5 cells) were sonicated and deproteinized by 100 μl of 10% TCA and 200 μl of ether were used to remove the lipids and TCA. Then an aliquot (containing 1-5 x 10^3 cells) of the aqueous solution was injected into HPLC, employing a fatty acid analysis column, and taurine concentrations were determined by a fluorometric detector (excitation - 325 nm, emission - 410 nm).

The post-column reagent was O-phthaldialdehyde (0.65%) (OPA,

Sigma Chemical Co.) with 0.2% 2-mercaptoethanol and borate buffer (0.5 M) which was adjusted to a pH of 10.4. The solvent was 0.5% acetic acid and 0.05% sodium lauryl sulfate in H_2O. Flow rate was 0.5 ml/min, and analysis time was 15 min.

^3H-thymidine Uptake by L1210 Cells

Cells were prepared in McCoy's 5A medium instead of HBSS. ^3H-Thymidine (23 Ci/mmole) was added to cell suspension fluid (1 ml, 0.1-80 x 10^5 cells) to give a concentration of 100 μCi/ml. The cell suspension was then incubated at 37° and aliquots were removed at 15, 30, 60, and 120 min. Unlabeled thymidine (84.3 μg/ml) was then added to wash the cells on the Millipore filter after incubation. The filter was placed in a scintillation vial and counted using a Packard scintillation spectrometer.

Cysteine Studies

After isolation, lymphocytes were diluted with the buffered salt solution to a concentration of 1.9 x 10^6 cells/ml. The cells were kept in suspension by constant mild agitation as 0.10 ml portions of the suspension were withdrawn into 40 T.B. syringes with 25 gauge needles. Forty mice were each injected ip with 0.10 ml of the cell suspension by randomly selecting from the prepared syringes in a total time interval for injection of 20 min. An aliquot of the cell suspension was checked for cell viability using a trypan blue exclusion test and recorded as greater than 90%. Four groups of 10 mice which had previously been innoculated with the L1210 cells or the equivalent volume of buffered salt solution (control) were injected ip with either 0, 25, 100 or 400 mg/kg L-cysteine HCl per day.

Cysteine hydrochloride solutions were prepared daily. The solutions were adjusted to pH 6.0 ± 0.1 with NaOH and to an osmolality of 315 ± 10 mosmole/kg-H_2O with NaCl. An acidic solution is necessary to prevent the oxidation of L-cysteine to L-cystine, which is much less soluble. The solutions were then passed through 0.22 μm Millipore filters into 30 ml sterile dosette vials.

L-cysteine administration began 24 hours after injection of the L1210 cells into the 40 mice allowing for nearly two doubling periods of the leukemic cells. The doubling period of this leukemic strain is 12.4 hr. Dosages were administered in 0.10 ml portions ip every 3 hr for 142 hr. This regimen was selected because of the short plasma half-life of L-cysteine (approximately 2 hr) following intraperitoneal injection in rats (38). Analogous studies were performed using Model 2002 Alzet osmotic minipump implants (Alza. Palo Alto, CA) to control for a possible injection artifact. Ketamine HCl

(180 mg/kg, ip) was used as an anesthetic agent for the implant studies.

RESULTS

Studies in L1210 Cells

L1210 Cell proliferation was examined in time sequence after innoculation of L1210 cells. This experiment was repeated seven times and a total of 152 mice were use. Table 1 shows mean cell number at each time point. The exponential increase of Bo-cell number occurred from 48 hr to 96 hr compared with 72-120 hr in L-cells. This may indicate that two populations of L1210 cells exist or that different phases of a cell cycle are being reflected. Cytological findings of L and Bo-cells with the aid of light microscopy did not show any detectable morphological difference.

Changes in cell volume distribution and mean cellular volume (MCV) after injection of L1210 cells are shown in Figure 1. The experiment was performed twice using the cells from 10 mice at each time point. A rapid shift in cell balance was observed in the Bo-cells; the time of greatest cell volume was at 72 hr, while that of L-cells was at 96 hr. Cell volume in L- and Bo-cells showed significant differences primarily at the last time points (e.g., 144 hr). However, the change in the MCV of Bo-cells occurred more rapidly than that of the L-cells.

Change in ^{3}H-thymidine uptake by L1210 cells are shown in Table 2. In this experiment, 5 mice were used per time point. The uptake of ^{3}H-thymidine by Bo-cells was greater than L-cells at all time points except at 168 hr. For example. the maximum uptake of Bo-cells at 24 hr was 30 fold higher than that of L-cells. Uptake by L-cells reached a maximum at 72 hr, but it is still less than 1/5 that of Bo-cells.

Experiments were performed (Table 3) to measure ^{14}C taurine uptake *in vitro* and at 25, 50, and 75 hr after *in vivo* transplantation. Taurine concentration in cells at 0, 25, 50, 75 and 100 hr were compared with the taurine uptake data. Table 3 shows an initial ten-fold increment in the uptake of ^{14}C-taurine followed by a decrease to levels similar to those observed for control lymphocytes. This was followed by another increase, but less than that observed at 25 hr. We observed a parallel increase and decrease in the cellular taurine content 25 hr after the changes which were observed in the uptake of taurine by L1210 cells.

Changes in the taurine/protein ratio in L and Bo-cells were examined following innoculation with L1210 cells (Table 4).

TABLE 1

L1210 CELL PROLIFERATION

HOURS AFTER INOCULATION	24	48	72	96	120	144	168
NUMBER OF TOP CELLS (L-CELLS) PER ML	1.6	2.5	5.1	35	120	330	230
NUMBER OF BOTTOM CELLS (BO-CELLS) PER ML	0.25	0.13	8.0	29	82	25	71

All cell counts are X 10^{-6}

TABLE 2

CHANGE IN ^{3}H-TDR UPTAKE IN L1210 CELLS AFTER INOCULATION[1]

HOURS AFTER INOCULATION	Uptake (dpm X 10^{-3}/2 hr/10^{5} cells)						
	24	48	72	96	120	144	168
TOP CELLS (L-CELLS)	1.6±0.3	1.3±0.2	7.9±1.6	7.1±1.8	1.7±0.3	1.1±0.3	1.4±0.3
BOTTOMS CELLS (BO-CELLS)	53.0±3.0	29.5±2.0	11.8±1.5	9.4±2.2	11.8±0 8	9.9±0.6	0.7±0.1

1 ^{3}H TDR uptake in normal lymphocytes ranged between 500 and 3,000 DPM/10^{5} cells

Values are means ± SD

TABLE 3

TAURINE UPTAKE COMPARED WITH TAURINE LEVELS

Time After Transplantation (hr)	0	25	50	75	100
^{14}C Taurine Uptake (mmole/10^6 cells X 10^{-9})	0.16±0.04	1.56±0.14	0.32±0.04	0.96±0.14	-
Taurine Content (μg/10^6 cells)	1.01±0.19	0.92±0.13	4.60±0.56	1.40±0.15	3.38±0.42

TABLE 4

TAURINE CONCENTRATION OF L1210 CELLS

ng Taurine/μg Protein[1]

Time Following Inoculation - (hrs)	24	48	72	96	120	144	168
L-Cell	81	100	71	29	27	32	47
Bo-Cell	99	-	23	29	21	20	12

[1] The mean taurine concentrations (ng taurine/μg Protein) for normal L and Bo cells respectively are 49 and 29. Data are means of 5 mice per time.

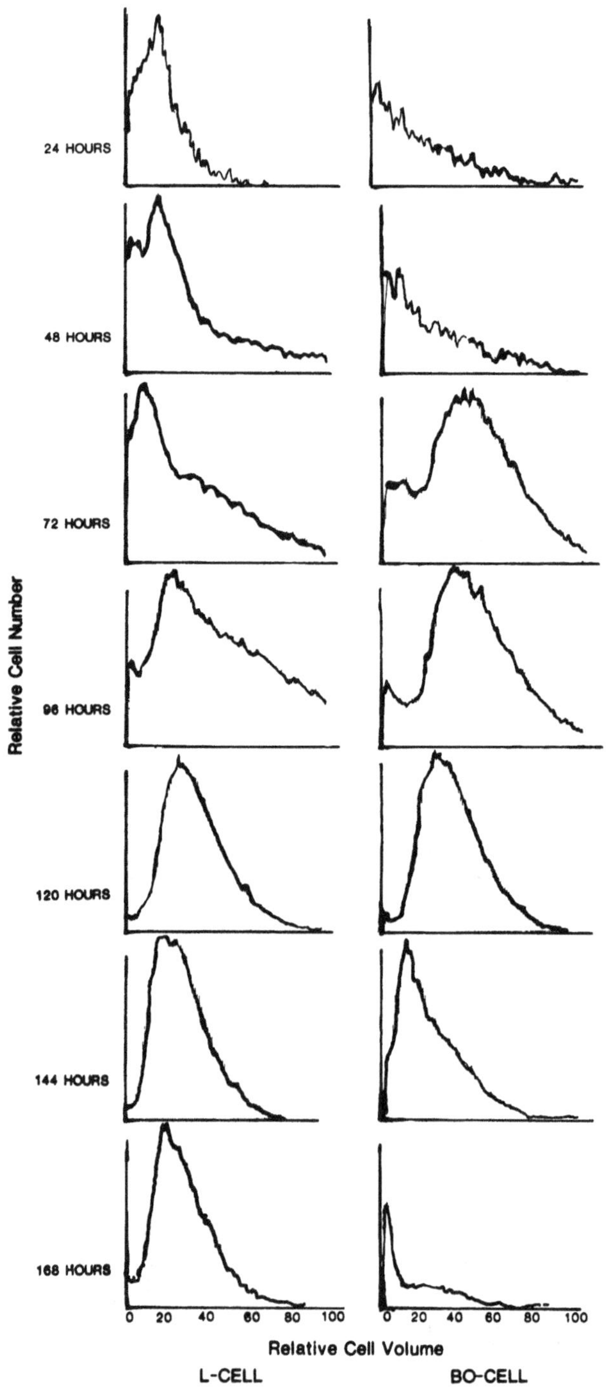

Change in Cell Volume Distribution of L1210 Cells After Inoculation

Fig. 1. The effect of time after inoculation on change in cell volume distribution of L1210 cells. The relative cell numbers for L and Bo-cells are illustrated on the ordinate. The relative cell volume for L and Bo-cells are illustrated on the abscissa. The technique is discussed in the Methodology Section.

The number of Bo-cells from one mouse are insufficient for analysis at 24 and 48 hrs, so Bo-cells from 5 mice were pooled and centrifuged to obtain enough cells. Therefore, taurine concentration of Bo-cells at 24 hr represented pooled value and no values were obtained at 48 hr. Although these data are preliminary, both L- and Bo-cells had high taurine concentrations during the early period of L1210 cell proliferation. These results are similar to those previously obtained in L-cells by the DNFB method (1). The taurine concentration of Bo-cells at 24 hr was 3.3 times greater than that of normal Bo-cells. At 48 hr, the taurine concentration of L-cells was 1.6 times greater than that of normal L-cells. Thereafter, taurine concentration in both L- and Bo-cells appeared to decrease slightly below the level of taurine as measured in normal lymphocytes.

In the *in vivo* studies performed using cysteine, the mean survival time of the group receiving 400 mg/kg/day of L-cysteine was extended by 14 hr over a control group not receiving L-cysteine. In a subsequent study using Alzet minipumps to eliminate possible injection artifacts, 6 DBA/2J mice receiving 750 mg/kg/day of L-cysteine showed no differences in mean survival time or mean WBC counts with respect to a control group (n=8) receiving either HCl solution or no implants.

DISCUSSION

Cellular growth and proliferation is known to be regulated and triggered by ionic factors (23). Therefore, it is not surprising that taurine, a major intracellular ion, shows marked differences within cells of altered proliferation rates. The importance of the differences in taurine observed may not be in the differences from control *per se* but may be in the fact that such differences occur very early in the course of the cellular proliferation. This may be of potential diagnostic or chemotherapeutic value. For example, we found that the changes in taurine uptake by the leukemic cells appear to precede changes in thymidine uptake.

Amino acid analysis of leukemic cells from patients with acute myelogenous and lymphoblastic leukemia as well as CML in blast crises revealed that abnormally low concentrations of taurine were often present in these clinical studies as well as in our present animal studies. In conjunction with this observation, the deficient α-cystathionase activity and/or concentration in human leukemic cells led us to develop an initial hypothesis that a taurine deficient state existed in the malignant cells. It was possible that taurine might have an inhibitory effect on differentiation of the leukopoietic cells and that a metabolic disorder in the sulfur-containing amino acid pathway might be responsible for maturation arrest. Supplementation of taurine or L-cysteine, however, produced no significant alterations of host survival time or leukopoiesis in mice.

The early increase in concentration of taurine in leukemic cells in the animal experiments supported an alternative hypothesis that taurine may be responsible for abnormal cellular proliferation. Other support for this hypothesis involved the cysteine analogue studies and cysteine depletion studies, which showed significant reduction of leukopoiesis (27,32,35). Our recent data indicate that an increase in taurine uptake is the primary event in cellular proliferation and that the decrease in sulfur-containing amino acid synthesis due to a decreased cystathionase concentration and/or activity is a secondary event. The depressed α-cystathionase activity that occurs following the initial taurine accumulation followed by decreased taurine levels may be due to inhibition by end product depression. The end product, taurine, therefore, may regulate a key synthetic enzyme (possibly α-cystathionase). Further experimentation needs to be performed to establish this hypothesis. Since taurine uptake is greater during the early log growth phase, it seems that taurine analogues may prove useful in chemotherapy of certain leukemic states.

Furthermore, it is interesting to note that in a recent study of ours examining the taurine distribution in cells using autoradiography taurine was preferentially localized in the nucleus of certain cells. While this fact alone does not provide one universal mechanism for the action of taurine in cells, it does suggest that one way this substance may act is through nuclear modulation of cellular function.

It is our hypothesis that taurine may act through mechanisms found by Berlowitz et al. (4) and others (7,13,16,19). These various authors have demonstrated that an anion with a very low pKa (1.5) could displace histone and possibly non-histone protein from DNA. Thus, there would be an exposed template and new RNA molecules could be transcribed. In these studies, synthetic anionic sulfone was used to bind to histones and activate RNA synthesis. We believe that taurine, an endogenous sulfonate, could be displacing histone or non-histone protein from DNA and thereby, influence cellular proliferation via this mechanism.

It is of interest that other investigators have found that the phosphorous analog of taurine, 2-amino ethyl phosphonic acid (AEP) concentration also changes with cell growth (18). However, unlike taurine, which decreases with cell growth, AEP increases with cell growth. This is exciting because we found (2) that the methyl analogue of AEP specifically depresses taurine accumulation in cells. Since we found that the change in taurine levels occurred just prior to the change (33) in rate of proliferation of leukemic cells, it is believed that taurine changes are the cause of changes in cellular kinetics rather than the effect of such changes.

ACKNOWLEDGMENTS

The authors thank Dr. Jay Roberts and Dr. Rosaline R. Joseph for their continued support and advice. The authors are indebted to Ms. Mary Montone for her typing skills. The authors thank Ms. Leslie Teeple, Ms. Rochelle Lanciano, Mrs. Vicky Nade-Hoffman, Mr. Adam Gutuntag, Mr. Phillip Smeraski and Dr. Harry Steinman, Dr. Richard Howell and Dr. John Claude Krusz for their technical assistance and also thank the support of the Hematology-Oncology Research Fund of the Medical College of Pennsylvania. Supported in part by NIH grant CA-22170.

REFERENCES

1. Baskin, S.I., Howell, R.R., Stinman, H., Smeraski, P., and Besa, E.C., 1978, Proliferation and growth-related changes in concentration of taurine in L1210 leukemia cells, Fed. Proc., 37:680.
2. Baskin, S.I., Zaydon, P.T., Kendrick, Z.V., Katz, T.C., and Orr, P.L., 1980, Pharmacokinetic studies of taurine in bovine Purkinje fibers, Circ. Res., 47:763-769.
3. Benson, J.R., and Hare, P.E., 1975, O-phthalaldehyde:fluorogenic detection of primary amines in the picomole range. Comparison with fluorescamine and ninhydrin, Proc. Natl. Acad. Sci., 72: 619-622.
4. Berlowitz, L., Kitchen, R., and Pallotta, D., 1972, Histone and RNA synthesis: selective binding of histones by a synthetic polyanion in calf thymus nuclei, Biochem. Biophys. Acta., 262: 160-168.
5. Bradford, M.M., 1976, A rapid and sensitive method for the quantitation of microgram quantities of protein utilizing the principle of protein-dye binding, Analytical Biochem., 72: 248-254.
6. Brigham, M.P., Stein, W.H., and Moore, S., 1960, The concentrations of cysteine and cystine in human blood plasma, J. Clin. Invest., 39:1633.
7. Chambon, P., Ramuz, M., Mandel, P., and Doly, J., 1968, The influence of ionic strength and a polyanion on transcription *in vivo*, I. Stimulation of the aggregate RNA polymerase from rat liver nuclei, Biochim. Biophys. Acta., 157:504.
8. Chatagner, F., Hirsch, A., Durieu-Trautmann, O., and Rairi,

M.C., 1969, Activities of cystathionase, cystine sulphinic acid decarboxylase and serine dehydrase in the liver of tumor-bearing rats, Experientia, 25:2077-2078.

9. Christensen, H.N., Hess, B., and Riggs, T.R., 1954, Concentration of taurine β-alanine and tri-iodothyronine by ascites tumor cells, Cancer Res., 14:124-127.
10. Foley, G.E., Barell, E.F., Adams, R.A., and Lazarus, H., 1969, Nutritional requirements of human leukemic cells, Exp. Cell. Res., 57:129.
11. Glade, L.M., Greene, H.L., and Bikel, L., 1979, α-cystathionase in normal and leukemic cells, Cancer Treat. Rep., 63:1081.
12. Green, J.P., and Day, M., 1963, Biosynthetic pathways in mastocytoma cells in culture and in vivo, Ann. N.Y. Acad. Sci., 103:334-350.
13. Holubek, V., Fanshier, L., and Crocker, T.T., 1966, The inhibition of nuclear RNA synthesis by added RNA, Exp. Cell. Res., 44:362.
14. Inglehart, J.D., York, R.M., Modes, A.P., Lazarus, H., and Livingston, D.M., 1977, Cystine-requirements of continuous human lymphoid cell lines of normal and leukemic origin, J. Biol. Chem., 252:7184-7191.
15. Jacobsen, G., and Smith, L.H., Jr., 1968, Biochemistry and physiology of taurine and taurine derivatives, Physiol. Rev., 48:424-511.
16. Kinoshita, S., 1971, Heparin as a possible initiator of genomic RNA synthesis in the early development of sea urchin embryos, Exp. Cell. Res., 64:403.
17. Kissane, J.M., and Robins, E., 1958, The fluorometric measurement of the deoxyribonucleic acid in animal tissues with special reference to the central nervous system, J. Biol. Chem., 233: 184-188.
18. Kittredge, J.S., and Roberts, E., 1969, A carbon-phosphorous bound in nature, Science, 164:37-42.
19. Kraemer, R.J., and Coffey, D.S., 1970, The interaction of natural and synthetic polyanions with mammalian nuclei. I. DNA synthesis, Biochim. Biophys. Acta., 224:553.
20. Kries, W., 1979, Tumor therapy by deprivation of L-methionine: rationale and results, Cancer Treat. Rep., 63:1069-1072.
21. Kries, W., and Hession, C., 1973, Biological effects of enzymatic deprivation of L-methionine in cell culture and an experimental tumor, Cancer Res., 33:1866-1869.
22. Lazarus, H., Burell, E., Oppenheim, S., and Krishan, A., 1974, Divergent properties of two human lymphocytic cell lines isolated from a single specimen of peripheral blood, In Vitro, 9:303.
23. Leffert, H.L., 1980, Growth regulation by ion fluxes, Ann. N.Y. Acad. Sci., 339:1-335.
24. Lin, J., and Chang, J., 1975, Chromophoric labeling of amino acids with 4-dimethylaminoazobenzene-4'-sulfonyl chloride, Anal. Chem., 47:1634-1638.
25. Livingston, D., Ferguson, C., Gollogly, R., and Lazarus, H.,

1976, Accumulation of cystine auxotrophic thymocytes accompanying type G viral leukemogenesis in the mouse, Cell., 7:41.
26. Metcalf, D., Brumby, M., 1967, Coulter counter analysis of lymphoma differentiation pattern in AKR mice in the lymphoid leukemia, Int. J. Cancer, 2:37.
27. Ohnuma, T., Waligunda, J., and Holland, J.F., 1971, Amino acid requirements *in vitro* of human leukemic cells, Cancer Res., 31:1640-1644.
28. Piez, K.A., and Eagle, H., 1958, The free amino acid pool of cultured human cells, J. Biol. Chem., 231:533-545.
29. Rouser, G., Samuels, A.J., Kinugasa, K., Jelinek, B., and Heller, D., Free amino acids in the blood of man and animals, *in*: "Amino Acid Pools," J.T. Holder, ed., Elsevier, New York (1962).
30. Schrier, B.K., and Thompson, E.J., 1974, On the role of glial cells in the mammalian nervous system; uptake, excretion, and metabolism of putative neurotransmitters by cultured tumor cells, J. Biol. Chem., 249:1769-1780.
31. Uren, J.K., and Jazama, H., 1975, Enzymatic approaches to cystine depletion therapy, Proc. Am. Assoc. Cancer Res., and ASCO, 16:144.
32. Uren, J.R., and Lazarus, H., 1979, L-cysteine requirements of malignant cells and progress toward depletion therapy, Cancer Treat. Rep., 63:1073.
33. Wakayama, K., Besa, E.C., Jepson, J.H., Nade, V., Melin, J., and Baskin, S.I., 1979, Cell volume and chemical analysis of L1210 cells, Expt. Hematol., 7(6):70.
34. Weisberger, A.S., and Levine, B., 1954, Incorporation of radioactive L-cystine by normal and leukemic leukocytes *in vivo*, Blood, 9:1082.
35. Weisberger, A.S., and Suhrland, L.G., 1956, The effect of selenium cystine on leukemia, Blood, 9:19.
36. Weisberger, A.S., Suhrland, L.G., and Griggs, R.C., 1954, Incorporation of radioactive L-cystine and L-methionine by leukemic leukocytes *in vitro*, Blood, 9:1095.
37. Weisberger, A.S., Suhrland, L.G., and Seifter, J., 1956, Some structural requirements for inhibiting the incorporation of radioactive L-cystine by leukemic leukocytes, Blood, 11:1.
38. Yamaguchi, K., Shigehisa, S., Sakakibara, S., Hosokawa, Y., and Ueda, I., 1975, Cysteine metabolism *in vivo* of vitamin B_6-deficient rats, Biochim. Biophys. Acta., 381:1.

TAURINE TRANSPORT BY RECONSTITUTED MEMBRANE VESICLES

S. W. Schaffer, E. C. Kulakowski and J. H. Kramer

University of South Alabama
Department of Pharmacology
Mobile, Alabama

Although taurine is found in very high concentration in the heart, a role for this amino acid has not been established. Read and Welty (24) initially reported in 1963 that taurine exhibited antiarrhythmic activity; they attributed this effect of taurine to changes in potassium flux. Since that study taurine has also been linked to changes in calcium transport (26). In addition, several investigators have described specific membrane-stabilizing effects of taurine (10,16).

Recent work has focused on the relationship between heart failure and myocardial taurine content (6,11,16,20,22). In experimentally induced cardiac hypertrophy (20,22) and congestive heart failure (11), myocardial taurine levels increase dramatically. The increase appears to be largely a result of an acceleration in taurine uptake by the heart (12). The view that taurine may play a role in the disease process has recently been supported by evidence that taurine treatment improves the state of the calcium overload myocardium (2,16).

Since taurine uptake into both the normal and diseased myocardium appears to be important we decided to examine the mechanism of this uptake process in more detail. In this chapter we describe the solubilization and reconstitution of the taurine transport protein.

EXPERIMENTAL PROCEDURES

Preparation of Ammonyx-Lo Heart Extract

Swine ventricular sarcolemma, prepared according to the method of Sulakhe et al. (28), were solubilized by adding sufficient Ammonyx-Lo to yield a final concentration of 2% detergent (w/v). The resulting solution was stirred for 90 min at 24°C and then centrifuged at 150,000 X g for 70 min at 4°C. The supernatant fraction was dialyzed for 12 hr at 4°C against 10 mM Tris HCl, pH 8.0, to reduce the detergent concentration to 0.02%. This preparation was subsequently used in the reconstitution and chromatographic studies.

Taurine Binding Assay

Following passage of the solubilized extract through the Sepharose 6B column, the appropriate samples were dialyzed, concentrated and assayed using a modified polyethylene glycol method. The taurine binding protein was first applied to a Concanavalin A (ConA) Sepharose column and then eluted from the column using 0.3% α-methyl mannoside; all α-methyl mannoside was subsequently removed by dialysis. Taurine binding was then assayed using the standard polyethylene glycol method (7).

Preparation of Proteoliposome Vesicles

Reconstituted taurine transport proteoliposomes were prepared from the Ammonyx-Lo heart extract and partially purified soybean phospholipids (14) using the procedure of Kasahara and Hinkle (15). To 40 mg of phospholipid was added 1.0 ml of nitrogen-flushed buffer (0.2 mM potassium phosphate, pH 7.4, containing 150 mM KCl). Following aeration with nitrogen, the liposomes were formed by sonicating the mixture until it became clear. These liposomes were then combined with an equal volume of solubilized heart extract (1.8 mg protein/ml) and quickly frozen in the acetone-ice bath. The temperature of the mixture was allowed to rise to 25°C and was then sonicated for 10 sec to disperse any aggregates. Electron micrographic examination of the proteoliposomes revealed vesicles composed of a single lipid bilayer and having an average diameter of 5 μm.

Reconstituted Taurine Transport Assay

The K^+ loaded proteoliposomes were diluted four-fold into an external 10 mM Tris buffer (pH 7.4) containing 120 mM NaCl,

1.2 mM $MgSO_4$ 1.25 mM $CaCl_2$ and 24.8 mM $NaHCO_3$. After a 7 min preincubation with ^3H taurine, ionophore-stimulated uptake was initiated by the addition of 11 μM valinomycin. The reaction was terminated at the appropriate time by rapid addition of 2 ml ice cold buffer and filtration on 0.2 μ filters. The filters were washed twice with 3 ml of ice cold buffer and then counted for radioactivity.

RESULTS

Membrane Solubilized Transport Protein

Kulakowski et al. (17) initially reported the existence of two taurine binding proteins in mammalian heart plasma membrane. As seen in Figure 1A these two proteins (designated I and II) can be extracted from the membrane with the detergent Ammonyx-Lo and separated from each other on Sepharose 6B. The chromatographic pattern shown in figure 1 reveals that the two components differ in molecular size; based on a standard calibration curve of the column, the Stokes radii of components I and II are 44 and 40 Å, respectively (Figure 1B).

The binding isotherm of component II is shown in Figure 2. A replot of these data according to the method of Scatchard yields a linear line and reveals that the protein contains a single binding site with a Kd of 150 μM (Figure 2, inset). Since this Kd value is similar to the Km of taurine uptake by the heart (1,3,8,12), it is likely that the 40 Å protein is associated with this transport process.

The specificity of the binding site also shows characteristics similar to the β-amino acid transport system (8,12). As seen in Table 1, the most potent inhibitors of taurine binding to this protein are β-amino acids. Based on competition studies, the Ki values of three of these β-amino acids (β-alanine, hypotaurine and taurocyamine) for the binding site are 15 μM, 30 μM and 75 μM, respectively (Figure 3). These values are virtually identical to the reported Km values for the uptake of these analogs by the perfused rat heart (12,13). Several α-amino acids were also examined for their ability to effectively compete with taurine for its binding site; however, none had a significant effect.

Reconstitution of Taurine Transport

Reconstitution of a transport system has generally been used to study the bioenergetics of an uptake process and to

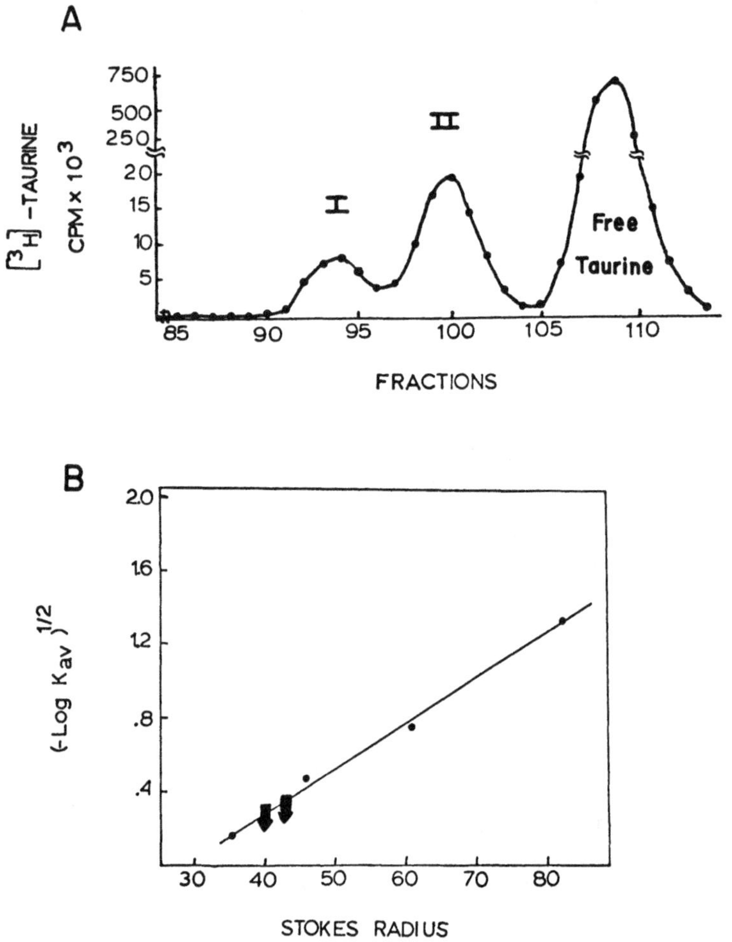

Fig. 1. Sepharose 6B chromatography of solubilized swine sarcolemma.

A. Membrane was incubated with 3.5 mM ^{14}C-taurine at 24°C and then placed on ice for 10 min. The samples were chromatographed at 4°C on a Sepharose 6B column (1.5 x 125 cm), which had previously been equilibrated with 10 mM Tris HCl, pH 8.0, containing 0.02% Ammonyx-Lo.

B. Determination of molecular size of taurine binding proteins by gel filtration.

The standard curve was constructed by passing ^{3}H-acetylated derivatives of thyroglobulin, apoferritin, aldolase and albumin through the Sepharose 6B column.

TABLE 1

Effect of Taurine Analogs on Taurine Binding to Component II[a]

Taurine Analog	% Taurine Binding
Control	100
β-alanine	8 ± 4[b]
Hypotaurine	6 ± 6
Homotaurine	43 ± 1
Taurocyamine	57 ± 7
Cysteine Sulfinic Acid	6 ± 5
L-Threonine	81 ± 10
Glycine	59 ± 17
α-L-Alanine	106 ± 15
L-Cysteine	113 ± 4

[a]All values represent the amount of taurine bound to component II following incubation with 50 μM ^{3}H-taurine and an equal concentration of the analog. Assays were performed as described in experimental procedures.

[b]Values shown are means ± SEM of 3 assays.

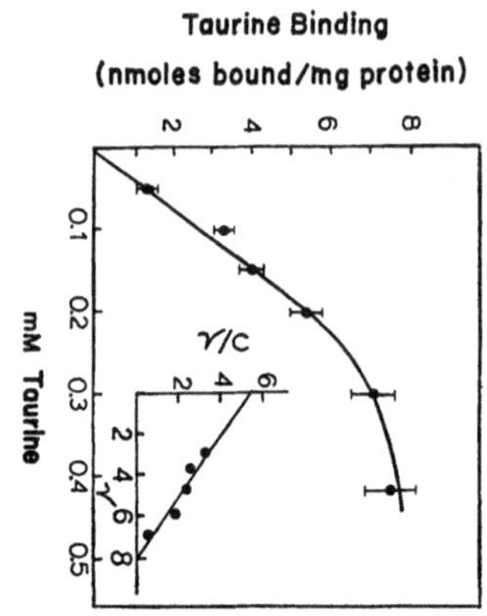

Fig. 2. Saturation isotherm of component II. Component II was assayed for taurine binding activity using the polyethylene glycol method as described in the methods. Values shown are means ± SEM of 3 assays. The inset shows a Scatchard plot of the data.

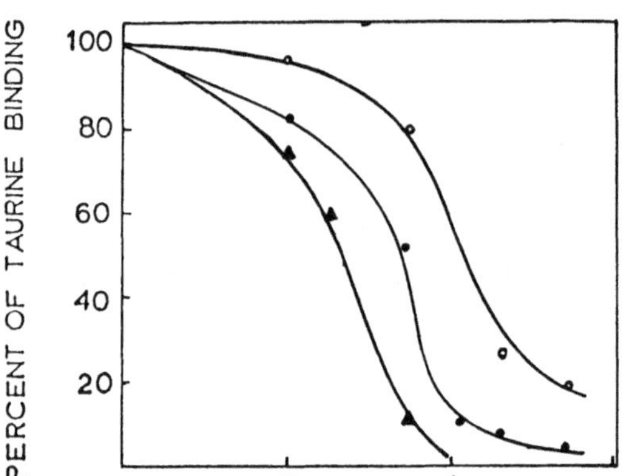

Fig. 3. Inhibition of taurine interaction with component II by various taurine analogs. Component II was assayed utilizing 100 μM ^{3}H-taurine and varying concentrations of β-alanine (▲), hypotaurine (●) and taurocyamine (○). Values shown are the means ± SEM of 3 assays.

establish the function of specific transport proteins. The most common approach has been to prepare K^+ loaded vesicles containing the transport protein and suspend the vesicles in K^+ free medium. The ionophore valinomycin can be used to induce efflux of this cation down a concentration gradient with concurrent generation of a membrane potential (interior negative). This membrane potential is then used to drive the accumulation of the substrate by the vesicles. Figure 4 shows the results of this approach on the uptake of ^{3}H-taurine by K^+ loaded, reconsituted vesicles prepared from the Ammonyx-Lo heart extract and soybean phospholipids. Addition of valinomycin to this system led to a rapid, but transient, uptake of taurine by the vesicles. This process was found to depend on the membrane potential generated by the K^+ gradient since elimination of this gradient also abolished the effect of the ionophore.

To verify that the transport was carrier-mediated, we first examined the effect of varying concentrations of heart extract on taurine uptake. It was found that in the absence of the extract no taurine was accumulated by the vesicles; however, as the amount of protein in the proteoliposomes increased, there was a progressive increase in taurine accumulation (Figure 5).

Another important criterion of a carrier-mediated process is the demonstration of saturation kinetics. In Figure 6A it is seen that taurine uptake by the vesicles follows classical Michaelis-Menton kinetics. The Lineweaver-Burke plot of these data is linear and yields a Km of 100 μM and a Vmax of 11 pmole/mg protein/min (Figure 6B). The near identity of the Km for taurine uptake by these vesicles and the Kd for taurine binding to the 40 Å protein is further evidence in support of a carrier-mediated process.

The last experiment employed to verify that the process is carrier-mediated was an examination of its substrate specificity. In agreement with the specificity of component II, it was found that the most potent inhibitors of taurine uptake by the vesicles are hypotaurine, β-alanine and cysteine sulfinic acid (Table 2). Previous studies have shown that both hypotaurine and β-alanine inhibit taurine uptake by the β-amino acid transport system in the fetal mouse heart (8) and can be transported by this system in the perfused rat heart (12). The myocardial β-amino acid system is highly specific for β-amino acids and is unable to transport α-amino acids (8). As seen in Table 2, none of the α-amino acids examined affected taurine accumulation by the vesicles indicating that the ionophore-stimulated process has a specificity identical to the β-amino acid transport system.

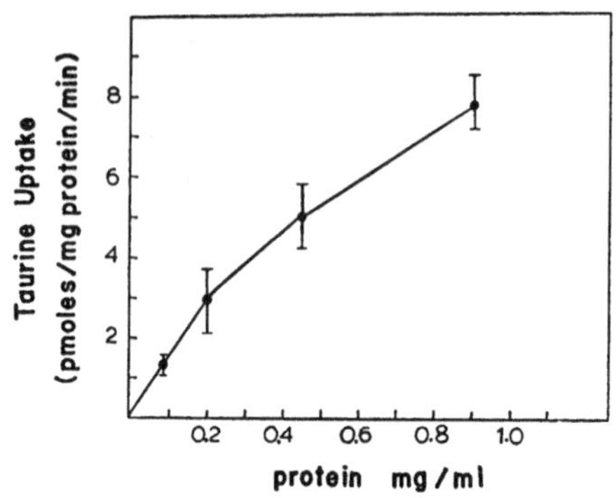

Fig. 4. Ionophore-stimulated uptake of taurine by myocardial proteoliposomes. Taurine transport proteoliposomes were prepared from the Ammonyx-Lo heart extract and soybean phospholipids using the procedure described in the methods. Following a 7 min preincubation with 200 μM ^{3}H-taurine, ionophore-stimulated uptake was initiated by the addition of 11 μM valinomycin. The amount of taurine accumulated by the vesicles was determined using procedures outlined in the methods. Values shown are the means ± SEM of 3 assays.

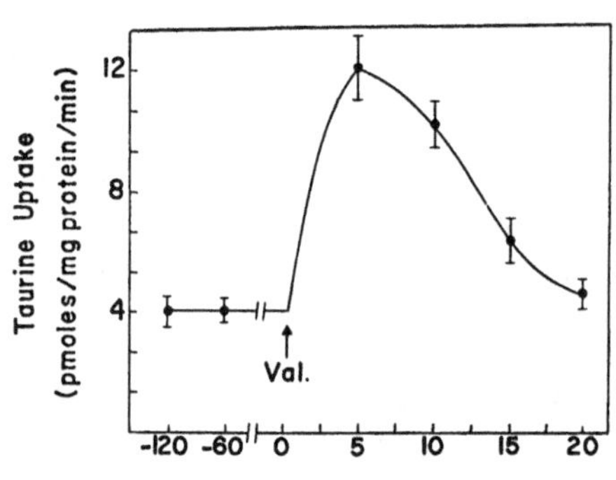

Fig. 5. Effect of protein concentration on taurine uptake by reconstituted vesicles. Proteoliposomes of varying protein concentration were prepared according to the procedure described in the methods. Taurine uptake into these vesicles was determined 7 sec after addition of the ionophore. The values shown represent ionophore-stimulated taurine uptake and were obtained by correcting for taurine uptake in the absence of the ionophore. Each point is the mean ± SEM of 3 assays.

TABLE 2

Effect of Taurine Analogs and Amino Acids on Ionophore - Stimulated Taurine Uptake[a]

Taurine Analog	Percent Taurine Uptake
Control	100
Beta-alanine	12 ± 3
Hypotaurine	9 ± 9
Homotaurine	25 ± 19
Taurocyamine	71 ± 11
Cysteine Sulfinic Acid	14 ± 10
Glycine	73 ± 8
L-alanine	96 ± 4
L-Threonine	91 ± 7

[a]Proteoliposomes (0.9 mg protein/ml) were preincubated with buffer containing 60 μM ^{3}H-taurine and an equal concentration of the analog. Ionophore-stimulated uptake was determined according to Figure 5. Values shown are means ± SEM of 3 assays.

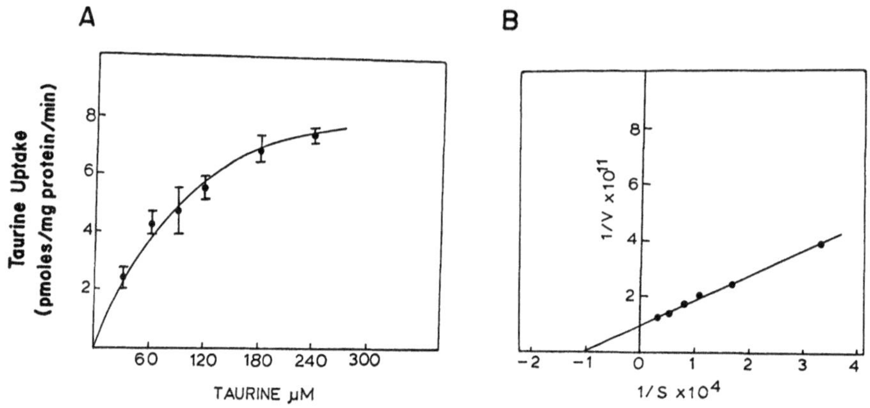

Fig. 6. Kinetics of taurine uptake by reconstituted system.

A. Ionophore-stimulated uptake at various taurine concentrations was determined as described in Figure 5. The final protein concentration of the proteoliposomes was 0.9 mg/ml. Values shown are means ± SEM of 3 assays.

B. Lineweaver - Burke plot of data from Figure 6A.

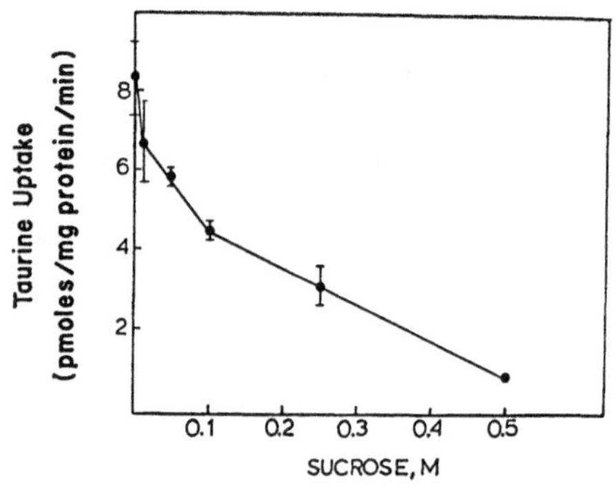

Fig. 7. Effect of sucrose on ionophore-stimulated taurine uptake. Proteoliposomes (0.9 mg protein/ml) were preincubated for 7 min in buffer containing 200 µM ^{3}H-taurine and the appropriate concentration of sucrose. Ionophore-stimulated uptake was calculated described in Figure 5. Values shown are means ± SEM of 3 assays.

To distinguish between transport into the intravesicular spaces of the reconsituted vesicle and binding to the carrier protein, experiments were performed to see if high concentrations of a nonpenetrating solute would diminish the extent of uptake. As noted in Figure 7, an increase in the osmolarity of the outside medium by addition of sucrose resulted in a decrease in taurine uptake. Since sucrose decreases the intravesicular volume, this observation establishes that taurine is being transported into the vesicles.

It is generally accepted that taurine transport by the cell is Na^+ dependent (4,8,9,19). This dependence on sodium is also apparent in the reconstituted system; partial substitution of Li^+ for Na^+ in the exterior medium dramatically reduces taurine uptake (Table 3). Also of interest is the observation that taurine uptake is highly dependent upon extravesicular bicarbonate, but not calcium content.

Neutral ionophores, such as valinomycin, stimulate solute transport by generating a membrane potential (23). On the other hand, carboxylic ionophores influence the pH gradient of the membrane without altering the membrane potential (23). An effect of these charged ionophores on solute transport would imply that a pH gradient may serve as an alternate energy source in driving solute transport. Pasantes-Morales et al. (21) have shown that one such charged ionophore, X-537A, causes the release of taurine from both mouse cerebral cortex slices and chick retina. They suggested that taurine release is related to the effects of the ionophore on calicum redistribution. In Table 4 it is seen that X-537A also stimulates taurine uptake by the K^+ loaded vesicles. However, this effect is more easily explained by a direct action of the ionophore on taurine transport rather than an indirect effect due to the stimulation of calcium uptake. As shown in Figure 8, X-537A accelerates the transfer of taurine from an aqueous compartment through a bulk organic phase into another aqueous compartment of a Pressman cell (23,27). Thus, it is likely that X-537A promotes taurine uptake by directly mediating the transport of this amino acid through the membrane. Monensin, the other carboxylic ionophore examined, also accelerates taurine uptake by the vesicles, although its modes of action is unknown.

DISCUSSION

Role of Component II

The results of this study suggest that the solubilized protein we designate as component II is associated with the

TABLE 3

Effect of Ions on Ionophore-Stimulated Taurine Uptake[a]

Ions	Percent Taurine Uptake
Control	100
less calcium	103 ± 7
less bicarbonate	6 ± 10
less magnesium	71 ± 5
Sodium, 25 mM	38 ± 17

[a]Proteoliposomes (0.9 mg protein/ml) were preincubated for 7 min in buffer containing 60 μM ^{3}H-taurine and the appropriate ions. Extravesicular sodium was reduced from 140 mM to 25 MM by replacement with LiCl, HCO_3^- was substituted by Cl^-, Ca^{++} was replaced by Mg^{++} and Na^+ was used in place of Mg^{++}. Ionophore-stimulated uptake was determined according to Figure 5. Values shown are means ± SEM of 3 assays.

TABLE 4

Effect of various ionophores on taurine uptake by reconstituted vesicles[a]

Ionophore	Taurine Uptake (pmoles/mg protein/min)
Valinomycin	6.3 ± 0.6
Gramicidin	6.4 ± 0.6
X-537A	7.4 ± 1.1
Monensin	3.0 ± 0.2
Nigericin	1.5 ± 0.3

[a]Proteoliposomes (0.9 mg protein/ml) were preincubated for 7 min with buffer containing 200 μM ^{3}H-taurine. Ionophore-specific uptake was initiated by addition of 11 μM valinomycin, 1 μM gramicidin, 1 μM X-537A, 1 μM monensin or 1 μM nigericin. The reaction was terminated after 10 sec according to the procedures described in the methods. Specific ionophore-stimulated uptake was determined from the difference between total uptake and non-ionophore mediated uptake. Values shown are means ± SEM of 3 assays.

β-amino acid transport system of the heart. In support of this hypothesis, it was found that the Kd for taurine binding to this protein (Kd = 150 μM) is virtually identical to the Km for taurine uptake by reconstituted vesicles (Km = 100 μM). Moreover, these values are comparable to the Km values reported for taurine uptake by rat heart slices (Km = 40 μM), perfused rat heart (Km = 45 μM), fetal mouse heart in culture (Km = 440 μM) and rat myocytes (Km - 133 μM) (1,3,8,12).

The other major line of evidence supporting this hypothesis is the effect of various analogs on taurine uptake. Both component II and the reconstituted transport system exhibit identical specificity patterns. In both systems, the analogs homotaurine, hypotaurine, β-alanine and cysteine sulfinic acid are potent inhibitors of taurine uptake. By contrast, taurocyamine (the guanidinium analog of taurine) displays much weaker affinity for the taurine binding site, while the α-amino acids are virtually without effect. A similar dependence of the myocardial β-amino acid transport system has been described (8,12). Huxtable et al. (12,13) examined the transport of β-alanine, hypotaurine and taurocyamine by the perfused rat heart and reported Km values for these transport processes of 18 μM, 33 μM and 65 μM, respectively. The striking similarity between these values and the corresponding Ki values reported here for the interaction of these analogs with component II (15 μM, 30 μM and 75 μM) is compelling evidence in favor of the view that component II must be an integral part of the myocardial β-amino acid transport system.

Mechanisms of Taurine Uptake

It is now generally accepted that the active transport of solutes by the cell can be envisaged in terms of Mitchell's chemiosmotic hypothesis (18). According to this theory the respiratory chain translocates protons across the membrane and thus generates a proton-motive force. It is assumed that this proton-motive force (Δ p) consists of two components, an electrical term (the membrane potential, $\Delta \psi$) and a concentration term (the pH gradient, Δ pH). Thus,

$$\Delta p = \Delta \Psi - \frac{2.3\ RT}{F} \Delta pH$$

where R is the gas constant, T the temperature and F in the Faraday. In prokaryotes the generation of the proton-motive force by proton translocating systems is coupled to either H^+/solute symports or to electrogenic uniports (5,18). By contrast, in eukaryotes the H^+ appears to be replaced by Na^+ and the electrochemical Na^+ gradient, established by the Na^+, K^+ ATPase, is utilized by a Na^+/solute symport (5).

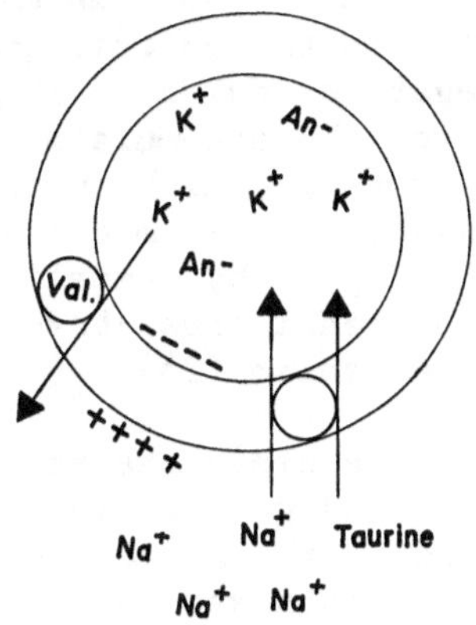

Fig. 8. Effect of X-537A on the passage of ^{3}H-taurine through a chloroform layer of a Pressman cell. ^{3}H-Taurine (0.65 μM, 10^7 cpm) was placed in 1 ml of buffer in the left hand side of a Pressman cell. This aqueous layer was separated from the right hand side of the Pressman cell (also containing 1 ml of buffer) by 30 ml of chloroform. Taurine transfer from the left hand to the right hand side of the cell was examined in the presence and absence of 175 μM X537A. The curve depicts specific ionophore-mediated transfer of ^{3}H-taurine across the chloroform layer. Transfer in the absence of the ionophore was less than 10% of the ionophore-stimulated process.

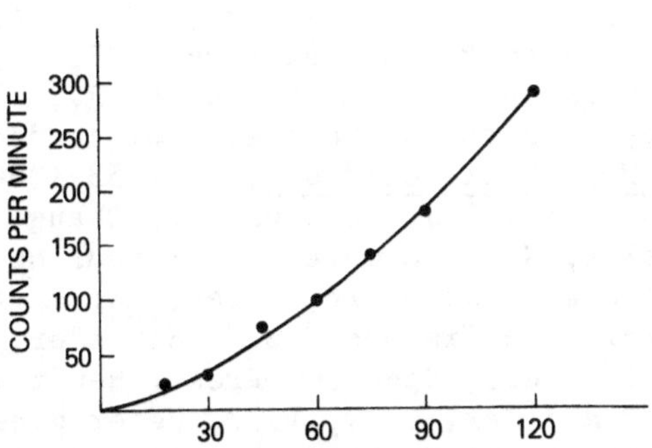

Fig. 9. Model of ionophore-stimulated taurine uptake by myocardial proteoliposomes.

A simpler method of energizing a transport system is to generate artifically an electrical potential by using the neutral ionophore valinomycin. The mechanism for this process is depicted in Figure 9. The valinomycin-mediated transport of K^+ from K^+ loaded vesicles to the exterior medium generates a membrane potential (negative inside). This electrical potential ($\Delta\ \psi$) then drives the accumulation of taurine and Na^+ via an electrogenic symport, which in this study is the β-amino acid transport system. This model seems to be generally applicable for taurine transport by mammalian systems since the accumulation of taurine by renal brush-border membrane vesicles also involves a similar mechanism (25).

Although our model may explain the stimulation of taurine uptake by both valinomycin and gramicidin, a nonelectrogenic mechanism must be involved in the actions of the carboxylic ionophores. It is widely recognized that X-537A is capable of transporting biological amines, as well as divalent cations (23,27). In this study, we find that this ionophore is also capable of directly transporting taurine through a nonpolar medium. Thus, the accumulation of taurine by the proteoliposomes in the presence of X-537A may be ionophore rather than carrier-mediated. These results suggest that X-537A may be used as a general agent to modify tissue taurine content.

AKNOWLEDGMENT

Supported by NIH Grant HL 18554.

REFERENCES

1. Awapara, J., and Berg, M., Uptake of taurine by slices of rat heart and kidney, in: "Taurine," R. Huxtable and A. Barbeau, eds., Raven Press, New York (1976), pp.135-143.
2. Azari, J., Brumbaugh, P., and Huxtable, R., 1980, Prophylaxis by taurine in the hearts of cardiomyopathic hamsters, J. Mol. Cell. Cardiol., 12:1353-1366.
3. Bahl, J., Frangakis, C.J., Larsen, B., Chang, S., Grosso, D., and Bressler, R., Accumulation of taurine by isolated rat heart cells and rat heart slices, in: "Action of Taurine on Excitable Tissues," S.W. Schaffer, J.J. Kocsis and S.I. Baskin, eds., Spectrum Press, New York, in press.
4. Chesney, R.W., and Jax, D.K., 1977, Heterogeneity of the β-amino preferring transport system in rat kidney cortex, Biochim. Biophys. Acta., 466:84-96.
5. Christensen, H.N., 1979, Exploiting amino acid structure to learn about membrane transport, Adv. in Enzymology, 49:41-101.

6. Crass, III, M.F., and Lombardini, J.B., 1977, Loss of cardiac muscle taurine after left ventricular ischemia, Life Sciences, 21:951-958.
7. Cuatrecasas, P., 1972, Isolation of the insulin receptor of liver and fat-cell membranes, Proc. Nat. Acad. Sci., U.S.A., 69:318-322.
8. Grosso, D.S., Roeske, W.R., and Bressler, R., 1978, Characterization of a carrier-mediated transport system for taurine in the fetal mouse heart in vitro, J. Clin. Invest., 61:944-952.
9. Hruska, R.E., Huxtable, R.J., and Yamamura, H.I., High-affinity, temperature-sensitive, and sodium-dependent transport of taurine in rat brain, in: "Taurine and Neurological Disorders," A. Barbeau and R.J. Huxtable, eds., Raven Press, New York, (1978), pp.109-117.
10. Huxtable, R., and Bressler, R., 1973, Effect of taurine on a muscle intracellular membrane, Biochim. Biophys. Acta, 323:573-583.
11. Huxtable, R., and Bressler, R., 1974, Elevation of taurine in human congestive heart failure, Life Sciences, 14:1353-1359.
12. Huxtable, R.J., Chubb, J., and Azari, J., 1980, Physiological and experimental regulation of taurine content in the heart, Federation Proc., 39:2685-2690.
13. Huxtable, R.J., Laird, II, H.E., and Lippincott, S.E., 1979, The transport of taurine in the heart and the rapid depletion of tissue taurine content by guanidinoethyl sulfonate, J. Pharmacol. Exp. Ther., 211:465-471.
14. Kagawa, Y., and Racker, E., 1971, Partial resolution of the enzyme catalyzing oxidative phosphorylation. XXV Reconstitution of vesicles catalyzing ^{32}Pi - Adenosine Triphosphate exchange, J. Biol. Chem., 246:5477-5487.
15. Kasahara, M., and Hinkle, P.C., 1977, Reconstitution and purification of the D-glucose transporter from human erythrocytes, J. Biol. Chem., 252:7384-7390.
16. Kramer, J.H., Chovan, J.P., and Schaffer, S.W., 1981, The effect of taurine on calcium paradox and ischemic heart failure, Am. J. Physiol., in press.
17. Kulakowski, E., Maturo, J., and Schaffer, S.W., 1978, Identification of taurine binding receptors from rat heart sarcolemma, Biochem. Biophys. Res. Commun., 80:936-941.
18. Mitchell, P., 1979, Compartmentation and communication in living systems, ligand conduction: a general catalytic principle in chemical, osmotic and chemiosmotic reaction systems, Eur. J. Biochem., 95:1-20.
19. Nauss-Karol, Cheryl and Vanderwende, C., 1981, High affinity taurine uptake in human blood platelets, in: "Actions of Taurine on Excitable Tissues," S.W. Schaffer, J.J. Kocsis and S.I. Baskin, eds., Spectrum Press, New York, in press.

20. Newman, W.H., Frangakis, C.J., Grosso, D.S. and Bressler, R., 1977, A relation between myocardial taurine content and pulmonary wedge pressure in dogs with heart failure, Physiol. Chem. and Physics, 9:259-263.
21. Pasantes-Morales, H., Salceda, R., and López Colomé, A.M., Taurine in normal retina, in: "Taurine and Neurological Disorders", A. Barbeau and R.J. Huxtable, eds., Raven Press, New York, (1978), pp.265-279.
22. Peterson, M.B., Mead, R.J., and Welty, J.D., 1973, Free amino acids in congestive heart failure, J. Mol. Cell. Cardiol., 5:139-147.
23. Pressman, B.C., 1976, Biological applications of ionophores, Ann. Rev. Biochem., 45:501-530.
24. Read, W.O., and Welty, J.D., 1963, Effect of taurine on epinephrine and digoxin induced irregularities of the dog heart, J. Pharmacol. Exp. Ther., 139:283-289.
25. Rozen, R., Tenenhouse, H.S., and Scriver, C.R., 1979, Taurine transport in renal brush-border membrane vesicles, Biochem. J., 180:245-248.
26. Schaffer, S.W., Kramer, J., and Chovan, J.P., 1980, Regulation of calcium homeostasis in the heart by taurine, Federation Proc., 39:2691-2694.
27. Schaffer, S.W., Safer, B., Scarpa, A., and Williamson, J.R., 1974, Mode of action of the calcium ionophores X-537A and A-23187 on cardiac contractility, Biochem. Pharmacol. 23:1609-1617.
28. Sulakhe, P.V., Drummond, G.I., and Ng, D.C., 1973, Calcium binding by skeletal muscle sarcolemma, J. Biol. Chem., 248:4150-4157.

PART IV

TAURINE AND THE HEART

Ryan J. Huxtable

Department of Pharmacology
Health Sciences Center
University of Arizona
Tucson, Arizona

Taurine concentrations are higher in the heart than in other tissues in most mammals. Indeed, taurine is by far the most abundant free amino acid in the heart. Taurine concentration in rat heart, for example, is four times that of glutamate, the next most abundant amino acid (13). These levels can be lowered by gross insults such as anoxia (14), or myocardial infarction (4), and are raised in pathological states such as congestive heart failure (12). Interest in the cardiac actions of taurine stems first from these high levels combined with its numerous electrophysiological actions (10), and secondly from the protection that taurine affords in certain cardiopathic states (e.g. 1) and the association of a defect in the biochemistry of taurine with some clinical conditions (e.g. 5). Numerous studies over the past two decades have revealed the positive inotropic actions of taurine (6), its calcium modulatory effects (7), and its ability in striated muscle to shorten the action potential, increase the rate of both depolarization and repolarization, and its stabilizing effect on membrane potential (8). It is of practical importance that taurine, given orally, ameliorates the progession of disease in genetically cardiomyopathic hamsters (1), and that defects in taurine transport may be associated with the cardiomyopathy of Friedreich's ataxia (3). Furthermore, inability to maintain cardiac taurine concentrations is associated with genetic congestive cardiomyopathy in humans (5).

This synopsis summarizes why further investigations of taurine in the heart are of both scientific and clinical importance. It is a substance normally abundant in the diet, and normally well conserved within the body. However, under certain nutritional conditions, such as taurine deprivation in neonates, or with certain genetic conditions, in which resorption of taurine by the kidney may be impaired, clinical abnormalities associated with taurine may reveal themselves. There is little toxicity associated with taurine, although there is one report that in excess it may cause arrhythmias (9), and so the finding that relatively low amounts of taurine given orally retard the development of lesions in genetic cardiomyopathy in hamsters is of great relevance. In this regard, clinical trials have been proposed (2,11) for taurine in the cardiomyopathy of Friedreich's ataxia. The following three chapters report ongoing work on the biochemistry and electrophysiological of taurine in the heart, and some clinical implications and (possibly) applications. Nathan and Crass find sharp falls in the taurine content of the heart on ligation of a coronary artery. They also observe, as Gruener et al. found with skeletal muscle, that taurine shortens the duration of the action potential. Franconi and her coworkers establish in yet another system the biphasic and apparently contradictory effects of taurine on calcium. They find that taurine has a negative inotropic effect at high calcium concentration, but a positive effect at low calcium concentration. Taurine, in effect, acts as a homeostatic agent, antagonizing deviations in either direction from normalcy. This seems to be a general action of taurine as it has now been observed in a number of situations.

In response to stress of various kinds, the heart loses taurine. Schaffer has shown that hearts perfused in calcium-free medium and then exposed to calcium lose large amounts of taurine over a short period. Lombardini and his coworkers have found similar losses of taurine in response to coronary artery ligation or isoproterenol-induced infarction. Lombardini has explored clinical applications of this phenomenon, and reports some of his findings below. His work strongly suggests that measurements of blood taurine levels, which may be readily done, may be a useful addition to the battery of diagnostic tests for myocardial infarction.

REFERENCES

1. Azari, J., Brumbaugh, P., and Huxtable, R., 1980, Prophylaxis by taurine in the hearts of cardiomyopathic hamsters, J. Mol. Cell. Cardiol., 12:1353-1366.
2. Barbeau, A., 1976, Friedreich's ataxia 1976 - An overview, Can. J. Neurol. Sci., 3:389-397.
3. Barbeau, A., Melancon, S., Huxtable, R., and Lemieux, B., Taurine and Friedreich's Ataxia: An update, This volume, 1981.
4. Crass, M.F., and Lombardini, J.B., 1977, Loss of cardiac muscle taurine after acute left ventricular ischemia, Life Sciences, 21:951-958.
5. Darsee, J.R., and Heymsfield, S.B., 1981, Decreased myocardial taurine levels and hypertaurinurea in a kindred with mitral value prolapse and a congestive cardiomyopathy, New Eng. J. Med.,

304:129-135.
6. Dietrich, J., and Diacono, J., 1971, Comparison between ouabain and taurine effects on isolated rat and guinea pig hearts in low calcium medium, Life Sciences, 10:499-507.
7. Dolara, P., Agresti, A., Giotti, A., and Pasquini, G., 1973, Effect of taurine on calcium kinetics of guinea pig heart, Eur. J. Pharmacol., 24:352-358.
8. Gruener, R., Bryant, H., Markovitz, D., Huxtable, R., and Bressler, R., Ionic actions of taurine on nerve and muscle membranes: electrophysiologic studies, in: "Taurine," R. Huxtable and A. Barbeau, eds., Raven Press (1976), pp225-242.
9. Hinton, J.R., Dias Souza, J., and Gillis, R.A., 1975, Deleterious effects of taurine in cats with digitalis-induced arrhythmias, Eur. J. Pharmacol., 33(2):383-387.
10. Huxtable, R., Metabolism and function of taurine in the heart, in: "Taurine," R. Huxtable and A. Barbeau, eds., Raven Press, New York (1976), pp99-120.
11. Huxtable, R.J., 1978, Cardiac pharmacology and cardiomyopathy in Friedreich's ataxia, Can. J. Neurol. Sci., 5:83-91.
12. Huxtable, R., and Bressler, R., 1974(a), Taurine concentration in congestive heart failure, Science, 184:1187-1188.
13. Huxtable, R.J., Laird, H.E., and Lippincott, S.E., 1979, The transport of taurine in the heart and the rapid depletion of tissue taurine content by guanidinoethyl sulfonate, J. Pharmacol. Exptl. Therap., 211:465-471.
14. Lombardini, J.B., 1980, Effects of isoproterenol and methoxamine on the contents of taurine in rat tissues, J. Pharmacol. Exptl. Therap., 213:399-405.

ELECTROPHYSIOLOGICAL EFFECTS OF TAURINE IN CARDIAC PURKINJE FIBERS AND MYOCARDIAL TAURINE LOSS DURING ISCHEMIA. IS THERE A RELATIONSHIP?

R.D. Nathan and M.F. Crass III

Department of Physiology
School of Medicine
Texas Tech University Health Sciences Center
Lubbock, Texas

Interest in a role for taurine in the maintenance of normal electrophysiology of the heart originated from the experiments of Read and Welty (16) and later, Chazov and colleagues (1). These workers showed that exogenous taurine was capable of causing the disappearance or reduction in severity of arrhythmias, principally premature ventricular contractions, in whole animals or isolated hearts administered toxic doses of catecholamines or glycosides. The proposed mechanisms for these antiarrhythmic effects of taurine have not received uniform acceptance by workers in the field, and the subject is clearly in need of much investigation.

Pursuant to certain questions related to ischemia-induced alterations in myocardial metabolism, we became intrigued with these early findings and decided to examine the possibility that myocardial ischemia, with its attendant abnormalities of electrical and pump function, might involve major perturbations, not only of taurine content but taurine function as well. This report will review our findings on content and distribution of taurine in non-ischemic and ischemic or oxygen-deficient heart preparations and overview our current studies aimed at discerning the effects of taurine on electrical activity of the heart.

MATERIALS AND METHODS

Animals

Mongrel dogs weighing 15-35 kg were used for taurine content and distribution as well as isolated Purkinje fiber studies. The dogs were anesthetized with sodium pentobarbital (30 mg/kg IV) and, in ischemic and non-ischemic studies, were tracheostomized and ventilated with room air using a positive pressure respirator. In perfused heart studies, male Sprague-Dawley rats, weighing 250-300g and given free access to Purina rat chow and water, were used. The rats were anesthetized with ether.

Ischemic and Non-ischemic Dog Preparation

A detailed methodology has been published (3). Briefly, a left thoracotomy or midline sternotomy was performed. Transmural biopsy punctures of the left and right ventricles were taken after 4 hr with or without left circumflex coronary artery ligation using a stainless steel boring device attached to an electric hand drill. The transmural tissue plugs along with pieces of right and left atria were rinsed and frozen in liquid nitrogen. The frozen ventricular tissues were then divided into outer, mid and inner sections and taurine was determined on an amino acid analyzer.

Isolated Perfused Rat Heart Preparation

A detailed methodology has been published previously (4). In brief, after ether anesthesia, the chest of the rat was opened and the heart excised and mounted on the perfusion apparatus. One group of hearts was perfusion-washed for 5 min only with Krebs-Henseleit bicarbonate buffer containing 5.5 mM glucose. The ventricles were freeze-clamped and taurine ("initial") analyzed as before. A second group of hearts was perfused for an additional 30 min followed by freeze-clamping of the ventricles and taurine determination. The two groups cited above were each subdivided into two groups, one perfused with buffer equilibrated with 95% O_2 - 5% CO_2 and the other perfused with buffer equilibrated with 95% N_2 - 5% CO_2. Samples of perfusate were taken at 10, 20, and 30 min intervals for taurine analyses.

Isolated Dog Purkinje Fibers

The dog hearts were excised through an intercostal incision. Bundles of Purkinje fibers, removed from either ventricle, were incubated for 1 hr in oxygenated Tyrode's solution containing 5.4 mM $CaCl_2$ to promote "healing over" (Dr. Ira Cohen, personal communication). The fibers were then secured with two stainless steel clamps and bathed in 3.0 ml of oxygenated Tyrode's solution composed of (mM): NaCl 137, KCl 2.7, $NaHCO_3$ 11.9, NaH_2PO_4 0.45, $MgCl_2$ 0.5, $CaCl_2$ 2.7, and glucose 5.5. Temperature (36-38°C) and pH (7.3-7.4) were monitored continuously, and the concentrations of Na^+ and K^+ were verified by flame photometry at the end of most experiments.

Transmembrane potentials were measured as the potential difference between an intracellular micropipette filled with 3M KCl (20-30 MΩ) and the Tyrode's solution which was coupled to a virtual ground by a chlorided silver wire (14). The maximum upstroke velocity of the action potential (Vmax) was determined by active electronic differentiation (time constant 25 μsec, linear to 800 V/sec) and displayed on a storage oscilloscope (some of the values shown in Table 4 might have been underestimated because of inadequate electrode capacitance compensation). Extracellular stimuli (2 msec duration) were delivered at a rate of 120/min through an extracellular bipolar electrode made of Teflon-insulated silver wires.

Action potentials and their derivatives were recorded on FM tape at a speed of 3.75 in/sec. The taped data were played back at a later time through a transient waveform recorder, digitized, and plotted at greatly reduced speed on a strip chart recorder (15). From these expanded traces, a graphic digitizer was used to measure such parameters as the maximum diastolic potential (MDP), overshoot potential (OS), and the duration of the action potential (DUR) at -20, -40, -60, and -80 mV. Values obtained from two non-consecutive action potentials recorded in the same fiber were then averaged, and the means were used to construct Table 4.

Taurine or hypotaurine was dissolved in Tyrode's solution (3 mM or 12 mM) and added to the bathing medium in volumes of 1-100 μl. Treatment of Purkinje fibers with equivalent amounts of Tyrode's solution produced no significant effects.

RESULTS AND DISCUSSION

Content and Distribution of Taurine in the Non-ischemic and Ischemic Dog Heart

The content and distribution of taurine in the non-ischemic canine heart can be seen in Table 1. Total right atrial taurine content was substantially greater than that found in the right ventricle or in any region of the left heart. Furthermore, left atrial taurine also exceeded ventricular taurine content. Total right and left ventricular taurine contents were not significantly different ($P > 0.05$). Somewhat surprising was the finding that taurine content in the left ventricle was clearly not uniform transmurally. Table 1 shows distribution of taurine across the posterolateral left ventricular wall midway between base and apex, indicating an increasing outer to inner gradient; the inner or subendocardial content being significantly

Table 1. Content and distribution of taurine in the dog heart with and without 4 hours of left circumflex coronary artery occlusion

Tissue Region	Taurine Content	
	Without Occlusion	With Occlusion
	μmole/g dry weight	
Right atrium	96.8 ± 3.5	—
Left atrium	72.9 ± 7.2	—
Right ventricle	47.1 ± 2.7	—
Subendocardium	48.2 ± 3.3	—
Subepicardium	46.0 ± 5.5	—
Left ventricle	55.5 ± 3.0	29.5 ± 2.1
Subendocardium	68.0 ± 2.0	35.5 ± 4.3
Midmyocardium	53.7 ± 3.1	27.7 ± 2.5
Subepicardium	44.9 ± 3.7	25.2 ± 2.6

Each value represents the mean ± SEM of 5 dogs

greater than the content of other regions ($P < 0.01$ vs midmyocardium; $P < 0.001$ vs subepicardium). This agrees, at least on an anatomical basis, with the data of Kocsis and colleagues (13) who showed earlier that Purkinje fibers, with the greatest density in the subendocardial zone, contained the highest taurine content of heart tissue. The right ventricular taurine profile expectedly did not show a similar transmural taurine gradient. Thus, the content of taurine in the heart is not uniformly distributed.

Next, the effects of acute coronary artery ligation on myocardial taurine content and distribution were assesed. The *in vivo* model was the same as that used in non-ischemic dogs with the exception that a length of the left circumflex coronary artery was dissected to permit placement of an occluder. In terms of the performance of the dogs during 4 hr of coronary occlusion, the only parameter to change, aside from electrical events, was a statistically significant increase ($P < 0.05$) in the left ventricular end-diastolic pressure amounting to an increase of 3-5 mm Hg after the 4 hr occlusion period. Basically, the experimental approach was similar to the model for acute canine myocardial ischemia described by Jennings et al. (10).

Table 1 shows total left ventricular taurine with and without 4 hr of ischemia. As can be seen, taurine decreased by approximately 50% ($P < 0.001$). In addition to the already described non-uniform tissue distribution of taurine in the non-ischemic heart, data in Table 1 show that loss of taurine was similarly non-uniform, being greatest in the inner region. Disappearance of taurine in this region amounted to 32 μmole/g dry weight or almost twice that seen in the subepicardial region. Since the Purkinje system is localized primarily in the subendocardium, and since Purkinje fibers probably account for much of the taurine in this region (13), these results would suggest a major taurine-deficiency of the Purkinje conduction tissue in the ischemic zone. To our knowledge, this taurine deficiency has not been assessed by direct quantitation. Much importance must be attached to such determinations in the context of the possible importance of taurine deficiency to ischemia-induced abnormal electrical activity.

In order to account for the disappearance of taurine in the acutely ischemic myocardium, we considered the possibilities that taurine could have been metabolized to another compound or simply released into the extracellular fluid compartment. The first possibility is unlikely from the present literature (9). To test the second possibility, that of release into the circulation, we employed an isolated rat heart preparation perfused with a taurine-free, oxygen-deficient Krebs-Henseleit buffer. This approach provided a simpler approach vis-a-vis the _in vivo_ setting, since any taurine present in the perfusate must have been released and of endogenous origin. As can been seen in Table 2, tissue taurine levels were unchanged ($P > 0.10$) after 30 min of oxygenated perfusion. On the other hand 30 min of oxygen-deficient perfusion resulted in a significant ($P < 0.02$) tissue taurine loss. Upon examination of the heart perfusates, which contained no taurine at 0 time, the following observations were made: 1) oxygenated perfusates contained a small quantity of taurine after 10 min which remained unchanged after 30 min (this taurine could have been released during the transient hypoxia of heart extirpation); 2) taurine presence in oxygen-deficient perfusates was increased 3- to 4-fold after 10 min and continued to increase over the remaining 30 min of perfusion; and 3) recovery of taurine in the perfusates was, in all cases, stoichiometric with tissue taurine loss. Thus, cellular hypoxia, the principal manifestation of ischemia, resulted in the efflux of intracellular taurine into the extracellular compartment.

Table 2. Taurine content of rat heart and medium during perfusion for 30 minutes with or without oxygenation

	Perfusion Time (min)			
	Initial	10	20	30
	Taurine (μmole/g dry weight)			
Tissue				
oxygenated (95% O_2 - 5% CO_2)	154.4 ± 2.6 (8)	—	—	159.6 ± 2.6 (11)
oxygen-deficient (95% N_2 - 5% CO_2)	146.3 ± 3.6 (4)	—	—	132.3 ± 3.6 (9)
Perfusate				
oxygenated (95% O_2 - 5% CO_2)	0 (6)	3.8 ± 0.2 (6)	4.2 ± 0.3 (6)	3.9 ± 0.2 (6)
oxygen-deficient (95% N_2 - 5% CO_2)	0 (6)	12.9 ± 1.6 (6)	16.6 ± 2.9 (6)	19.4 ± 2.5 (6)

Perfusate taurine values are cumulative with time. Each value represents the mean ± SEM of the number of hearts or perfusates shown in parenthesis.

The next question we posed was: "Is the loss of taurine during myocardial ischemia specific vis-a-vis other cellular amino acids?" Taurine comprises roughly 50 percent of the free amino acids of the myocardial cell (8) and presumably is not a component of cellular proteins, structural or otherwise. Thus, this sulfonic amino acid could possibly be released from the myocardial cell without the prerequisite of lysosomal or other proteolytic activity. To test the specificity of taurine release, the same *in vivo* ischemic dog model was employed. Serial sampling of blood was performed simultaneously from an artery and from the great cardiac vein. Thus, any changes in blood amino acid composition by this coronary arteriovenous technique represented the myocardial contribution exclusively, independent of changes in the systemic circulation. Fig. 1 shows the ECG tracing (lead II) before and after ligation of the circumflex artery. Unexpectedly, in this particular dog, the heart began to fibrillate at approximately 5 minutes post-ligation. At this time, a blood sample was taken and later analyzed for taurine as well as the acidic, basic and neutral amino acids. Table 3 shows the whole blood amino acid profile of blood withdrawn from the great cardiac vein before circumflex ligation and after onset of fibrillation. All but four of the acidic and neutral amino acids showed a decrease or no change in concentration. The combined asparagine/glutamine fraction and isoleucine concentrations were slightly elevated (7 and 4% respectively). Taurine concentration showed the largest increase at 24%. The only other amino acid to increase was aspartic acid (18%). In another experiment, representative of several, arterial and coronary venous blood samples were taken before and after 4 hr of left circumflex ligation. After four hours of ischemia, the coronary venous plasma taurine concentration had increased 5.4-fold. No change was found in aspartic acid, phosphoserine or phosphoethanolamine levels. All other venous plasma amino acid concentrations increased, but none greater than the 2.2-fold change exemplified by tyrosine. This data on specificity of taurine release is rather preliminary, but hints strongly at early selective release of taurine. Establishing a correlation between ischemia-induced electrophysiological changes and taurine loss would be of considerable interest. Furthermore, evidence which would establish a positive correlation between extent and duration of ischemia and/or severity of electrophysiological disturbances and the level of reduction in cellular taurine would also be of interest. So far, it would appear that longer periods of ischemia result in a more generalized, less specific rise in coronary venous plasma amino acids. But selective taurine loss may predominate early or shortly after occlusion, a time which would coincide with a period of serious and oftimes fatal arrhythmias. Finally, in terms of directional changes in tissue amino acids, our findings indicate that cellular necrosis is probably not a prerequisite for taurine release.

The loss of tissue taurine and the coincident appearance of arrhythmias does little to suggest a direct correlation between these

Table 3. Whole blood* acidic, neutral and basic amino acid profile before and after left circumflex coronary artery ligation

Amino Acid	Before Ligation of Artery	Immediately After Onset of Fibrillation (5 min post-ligation)	Percent Change
	μmole/ml whole blood		%
P-Ser	0.125	0.120	- 4
P-$EtOHNH_2$	0.015	0.013	-13
Urea	4.930	4.800	- 3
Aspartic Acid	0.032	0.038	+18
Threonine	0.205	0.177	-14
Serine	0.322	0.204	-37
$AspNH_2$+$GluNH_2$	0.734	0.789	+ 7
Glutamic Acid	0.188	0.175	- 7
Citrulline	0.044	0.044	± 0
Glycine	0.137	0.122	-11
Alanine	0.282	0.250	-11
α-Amino Butyric Acid	0.012	0.010	-17
Cystine	0.021	0.021	± 0
Valine	0.130	0.125	- 4
Methionine	0.017	0.015	-12
Isoleucine	0.050	0.052	+ 4
Leucine	0.098	0.093	- 5
Tyrosine	0.067	0.056	-16
Phenylalanine	0.048	0.039	-19
Taurine	0.139	0.172	+24
Ornithine	0.021	0.021	± 0
Lysine	0.227	0.114	-50
Histidine	0.065	0.050	-23

* Venous blood was obtained from the great cardiac vein 4.5 cm from the apex in the lower half of the anterior interventricular sulcus

two events. However, electrophysiological experiments involving normal and ischemic cardiac muscle or conduction tissue, in the presence or absence of taurine and/or its analogs, may provide such a correlation. Only a few reports, for example those by Read and Welty (17) using guinea pig papillary muscle, Dolara et al. (6) using sheep Purkinje fibers and guinea pig ventricular muscle and Diacono et al. (5) using rat and guinea pig ventricular muscle, have addressed the question of taurine effects on cardiac electrophysiology. Recently, we have begun to assess this question using 7-day-old chick heart cell aggregates and canine Purkinje fibers. Only the Purkinje fiber studies, which may be more relevant to the overall question of ischemia-induced arrhythmogenesis, will be discussed.

Continuous intracellular recordings were obtained from seven Purkinje fibers during addition of taurine in varying doses over the range 0.01-0.4 mM. These represent lower concentrations than employed previously (5, 6, 17). Table 4 shows characteristics of action potentials recorded from Purkinje fibers driven at 120 beats/min. The two most significant effects after taurine addition were reductions in the duration and overshoot of the action potential. Duration was assessed at multiple membrane potentials to determine more accurately changes in phases 2 and 3 of the action potential. Reductions in duration up to 53 msec were observed. Overshoot decreased by as much as 10 mV. Though the changes in maximum diastolic potential (MDP) and upstroke velocity (Vmax) lacked statistical significance, taurine produced an apparent biphasic shift which was dose-dependent (Table 4). As shown in the actual tracings, the MDP and Vmax increased by as much as 7 mV and 35 V/sec, respectively, in the range of 0.01-0.08 mM (Fig. 2A). At higher taurine concentrations, both parameters declined by as much as 12 mV and 100 V/sec, respectively, from control values (Fig. 2B). These alterations in action potential configuration were apparent within 2 min after the addition of taurine, were maximal within 5-10 min (Fig. 3A) and were dose-dependent (Fig. 3B).

Interestingly, the initial condition of the fibers seemed to influence the amplitude of these effects. For example, in a "healthy" fiber (MDP = -91 mV; OS = +33 mV; Vmax = 355 V/sec), 0.08 mM taurine shortened the action potential by 21 msec but had little effect on the other parameters (Fig. 4A). On the other hand, a similar concentration significantly enhanced the MDP and OS (by 11 mV and 5 mV, respectively) of a "damaged" fiber (MDP = -79 mV; OS = +21 mV; Vmax = 280 V/sec) (Fig. 4B). The so-called "damage" seen in certain fibers probably arose from unintentional stretching.

Spontaneously-beating Purkinje fibers also responded differently to taurine depending upon their condition. These fibers depolarized in 0.4 mM taurine if they were "healthy" (Fig. 5A, MDP = -93 mV; Fig. 5B, MDP = -91 mV), but hyperpolarized if they were "damaged" (Fig. 5C, MDP = -48 mV; Fig. 5D, MDP = -55 mV). The slowing of beat rate in

Table 4. Effects of taurine on characteristics of canine Purkinje fiber action potentials

Conc. (mM)	N[a]	MDP (mV)	OS (mV)	DUR(-20mV) (msec)	DUR(-40mV) (msec)	DUR(-60mV) (msec)	DUR(-80mV) (msec)	Vmax (V/sec)
0.00	7	-85.8 ±1.4	+32.1 ±0.5	126 ±4	160 ±4	192 ±6	259 ±12	324 ±20
0.01	5	-87.2 ±2.9	+31.2 ±1.2	121 ±7	154[b] ±5	183[c] ±6	246 ±10	325 ±32
0.04	5	-89.2 ±3.0	+29.4 ±2.0	111 ±13	144 ±8	171 ±6	221[e] ±7	351 ±21
0.08	5	-88.1 ±2.9	+29.2 ±3.0	103 ±13	140 ±8	170 ±6	226[c] ±12	340 ±20
0.20	5	-85.3 ±2.4	+28.2[b] ±2.1	111[c] ±6	146[d] ±7	182 ±13	259 ±35	316 ±25
0.40	5	-83.2 ±3.3	+27.2[c] ±1.9	105[d] ±6	144[d] ±5	185 ±16	229[b] ±7	299 ±34

The maximum diastolic potential (MDP), overshoot potential (OS), durations (DUR) measured at -20, -40, -60 and -80 mV, and the maximum upstroke velocity (Vmax) were determined during continuous recordings from canine Purkinje fibers driven at 120 beats/min. Values representing means ± SE were measured within 3-10 min after addition of taurine. Significance of differences between experimental and control values was calculated by Student's paired t-test.

[a]N represents the number of Purkinje fibers, which was one less than that indicated for DUR (-80mV); [b]$p < 0.05$; [c]$p < 0.02$; [d]$p < 0.01$; [e]$p < 0.005$.

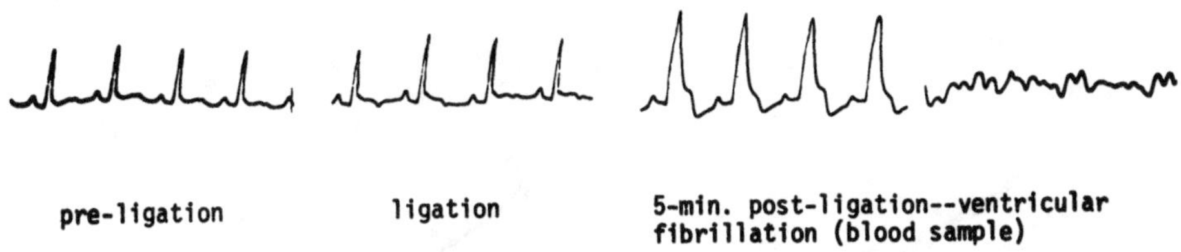

Fig. 1. Electrocardiogram (Lead II) of dog before and after ligation of the left circumflex coronary artery. Blood samples were withdrawn from the great cardiac vein before ligation and after onset of ventricular fibrillation.

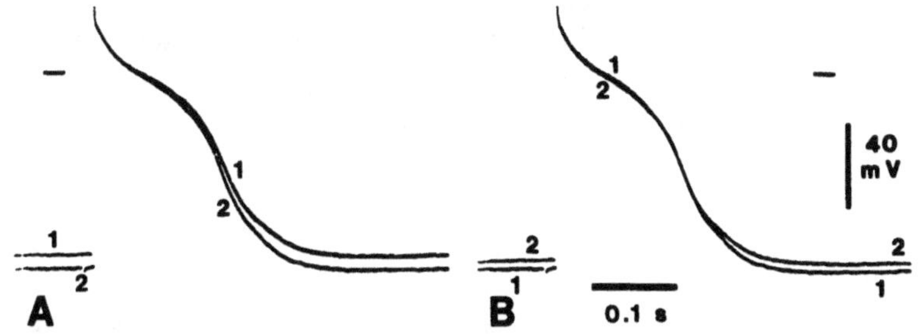

Fig. 2. A. Control (trace 1) and 10 min after addition of 0.04 mM taurine (trace 2). B. Ten min after addition of 0.04 mM taurine (trace 1) and 6 min after addition of 0.4 mM taurine (trace 2). Purkinje fiber driven at 120 beats/min. Horizontal lines at upper left- and right-hand margins represent 0 mV.

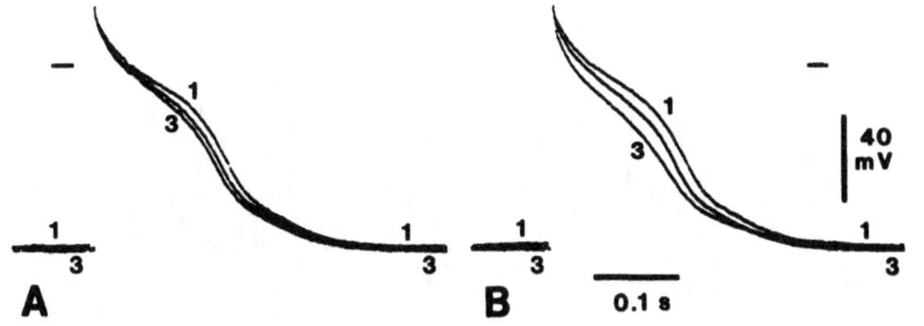

Fig. 3. A. Control (trace 1), 4 min (middle trace), and 10 min (trace 3) after addition of 0.01 mM taurine. B. Control (trace 1) and 7 min after addition of 0.01 mM (middle trace) and 0.04 mM taurine (trace 3).

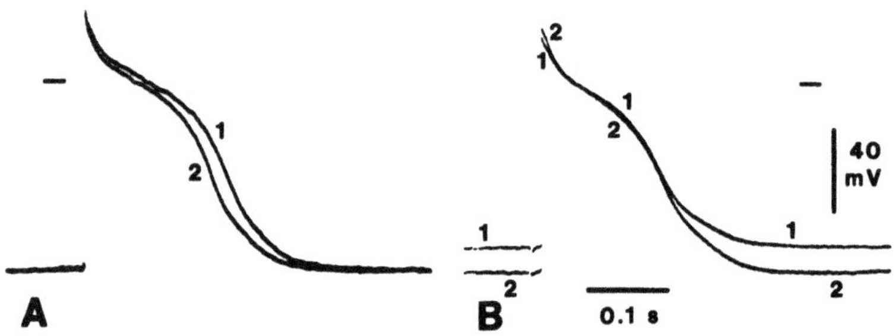

Fig. 4. A. Control (trace 1) and 6 min after addition of 0.08 mM taurine (trace 2). B. Control recording from a "damaged" fiber (trace 1) and 6 min after addition of 0.08 mM taurine (trace 2).

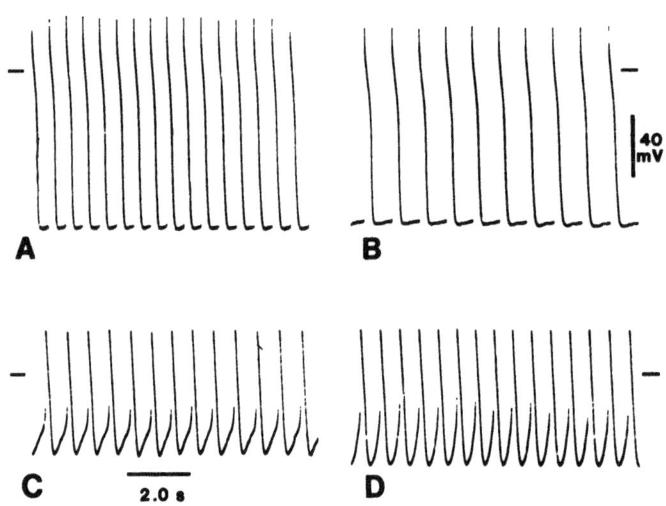

Fig. 5. A. Control recording from a spontaneous Purkinje fiber; 107 beats/min. B. Five min after addition of 0.4 mM taurine; 68 beats/min. C. Control recording from a "damaged" spontaneous Purkinje fiber; 87 beats/min. D. Five min after addition of 0.4 mM taurine; 98 beats/min.

"healthy" fibers was probably a result of the increased distance from MDP to threshold; whereas in the "damaged" fibers, beating accelerated due to a more rapid rate of diastolic depolarization.

From these preliminary experiments, the electrophysiological effects of taurine (0.01-0.4 mM) on canine Purkinje fibers may be summarized as follows: a) significant reductions in the overshoot and duration of the action potential; b) hyperpolarization of the maximum diastolic potential and an increase in upstroke velocity at concentrations between 0.01 and 0.08 mM; c) depolarization of the MDP and a decline in Vmax at higher levels of taurine; and d) an enhancement of these changes in "damaged" fibers, as well as the requirement for more taurine to depolarize the MDP. Our observations with respect to the MDP, Vmax and DUR of the action potential in the presence of low levels of taurine (0.01-0.08 mM) are directionally similar to those of Dolara et al. (6), who used much larger doses (1.0 mM taurine) in studying sheep Purkinje fibers and guinea pig ventricular muscle. Also, these workers found an increase rather than a decline in the OS of their muscle preparation. Higher concentrations have been reported to reduce the OS and Vmax of action potentials recorded from guinea pig and rat ventricular trabeculae, without additional modifications (5), and to reduce the OS and resting potential of guinea pig papillary muscle (17). The results of the former study (5) were explained by voltage clamp experiments in which the fast inward sodium current (i_{Na}) declined, and the slow inward current (i_{si}) and outward plateau current (i_x) were unchanged by 8 mM taurine. In contrast, 20 mM taurine reduced the plateau and shortened the action potential at potentials positive to -54 mV, while increasing the duration at more negative potentials (17). When taurine was added in the presence of reduced calcium (1.8 mM), the DUR was shortened at all potentials, and the resting potential was not changed significantly. Because of the biphasic effects of taurine observed in the present study (Table 4 and Fig. 2), we suggest that the conflicting results reported for guinea pig ventricular tissue, i.e., an enhancement of the MDP, OS and Vmax in 1.0 mM taurine (6) and a reduction in these parameters at higher levels (5,17), may simply be due to differences in concentration.

Without voltage clamp data, we can only speculate on the ionic mechanisms underlying the electrophysiological effects of taurine in canine Purkinje fibers. Hyperpolarization of the MDP and shortening of the action potential, which occurred at concentrations between 0.01 and 0.08 mM, could have resulted from: a) an increase in membrane conductance(s) to K^+ (11); b) elevation of intracellular K^+; c) a reduction in slowly inactivating inward current(s) carried by Ca^{++} and/or Na^+ (2); d) stimulation of the electrogenic Na-K pump (7); or e) a negative shift of voltage-dependent conductance parameters as a result of increased positive intracellular surface charge (15). The rise in Vmax seen at these concentrations might also have been derived from the last mechanism or as a consequence of hyper-

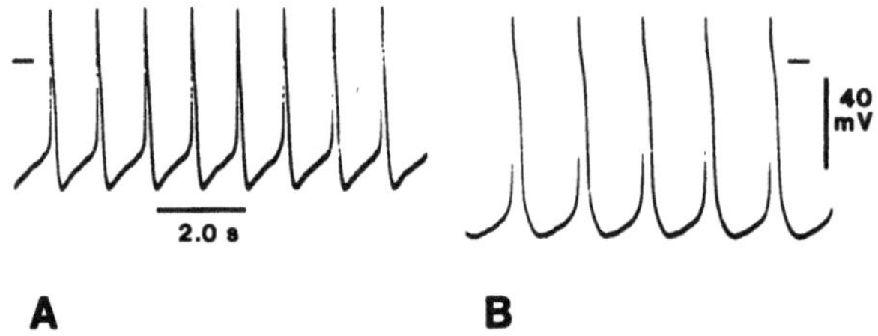

Fig. 6. A. Control recording from a "damaged" spontaneous Purkinje fiber; 56 beats/min. B. Five min after addition of 0.4 mM hypotaurine; 42 beats/min.

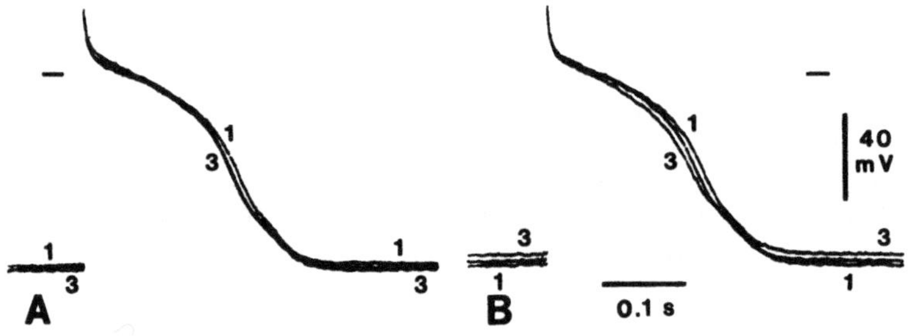

Fig. 7. A. Control (trace 1) and 6 min after addition of 0.001 mM (trace 2) and 0.005 mM hypotaurine (trace 3). Six min after addition of 0.005 mM (trace 1), 0.02 mM (trace 2), and 0.18 mM hypotaurine (trace 3).

polarization of the MDP (by removal of inactivation). One explanation for depolarization of the MDP at higher levels of taurine is accumulation of K^+ in the restricted extracellular space surrounding the Purkinje fibers (12). Such accumulation could arise if the sarcolemma were to become increasingly permeable to K^+. Elevation of extracellular K^+ would shift the potassium equilibrium potential to more positive values and lead to depolarization of the MDP. Additional experiments will be necessary to test these hypotheses.

Finally, Purkinje fibers appeared to be more sensitive to hypotaurine. For example, hypotaurine enhanced the MDP of a "damaged" spontaneous fiber by 20 mV (from -56 mV to -76 mV) and increased the Vmax from <10 V/s to 32 V/s (Fig. 6). These changes were much greater than those produced with an equivalent dose of taurine (Fig. 5). In drive fibers, the addition of only 0.001-0.005 mM hypotaurine was necessary to hyperpolarize the MDP by 2-4 mV (Fig. 7A). Higher concentrations (0.01-0.4 mM) resulted in depolarization (Fig. 7B).

The subject of ischemia-induced electrical abnormalities in cardiac muscle has been, and continues to be, an important area of research. This report brings together a number of observations and results of preliminary experiments which support the inclusion of changes in heart tissue taurine in the myriad of ischemia-induced events. The marked loss of myocardial cell taurine after coronary occlusion, particularly in the vulnerable subendocardial region with its high density of Purkinje conducting tissue, and the proposed relative specificity of this loss vis-a-vis other amino acids, do not necessarily imply a commensurate loss of a key functional role for taurine in myocardial cell electrical stability. However, the effects of added taurine (or hypotaurine) in low (approximately physiologic) extracellular concentrations on Purkinje fiber action potential configuration lends support to a functional role in this area. The greater effect of taurine and hypotaurine on "damaged" Purkinje fibers may add to the significance of these findings. Development of a possible correlation between ischemia-induced loss of tissue taurine with this apparent functional role constitutes the focus of our continuing investigations.

ACKNOWLEDGMENTS

We thank Dr. M.L. Bhattacharyya for his assistance in the removal of canine Purkinje fibers, and Mrs. Virginia Shelton for her technical assistance. Supported, in part, by grant HL 20708.

REFERENCES

1. Chazov, E.I., Malchikova, L.S., Lipina, N.V., Asafov, G.B., and

Smirnov, V.N, 1974, Taurine and electrical activity of the heart, Circ. Res., 34 & 35 (Suppl. III):11-21.
2. Coraboeuf, E., Deroubaix, E., and Coulombe, A., 1979, Effect of tetrodotoxin on action potentials of the conducting system in the dog heart, Am. J. Physiol., 236:H561-H567.
3. Crass, M.F., III, and Lombardini, J.B., 1977, Loss of cardiac muscle taurine after acute left ventricular ischemia, Life Sci., 951-958.
4. Crass, M.F., III, and Lombardini, J.B., 1978, Release of tissue taurine from the oxygen-deficient perfused rat heart, Proc. Soc. Exptl. Biol. Med., 157:486-488.
5. Diacono, J., and Dietrich, J., 1976, Effects of taurine on the transmembrane ionic currents of rat and guinea pig ventricular fibers, Il. Farmaco-Ed. Sci., 31:277-283.
6. Dolara, P., Ledda, F., Mugelli, A., Mantelli, L., Zilletti, L., Franconi, F., and Giotti, A., 1978, Effect of taurine on calcium, inotropism and electrical activity of the heart, in: "Taurine and Neurological Disorders," A. Barbeau and R.J. Huxtable, eds., Raven Press, New York.
7. Gadsby, D.C., and Cranefield, P.F., 1979, Direct measurement of changes in sodium pump current in canine cardiac Purkinje fibers, Proc. Natl. Acad. Sci., 76:1783-1787.
8. Grosso, D.S., and Bressler, R., 1976, Taurine and cardiac physiology, Biochem. Pharmacol., 25:2227-2232.
9. Huxtable, R.J., Chubb, J., and Azari, J., 1980, Physiological and experimental regulation of taurine content in the heart, Federation Proc., 39:2685-2690.
10. Jennings, R.B., Wartman, W.R., and Zudyk, Z.E., 1957, Production of area of homogenous myocardial infarction in dog, AMA Arch. Path., 63:580-585.
11. Kass, R.S., and Tsien, R.W., 1976, Control of action potential duration by calcium ions in cardiac Purkinje fibers, J. Gen. Physiol., 67:599-617.
12. Kline, R.P., Cohen, I., Falk, R., and Kupersmith, J., 1980, Activity-dependent extracellular K^+ fluctuations in canine Purkinje fibers, Nature, 289:68-71.
13. Kocsis, J.J., Kostos, V.J., and Baskin, S.I., 1976, Taurine levels in the heart tissues of various species, in: "Taurine," R. Huxtable and A. Barbeau, eds., Raven Press, New York.
14. Nathan, R.D., and DeHaan, R.L., 1979, Voltage clamp analysis of embryonic heart cell aggregates, J. Gen. Physiol., 73: 175-198.
15. Nathan, R.D., Fung, S.J., Stocco, D.M., Barron, E.A., and Markwald, R.R., 1980, Sialic acid: regulation of electrogenesis in cultured heart cells, Am. J. Physiol., 239:C197-C207.
16. Read, W.O., and Welty, J.D., 1963, Effect of taurine on epinephrine and digoxin-induced irregularities of the dog heart, J. Pharmacol. Exp. Ther., 139:283-289.
17. Read, W.O., and Welty, J.D., 1976, Effect of taurine on the action potential of guinea pig papillary muscle, in: "Taurine," R. Huxtable and A. Barbeau, eds., Raven Press, New York.

OBSERVATIONS ON THE ACTION OF TAURINE AT ARTERIAL AND CARDIAC LEVELS

F. Franconi[1], A. Giotti[1], S. Manzini[2], F. Martini[1], I. Stendardi[1] and L. Zilletti[1]

[1]Istituto Interfacolta' di Farmacologia e Tossicologia dell'Universita' di Firenze
[2]Reparto Farmacologico
Laboratorio Ricerche "A. Menarini" Firenze

Taurine exerts a positive inotropic effect in guinea-pig hearts (1,2,3,4,5). This effect seems to be mediated by calcium. In fact, Dolara et al. (2,6) suggest that taurine modifies calcium exchange, and Huxtable and Bressler (7) demonstrate that taurine modifies calcium binding and uptake in rat muscle endoplasmic reticulum. As regards calcium binding to rat myocardial plasma membranes the data are in contrast, indicating that taurine both increases (8) and decreases (9) this binding. In order to clarify the action of taurine on cardiac contraction we studied the effect of taurine on guinea-pig ventricular strips at different calcium concentrations, and the interaction between taurine and calcium antagonists on the same preparations. At the same time, since it is likely that the key factor in the activation of contractile response in smooth muscle is a rise in the cytoplasmic free calcium concentration (10), it seemed of interest to study also the effect of this sulfonic amino-acid on vascular smooth muscle.

EFFECT OF TAURINE ON CONTRACTILITY OF GUINEA-PIG VENTRICULAR STRIPS

From guinea-pig ventricle two strips are dissected, one of which is treated and the other used as control. The strips are placed in separate organ baths into a solution with the following mM composition: NaCl 115, KCl 4.7, KH_2PO_4 1.2, $NaHCO_3$ 25, Glucose 10 and $CaCl_2$ 1.8. The solution is bubbled with a mixture of 95% O_2 and 5% CO_2 and maintained at 30°C. The stimulation frequency is 120 beats/min. After 1 hr equilibration, the ventricular strips are incubated in a medium containing the test calcium concentration; taurine is added to the treated preparation while in the control the osmolarity of the medium is adjusted with sucrose. As shown in Fig. 1, taurine exerts a positive or negative inotropic effect depending on the calcium concentration in the external medium. When calcium is 0.9 mM, it is possible to observe at the three taurine concentrations studied a statistically significant increase in contraction (Fig. 1); this effect is particularly evident when 20 mM taurine is administered. At 1.8 mM calcium, on the other hand, the positive inotropic effect of taurine is evident only when taurine concentration is 20 mM. When the calcium concentration is increased to 2.7, 20 mM taurine exerts a negative inotropic effect, while 10 and 4 mM taurine do not modify contractility (Fig. 1). At 3.6 mM calcium, 10 and 4 mM taurine also produce a slightly negative inotropic effect.

In the same preparations under the same conditions, we studied the action of taurine on the negative inotropic effect induced by D600. The external calcium concentration used in these studies is 1.8 mM. As shown in Fig. 2, taurine slowly antagonizes the negative inotropic effect of D600. The slowness might be explained by the slowness of taurine uptake (11). It is however, important to point out that after one hour the D600-induced fall in contractility is completely reversed with 20 mM taurine.

In the strips treated with D600 alone there is a significant decrease in calcium levels (Table 1), while in those treated with both D600 and taurine, the level of total calcium is almost equivalent to the control value. The cumulative D600 dose-effect curve is shifted to the right in the presence of taurine, suggesting

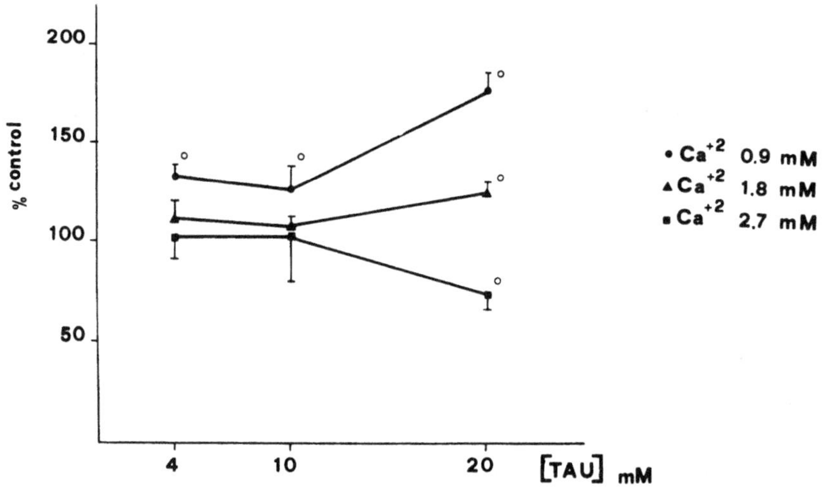

Fig. 1. The effect of taurine on cardiac contraction at different Ca^{++} concentrations. Results are mean ± SE of 6-8 experiments. The small circles (o) denote a significant difference.

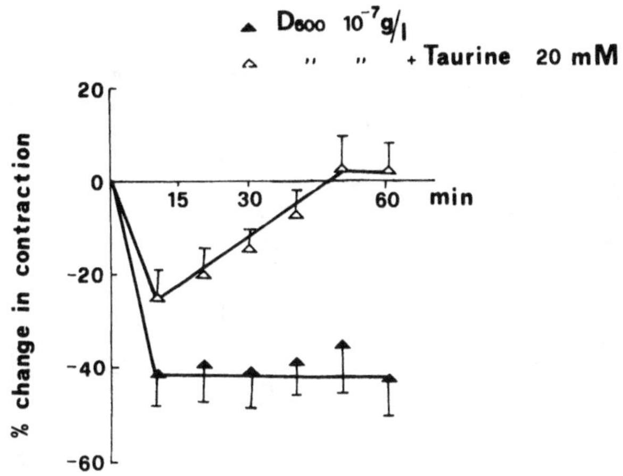

Fig. 2. Effect of taurine ana D600 on contraction in guinea pig right ventricle strips. Results are mean ± SE of 6 experiments. Closed symbols: 0.1 μg/l D600. Open symbols: 0.1 μg/l D600 + 20 mM taurine.

competitive antagonism between the two molecules (Fig. 3). Chovan et al. (12) also demonstrate antagonism between taurine and calcium antagonists on the recovery of contractility in rat heart. Taurine, as shown in Fig. 1, exerts a positive inotropic effect when the strips are incubated in a medium containing 0.9 mM calcium. Kinetic analysis of the calcium wash-out data demonstrate that taurine increases total calcium content, and greater amounts of this ion are washed out from calcium exchange compartments. It seems also that most of the calcium retained in the presence of taurine is bound to tissue structure, but this bound calcium can be partially mobilized when the concentration of easily exchangeable calcium falls (2).

The antagonism (Fig. 3) by taurine of the D600 negative inotropic effect could be explained by the fact that taurine prevents the decrease in calcium content observed in the D600 - treated strips (Table 1). Moreover, electrophysiological studies (13,14) have shown that D600 reduces the number of slow channels activated at a given voltage per unit area of cell membrane in the heart. Calcium antagonists have a negative inotropic effect (15) presumably because of their ability to restrict the excitation-induced inward displacement of calcium (16). Nayler et al. (17,18) exclude verapamil acting either on the calcium sensitivity of the myofibrils or on the calcium uptake and releasing activity of sarcoplasmic reticulum. Moreover Dolara et al. (4) demonstrate with electrophysiological studies that taurine modifies the action potential in guinea-pig ventricle cells perfused with 1.8 mM calcium. Decrease in action potential duration, overshoot and maximum diastolic potential are observed and these modifications can be attributed in turn to an

Table 1

Effect of Taurine and D600 on Contractility and Calcium Content of Guinea Pig Ventricular Strips

	CONTROL	TAURINE	D600	TAURINE + D600
Number of experiments	(25)	(8)	(7)	(7)
Calcium[a] (mEq/1)	13.01 ± 1.04	17.91 ± 3.19[b]	10.57 ± 1.55[c]	14.37 ± 1.06[d]
% Contraction	100	126.3 ± 4.0[b]	58.3 ± 6.5[c]	106.6 ± 8.9[b]

Data are means ± SE for indicated number of experiments.

Determinations were made 1 hr after taurine (20 mM) and/or D600 (0.1 μg/1) were administered. Calcium was determined by atomic spectrophotometry (2).

[a] mEq/1 Intracellular water; [b] $p<.02$ versus control; [c] $p<.001$ versus control; [d] $p<.001$ versus treatment with D600 alone.

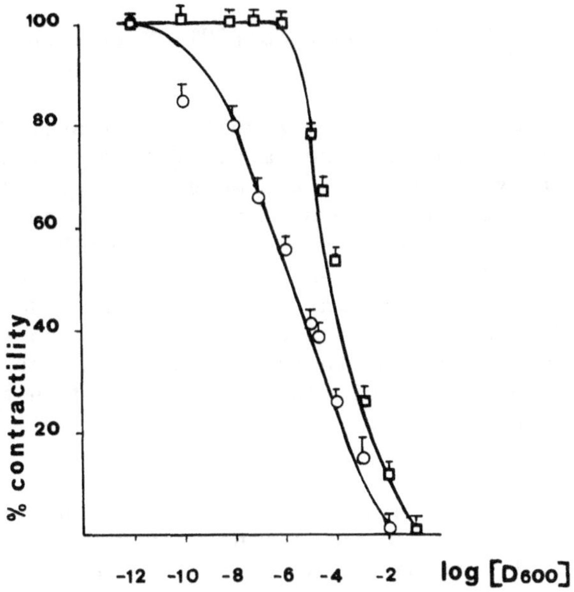

Fig. 3. Effect of taurine on log dose-response curve of D600. Circles: D600 alone; squares; D600 and taurine (20 mM). Calcium concentration is 1.8 mM. Each point is mean ± SE of 6 experiments.

increase in the calcium content. Taurine also increases the calcium binding capacity of sarcolemma (8,19,20). With the above data the ability of taurine to induce a positive inotropic effect becomes clear.

On the other hand, it is more difficult to explain the negative inotropic effect of taurine exerted when the external calcium concentration is 2.7 mM. It is known that taurine treatment reduces the calcium level in some diseases characterized by myocardial calcium overload. In fact, taurine reduces calcium content in cardiomyopathic hamster hearts (21) and also protects against the calcium paradox (22). The inhibitory effect of taurine which we observed can not be attributed to calcium toxicity, since in this case we would have noted the same maximum positive inotropic effect due to calcium increase in the taurine treated strips as in controls, and a subsequent fall in this effect. Taurine also increases rat liver and guinea-pig heart mitochondrial calcium uptake (23,19). This effect on rat liver mitochondria is evident only in a high-calcium medium, and could cause a reduction in cytoplasmic calcium which might explain the negative inotropic effect observed with 20 mM taurine at 2.7 mM calcium. It is further known that taurine also has a biphasic behavior on digitalized heart. In fact, when the cardioactive steroids are in the therapeutic range this sulfonic amino-acid exerts an agonistic action (24,25), while it antagonizes the effects of a toxic dose of digitalis (26,27).

RELAXING EFFECT OF TAURINE ON VASOCONSTRICTION INDUCED BY 54 mM POTASSIUM

Central ear artery segments of rabbit are dissected, cannulated at both ends and placed in an organ bath. The apparatus is outlined in Fig. 4. The artery is perfused and superfused with Krebs solution, delivered at a constant rate of 5 ml/min. After an hour of stabilization, the artery is perfused and superfused with Krebs solution containing 54 mM potassium; the osmolarity of the medium is maintained with a parallel decrease in NaCl. Drugs are administered intraluminally. The potassium-induced contraction has two components (Fig. 5): phasic and tonic, the tonic component remains stable for at least 120 min. The action of taurine is studied on the tonic component, which is decreased by about 50% by verapamil (25 μg/ml) and $NaNO_2$ (30 mM).

Taurine has a relaxing effect (Fig. 6) upon depolarization-induced contraction at all concentrations tested. This effect is dose-dependent in a linear manner and is selective, β-alanine and homotaurine not exerting any vasodilatory action (Table 2).

The effect of taurine is quite rapid, reaching the maximum after about 2 min and remaining constant for at least an hour. When the taurine is removed, the tonic contraction component goes back to the

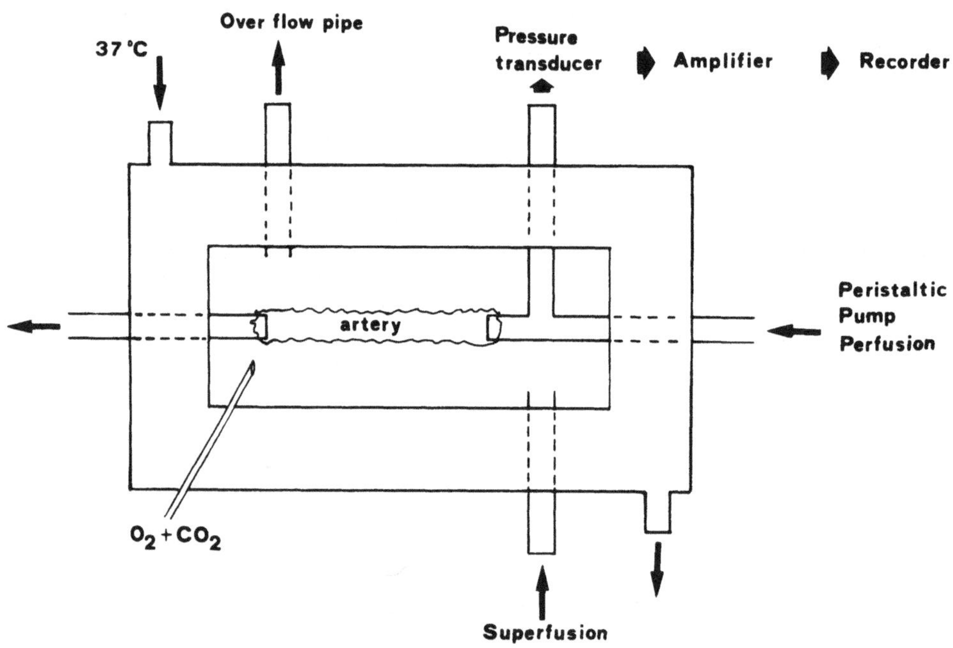

Fig. 4. Diagram of apparatus for ear artery perfusion.

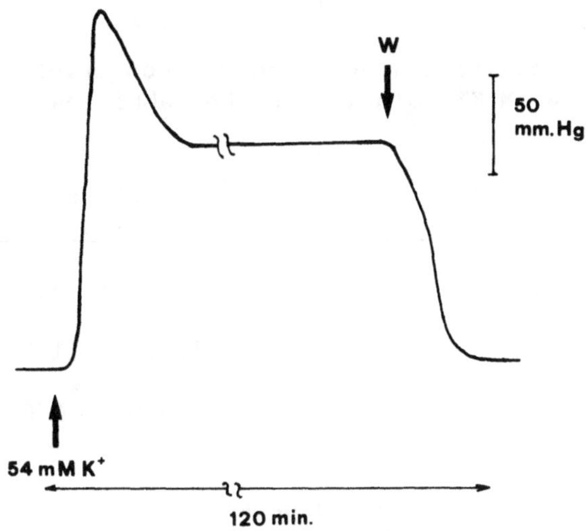

Fig. 5. Biphasic response induced by 54 mM KCl in the rabbit ear artery.

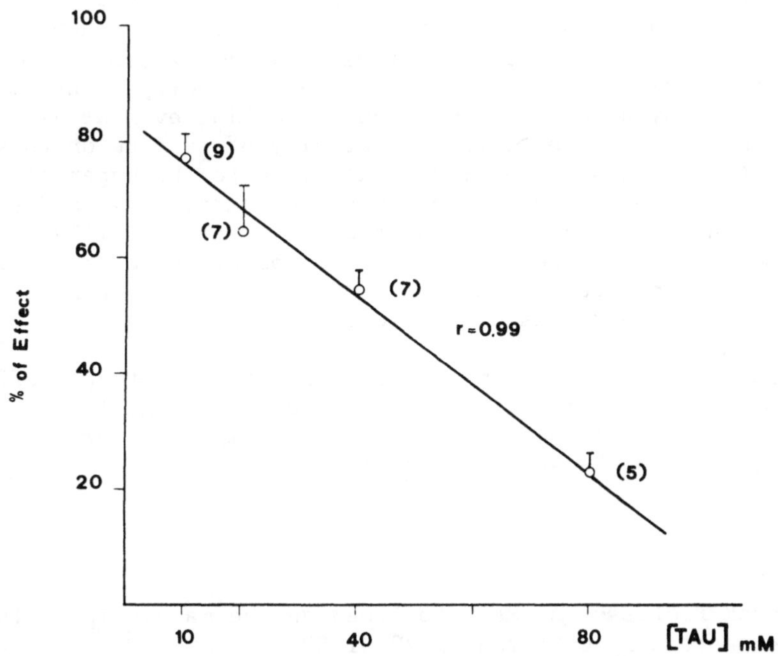

Fig. 6. Dose response relationship between taurine concentration and relaxing effect of taurine on vasoconstriction induced by 54 mM K^+ on the rabbit ear artery. Results are means ± SE; for the number of experiments indicated in parentheses.

Table 2

Effect of taurine and its analogues on vasoconstriction induced by perfusion with a 54 mM K^+ solution in the rabbit ear artery.

Treatment	Effect
none	100%
Taurine	54.1 ± 4.2 (7)°
β-alanine	117.5 ± 11.5 (4)
Homotaurine	100.0 ± 0.0 (3)

Results are mean ± SE, for the indicated number of experiments. Amino acids were used at 40 mM concentration. °$P<0.001$

value found before taurine administration. Preliminary data indicate that taurine does not influence norepinephrine-induced contraction. The different action of taurine on norepinephrine- and potassium-induced contraction might be explained by the fact that administration of a high potassium solution leads to a depolarization of the cell membrane, while the mechanical response to norepinephrine is not accompanied by a change in membrane potential even at high concentrations (28). The different activity of taurine on these two contraction models could also be ascribed to the hyperpolarizing activity of taurine. Taurine hyperpolarizes cardiac (4) and striated muscle (29) and it is possible that taurine exerts a similar action also in smooth muscle. We can not at the moment, however, exclude the possibility that taurine also interferes in vascular smooth muscle with the homeostasis of calcium.

In conclusion some taurine actions (positive inotropic effect) would seem to be calcium mediated whereas others (negative inotropic effect and relaxing effect) point to a more complex interaction with cation movement and the subcellular component.

ACKNOWLEDGMENT

We thank Professor A. Meli for reading the manuscript. This work was supported by CNR grant # 79.01950.

REFERENCES

1. Dietrich, J., and Diacono, J., 1971, Comparison between ouabain

and taurine on isolated rat and guinea-pig hearts in low calcium medium, Life Sci., 10:499-507.

2. Dolara, P., Agresti, A., Giotti, A., and Pasquini, G., 1973, Effect of taurine on calcium kinetics of guinea-pig heart, Eur. J. Pharmacol., 24:352-358.
3. Chubb, J., and Huxtable, R., Transport and biosynthesis of taurine in the stressed heart, in: "Taurine and Neurological Disorders," A. Barbeau and R. Huxtable, eds., Raven Press, New York (1978), pp. 161-178.
4. Dolara, P., Ledda, F., Mugelli, A., Mantelli, L., Zilletti, L., Franconi, F., and Giotti, A., Effect of taurine on calcium, inotropism and electrical activity in the heart, in: "Taurine and Neurological Disorders," A. Barbeau and R. Huxtable, eds., Raven Press, New York (1978), pp. 151-159.
5. Guidotti, A., and Giotti, A., 1970, Taurine antagonizes the negative effect of low calcium perfused hearts, Rec. Prog. Med. (Roma), 49:61-67.
6. Dolara, P., Agresti, A., Giotti, A., and Sorace, E., 1976, The effect of taurine on calcium exchange of sarcoplasmic reticulum of guinea-pig heart studied by means of dialysis kinetics, Can. J. Physiol. Pharmacol., 54:529-533.
7. Huxtable, R, and Bressler, R., 1973, Effect of taurine on a muscle intracellular membranes, Biochim. Biophys. Acta., 323: 573-583.
8. Chovan, J., Kulakowski, E.C., Benson, B.W., and Schaffer, S.W., 1979, Taurine enhancement of calcium binding to rat heart sarcolemma, Biochim. Biophys. Acta., 551:129-136.
9. Azari, J., and Huxtable, R.J., 1980, The mechanism of the adrenergic stimulation of taurine influx in the heart, Eur. J. Pharmacol., 61:217-223.
10. Perry, S.V., and Grand, R.J., 1979, Mechanism of contraction and specialized protein components of smooth muscle, Brit. Med. Bull., 35:219-226.
11. Franconi, F., Martini, F., Manghi, N., Galli, A., Bennardini, F., and Giotti, A., 1980, Uptake of ^{3}H taurine into myocardial membranes, Biochem. Pharmacol., in press.
12. Chovan, J.P., Kulakowski, E.C., Sheakowski, S., and Schaffer, S.W., 1980, Calcium regulation by the low affinity taurine binding sites of cardiac sarcolemma, Mol. Pharmacol. 17:295-300.
13. Kass, R.S., and Tsien, R.W., 1975, Multiple effects of calcium antagonist on plateau currents in cardiac Purkinje fibers, J. Gen. Physiol., 36:169-192.
14. Kohldart, M., and Mnich, Z., 1978, Studies on the inhibitory effect of verapamil and the slow inward current in mammalian ventricular myocardium, J. Mol. Cell. Cardiol., 10:1037-1052.
15. Fleckestein, A., Specific inhibitors and promotors of calcium action in the excitation-contraction coupling of heart muscle and their role in the prevention of production of myocardial lesions, in: "Calcium and Heart," L. Opie and P. Harris, eds., Academic Press, London, New York (1971), pp. 135-188.

16. Nayler, W.G., Mas-Oliva, J., and Williams, J.A., 1980, Cardiovascular receptors and calcium, Circ. Res. Supp. I, 46:161-166.
17. Nayler, W.G., and Szeto, J., 1972, Effects of verapamil on contractility, oxygen utilization and calcium exchangeability in mammalian heart muscle. Cardiovasc. Res., 6:120-128.
18. Dunnett, J., and Nayler, W.G., 1978, Calcium efflux from cardiac sarcoplasmic reticulum: effect of calcium and magnesium, J. Mol. Cell. Cardiol., 10:487-498.
19. Khatter, J.C., Soni, P.L., Hoeschen, R.J., Alto, L.E., and Dhalla, N.S., Subcellular effects of taurine on guinea-pig heart, in: "Actions of taurine on excitable tissues, S.W. Schaffer, J.J. Kocsis and S.I. Baskin, eds., Spectrum Press, New York (1981), in press.
20. Welty, J.D., and Welty, M.C., The effects of taurine on subcellular calcium dynamics in the normal and cardiomyopathic hamster heart, in: "Actions of taurine on excitable tissues, S.W. Schaffer, J.J. Kocsis and S.I. Baskin, eds., Spectrum Press, New York (1981), in press.
21. Mc Broom, M.J., and Welty, J.D., 1977, Effects of taurine on heart calcium in the cardiomyopathic hamster, J. Mol. Cell. Cardiol., 9:853-858.
22. Chovan, J., Kramer, J., and Schaffer, S., 1979, Protection by taurine of calcium paradox damage, J. Mol. Cell. Cardiol. II (SI):15.
23. Dolara, P., Marino, P., and Buffoni, F., 1973, The effect of taurine on calcium uptake and oxygen consumption of rat liver mitochondria, Biochem. Pharmacol., 22:2085-2094.
24. Guidotti, A., Badiani, G., and Giotti, A., 1971, Potentiation by taurine of inotropic effect of strophantin-K on guinea-pig isolated auricles, Pharmac. Res. Comm., 3:29-38.
25. Iwata, H., and Fujimoto, S., 1976, Potentiation by taurine of the inotropic effect of ouabain and the content of intracellular Ca^{2+} and taurine in the heart, Exp., 32:1559-1560.
26. Read, W.O., and Welty, J.D., 1963, Effect of taurine on digoxin and epinephrine irregularities of the dog heart, J. Pharmacol. Exp. Ther., 139:283-289.
27. Fujimoto, S., 1977, Inhibitory effect of taurine on decrease in the inotropic action of ouabain at high concentrations in isolated atria, 33:1350-1351.
28. Droogmans, G., Raeymaekers, and Casteels, R., 1977, Electro- and Pharmacomechanical coupling in the smooth muscle cells of the rabbit ear artery, J. Gen. Physiol., 70:129-148.
29. Gruener, R., Markovitz, D., Huxtable, R., and Bressler, R., 1975, Excitability modulation by taurine, J. Neurol. Sci., 24:351-360.

ELEVATED BLOOD TAURINE LEVELS AFTER MYOCARDIAL INFARCTION OR CARDIOVASCULAR SURGERY: IS THERE ANY SIGNIFICANCE?

M. Wayne Cooper and J. Barry Lombardini

Departments of Internal Medicine and Pharmacology and Therapeutics
Texas Tech University
Health Sciences Center
Lubbock, Texas

Though it is well known that cardiac tissue contains large quantities of taurine the function of taurine in the myocardium has not been defined. In rat cardiac tissue the content of taurine comprises over 50% of the total free amino acids (9) with a level of approximately 100 μmoles/g dry weight (12). Not only is cardiac tissue high in taurine content but there is also an unequal distribution throughout the myocardium. For example, it has been reported by both Kocsis and colleagues (10) and Crass and Lombardini (1) that atrial taurine content is significantly higher than ventricular content in the dog heart. Furthermore, both laboratories reported that a transmural gradient existed in the dog left ventricle with the highest taurine content being present in the subendocardium (1,10). While there are few reports of taurine content in human heart it appears that our species is no different from the rest of the animal kingdom in that we also harbor high levels of taurine in our cardiac tissue (7,8). Furthermore, the disparity in regional distribution of taurine in the human heart has been noted in our laboratory. However, the human heart differs from the dog heart in that the ventricular content of taurine is higher than the atrial content (unpublished observations).

Since cardiac tissue contains high levels of taurine we wondered whether myocardial damage in patients due to acute myocardial infarction or cardiovascular surgery would release taurine from cardiac tissue and subsequently be observed as elevated taurine concentrations in the blood. Three animal models suggested this possibility. First, it was reported by Crass and Lombardini (1) that dog cardiac muscle released significant quantities of taurine after acute left ventricular ischemia. In these experiments the circumflex branch of the left main coronary artery was ligated in a series of animals and tissue biopsies were obtained from four regions of the heart (right and left atria, right and left ventricles) four hours later. Taurine levels were determined in these regions and in transmural sections of both the right and left ventricles (Fig. 1).

The largest decrease in taurine content was in the immediate ischemic zone of the left ventricle. However, it is very interesting that atrial taurine content decreased following left ventricular ischemia. The right ventricle also revealed similar decreases in taurine content (data not shown in Fig. 1). No ready satisfactory reason can be offered in explanation of this decrease in taurine content in the non-ischemic regions.

In the second animal model for determining the potential significance of taurine release or leakage from cardiac tissue as an indicator of myocardial damage, rat hearts were perfused in oxygen-deficient media (2). In Fig. 2 it is observed that perfusion with either oxygen or nitrogen for the initial 5 min washout period had no effect on taurine content. However, 30 min of recirculation in buffer equilibrated with 95% nitrogen -5% CO_2 produced a 20% decrease

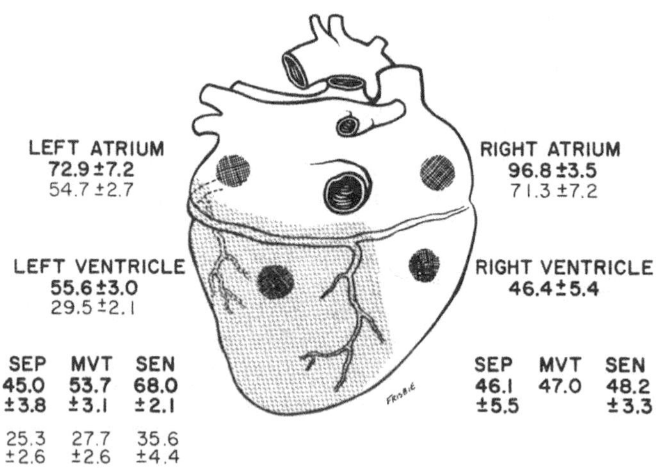

Fig. 1. Posterior surface of the canine heart. Schematic diagram of area of myocardial ischemia (hatched area) after ligation of circumflex branch of the left main coronary artery. Taurine levels (μmoles/g dry wt) of cardiac regions (cross-hatched circles) obtained from hearts which were non-ligated (controls) are shown in bold type; taurine levels from ischemic heart regions are shown in light type. SEP = subepicardium; MVT = midventricle; SEN = subendocardium. Each value represents the mean ± SE of samples from 4 or 5 animals. (Figure reproduced from Lombardini, 1980, ref. 13, with permission from Plenum Publishing Corporation; data originally from Crass and Lombardini, 1977).

Taurine Content in Rat Heart after Perfusion with or without Oxygen

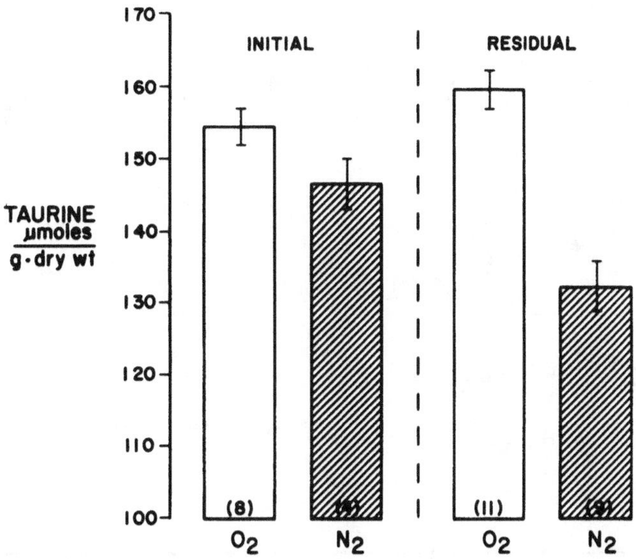

Fig. 2. Taurine content in rat heart after perfusion with or without oxygen. Hearts were perfused with Krebs-Henseleit bicarbonate buffer containing 5.5 mM glucose for 5 min ("initial") or for an additional 30 min ("residual"). Clear bars and hatched bars show values (μmole/g dry wt) for tissue taurine in hearts perfused with buffer equilibrated with 95% O_2-5% CO_2 and 95% N_2-5% CO_2, respectively. Each value represents the mean ± SE of the number of hearts indicated in parenthesis. (Figure reproduced from Crass and Lombardini, 1978, ref. 2, with permission from the Society for Experimental Biology and Medicine).

in taurine content ($p < 0.02$). All the taurine released from the heart was recovered in the perfusate.

The third animal model that established a relationship between cardiac taurine content and myocardial damage is the administration of high doses of sympathomimetic agents (12,13). It is well established that the subcutaneous administration of a single dose of isoproterenol (80 mg/kg) produces myocardial necrosis in rats (11, 15). The mechanism for the development of the necrotic lesions in the cardiac tissue are thought to be due to a lowering of coronary perfusion pressures (4,6,14).

The effects of DL-isoproterenol on levels of taurine in rat heart and blood as function of time are shown in Fig. 3. Isoproterenol reduced the ventricular tissue content of taurine by 5 hr and the taurine levels did not return until 48 hr. The pattern of

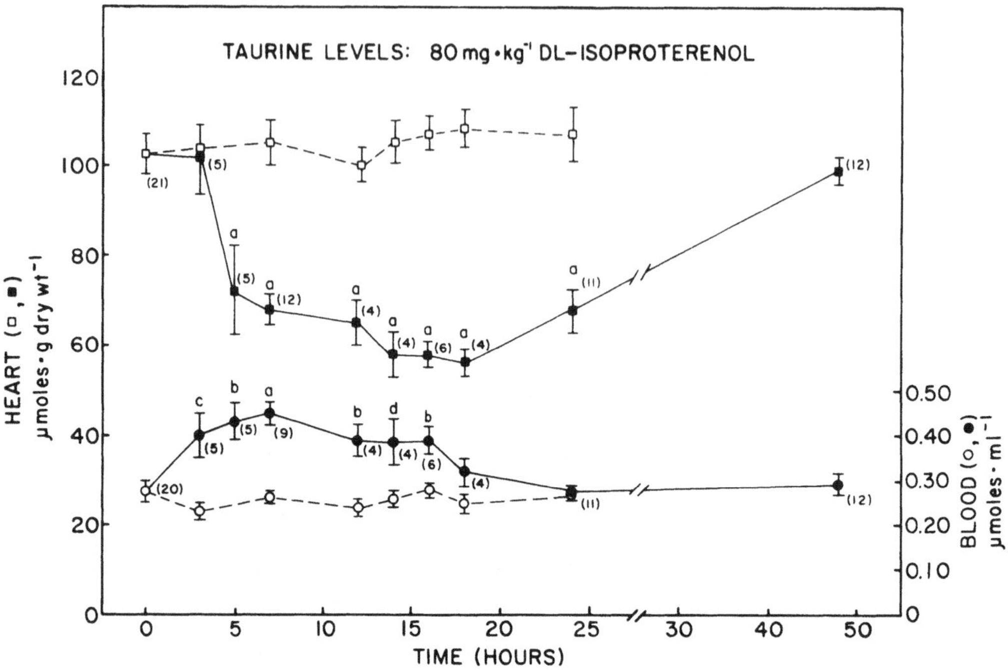

Fig. 3. Effects of DL-isoproterenol (subcutaneous administration) on levels of taurine in rat ventricles and blood. Animals were killed and tissues analyzed for taurine content at indicated times after administration of 80 mg/kg of DL-isoproterenol or saline solutions. Taurine levels are expressed as µmole/g dry wt for ventricular tissue and µmole/ml for blood. The values are reported as means ± SE. The values for P were calculated by the Wilcoxon 2-sample test (^{a}p <0.001; ^{b}p <0.005; ^{c}p <0.02; ^{d}p <0.05). Open symbols denote control animals (N=4); solid symbols denote treated animals. The number of animals per group is shown in parentheses. (Figure reproduced from Lombardini, 1980, ref. 12, with permission from the American Society for Pharmacology and Experimental Therapeutics).

taurine concentrations in the blood was inversely proportional to that in the cardiac tissue but with some minor differences; for example, blood taurine concentrations were elevated by 3 hr and had returned to control values by 18 hr.

The above information derived from animal models that taurine is indeed released from cardiac tissue and elevated in the blood after cardiac insult suggested that taurine concentrations might also rise in patients having acute myocardial infarctions (AMI) or undergoing cardiovascular surgery.

METHODS

A series of blood samples were obtained from patients admitted to the coronary care unit with complaint of chest pain and possible AMI. Patients were subsequently diagnosed as having had an AMI by standard criteria, i.e., EKG changes in the QRS complex, creatine kinase (CK) isoenzyme fraction B greater than 3% of the total CK, lactic dehydrogenase (LDH) band 1 higher than LDH band 2, and positive technetium-99M pyrophosphate myocardial scintigrams (PYP). A series of blood samples were also obtained from patients who entered the hospital for coronary artery bypass (CAB) surgery. Taurine concentrations were measured on either deproteinized (equal volume of 5% perchloric acid), centrifuged whole blood or plasma by amino acid analyzer.

The sensitivity and specificity of each diagnostic test for AMI were calculated by the following equations:

$$\text{Sensitivity} = \frac{\text{True positive X 100}}{\text{True positive + False Negative}}$$

$$\text{Specificity} = \frac{\text{True negative X 100}}{\text{True negative + False positive}}$$

RESULTS

1. Acute Myocardial Infarction

Concentrations of taurine were measured in a series of blood and plasma samples obtained from patients with a complaint of chest pain who entered the emergency room (ER) and then the coronary care unit (Table 1). Admittance to the ER was used as the zero time for determining the time sequence of all subsequent samples. No differences in blood concentrations of taurine were observed during hospitalization of patients who were diagnosed as not having an AMI. However, patients who had an AMI diagnosed by standard criteria had elevated concentrations of taurine by the time they presented at the ER. The taurine levels remained elevated for the next 50 to 60 hr and then returned to non-AMI values by 70 to 120 hr. There were no differences in the plasma levels of taurine between the AMI and non-AMI patients.

In the course of hospitalization the patients who did not have an AMI were classified according to their specific medical diagnoses (Table 2). Blood levels of taurine in these patients were compared to control concentrations of taurine obtained from medical school personnel. No differences were noted in patients who had chest pain of unknown etiology, or skeletal muscle trauma, G.I. pain, pulmonary pain, supraventricular/ventricular tachycardia plus cardioversion, pericardial disease or unstable angina. However, the five patients with congestive heart failure had elevated taurine levels when analyzed by Student's t-test (parametric criteria were satisfied).

Table 1

Time course of changes in taurine concentrations (μmoles/liter) in blood and plasma of patients entering the emergency room and coronary care unit with chest pain. Patients were subsequently diagnosed by standard criteria as AMI or non-AMI.

Time (hr)	Whole Blood Non-AMI	Whole Blood AMI	Plasma Non-AMI	Plasma AMI
ER(0)	222 ± 9 (24)	318 ± 44 (7)[a]	70 ± 8 (16)	88 ± 16 (4)
5-10	226 ± 9 (40)	340 ± 28 (13)[c]	72 ± 7 (27)	83 ± 15 (10)
15-20	222 ± 8 (40)	346 ± 26 (12)[c]	75 ± 9 (33)	69 ± 8 (6)
30-40	220 ± 8 (32)	287 ± 14 (11)[b]	63 ± 8 (25)	73 ± 15 (9)
50-60	206 ± 7 (28)	291 ± 15 (13)[c]	62 ± 6 (23)	77 ± 22 (10)
70-120	214 ± 11 (18)	251 ± 18 (10)	69 ± 9 (14)	63 ± 18 (7)

Data are represented as means ± SE. Number in parenthesis is number of samples in each group. Statistical differences of data groups within time periods were determined by the Wilcoxon 2-sample test ([a]$p<0.05$; [b]$p<0.01$; [c]$p<0.001$). No differences were observed for non-AMI data between time periods (Kruskal-Wallis test). (Unpublished data of Lombardini and Cooper).

On the contrary, the Wilcoxon 2-sample test for non-parametric data did not demonstrate any statistical differences.

The value of measuring taurine levels compared to known diagnostic parameters for AMI is presented in Fig. 4. No single standard diagnostic procedure was 100% reliable. For instance, serum from 8 out

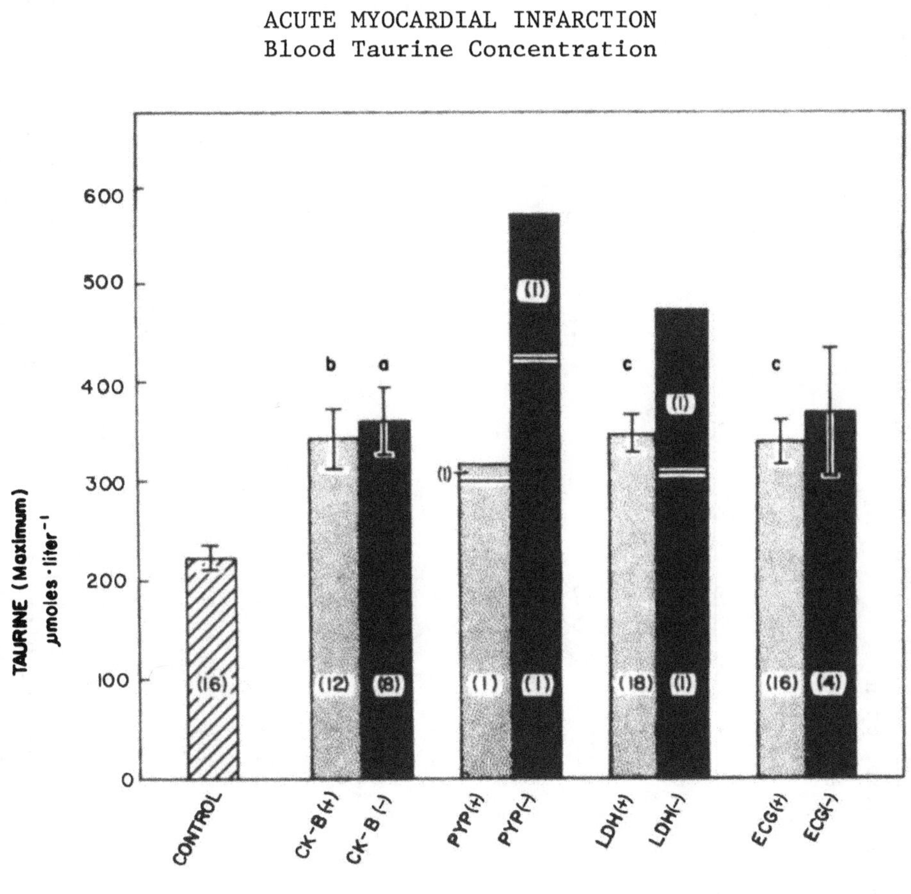

Fig. 4. Diagnostic value of measuring maximum taurine concentrations in blood compared to standard criteria of evaluating AMI. The twenty AMI patients were classified according to whether the laboratory results of CK-B isoenzyme activity (CK-B), pyrophosphate test (PYP), elevated isoenzyme band 1 of LDH (LDH), or electrocardiogram scan (ECG) were positive (+) or negative (-). Taurine content of blood samples are reported as means ± SE in µmole/liter. Value of control taurine concentrations were obtained from blood samples of medical school personnel. The number of patients per groups is shown in parentheses. Statistical differences as compared to control values were determined by the Wilcoxon 2-sample test ($^{a}p<0.01$; $^{b}p<0.01$; $^{c}p<0.001$). (Unpublished data of Lombardini and Cooper).

Table 2

Maximum concentrations of taurine in blood of Non-AMI patients who were admitted to the coronary care unit because of chest pain.

Diagnosis	μmole/liter	P value (non-parametric)	P value (parametric)
Control	226 ± 12 (16)	--	--
Chest pain of unknown etiology	231 ± 5 (28)	0.64	--
Skeletal muscle trauma	238 ± 13 (18)	0.62	--
G.I. pain	244 ± 36 (4)	0.27	--
Pulmonary pain	235 ± 8 (4)	0.75	--
SVT, VT + CV	241 ± 15 (9)	0.52	--
Pericardial Disease	253 (1)	--	--
Congestive heart	284 ± 30 (5)	0.13	0.05
Unstable angina	253 ± 28 (5)	0.68	--

Data are represented as means ± SE. Number of patients in group is shown in parentheses. SVT = supraventricular tachycardia; VT - ventricular tachycardia; CV = cardioversion. Non-parametric p value was obtained by analysis of variance. (Unpublished data of Lombardini and Cooper).

of 20 AMI patients did not demonstrate a positive CK-B isoenzyme fraction. (i.e. CK-B greater than 3% of total CK). However, taurine levels were elevated in both CK-B positive and CK-B negative groups when compared statistically to the control group. Similarly, reliance on elevation of LDH band 1, PYP, or ECG changes as solitary diagnostic procedures would have resulted in overlooking seriously ill patients. In contrast, taurine concentrations in blood were elevated in patients who had a negative PYP (424, 571 μmole/liter), in the two patients who did not have an elevated band 1 of LDH (308, 474 μmole/liter) and in 3 out of 4 patients who did not have ECG changes (411, 428, 465 μmole/liter). (Statistical evaluation of some of these data cannot be performed due to the small number of samples.)

2. Coronary Artery bypass surgery

The above observations that the concentrations of taurine are elevated in the blood of AMI patients suggested that such measurements might be utilized as an indicator of cardiac damage after coronary artery surgery. Thus, concentrations of taurine were measured in the blood and plasma of a series of 15 patients who were undergoing coronary artery bypass surgery (Table 3). Taurine levels increased immediately post-surgery (0-time) when compared to pre-surgery values and remained elevated for 48 hr reaching a maximum at 24 hr. No differences were noted in the plasma concentrations of taurine.

A representative sampling of the concentrations after CAB surgery of both the essential and non-essential amino acids is shown in Table 4. The levels of the essential aliphatic amino acids (isoleucine shown) in general decreased after surgery while phenylalanine increased. The concentrations of the non-essential amino acids remained unchanged.

DISCUSSION

It was previously observed in studies utilizing animal models for either ischemia or hypoxia that the taurine contents of cardiac tissue decreased (1,2) while blood concentrations increased in an inverse relationship (12). These observations prompted us to extend our investigations to clinical studies involving patients with an AMI or undergoing cardiac surgery. The rationale for these studies was that if mechanical or drug induced infarction could alter tissue levels of taurine then perhaps the same changes could be observed, with potential diagnostic value, in a pathological process such as AMI or in cardiac surgery.

1. Acute Myocardial Infarction

In the first studies concentrations of taurine in blood or plasma were obtained from a series of patients admitted to the coronary care

Table 3

Concentrations of taurine in blood and plasma of coronary artery bypass patients.

Time		Blood (µmole/liter)	Plasma
Pre-surgery		216 ± 13	65 ± 4
Post-surgery	0 hr	264 ± 15[a]	76 ± 6
	12 hr	362 ± 24[c]	74 ± 5
	24 hr	386 ± 23[c]	79 ± 7
	48 hr	289 ± 16[b]	59 ± 3

Fifteen patients are in each group. Data are represented as means ± SE. Taurine concentrations obtained at post-surgery time points are compared to pre-surgery values. (Arterial blood: pre-surgery, post-surgery 0, 12 and 24 hrs. Venous blood: post-surgery 48 hrs). Wilcoxon 2-sample tests were utilized to calculate p-values ([a]$p<0.02$; [b]$p<0.01$; [c]$p<0.001$). (Unpublished data of Lombardini and Bricker).

unit with a complaint of chest pain. Blood concentrations of taurine were elevated in the AMI patients by the time that they reached the ER compared to values obtained from non-AMI patients. Plasma levels were not elevated thus indicating that a component of the blood, most probably the platelets (5), was sequestering taurine.

Diagnostic Value of Taurine Measurements in Blood After Acute Myocardial Infarction

If a taurine concentration of 275 µmole/liter, which is one standard deviation above the control group (226 ± 49 µmole/liter), is assumed to be diagnostic of a pathologic condition then 15 of the 20 AMI patients fit this criteria. Two of the five patients missed had only a single blood taurine determination. Two had uncomplicated, presumably small myocardial infarctions. Only one of the twenty AMI patients had several blood taurine determinations below 275 µmole/liter, yet had other laboratory evidence of an extensive AMI.

Sensitivity and specificity of the various tests for AMI are shown in Table 5. By comparing peak blood taurine concentrations to other standard tests routinely used to diagnose acute myocardial infarction, we found the test to be of equal value to CK-B, better

Table 4

Concentrations of amino acids in blood of coronary artery bypass patients.

		Essential Amino Acids			Non-essential Amino Acids		
		µmole/liter					
		Ile	Phe	Lys	Ala	Asp	Ser
Pre-surgery		48 ± 3	47 ± 3	269 ± 15	410 ± 12	141 ± 10	147 ± 19
Post-surgery	0	22 ± 3^c	41 ± 2	204 ± 16	443 ± 21	127 ± 16	136 ± 15
	12	34 ± 2^c	64 ± 4^b	192 ± 20	438 ± 41	130 ± 16	125 ± 15
	24	48 ± 4^a	89 ± 5^c	251 ± 17	489 ± 49	130 ± 16	157 ± 33
	48	66 ± 5	88 ± 6^c	234 ± 26	411 ± 49	118 ± 17	159 ± 35

Data are represented as means ± SE from 11 patients. Arterial blood samples were obtained at the following time periods: Pre-surgery 0, 12 and 24 hr. Venous blood was obtained at the 48 hr time period. Statistical differences between pre-surgery and post-surgery time periods were obtained by the Wilcoxon 2-sample test ($^a p<0.05$; $^b p<0.01$; $^c p<0.001$). Unpublished data of Lombardini and Bricker).

Table 5

Sensitivity and Specificity of Various tests for Acute Myocardial Information

	ECG	CPK-MB	SGOT	LDH isoenzymes	Taurine µmole/liter >300	>290	>275
Sensitivity	80%	60%	95%	90%	65%	70%	75%
Specificity	97%	93%	62%	97%	91%	91%	86%

than SGOT but inferior to electrocardiogram and LDH isoenzyme activities. The sensitivity of taurine determinations was slightly better than CK-B and specificity was slightly less. SGOT activity, because of its widespread distribution in noncardiac tissue, was found to be the least specific of the blood tests. The isoenzyme pattern of LDH characteristic of an AMI was found to be the most sensitive and specific test. However, its usefulness, like SGOT, is limited by its widespread distribution. In addition, hemolytic anemia or renal infarction may give an LDH isoenzyme pattern similar to AMI.

If blood taurine concentrations of greater than 300 µmole/liter are used as the diagnostic cut-off point for AMI, 7 of the remaining 77 non-AMI patients could be classified as false positives. Four of these patients had a past history of cardiovascular disease and three of these a history of a previous AMI.

If blood taurine concentrations above 275 µmole/liter are used as the cut-off point for diagnosis of AMI, an additional 4 patients in the non-AMI category would be classified as false positives. One of these was admitted with atrial flutter with a rapid ventricular response and was cardioverted. Subsequently a modest rise in blood taurine concentration was noted (280 µmole/liter). One patient with chest pain of unknown etiology had a modest elevation of blood taurine concentration to 286 µmole/liter. Two of the five patients with G.I. pain had modest elevations of taurine levels to 286 and 289 µmole/liter. Both patients had acute cholecystitis and underwent cholecystectomy.

Total CK activity frequently rises after skeletal muscle trauma, prolonging hospitalization until CK-B is determined. Our demonstration that taurine levels fail to rise after skeletal muscle trauma suggests an advantage over measurement of the CK-B fraction. Failure of taurine concentrations in blood to rise in 3 of our 4 patients after cardioversion which is known to increase total CK levels but not CK-B activities is further confirmation that there was no myocardial injury as a result of the countershock.

From these data it is evident that AMI produces a significant increase in taurine concentrations in blood. However, the source of the elevated taurine levels can only be inferred from our previous work (1,2,12,13).

Clinical Implications

Although our series of AMI patients is relatively small and the number of patients with noncardiac chest pain is also small, the finding that taurine concentrations in whole blood are frequently elevated in patients with an AMI suggests that this relatively simple analysis may contribute significantly to the diagnosis of AMI. Further studies are warranted to prove this hypothesis.

2. Coronary artery bypass surgery

Concentrations of taurine increased in the blood but not in plasma of patients undergoing coronary artery bypass surgery. However, in subsequent studies similar results, i.e. elevated blood taurine levels, were obtained from patients undergoing repair of abdominal aneurysms and aorto-femoral bypass procedures (unpublished observations of Lombardini and Bricker). In these later studies there was no evidence that the patients had any myocardial necrosis and thus the source of the taurine is unknown.

One hypothesis is that muscle damage and/or hemolysis of platelets may be the source of the elevated concentrations of taurine in the blood of surgical patients. Alternatively, it may be postulated that stress modified cell membrane permeability thereby allowing taurine to leak out of the cell. It is well known that plasma catecholamines are markedly elevated during surgery and AMI (3) and thus tissue fluctuations of taurine may also be a function of elevated catecholamine content due to stress rather than overt tissue damage. In this context we have demonstrated that very high doses of epinephrine increased blood concentrations of taurine and decreased cardiac tissue levels of taurine in rats (Fig. 5).

Further study into the function of taurine in the heart is warranted and additional investigation of the possibility of using taurine as a marker of myocardial infarction appears justified.

ACKNOWLEDGMENTS

This study was supported in part by a grant from the Texas Affiliate of the American Heart Association. We acknowledge the skillful technical assistance of Mrs. S. Paulette Decker. Mr. Harvey O. Olney, Department of Biochemistry, is thanked for his help in

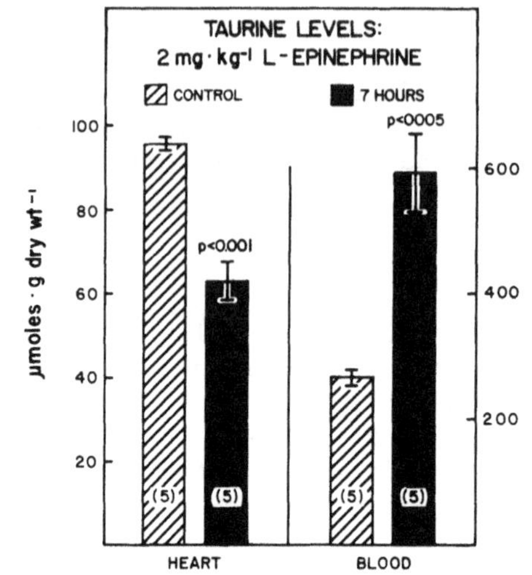

Fig. 5. Effects of L-epinephrine (subcutaneous: 2 mg/kg) on levels of taurine in rat ventricles and blood. Tissues were analyzed for taurine content seven hours after administration of L-epinephrine. Taurine levels are expressed as means ± SE in μmole/g dry wt for ventricular tissue and μmole/ml blood. The number of animals per group is shown in parentheses. The Wilcoxon 2-sample test was utilized to calculate significance (p<0.005) (Unpublished data of Lombardini).

performing the amino acid analyses. We express our appreciation to Mr. Raymond Lamont and Mr. Mark Hughes, Information Systems, for performing the statistical analyses. We are grateful to Dr. James A. Balch, Clinical Pathology Laboratory, TTUHSC for coordinating the laboratory staff who collected the blood samples from the AMI and non-AMI patients. The cooperation of the administration and clinical laboratory staff of St. Mary of the Plains Hospital & Rehabilitation Center is also acknowledged. The study involving the surgical patients could have been neither undertaken nor successfully completed without their assistance.

REFERENCES

1. Crass, M.F., III and Lombardini, J.B., 1977, Loss of cardiac muscle taurine after acute left ventricular ischemia, Life Sciences 21, 951-958.
2. Crass, M.F., III and Lombardini, J.B., 1978, Release of tissue taurine from the oxygen-deficient perfused rat heart, Proc. Soc. Exp. Biol. Med. 157, 486.
3. Cryer, P.E., 1980, Physiology and pathophysiology of the human sympathoadrenal neuroendocrine system, New Eng. J. Med. 303, 436-444.
4. Daniel, H.B., Bagwell, E.E. and Walton, R.P., 1967, Limitation

of myocardial function by reduced coronary blood flow during isoproterenol action, Circ. Res. 21, 85-98.

5. Gaut, Z.N. and Nauss, C.B., 1976, Uptake of taurine by human blood platelets: a possible model for brain, In Taurine, edited by Huxtable, R. and Barbeau, A., New York, Raven Press, pp 91-98.
6. Handforth, C.P., 1962, Isoproterenol-induced myocardial infarction in animals, Arch Pathol. 73, 161-165.
7. Huxtable, R. and Bressler, R, 1974, Elevation of taurine in human conjestive heart failure, Life Sciences 14, 1353-1359.
8. Huxtable, R. and Bressler, R., 1974, Taurine concentrations in congestive heart failure, Science 184, 1187-1188.
9. Jacobsen, J.G. and Smith, L.H., Jr., 1968, Biochemistry and physiology of taurine and taurine derivatives, Physiol. Reviews 48, 424-511.
10. Kocsis, J.J., Kostos, V.J. and Baskin, S.T., 1976, Taurine levels in the heart tissues of arious species, In Taurine, edited by Huxtable, R. and Barbeau, A., New York, Raven Press, pp 145-153.
11. Kutsuna, F., 1972, Electron microscopic studies on isoproterenol-induced myocardial lesions in rats, Jp. Heart J. 13, 168-175.
12. Lombardini, J.B., 1980, Effects of isoproterenol and methoxamine on the contents of taurine in rat tissues, J. Pharmacol. Exp. Ther. 213, 399-405.
13. Lombardini, J.B., 1980, Effects of ischemia on taurine levels, In Natural Sulfur Compounds, edited by Cavallini, D., Gaull, G.E. and Zappia, V., Plenum Press, New York, pp 295-306.
14. Rona, G., Bontet, M., Huttner, I. and Peters, K., 1973, Pathogenesis of isoproterenol-induced myocardial alterations: Functional and morphological correlates, In Recent Advances in Studies on Cardiac Structure and Metabolism, edited by Dhalla, N.S., vol. 3, University Park Press, Baltimore, pp 507-525.
15. Rona, G., Chappel, C.K., Balazs, T. and Gaudry, R, 1959, An infarct-like myocardial lesion and other toxic manifestations produced by isoproterenol in the rat, Arch. Pathol. 67, 443-455.

PART V

THE NEUROCHEMISTRY OF TAURINE

Ryan J. Huxtable

Department of Pharmacology
Health Sciences Center
University of Arizona
Tucson, Arizona

The progress in taurine research over the past decade is well illustrated by the following four chapters. All are concerned with the neurochemical relationships between taurine and its precursors, particularly cysteine sulfinic acid (CSA), and the relationship between the metabolism of CSA and glutamic acid. The dissection of these relationships is integral to our understanding of the central nervous system effects of taurine. Decarboxylases are importantly involved in the regulation of excitability in the brain, in that their substrates are neuroexcitatory amino acids while their products are neuroinhibitory amino acids. A major problem is the degree of identity between the metabolism of CSA and glutamic acid. In its most basic form, it poses the question: Are GAD and CSAD activities expressed by one or two enzymes?

Another major problem in the neurochemistry of taurine is the differentiation of its actions from those of GABA. GABA is accepted by many to be an inhibitory neurotransmitter. Are the central nervous system effects of taurine due to an action of taurine at GABA receptors, a modulation of GABA release or reuptake, neurotransmitter actions of taurine at a site separate from sites for GABA, or neuromodulatory actions of taurine not involving GABA?

In the first chapter, Iwata and Baba explore the neuroexcitatory properties of CSA, and suggest that the CSA: taurine ratio in the brain may be a modulator of brain excitability, and may be a determinant in the production of seizures. This is similar to a hypothesis I advanced in a recent review (1), and discuss elsewhere in this volume. In many ways, CSA fulfilled the requirements for an excitatory neurotransmitter. Its release is calcium modulated and potassium stimulated, it stimulates formation of cyclic AMP, and it is taken up by a transport system specific for acidic amino acids. Taurine protects against CSA-induced seizures, and taurine also antagonizes the CSA-stimulated formation of cyclic AMP.

Certain comparisons (and differences) may be drawn between the pair CSA: taurine and glutamate: GABA. Furthermore, is it of any significance or merely happenstance that excitatory amino acids produce inhibitory ones, and not the reverse? Does this mean that excitation by amino acids is the "normal", unregulated state, and that production of inhibitory amino acids serves a modulatory function?

The neurochemistry of taurine has been particularly well studied in the retina, although this area still needs intensive investigation. Kuriyama and his colleagues have compared the neurochemistry of taurine and GABA in the retina, and have studied the differential actions of light on these substances. In brief, they find that GABA levels and GABA metabolism alter rapidly in response to light and dark, whereas the corresponding changes in taurine occur only after prolonged light and dark exposure for periods of up to three weeks. These findings reinforce the feeling that these substances fulfill separate neurotransmitter and neuromodulator functions. Prolonged light exposure decreases taurine concentration in the retina, whereas prolonged dark exposure has the opposite effect. Coeval with this are corresponding changes in CSAD activity. GAD activity, on the other hand, increases after short light exposure. This is further evidence that CSAD and GAD activities are mediated by two distinct enzymes.

An indication of the increasing sophistication in this area is the use of biochemical markers of changes in physiological state. Van Gelder has investigated changes in amino acid levels and ratios as indices of cyto-architectural changes in the central nervous system. This approach is based on the localization or enrichment of certain amino acids in particular structures. Thus, changes in these amino acids, not only in absolute levels but in relationship to other amino acids, can indicate microscopic structural changes in the brain, such as increased gliosis at an epileptogenic focus.

Both Van Gelder and Rassin find strong correlation between glutamate and taurine content in the brain. These correlations have now been observed in a number of species, and have been interpreted as indicating a regulation of glutamate metabolism by taurine. Some caution may be needed here, however. Correlations between taurine and glutamate may indicate some overlap in physical location without necessarily implying that the one substance is directly affecting or modifying the concentration of the other. Furthermore, such correlations must be considered in context. For example, we find it to be the rule rather than the exception that amino acid pairs

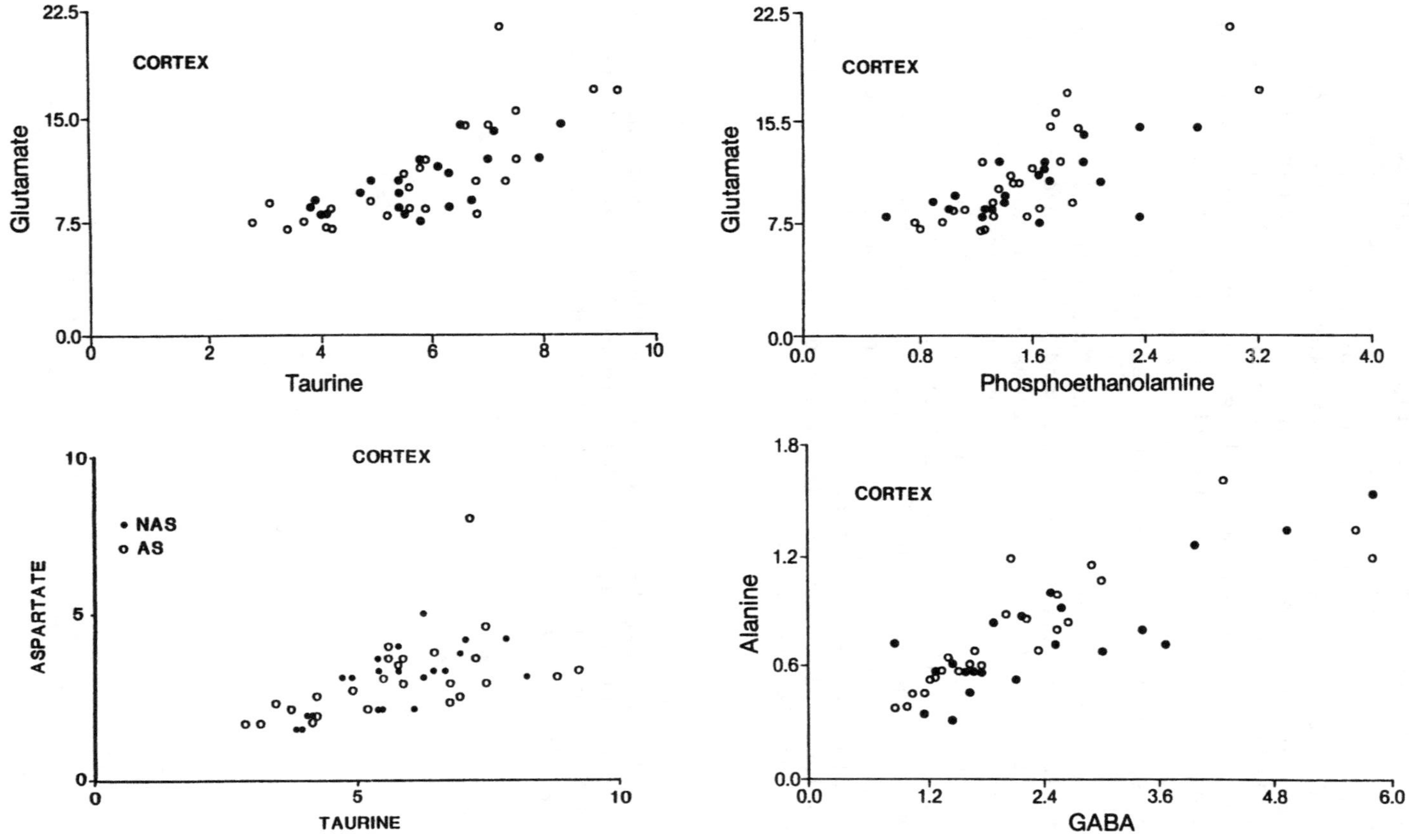

Fig. 1. Examples of amino acid correlations in the cortex. Filled symbols are seizure-resistant and open symbols are seizure-resistant brains. Correlation coefficients for resistant and susceptible animals respectively are: glutamate-taurine, .69,).76; glutamatephosphoethanolamine 0.66, 0.83 $p<.001$; aspartate-taurine 0.67, 0.56[a]; alanine-GABA 0.85, 0.84.

are strongly correlated in the brain. Fig. 1, for example, shows the correlation between four such pairs in rat cortex. Glutamate and taurine are strongly correlated, but glutamate and phosphoethanolamine show a slightly stronger correlation. Furthermore, taurine is also strongly correlated with aspartate. As might be expected, metabolically related substances such as glycine and serine are also strongly correlated. My feeling is that these relationships may be of "diagnostic" value in unraveling cyto-architectural changes in various states, but that they do not necessarily indicate a metabolic or regulatory relationship between the two substances.

In the final chapter of this section, Rassin incisively reviews the relationship of taurine to glutamate and GABA. Once again, the relationship between CSAD and GAD activities is addressed. He points out the lack of information concerning concentrations of CSA, CA and other sulfur metabolites in the brain. From the little information available, it appears that the concentrations of these substances are far below the concentration of glutamate. Therefore, even if one form of GAD could use CSAD as a substrate, it would be unlikely to be of much biochemical significance as the enzyme is exposed to CSA concentrations well below the Km, and glutamate would effectively compete with these levels of CSA.

REFERENCE

1. Huxtable, R.J., Insights on function: metabolism and pharmacology of taurine in the brain, in: "The Role of Peptides and Amino Acids as Neurotransmitters," J.B. Lombardini and A. Kenney, eds., J.R. Liss, New York (1981).

INTERACTION OF TAURINE WITH ITS PRECURSOR, CYSTEINE SULFINIC ACID, IN THE CENTRAL NERVOUS SYSTEM

Heitaroh Iwata and Akemichi Baba

Department of Pharmacology
Faculty of Pharmaceutical Sciences
133-1 Yamada-kami
Suita, Osaka, Japan

Among the amino acid neurotransmitter candidates, emphasis has been recently focused on the role of taurine in the central nervous system (CNS). It is thought that taurine may function as an inhibitory neurotransmitter or neuromodulator in the CNS.

Cysteine sulfinic acid (CSA) is an intermediate in the biosynthesis of taurine from cysteine. Curtis and Watkins (5) showed a strong excitatory action of CSA in the spinal neuron. Moreover, CSA stimulates the formation of cyclic AMP in guinea pig cerebral cortex *in vitro* (7). However, its possible physiological function and even its actual existence in the CNS had long been uncertain. Recently, we demonstrated the existence and the distribution of CSA in rat brain using a specific microassay for this substance (3).

If CSA is excitatory and taurine is inhibitory in the CNS, the ratio of the concentrations of CSA and taurine, or a possible interaction between CSA and taurine, should be considered in certain neuronal functions, such as the development of seizures. We report here the neurochemical characterization of CSA and possible functional interactions between CSA and taurine.

METHODS

Male Sprague-Dawley rats, weighing about 200 g, were used for the uptake, release, binding and EEG studies. The effect of CSA on cyclic AMP level was determined in the slices of guinea pig brain. Mice were used for a behavioral study.

Synaptosomal fraction (P_2) and purified synaptosomes were obtained from rat cerebral cortex by the method of Cotman (4). Slices (250 μm) of cerebral cortex, hippocampus and striatum were prepared by using a mechanical tissue chopper. The Krebs-Ringer bicarbonate buffer (KR) contained 138 mM NaCl, 5.6 mM KCl, 1 mM $MgCl_2$, 1 mM $CaCl_2$, 11 mM $NaHCO_3$, 1 mM NaH_2PO_4, 20 mM HEPES buffer (pH 7.4) and 10 mM glucose and was gassed with 95% O_2-5% CO_2.

Uptake of ^{14}C-CSA

The P_2 suspension in KR was preincubated at 37° for 30 min. After centrifugation at 10,000 × g for 10 min, the pellets were suspended in new KR. After 5 min of preincubation, the reaction was started by adding 200 μl of P_2 into 0.8 ml of KR medium and carried out for 5 min. Unless otherwise indicated, KR medium contained 10 μM CSA and 0.02 μCi/tube of ^{14}C-CSA. The uptake was stopped by rapid filtration through Whatmann GF/C filter as described previously (1).

Release of ^{14}C-CSA

Brain slices, suspended in KR, were preincubated for 20 min at 37°. After the medium was changed to new KR (containing 10 μM ^{14}C-CSA), the suspensions were incubated for 10 min. Then, the slices were washed several times, and moved to superfusion chambers (volume 0.4 ml) connected to a peristaltic pump (Minipuls 2, Gilson). Superfusion was performed with a constant flow rate of 2.0 ml/min at 37° under gentle shaking. Eluate (1.0 ml each) was collected in a fraction collector and its radioactivity was determined.

Binding of ^{35}S-Cysteic Acid

Crude synaptic membranes were obtained by the method of Simon et al. (8). ^{35}S-L-Cysteic acid (17.5 Ci/mmole) was synthesized from ^{35}S-L-cystine and purified by high-voltage paper electrophoresis. After membranes were frozen for 16 hr, the binding of ^{35}S-cysteic acid was examined at 37° as originally described for the binding of kainic acid (8).

Effect of CSA on Cyclic AMP Levels in the Slices

Slices from cerebral cortex, hippocampus and striatum of guinea pig were prepared as described for rat brain. The slices were preincubated in modified KR at 37° for 60 min. Modified KR contained 122 mM NaCl, 3 mM KCl, 1.2 mM $MgSO_4$, 1.3 mM $CaCl_2$, 0.4 mM KH_2PO_4, 25 mM $NaHCO_3$ and 10 mM glucose. After another preincubation for 10 min, test agents were added to the medium. The incubation was carried out for 10 min at 37° and stopped by placing the tubes in an ice-cold bath. Slices were collected by centrifugation and homogenized with 0.5 M perchloric acid. After neutralization of the extract, cyclic AMP content was determined as described previously (2).

EEG Recordings

Surgical procedures were carried out under ether anesthesia. Animals were fixed in a stereotaxic apparatus, immobilized with d-tubocurarine chloride, and artificially ventilated with room air at a rate of 90 strokes/min. Respiratory volume was adjusted properly. Surface activities were recorded from the left motor cortex and the right limbic cortex by bipolar silver ball electrodes, 0.5 mm apart, according to the map of Terzuolo and Adey (9). A stainless steel needle electrode (external diameter 0.3 mm), insulated except at the tip, was implanted into the left hippocampus according to the atlas of König and Klippel (6): A 1.2, L 4.2, H 1.0. A guide cannula (external diameter 0.6 mm) for intracerebroventricular (icv) injection of amino acid was implanted into the left posterior horn of the lateral ventricle: A 3.3, L 4.5, H -1.0. A reference electrode was inserted into a neck muscle. Amino acid solution (10 μl) was infused into the brain over a 10 sec period through the guide cannula.

RESULTS

Distribution of CSA in Rat Brain

CSA content in various regions of rat brain was determined by a specific microassay (3). The principle of this assay system is illustrated in Fig. 1. The concentration of CSA was high in the cerebral cortex and cerebellum and lower in the midbrain and olfactory bulbs. In the subcellular distribution of CSA in a whole brain, about 90% of the CSA was found in the supernatant fraction, and about 10% was detected in the P_2 fraction.

Uptake of CSA into the P2 fraction

Uptake of CSA into the P_2 fraction of rat cerebral cortex showed two systems: one is high affinity (Km 12 μM) and the other is low affinity (Km 155 μM). High affinity uptake of CSA was completely dependent on Na^+. On further subfractionation of the P_2 fraction, synaptosomes but not mitochondria had Na^+-dependent high affinity uptake of CSA. Aspartate, glutamate and cysteic acid competitively inhibit the uptake of CSA.

Release of CSA from the Slices

As shown in Fig. 2, change of the superfusion medium (KR) to high K^+-KR was followed by a sharp increase in the release of CSA. In Ca^{++}-free EGTA-containing KR, high K^+-induced release of CSA was markedly reduced. In these experiments, apparent reuptake of CSA into the slices was negligible. Similar results were obtained in the P_2 fraction of cerebral cortex and hippocampal slices. In hippocampal slices, Ca^{++}-dependency of the release was higher than in cerebral cortical slices. Among the various centrally acting reagents, diazepam markedly inhibits high K^+-induced release of CSA from hippocampal slices at concentration of 5×10^{-6} to 10^{-4} M. These doses of diazepam also inhibit high K^+-induced release of glutamate but not of norepinephrine, dopamine, acetylcholine and GABA.

Binding Site for CSA

The binding of ^{35}S-cysteic acid to crude synaptic membranes showed two different systems: one is Na^+-dependent and the other is Na^+-independent. Na^+-independent specific binding of ^{35}S-cysteic acid is saturable with a Kd of 358 nM and Bmax of 2.88 pmole/mg protein. The IC_{50} (nM) of L-CSA, L-glutamate, L-cysteic acid, L-aspartate and kainate for Na^+-independent specific binding of ^{35}S-cysteic acid were 9.6, 107.4, 704.3, 912 and more than 20,000, respectively.

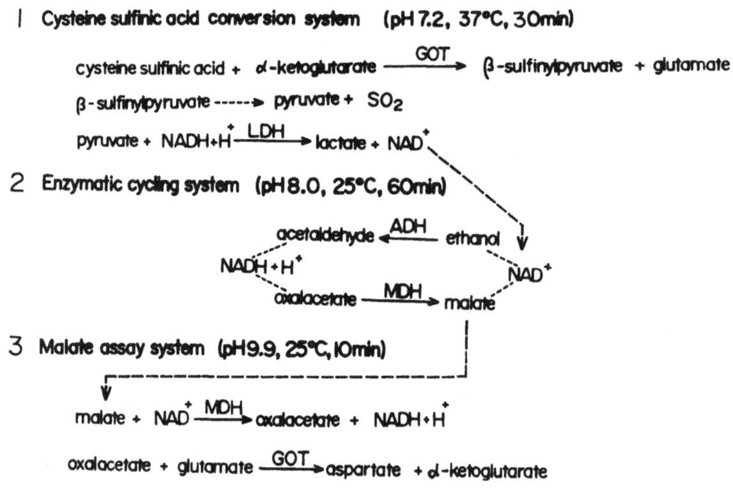

Fig. 1. Principle of cysteine sulfinic acid assay system.

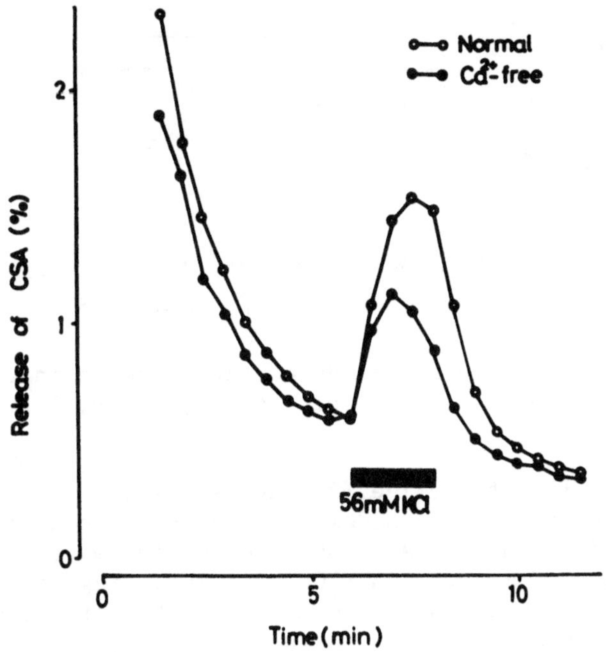

Fig. 2. Release of ^{14}C-cysteine sulfinic acid from preloaded cerebral cortical slices by a superfusion method.

Effect on Cyclic AMP System

As described previously (7), CSA markedly stimulates the formation of cyclic AMP in the brain slices. In the present study, hippocampus showed higher sensitivity to CSA than cortex and striatum: CSA at concentrations above 0.1 mM markedly stimulates cyclic AMP formation in the hippocampal slices. Addition of taurine (0.1 to 10 mM) caused 13 to 40% inhibition of CSA-induced elevation of cyclic AMP in hippocampal slices. On the other hand, taurine, up to 30 mM, did not affect the CSA effect in cortical slices. In addition, taurine (0.1 to 30 mM) did not inhibit glutamate- and aspartate-induced elevation of cyclic AMP in the slices.

Effect on EEG and Behavior

The effects of cysteine, CSA, cysteic acid, hypotaurine and taurine were examined in rats at a dose of 100 μg. Whereas cysteine, hypotaurine and taurine had no effect on the spontaneous EEG patterns, CSA and cysteic acid produced EEG seizures. These seizures were initiated in the hippocampus and propagated to the cortex (Fig. 3).

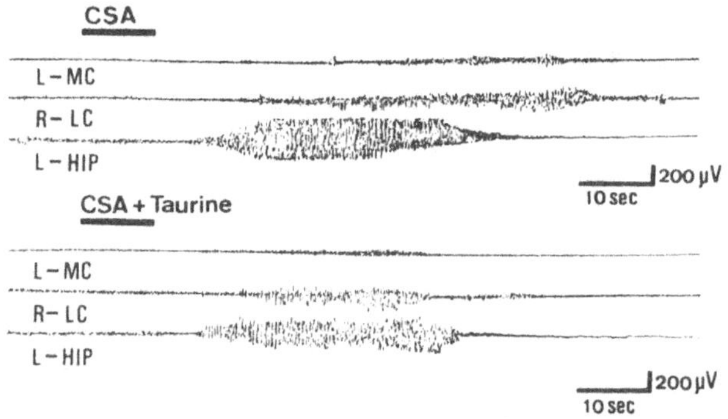

Fig. 3. Cysteine sulfinic acid-induced EEG seizures in rats. L-MC; left motor cortex, R-LC; right limbic cortex, L-HIP; left hippocampus.

Simultaneous addition of taurine (200 μg) significantly reduced the duration of seizures in motor and limbic cortex without affecting the occurrence of the seizures. The same results were obtained in a behavioral study of mice. CSA (100 μg/animal, icv) induced severe clonic and tonic convulsions within a few minutes of injection which continued for about 5 to 10 sec. Simultaneous injection of taurine by an intraventricular route markedly reduced the CSA-induced convulsions (Table 1).

DISCUSSION

In the present study, we have examined the neurochemistry of CSA in the CNS. Three findings should be mentioned: 1) Transport of CSA into synaptosomal fraction was via a Na^+-dependent high affinity system. 2) CSA was released by high K^+ in a partly Ca^{++}-dependent manner. 3) Na^+-independent specific binding sites for CSA were found in the synaptic membranes.

As is the case with many other substances, CSA was transported into synaptosomes by a Na^+-dependent high affinity system (Km 12 μM). Competitive inhibition of the uptake of CSA by glutamate and aspartate indicates that these three amino acids may be transported into synaptosomes by the same carrier system. Release of CSA from preloaded slices was immediately enhanced by the addition of high K^+ in a partly Ca^{++}-dependent manner. This result was the same as observed with other amino acid neurotransmitter candidates. It is interesting that diazepam inhibits K^+-induced release of CSA and glutamate, but not of norepinephrine, dopamine, GABA, or acetylcholine at low concentrations in the hippocampal slices. At present, it is not clear how these results correlate with the pharmacological

Table 1. Effect of taurine on CSA-induced behavioral changes of mice

	CSA[a]	CSA[a] + Taurine[b]	CSA[a] + Taurine[b]
No. of animals	45	19	20
Increase of touch response	45	19	0
Circling movement	6	2	0
Convulsion clonic	29	12	0
tonic	5	5	0
Mortality	6/45	4/19	0/20

[a] CSA (100 μg/animal) was injected intracerebroventricularly.

[b] Taurine (50 μg/animal) was injected simultaneously with CSA.

[c] Taurine (75 μg/animal) was injected simultaneously with CSA.

actions of diazepam. However, the specificity of this effect of diazepam may reflect functional roles of CSA or glutamate.

In the studies of uptake and release, we could not find any specificity for CSA; glutamate and aspartate sharing the same transport mechanisms. Therefore it is hard to characterize the neurochemical significance of CSA from these studies. However, the binding study of cysteic acid clarified the specificity of CSA. Synaptic membranes have a Na^+-independent specific binding site for cysteic acid. This binding site seems to be different from that for glutamate, since glutamate has one-tenth the affinity of CSA for this binding site and freezing the membranes does not reduce the binding capacity. CSA has the lowest IC_{50}, suggesting CSA to be an endogenous ligand for this binding site.

A previous study (7) indicates that CSA, glutamate and aspartate elevate cyclic AMP in brain slices by the same mechanisms. However, this is not the case, because taurine specifically antagonized the CSA-induced elevation of cyclic AMP. Furthermore, it should be mentioned that stimulation of cyclic AMP by CSA was highest in the hippocampus and that taurine inhibited the CSA effect in the hippocampus but not in the cortex. The regional specificity of the effects of CSA and taurine suggest functionality. These sulfur-containing amino acids may function in the hippocampus.

In a separate experiment, we found that intracerebroventricular injection of CSA evokes EEG seizures and convulsions. Simultaneous addition of taurine antagonized the seizures induced by CSA.

Our study indicates that CSA has various neurochemical aspects which have been observed in other neurotransmitters. Expecially, the presence of Na^+-independent specific binding sites for CSA in the synaptic membranes strongly suggest a functional role, since none of the other sulfur-containing amino acids have Na^+-independent specific binding sites. Specific interactions between CSA and taurine, which were observed for CSA-induced seizures and elevation of cyclic AMP, indicate the significant roles of these amino acids in neuronal function.

REFERENCES

1. Baba, A. and Cooper, J.R., 1980, The action of black widow spider venom on cholinergic mechanisms in synaptosomes, J. Neurochem. 34, 1369-1379.
2. Baba, A., Lee, E., Matsuda, T., Kihara, T. and Iwata, H., 1978, Reversible inhibition of adenylate cyclase activity of rat brain caudate nucleus by oxidized glutathione, Biochem. Biophys. Res. Comm. 85, 1204-1210.
3. Baba, A., Yamagami, S., Mizuo, H. and Iwata, H., 1980, Microassay of cysteine sulfinic acid by an enzymatic cycling method, Anal. Biochem. 101, 288-293.
4. Cotman, C.W., 1974, Isolation of synaptosomal and synaptic membrane fractions, Method in Enzymol. Vol. 31, p.445-452, Academic Press, New York.
5. Curtis, D.R. and Watkins, J.C., 1960, The excitation and depression of spinal neurones by structually related amino acids, J. Neurochem. 6, 117-141.
6. König, J.F.R. and Klippel, A.R., 1963, The rat brain: A stereotaxic atlas of the forebrain and lower parts of the brain stem, William and Wilkins Co., Baltimore.
7. Shimizu, H. and Yamamura , Y., 1977, Effects of diaminopropionate, deoxyadenosine, and theophylline on stimulated formation of cyclic AMP and GMP by depolarizing agents in slices of guinea pig cerebral cortex, J. Neurochem. 28, 383-388.
8. Simon, J.R., Contrena, J.F. and Kuhar, M.J., 1976, Binding of (^3H) kainic acid, an analogue of L-glutamate, to brain membranes, J. Neurochem. 26, 141-147.

9. Terzuolo, C.A. and Adey, W.R., 1960, Sensorimotor cortical activities, Handbook of Physiology, Sec. I: Neurophysiology, Vol. II, p.797, Amer. Physiol. Soc., Washington, D.C.

DIFFERENTIAL EFFECTS OF LIGHT AND DARK ADAPTATIONS ON FUNCTION AND METABOLISM OF RETINAL TAURINE AND γ-AMINOBUTYRIC ACID (GABA)

Kinya Kuriyama, Shuji Ida and Chihiro Nishimura

Department of Pharmacology
Kyoto Prefectural University of Medicine
Kawaramachi-Hirokoji
Kamikyo-Ku, Kyoto 602
Japan

It has been well established that taurine (2-aminoethanesulfonic acid) is present at a high level in the vertebrate retina, especially in photoreceptor cells (9). In addition, evidence suggesting possible roles of taurine as a neuromodulator or neurotransmitter in the retina, such as the presence of Na^{+}-dependent high affinity uptake of taurine (10), release of taurine by light stimulation (20), and suppressive effect of exogenously applied taurine on the b wave of the electroretinogram (ERG) (1), has been reported. On the other hand, it has been found that the highest levels of γ-aminobutyric acid (GABA) and l-glutamate decarboxylase (GAD) activity are present in the inner plexiform layer of frog retina (7). Moreover, various pieces of evidence indicating GABA may act as an inhibitory neurotransmitter in vertebrate retina, such as specific Na^{+}- and energy-dependent high affinity uptake (5), inhibition of the b wave of the ERG (22), and binding of (^{3}H)-muscimol (a GABA agonist) to GABA receptor (21) have been presented.

As exogenously applied taurine and GABA depress the b wave of the ERG, the b wave originates from bipolar cells, and the bipolar cell layer of retina is a major site of visual adaptation (3), it becomes important to determine whether or not the function and metabolism of taurine and GABA in the retina are altered following light and dark adaptations. It has been reported that the GABA contents decrease in dark adapted (for 2 hr) frog (6) and goldfish (14) retinae. Although it is known that taurine content in the dark-adapted retina of 6 days old chicken shows an increase (19), little is known on the alterations of function and metabolism of taurine and GABA in prolonged light and dark adapted retinae.

The present study aims to determine the differential effects of short and prolonged light or dark adaptations on the metabolism of taurine and GABA in the frog retina, and to elucidate possible physiological roles of these compounds therein.

MATERIALS AND METHODS

Frogs (Rana catesbiana) were used throughout this study. For studies on short light and dark adaptations, frogs were dark-adapted overnight (12 hr) prior to all experiments (Dark 1). To reach a light-adapted state, frogs were exposed to light for 1 hour under normal overhead fluorescent lights (Light). Subsequently, the frogs were returned to darkness for 4 (Dark 2) or 8 hr (Dark 3). For prolonged light and dark adaptations, frogs were divided into 3 groups: (1) light-adapted; (2) dark-adapted; (3) normal dark: light cycle. Each group was maintained in such a condition for 18 hr, and 1,2,3, and 6 weeks, respectively.

For the dark-adapted group, dissection and weighing of dissected retina were carried out with the aid of a safelight, but otherwise whole procedures were performed in complete dark. Protein content in the retina was assayed by the method of Lowry et al. (16).

Measurement of uptakes of (^{3}H)-taurine and (^{3}H)-GABA by retina

Following the dissection of retina, each retina was preincubated for 15 min at 25°C in 10 ml of Krebs-Ringer bicarbonate buffer (pH 7.4, KRB), gassed with 95% O_2 - 5% CO_2 gas mixture, to remove endogenous taurine. After preincubation the retinal tissues were centrifuged at 500 rpm for 1 min, and the sedimented tissues were further incubated for 20 min at 25°C with 2 ml of the KRB containing 12.5 μM (^{3}H)-taurine to measure the uptake. At the end of incubation, 10 ml of ice-cold KRB was quickly added and centrifuged at 500 rpm for 1 min. After washing twice using the same procedure, the washed tissues were placed into a scintillation counting vial containing 0.3 ml hyamine hydroxide and left several hours to dissolve the tissues. The radioactivity was measured by a Packard 3255 liquid scintillation spectrometer. The uptake of (^{3}H)-taurine observed in the presence of non-radioactive 10 mM taurine was considered to be due to non specific binding to the tissues, and subtracted from each experimental value to obtain the amount of specifically bound (^{3}H)-taurine. The measurement of (^{3}H)-GABA uptake was employed in the same manner as that of (^{3}H)-taurine, except 10^{-4}M of aminoxyacetic acid (AOAA) was added to the incubation medium for inhibiting the metabolism of GABA by retinal GABA-transaminase-succinic semialdehyde dehydrogenase (GABA-T:SSADH).

Determination of Release of (^{3}H)-taurine and (^{3}H)-GABA

After dissection of retina and placement of them in a small glass container with nylon mesh at the bottom, the container was preincubated for 15 min in a small glass vessel containing 10 ml of the KRB at 25°C. The preincubated retina was then transferred to KRB containing 12.5 μM of (^{3}H)-taurine and further incubated for 20 min to preload the tissues with (^{3}H)-taurine. After being washed twice with the KRB, the tissues were transferred at 4 minute intervals to a vial containing 2 ml of the KRB. For the application of high potassium stimulation, the tissues were put into a vial containing 60 mM of KCl (54.8 mM of NaCl was substracted from the KRB) at 20 to 24 min after the initiation of incubation, and subsequently returned to normal KRB. The spontaneous release was also measured with normal concentration of KCl in the medium. The radioactivity in each vial was counted to determine the release of (^{3}H)-taurine. At the end of the release measurements, radioactivities remaining in the tissues were also determined after solubilizing with hyamine hydroxide. The rate of release was expressed as percentage release and was calculated by the following equation:

$$\% \ (^3\text{H})\text{-Taurine released} = 100 \times \frac{\text{Radioactivity released for 4 min}}{\text{Summed radioactivity released + Residual radioactivity in the tissue.}}$$

For the measurement of (^{3}H)-GABA release, experimental procedures identical with those of (^{3}H)-taurine release were employed except 20 mM KCl was used for measuring the evoked release of (^{3}H)-GABA.

Composition of Reaction Media

The incubation medium (KRB) used for experiments measuring the uptake and release contained NaCl, 105 mM; KCl, 4.2 mM; $CaCl_2$, 2.2 mM; KH_2PO_4, 1.0 mM; $MgSO_4$, 1.0 mM; $NaHCO_3$, 22 mM; glucose, 5.0 mM. The medium was gassed with 95% O_2 - 5% CO_2 gas mixture and pH of the solution was adjusted to 7.4. The incubation medium depleted of sodium ions was prepared by the replacement of various proportions of sodium chloride with equimolar concentrations of lithium chloride, and 0.05 M-Tris-HCl buffer was substituted for $NaHCO_3$.

Chemical and Enzymatic Assays

Taurine content was measured by a column chromatographic method (25). GABA content was determined according to the enzymatic method (25) or enzymatic cycling procedures (25). The activity of CSAD, the enzyme involved in the synthesis of taurine, was determined by measuring the formation of $^{14}CO_2$ from DL-(1-^{14}C)-cysteine sulfinic acid (25). The GAD activity was measured by the radiometric method of Kimura and Kuriyama (11). The activity of GABA-T:SSADH was assayed by the measurement of the fluorescent substance NADH produced by the coupled reaction with this enzymatic system (25).

Measurement of Specific Bindings of (^{3}H)-taurine and (^{3}H)-muscimol

The specific binding of (^{3}H)-taurine was determined by a modification of the procedures of López-Colomé and Pasantes-Morales (15). Dissected retina was homogenized with 20 vol of distilled water. The homogenate was centrifuged for 15 min at 45,000 x g. The resulting sediment was washed once with water by resuspending and recentrifuging the pellet, and then resuspended in Krebs-Tris buffer (pH 7.4) containing NaCl: 105 mM, KCl: 4.2 mM, $CaCl_2$: 2.2 mM, KH_2PO_4: 1.0 mM, $MgSO_4$: 1.0 mM and Tris-HCl: 26.0 mM. The membrane suspension (0.4 - 0.6 mg protein) was incubated at 25°C for 5 min with Krebs-Tris buffer containing 10 nM (^{3}H)-taurine. The binding was terminated by centrifuging at 45000 x g for 15 min. The supernatant was decanted and the pellet was superficially rinsed twice with Krebs-Tris buffer, and then the pellet was solubilized with hyamine hydroxide to measure the bound radioactivity.

The specific binding of (^{3}H)-muscimol was determined by a modification of the procedures of Enna and Snyder (4). Briefly, dissected retina was homogenized using a Polytron (setting 6) for 30 sec, with 20 vol of distilled water, and centrifuged at 45,000 x g for 10 min. The pellet was resuspended in the same volume of H_2O and recentrifuged at 45,000 x g for 10 min.

After storing for 18 hr at -20°C, the pellet was resuspended with 100 vol of 0.05 M Tris-citrate buffer (pH 7.1, 4°C) and incubated for 30 min at 25°C in the presence of 0.05% Triton X-100. The suspension was then centrifuged at 45,000 x g for 10 min and the resultant pellet was diluted by Tris-citrate buffer to a final protein concentration of 2.0 - 3.0 mg/ml. For the binding assay, the tissue samples representing 200 - 300 μg protein/ml Tris-citrate buffer were incubated for 30 min at 4°C with 2 nM (^{3}H)-muscimol in the presence or absence of 1 mM GABA. For termination of the reaction, 2 ml of ice-cold buffer was added to the reaction mixture and subsequently filtered through a membrane filter (TM-2, Toyo Roshi) under vacuum. After being washed with 6 ml of Tris-citrate buffer, the filter was transferred into a counting vial containing 10 ml of Triton-toluene, and the radioactivity was measured by a liquid scintillation spectrometer.

Chemicals Used

DL-(1 - ^{14}C) Glutamic acid (51.02 mCi/mmole, New England

Nuclear), DL-(1 - ^{14}C) cysteine sulfinic acid (11 mCi/mmole, Centre D'Etudes Nucleaires de Saclay), (1,2 - ^{3}H) taurine (17.6 Ci/mmole, The Radiochemical Centre), (2,3 - ^{3}H) aminobutyric acid (28.2 Ci/mmole, New England Nuclear) and (^{3}H)-muscimol (methylene-^{3}H(N)-3-hydroxy-5-aminomethylizoxazole, 10.3 Ci/mmole, New England Nuclear) were used. Other chemicals used were reagent grade and obtained locally from commercial sources.

RESULTS

Effects of Light and Dark Adaptations on Retinal Taurine and GABA Contents

Although short light and dark adaptations had no significant effect on taurine content in the retina (Table 1), short light exposure (L) induced a significant increase in retinal GABA content (Fig. 1). This rapid increase in retinal GABA was reversed (D_1) within 4 (D_2) to 8 (D_3) hr if animals were maintained again under complete darkness. On the other hand, in the 3 weeks dark-adapted retina, a statistically significant increase was observed in taurine levels as compared with controls maintained in the routine dark: light cycle ($P < 0.05$). Furthermore a statistically significant decrease of taurine (16%) was observed in the group continuously

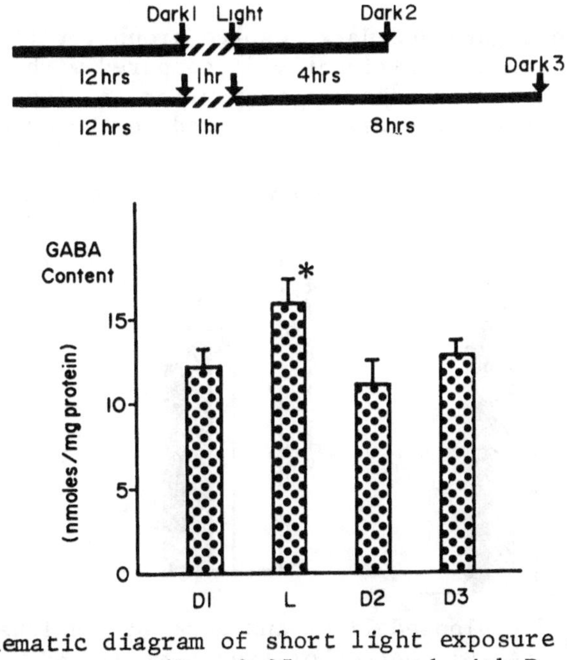

Fig. 1. Schematic diagram of short light exposure and change in retinal GABA content. *$P < 0.05$, compared with D_1.

exposed to light (Fig. 2). Similar alterations were also obtained when the results were expressed as nmole/g wet weight. As shown in Fig. 3, alterations of retinal taurine following prolonged light and dark adaptations were found at 3 and 6 weeks, but not in experimental groups having light or dark adaptation for 1 and 2 weeks. Under the same experimental conditions, no significant change in retinal GABA was detected (Fig. 2). These results clearly indicate, contrary to the results obtained in short light and dark adapted states, that only prolonged adaptations induce the changes in retinal taurine but not in retinal GABA, furthermore, the present results also indicate that these induced changes in retinal taurine only occur after 3 weeks on none of continuous light or dark.

Effect of Light and Dark Adaptations on Retinal Activities of CSD, GAD and GABA-T:SSADH

After short light exposure, it was found that the retinal CSAD activity, an enzyme involved in the biosynthetic pathway of taurine, remained essentially unchanged in each state (Table 1). These results are in essential agreement with the lack of changes in retinal taurine observed in these animals. On the other hand, retinal GAD activity increased significantly only in the Light when compared with Dark 1 (Fig. 4). Furthermore, a significant increase of GABA-T:SSADH activity was also found in the Dark 2 and Dark 3 (Fig. 4), when the light-exposure induced increase in retinal GABA has been ceased.

In prolonged light- and dark-adapted groups, retinal CSAD activity was unaltered significantly as compared with control group until 2 weeks. At three weeks and thereafter, a significant increase of retinal CSAD activity, however, occurred in the dark, whereas the

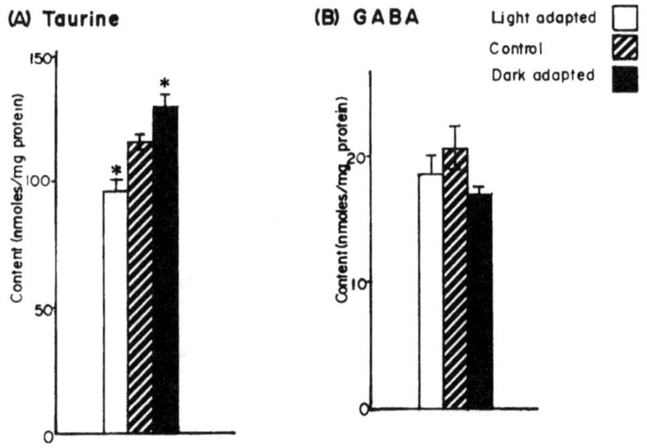

Fig. 2. Effect of prolonged light and dark adaptations (3 weeks) on taurine (A) and GABA (B) content of frog retina. *P < 0.05, compared with (A) taurine control.

activity in the light was reduced significantly (Figs. 5 & 6). On the other hand, the activity of GAD, the rate limiting enzyme of GABA biosynthesis, was not altered in the 3 weeks-adapted group (Fig. 5). Similarly, retinal GABA-T:SSADH activity was not altered significantly following prolonged light and dark adaptations. These results indicate again that prolonged light and dark adaptations selectively facilitate the retinal metabolism of taurine, but no such a change occurs in the case of metabolism of GABA.

Effect of Light and Dark Adaptations on Retinal Uptakes of (^{3}H)-taurine and (^{3}H)-GABA

Frog retinae incubated at 25°C in the presence of (^{3}H)-taurine or (^{3}H)-GABA accumulated radioactivity at a constant rate for 30 min. The rates of the accumulation, at longer times decreased gradually and reached a steady state level at 60 min. A linear relationship between tissue weight used and the uptake of (^{3}H)-taurine or (^{3}H)-GABA was observed. In addition it was found that the uptakes of both (^{3}H)-taurine and (^{3}H)-GABA were temperature and sodium dependent. In the complete absence of added sodium, the uptakes of (^{3}H)-taurine and (^{3}H)-GABA were reduced to less than 20% of those in the normal media. The Lineweaver-Burk plots of the uptakes of (^{3}H)-taurine and (^{3}H)-GABA in frog retina also revealed the presence of high and low affinity uptake systems (Fig. 7). On the basis of these preliminary experiments, subsequent experiments using light- and dark-adapted retinae were carried out in the presence of 12.5 μM (^{3}H)-taurine or (^{3}H)-GABA. No alteration of the uptakes of (^{3}H)-taurine and (^{3}H)-GABA, however, was found in frog retina following short as well as prolonged light and dark adaptations (Table 2 and Fig. 8).

Effect of Light and Dark Adaptations on Releases of (^{3}H)-taurine and (^{3}H)-GABA

The spontaneous releases of (^{3}H)-taurine and (^{3}H)-GABA from the preloaded retinal tissues were composed of the rapid and slow components. In the short light-adapted retina, the high K^+ (20 mM)-evoked release of (^{3}H)-GABA showed a significant increase as compared with the Dark 1 (Table 2). No significant alterations were found, however, in the releases of (^{3}H)-taurine and (^{3}H)-GABA from prolonged light-adapted and dark-adapted retinae. Similarly, no alteration of the retinal release of (^{3}H)-taurine was noted following short light and dark adaptations. These results indicate that an increase of the release from preloaded retinal tissues occurs only in the case of (^{3}H)-GABA, especially during the application of short light exposure.

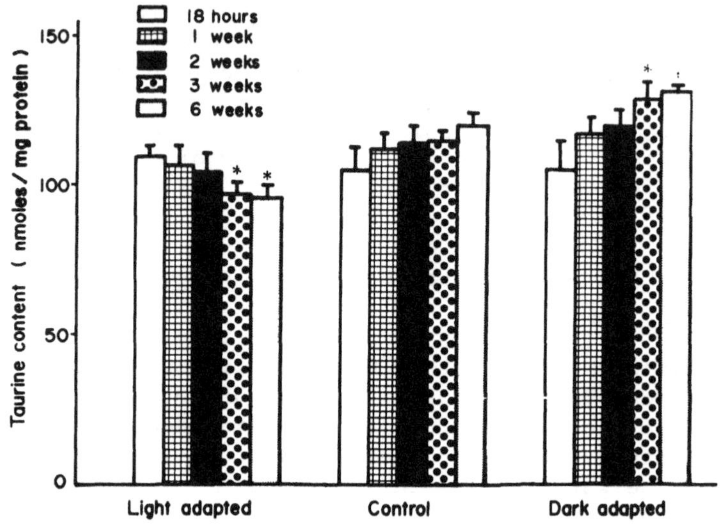

Fig. 3. Effect of prolonged light or dark adaptation on retinal taurine content. Each value represents the mean ± SEM obtained from 6 - 8 separate experiments. *P < 0.05, compared with each control.

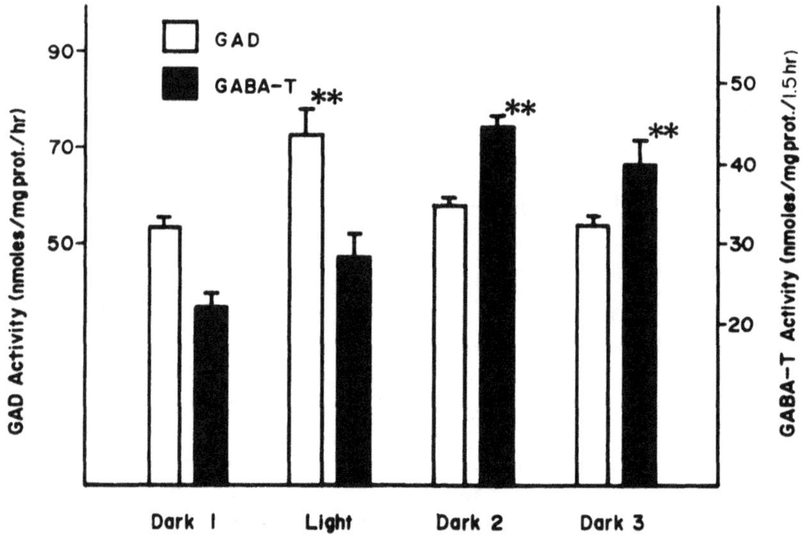

Fig. 4. Effect of short light exposure on retinal GAD and GABA-T activities. Each dissected retina was immediately homogenized with 20 vol of distilled water under either dark or light conditions as appropriate. Each value represents the mean ± SEM obtained from 4 - 6 separate experiments. **P < 0.01, compared with each value in Dark 1.

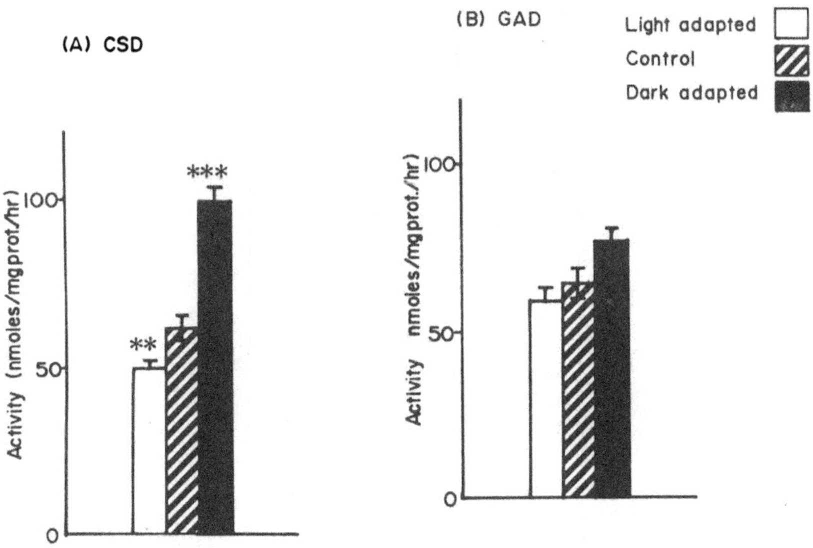

Fig. 5. Effect of prolonged light and dark adaptations (3 weeks) on retinal CSAD (A) and GAD (B) activities. Dissected retina was immediately homogenized with 20 vol of distilled water under either dark or light condition as appropriate. $**P < 0.01$, $***P < 0.001$, compared with each control group (CSAD). Each value represents the mean ± SEM obtained 6 - 8 separate experiments.

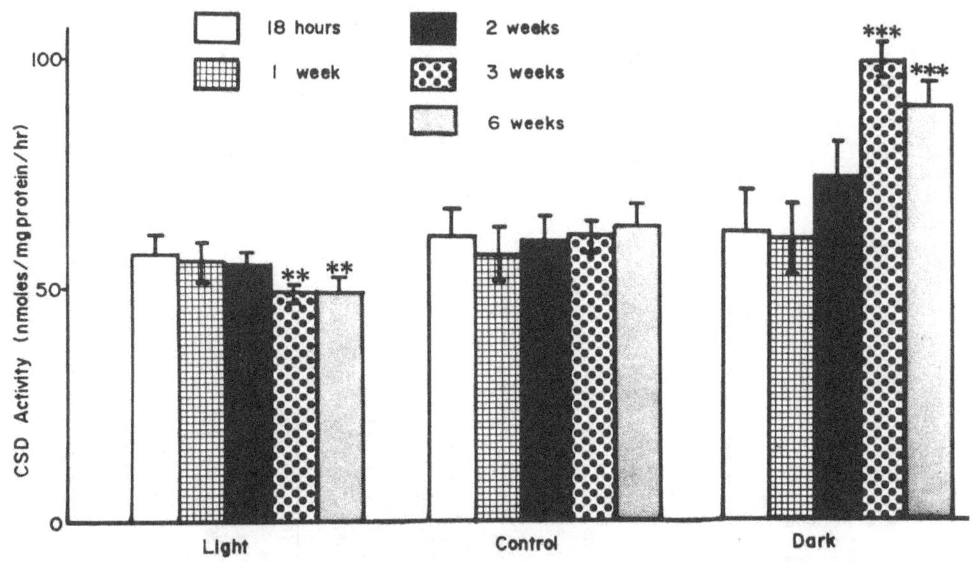

Fig. 6. Effect of duration of light and dark adaptations on retinal CSAD activity. The mean ± SEM obtained from 6 - 8 separate experiments is shown. $**P < 0.01$, $***P < 0.001$, compared with each control.

Table 1

Effect of Short Light Exposure on Retinal Taurine Content and CSAD Activity

	Dark 1	Light	Dark 2	Dark 3
Taurine (nmole/mg protein)	116.7 ± 3.6	111.7 ± 4.5	114.3 ± 4.4	117.9 ± 3.2
CSAD Activity (nmole/mg prot./hr)	56.8 ± 4.5	58.2 ± 3.1	55.1 ± 6.2	56.3 ± 5.2

Each value represents the mean ± SEM obtained from 3 - 6 separate experiments.

Table 2

Effect of Short Light Exposure on (^{3}H)-GABA Uptake and High K^+-evoked (^{3}H)-GABA Release in Frog Retina

	(^{3}H)-GABA Uptake (nmoles/g wet wt/min)	(^{3}H) GABA Release (% Increase by 20 mM KCl)
Dark 1 (12 hr)	1.32 ± 0.12	123.8 ± 5.7
Light (1 hr)	1.20 ± 0.15	184.9 ± 12.4**
Dark 2 (4 hr)	1.20 ± 0.13	141.0 ± 29.1
Dark 3 (8 hr)	1.15 ± 0.18	125.1 ± 10.9

Each dissected retina was incubated at 25°C for 20 min with Krebs-Ringer bicarbonate buffer containing 12 μM of (^{3}H)-GABA to measure the uptake. For the measurements of uptake and release of (^{3}H)-GABA, 0.1 mM aminoxyacetic acid (AOAA) was added for inhibiting GABA-T activity in the tissues.

Each value represents the mean ± SEM of 4 - 8 separate experiments. **$P < 0.01$, compared with Dark 1.

Table 3

Effect of Prolonged Light and Dark Adaptations on Specific Bindings of (^{3}H)-taurine and (^{3}H)-muscimol to Retinal Membrane.

Duration of Adaptation	(^{3}H)-Taurine			(^{3}H)-Muscimol		
	Specific binding (fmole/mg protein)					
	Light	Control	Dark	Light	Control	Dark
18 hours	19.6 ± 2.2	22.7 ± 3.0	22.6 ± 1.3	180.6 ± 15.8	138.6 ± 16.9	157.3 ± 8.1
3 weeks	20.4 ± 1.1	20.8 ± 1.3	21.5 ± 1.5	100.4 ± 5.8	121.0 ± 7.3	118.4 ± 8.7

Final concentrations of (^{3}H)-taurine and (^{3}H)-muscimol in the incubation medium were 10 and 2 nM, respectively.

Each value represents the mean ± SEM obtained from 6 separate experiments.

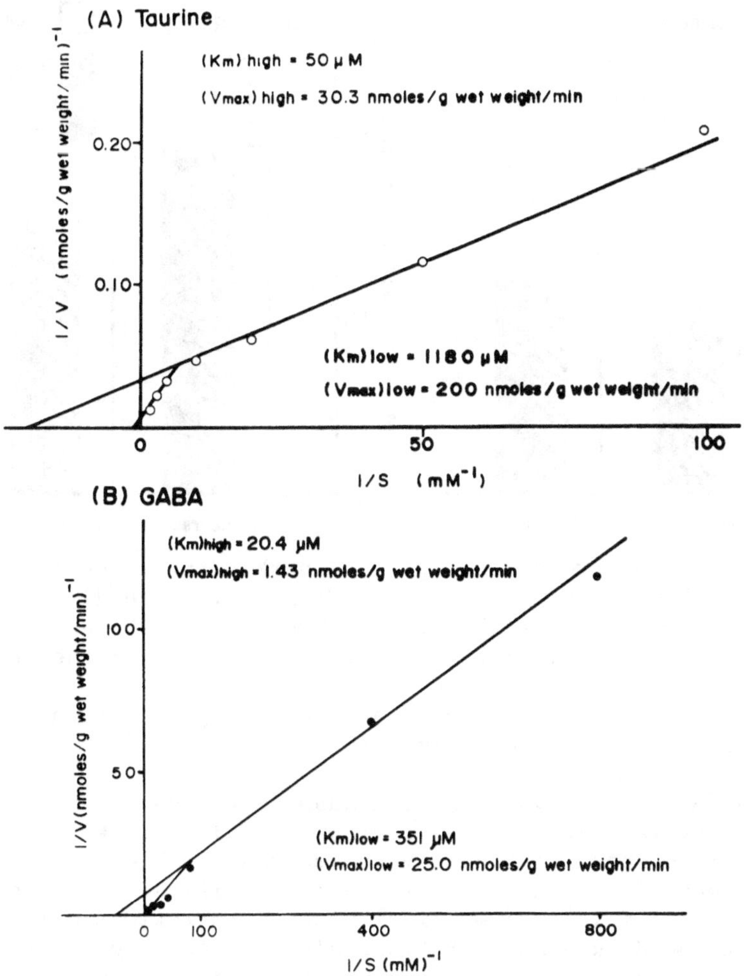

Fig. 7. Lineweaver-Burk plots for uptakes of taurine (A) and GABA (B). V = rate of (^{3}H)-taurine or (^{3}H)-GABA uptake (nmoles/g wet weight/min). S = (^{3}H)-taurine or (^{3}H)-GABA concentration (mM). Each value represents averages of 3 separate experiments.

Effect of Light and Dark Adaptations on Specific Bindings of (^{3}H)-taurine and (^{3}H)-muscimol to Retinal Membranes

The specific binding of (^{3}H)-taurine to retinal membrane was unchanged in both light and dark adapted groups for 18 hours and 3 weeks, respectively (Table 3). Similarly no alteration was found in the specific (^{3}H)-taurine binding to retinal membranes following short light and dark adaptations. As to the specific binding of (^{3}H)-muscimol to retinal GABA receptors, the amounts of specific

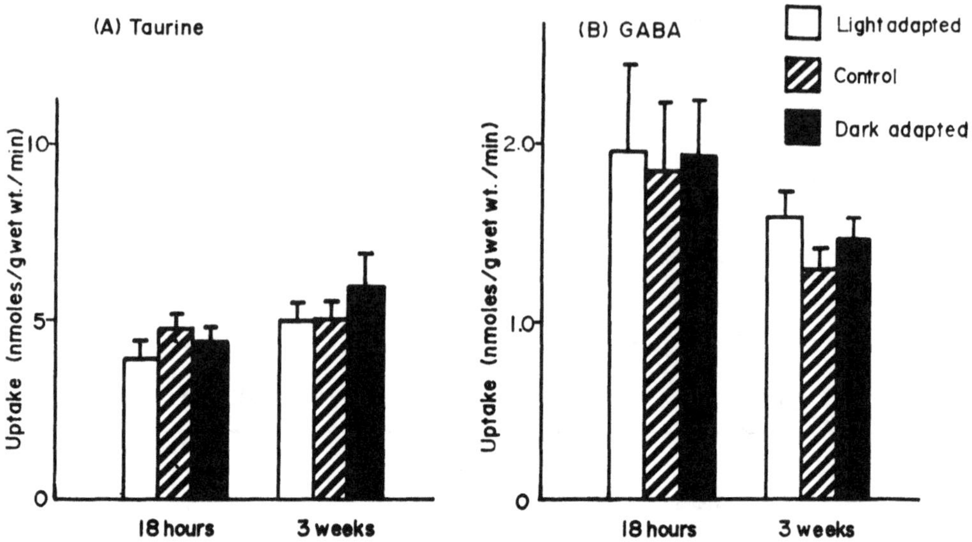

Fig. 8. Effect of prolonged light and dark adaptations on retinal uptakes of (^{3}H)-taurine (A) and (^{3}H)-GABA (B). Each dissected retina was incubated at 25°C for 20 min with Krebs-Ringer bicarbonate buffer containing 12.5 μM of (^{3}H)-taurine or (^{3}H)-GABA. For the measurements of uptake of (^{3}H)-GABA, 0.1 mM of AOAA was added for inhibiting GABA-T activity in the tissues. Each value represents the mean ± SEM obtained from 4 separate experiments.

binding in the 18 hr light and dark adapted retinae tended to increase compared with control, whereas that in 3 weeks light adapted animals showed a tendency to decrease (Table 3). These changes, however, were not statistically significant. As shown in Table 4, the specific (^{3}H)-muscimol binding to retinal membranes from short light and dark adapted retinae also showed a tendency of decrease, but these changes were again not statistically significant. These results suggest that the specific bindings of both (^{3}H)-taurine and (^{3}H)-GABA to retinal membrane, possibly to each receptor site respectively, may not be readily altered under light and dark adapted conditions.

DISCUSSION

It has been reported that taurine is present in a large quantity in vertebrate retina, especially at the inner segment of photoreceptor cells, outer nuclear layer and outer plexiform layer (18). Although the origin of taurine in these different retinal layers has not been fully elucidated, it is generally considered that the level of retinal taurine may be regulated by both biosynthesis and plasma-retinal exchange (13).

In the vertebrate retina the presence of cysteine oxidase and

Table 4

Effect of short light exposure on specific (^{3}H)-muscimol binding to retinal membrane.

	Dark 1 (12 hr)	Light (1 hr)	Dark 3 (8 hr)
(^{3}H)-Muscimol Binding (fmole/mg protein)	107.8 ± 9.1	89.0 ± 5.9	79.6 ± 10.7

Final concentration of (^{3}H)-muscimol in the incubation medium was 1.2 nM. Each value represents the mean ± SEM obtained from 6 separate experiments.

CSAD, enzymes involved in the biosynthesis of taurine from cysteine, has been demonstrated. Characteristics of retinal cysteine oxidase considered to be similar to those in the brain (2). Furthermore, studies on the distribution of retinal CSAD activity indicated that approximately 30 and 65% of the activity was localized in the inner segment of photoreceptor cells and inner nuclear and plexiform layers, respectively (17). The results obtained in this study using frog retina also indicated that a relatively high CSAD activity, which altered significantly following prolonged light and dark adaptations, was present in the frog retina.

Considering these results it is most reasonable to assume that a major pathway of taurine biosynthesis in retinal tissues may involve the enzymatic step catalyzed by CSAD. The present results also suggest that CSAD and GAD may be two distinct enzymes, only the latter enzyme being activated by a short light exposure.

One of the important findings in this study is that short light exposure induces a rapid increase of retinal GABA, which is readily reversed by returning animals to complete dark, whereas no such change occurs in the case of taurine. The short light exposure-induced increase of retinal GABA is in essential agreement with previous reports by Graham et al. (6) and Lam (14), respectively. Since the GABA increase found in retina following a short light exposure was accompanied with a significant activation of GAD as well as an increase of evoked release of (^{3}H)-GABA, it was assumed that this increase in retinal GABA might be an indicative measure of the augmented functional state of GABA-ergic neurons. Furthermore the activation of retinal GABA-T:SSADH, enzymes responsible for the degradation of GABA, followed the activation of GAD, and this GABA-T:SSADH activation in the retina coincided with the declining phase of short light exposure-induced increase of GABA. These results suggest that alterations of retinal GABA induced by short light exposure may be regulated, at least in part, by the differential activation of GAD and GABA-T:SSADH therein.

In contrast with the results obtained in short light and dark adapted retinae, prolonged light and dark adaptations induced respectively statistically significant decrement and increment in retinal taurine contents and CSAD activities following 3 and 6 weeks adaptations. Under the same experimental conditions, retinal GABA content and GAD activity exhibited no significant changes. These results indicate clearly that light and dark adaptations induce differential effects on the retinal metabolism of GABA and taurine if short and prolonged light exposures are applied. Although it has been reported that a dark adaptation applied for 6 days to newly hatched chicken induced an increase of retinal taurine (19), no such alterations were found in the retinae of rat, frog, chicken and goldfish maintained for at least 18 hr in continuous light or total dark (23). In this study we have also found that statistically significant alterations in retinal taurine and CSAD activity, an increase in dark-adapted and decrease in light-adapted states respectively, occur in frogs adapted continuously to light or dark more than 3 weeks. Considering that such a long dark or light adaptation was necessary for inducing changes in retinal taurine, and the observed increase and decrease of retinal taurine contents were not coincident with the occurrence of alterations of any possible functional parameters for neurotransmitter substances such as changes of the high K^+ evoked release, high affinity uptake and specific binding to synaptic membrane preparations, it can be said that these changes may not correspond to rapid changes induced or associated with the alterations of synaptic activity within the retina. Although the mechanisms and functional significances underlying these changes remain to be elucidated, the possible importance of retinal taurine for maintaining cellular integrity under pathophysiological conditions may be suggested. In this regard, it is noteworthy that taurine deficiency leads to photoreceptor cell death in cat (8), and pancreatic taurine may play important physiological roles in maintaining the function and/or integrity of β-cells in the islets of Langerhans and protecting these structures from toxic substances such as streptozotocin (24).

SUMMARY AND CONCLUSION

Short light exposure on dark adapted frog retina induced significant increases of retinal GABA levels via the activation of l-glutamate decarboxylase (GAD) and increased the evoked release of (^{3}H)-GABA. GABA-transaminase:Succinic semialdehyde dehydrogenase (GABA-T:SSADH) activities in the retina were also activated, but these activations occurred later than those of GAD, and coincided with the recovery of GABA to the level found in dark adapted state. Under the same experimental conditions, no alterations in retinal taurine and cysteine sulfinate decarboxylase (CSAD) activity were noted.

Contrary to the results obtained in short light and dark adapted retinae, prolonged light and dark adaptations for 3 - 6 weeks induced significant decreases and increases respectively in retinal taurine and CSAD activity without altering the GABA system therein. These alterations in retinal taurine were, however, unaccompanied with any changes in factors possibly responsible for transmitter actions such as evoked release, high affinity uptake and specific binding to synaptic membranes.

The present data suggest that, in distinction to GABA as a potential candidate for inhibitory neurotransmitter, retinal taurine may act as a neuromodulator (12) and/or may play an important role as a basic factor for maintaining cellular integrity under certain pathophysiological conditions.

REFERENCES

1. Bonaventure, N., Wioland, N., and Mandel, P., 1974, Antagonists of the putative inhibitory transmitter effects of taurine and GABA in the retina, Brain Res., 80:281-289.
2. Di Giorgio, R.M., Tucci, G., and Macaione, S., 1975, Cysteine oxidase activity in rat retina during development, Life Sci., 16:429-436.
3. Dowling, J.E., 1967, The site of visual adaptation, Science, 155:273-279.
4. Enna, S.J., and Snyder, S.H., 1976, Gamma-aminobutyric acid (GABA) receptor binding in mammalian retina, Brain Res., 115:174-179.
5. Goodchild, M., and Neal, M.J., 1970, Uptake of ^{3}H-gamma-aminobutyric acid (GABA) by rat retina, J. Physiol., 210:182-183p.
6. Graham, L.T., Jr., Baxter, C.F., and Lolley, R.N., 1970, In vivo influence of light or darkness on the GABA system in the retina of the frog (Rana pipiens), Brain Res., 20:379-388.
7. Graham, L.T., Jr., 1972, Intraretinal distribution of GABA content and GAD activity, Brain Res., 36:476-479.
8. Hayes, K.C., Carey, R.E., and Schmidt, S.Y., 1975, Retinal degeneration associated with taurine deficiency in the cat, Science, 188:949-951.
9. Kennedy, A.J., and Voaden, M.J., 1974, Free amino acids in the photoreceptor cells of the frog retina, J. Neurochem., 23: 1093-1095.
10. Kennedy, A.J., and Voaden, M.J., 1976, Studies on the uptake and release of radioactive taurine by the frog retina, J. Neurochem., 27:131-137.

11. Kimura, H., and Kuriyama, K., 1975, A new microassay method for L-glutamic acid decarboxylase (GAD) activity, Jap. J. Pharmacol., 25:189-195.
12. Kuriyama, K., (1980), Taurine as a neuromodulator, Federation Proc., 39:2680-2684.
13. Lake, N., Marshall, J., and Voaden, M.J., 1977, The entry of taurine into the neural retina and pigment epithelium of the frog, Brain Res., 128:497-503.
14. Lam, D.M.K., 1972, The biosynthesis and content of gamma-aminobutyric acid in the goldfish retina, J. Cell. Biol., 54:225-231.
15. López-Colomé, A.M., and Pasantes-Morales, H., 1980, Taurine interactions with chick retinal membranes, J. Neurochem., 34: 1047-1052.
16. Lowry, O.H., Rosebrough, N.J., Farr, A.L., and Randall, R.J., 1951, Protein measurement with the Folin phenol reagent, J. Biol. Chem., 193:265-275.
17. Mathur, R.L., Klethi, J., Ledig, M., and Mandel, P., 1976, Cysteine sulfinate decarboxylase in the visual pathway of adult chicken, Life Sci., 18:75-80.
18. Orr, H.T., Cohen, A.I., and Lowry, O.H., 1976, The distribution of taurine in the vertebrate retina, J. Neurochem., 26:609-611.
19. Pasantes-Morales, H., Klethi, J., Ledig, M., and Mandel, P., 1973, Influence of light and dark on the free amino acid pattern of the developing chick retina, Brain Res., 57:59-65.
20. Pasantes-Morales, H., Urban, P.F., Klethi, J., and Mandel, P., 1973, Light stimulated release of (^{35}S) taurine from chicken retina, Brain Res., 51:375-378.
21. Redburn, D.A., Kyles, C.B., and Ferkany, J., 1979, Subcellular distribution of GABA receptors in bovine retina, Exp. Eye Res., 28:525-532.
22. Scholes, N.W., and Roberts, E., 1964, Pharmacological studies of the optic system of the chick: Effect of γ-aminobutyric acid and pentobarbital, Biochem. Pharmacol., 13:1319-1329.
23. Starr, M.S., 1973, Effect of dark adaptation on the GABA system in retina, Brain Res., 59:331-338.
24. Tokunaga, H., Yoneda, Y., and Kuriyama, K., 1979, Protective actions of taurine against streptozotocin-induced hyperglycemia, Biochem. Pharmacol., 28:2807-2811.
25. Yoneda, Y., 1978, Studies on functional roles of γ-aminobutyric acid (GABA) and taurine in the spinal cord and thalamus. I. Microdistributions of GABA, Taurine and their metabolically related enzymes, J. Kyoto Pref. Univ. Med., 87:675-694. (Japanese)

CHANGED TAURINE-GLUTAMIC ACID CONTENT AND ALTERED NERVOUS TISSUE CYTOARCHITECTURE

Nico M. Van Gelder

Centre de recherche en sciences neurologiques
Département de physiologie
Faculté de médecine
Université de Montréal
C.P. 6128, Succursale A
Montréal, Québec H3C 3J7 (Canada)

Despite the efforts of numerous investigators, it appears difficult at this time to assign in mammals any definite biological role to taurine. Nevertheless, the special association of taurine with electrically excitable tissue (heart, muscle and CNS) supposes its importance in the process of conduction and/or synaptic transmission. In these organs taurine seems involved in, among other phenomena, the regulation of calcium metabolism, temperature control as well as in the maintenance of the compartmentalized metabolism of glutamic acid, and other amino acids, whose metabolism and sequestration require a combination of neuronal and glial elements. Other findings suggest that the amino acid may perform a critical function in neonatal and, perhaps, embryonic development of the central nervous system (2,15). The latter suggestion is further supported by the fact that in women, taurine excretion becomes sharply curtailed during the period of pregnancy and lactation (1). That its role in assuring proper cellular "development" is not confined to nervous tissue may be extrapolated from findings that a similar diminished excretion of taurine occurs during wound healing (36). Other conditions such as radiation damage (14), hormonal imbalance (7,8), the viability of tissue cultures (13) and certain disease states (16,17) also appear reflected by alterations in taurine excretion or tissue taurine content. The possible function of taurine in such critical biological phenomena assumes even greater significance when it is considered in the context of findings indicating that in man and a number of other species the amino acid probably remains essential nutrient throughout life (19). Temporary or chronic dietary deficiencies as well as certain drugs, an acquired or familial tendency for high taurine excretion, and the physical condition of the individual, may therefore all represent important factors in the prevention or reversal by taurine of cytoarchitectural or functional damage in organs and tissues (21,22,23,33,35).

Alterations in the physiological state of nervous tissue, such as hyperirritability, epilepsy, chronic activation, etc., often can be ascertained by changes of the EEG or other electrophysiological parameters used to monitor the level of nervous tissue excitation. However, shifts of intercellular anatomical relationships and the exact nature of the cyto-architectural derangements often may be so subtle that they appear undetectable on routine anatomical examination of the tissues. Yet, when cytoarchitectural changes are known to have occurred in a certain brain region, they are often accompanied by a permanent change in the physiological function of that region (12).

Work in this and other laboratories on epilepsy (6,26) have indicated quite clearly that this disorder may arise from a variety of environmental factors or may be the result of a genetic predisposition (28,32). Not all forms of epilepsy, and especially not the inborn types, are accompanied by appreciable anatomic changes of the affected brain regions. Structural alterations when detectable in association with epilepsy often arise as secondary consequences of repeated seizure attacks (hypoxia, energy deficit, acidosis, falling, etc.). On the other hand, a number of studies over the past 10 years now suggest that the many divergent factors which may all give rise to an enhanced nervous tissue excitation (25,30,34) are almost invariably accompanied by subtle alterations in the metabolism of two specific groups of amino acids. The first group encompasses the amino acids associated with glutamic acid metabolism (pH regulation, excitation-inhibition, ammonia detoxification and energy metabolism, membrane structure). The second group of amino acids involves (so far) always one nutritionally essential amino acid (taurine, valine, etc.), in combination with their metabolically interrelated amino acids (structural repair, amino acid retention, enzyme synthesis, etc.). The explanation as to why such a multiplicity of different conditions give rise to an almost identical bio-chemical anomaly resides in the fact that the metabolism of these amino acids is divided between neuronal elements and adjacent satellite cells. Metabolic communication between the various cytoarchitectural elements (metabolic compartments) requires a host of complex interconnecting transport and uptake systems which demands that the precise relationships between the various anatomical structures is exactly maintained (see Fig. 1 of 31). This type of metabolism may thus represent one of the most sensitive indicators of either chemical or physical damage to the CNS. A derangement of such metabolism seems detectable long before either individual electrophysiological or anatomical parameters of brain function become appreciably altered (cf. predisposed individuals without clinical signs; 28).

It would seem desirable therefore to establish more precise biochemical indices of a specific <u>chronic</u> change in physiological function or in the cytoarchitecture of a particular brain region (5,18,20). Such indices would have to fulfill certain criteria, the most important of which would have to be that a specific biochemical event can be assigned to a defined anatomical structure. The detailed work on the compartmentalized metabolism of glutamic acid, initiated by Berl et al. (4) and subsequently extended by many other investigators (refs. cited in 3), has led this investigator in recent years into attempting to use that metabolism as one such indice. The investigations referred to above have suggested that most of the steady state glutamic acid content in nervous tissue is present throughout most neuronal elements, that of glutamine in glial structures, and that GABA and aspartic acid (ASP) as well as a small but readily releasable fraction of the total glutamic acid pool is more specifically situated in nerve terminals. Evidently, the resolution of the compartmentalized metabolism of glutamic acid into three distinct metabolic pools in association with three anatomic subdivisions of nervous tissue cytoarchitecture, cannot considered to be absolute. However, a combination of techniques such as specific enzyme localization in certain anatomical structures, tracer studies, the use of tissue culture, brain slices and differential centrifugation, have to a large extent validated the postulation of a compartmentalized metabolism for glutamic acid and its metabolites, divided among (at least) three cellular subdivisions of nervous tissue (24).

On the other hand, the metabolic mechanisms which assure interconnections between these compartments are very complex and entail a number of intracellular enzymic processes (11). Since all these phenomena, while interrelated, can vary to some extent independently from one another, it is difficult to ascribe changes in the steady state levels (metabolic pools) of the amino acids implicated to alterations of any one of these many mechanism. The recent observation that taurine, present in both glial and neuronal elements, could influence glutamic acid compartmentation, seemed to have obviated some of the problems.

The peculiarities of taurine metabolism, being extremely slow or absent in the CNS (15,19), suggests that changes in taurine content may be equated primarily with changes in the ability of certain CNS structures to retain (sequester) the amino acid. Moreover, to date taurine has not been demonstrated to effect directly any of the metabolic mechanisms involved in the compartmentalized metabolism of glutamic acid. The transport or uptake of the various glutamic acid cycle constituents appears to occur independently of taurine, with the exception possibly of GABA (weak transport competition). Neither are the enzymes of glutamic acid metabolism influenced appreciably by taurine (although considerable work still has to be done in this area, 22). Thus, the clearly demonstrable closely linked fluctuations of taurine content in a tissue with that of glutamic acid and/or glutamine, as well as of (GABA + ASP), would imply that the combined changes can be attributed to simultaneous alterations in the tissue retention capacity for taurine and of that constituent of the glutamic acid cycle with which taurine changes are associated. While the sum total of the various transport and enzymic mechanisms thus may determine the rates of synthesis and metabolism of those constituents, taurine retention may indicate their "storage" capacity in the various compartments. Hence, for taurine it should be possible to model a three compartment system (perikaryum-glia-nerve terminal), each equipped with an inflow valve and an outflow valve which interconnect the compartments. It then becomes fairly simple to deduce theoretically how the levels of taurine in one compartment can fluctuate either dependently or independently with the content of the other two compartments, conditional on whether in each compartment the inflow rate exceeds, is similar to, or is less than the rate of outflow towards the next compartment. Unfortunately, when measuring steady state tissue taurine content, no means are available to detect whether taurine is derived from one compartment or another, unless it were possible to tag the taurine content of each compartment by a different label. Nevertheless, an indirect "labeling" method seems provided if one adopts a convention which in fact is supported by experimental observations; that changes in the level of taurine in each compartment are accompanied by changes in the levels of another substance which presence in each case is predominantly confined to one particular compartment. For example, if taurine fluctuations in the large neuronal compartment is accompanied by fluctuations of glutamic acid, while in glia and nerve terminals with, respectively, changes in the steady state content of glutamine or (GABA + ASP), one can designate each of these amino acids as representing "marker" substances for the different taurine pools.

The notion that each amino acid of the glutamic acid cycle defines a specific metabolic-anatomic compartment appears quite simplistic. On the other hand, when the data from a number of laboratories were interpreted in terms of three taurine compartments, it became evident that in this manner it was possible to distinguish between physiological states of nervous tissue, differences in the cytoarchitecture of various brain regions, and also, apparently, to determine which of the major structural components of nervous tissue cytoarchitecture had undergone change following anatomic manipulation of the tissue (24,30, 31). It should be emphasized here that these studies attempted to provide an explanation for a set of data. A clear distinction must therefore be made between the data obtained, which require interpretation in terms of existing physiological or anatomical alterations, and an explanation which at this moment can only be considered as one of several possibilities (see 5). In order to test the plausibility of the explanation proposed, several different demonstrated physiological states were recently reexamined in more detail, this time by making use of only two compartments - the "large" neuronal glutamic acid compartment and the glial glutamine compartment (27). In this article such studies are continued, but in this instance by examining conditions in nervous tissue which represent known cytoarchitectural derangment (damage).

RESULTS

In Figs. 1 and 2 two sets of data are shown, obtained from representative species with either a low (man, cat, monkey) or a high (mouse, fish) nervous tissue taurine content. In terms of a hypothetical two compartment system interconnected by an adjustable flow through valve (valve a) and an outflow valve (valve b), the data can be interpreted as follow:

(I) Low Cerebral Taurine Levels. Normal Cortex

The neuronal compartment is not yet saturated with as consequence that varying influx of (exogenous) taurine into that compartment gives rise to intracompartmental fluctuations of the amino acid. Valve a regulates the transfer of taurine to the glutamine compartment at a steady rate. Since glutamic acid storage capacity appears to follow that of taurine (24,27), varying intraneuronal glutamic acid synthesis is also not translated by release and changes of its uptake into the glial structures. The six fold range of taurine values and three fold range of glutamic acid values in the normal cat cortex was

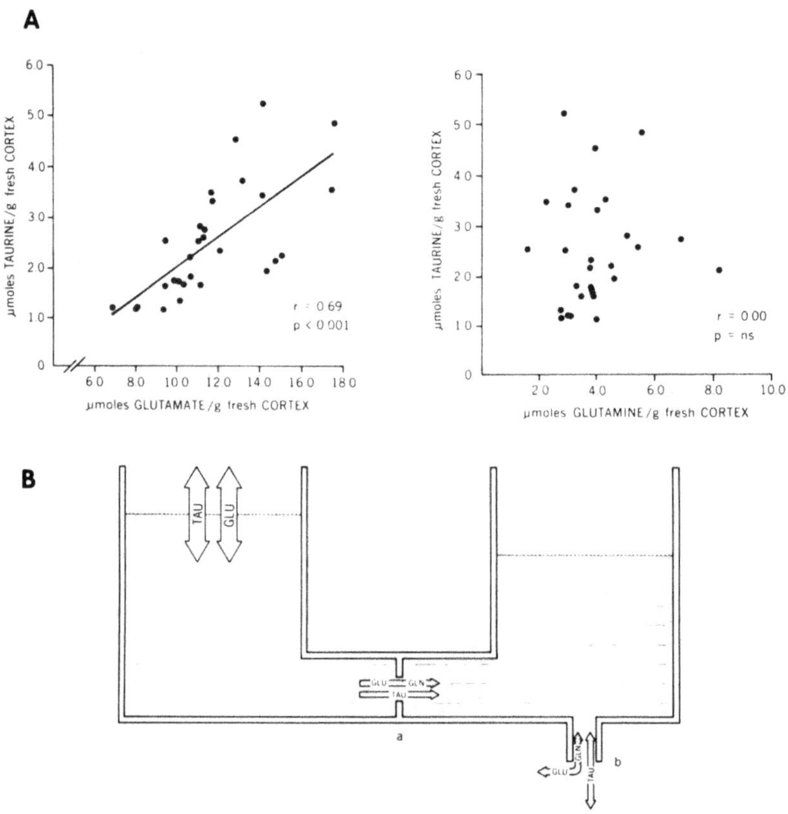

Fig. 1. A. Steady state amino acid content determined in the cerebral cortex of a number of cats. This situation may be considered analogous to repeatedly taking tissue samples from the same cortex over a period of time i.e. each animal represents one time internal during the sampling period. Glutamic acid and taurine contents vary interdependently while no such interrelationship exists for glutamine and taurine. Regression lines and statistics (two tail test of significance) by "Plot 50: Statics" Program of Tektronix, Inc. Combined data from (9,10,29) and unpublished.
B. A hypothetical two compartment system joined by an adjustable connecting valve (a) and an outflow valve (b). The GLU-TAU compartment may be analogous to cerebral neuronal elements while the GLN-TAU compartment may represent adjacent glial structures. Arrows indicate how combined neuronal GLU-TAU fluctuations do not necessarily relate to changes in the content of glutamine or taurine of the glial compartment, the GLU-TAU compartment has not yet reached maximum storage capacity for the two amino acids. With permission from Raven Press, (27).

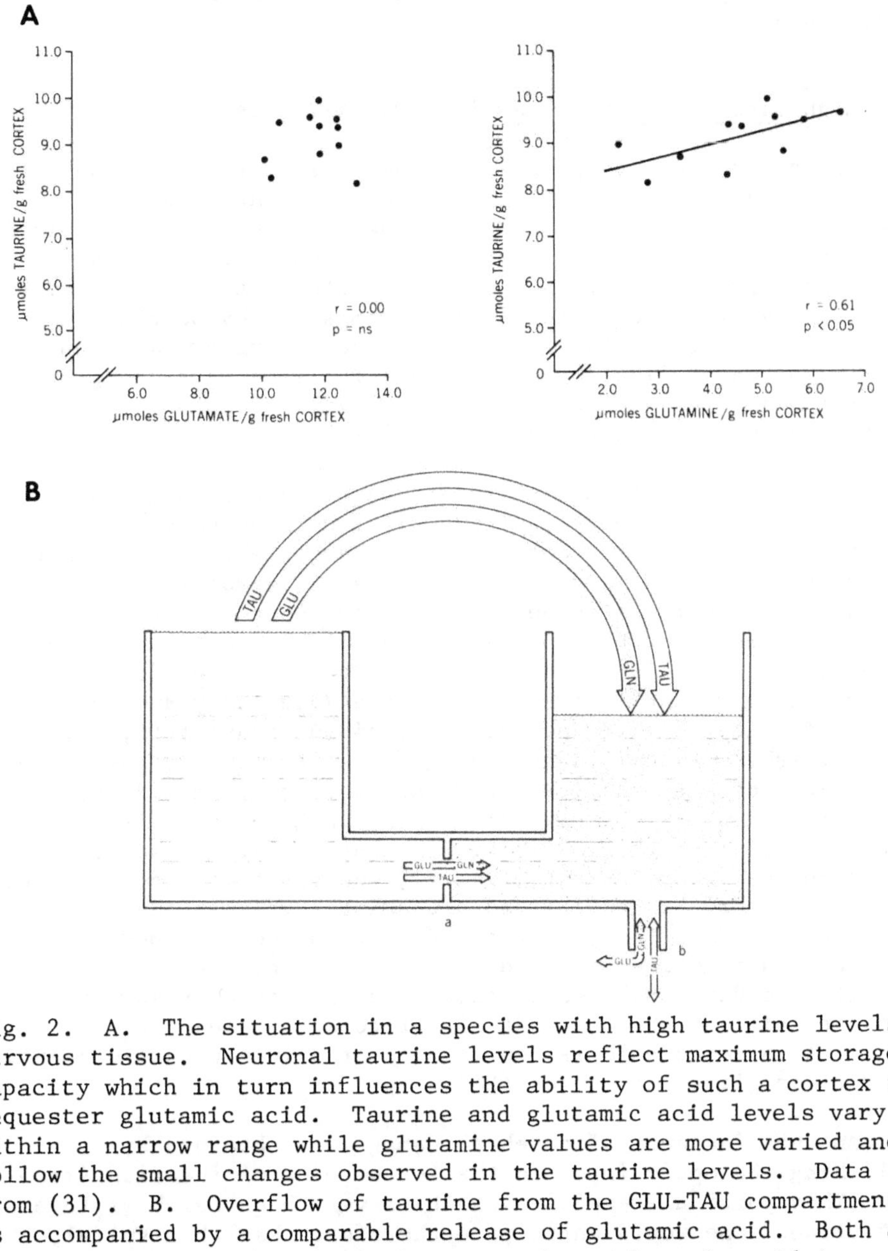

Fig. 2. A. The situation in a species with high taurine levels in nervous tissue. Neuronal taurine levels reflect maximum storage capacity which in turn influences the ability of such a cortex to sequester glutamic acid. Taurine and glutamic acid levels vary within a narrow range while glutamine values are more varied and follow the small changes observed in the taurine levels. Data from (31). B. Overflow of taurine from the GLU-TAU compartment is accompanied by a comparable release of glutamic acid. Both amino acids are taken up by glial elements, where glutamic acid is rapidly converted to glutamine. Hence steady GLU-TAU content and interrelating fluctuations of taurine and glutamine, reflecting primarily changing steady state levels of the amino acids in glial elements. With permission from Raven Press (27).

obtained by combining data derived from studies by Koyama (9); Koyama and Jasper (10); van Gelder and Courtois (29) and unpublished data.

(II) High Cerebral Taurine Levels. Normal Cortex

As the neuronal compartment becomes saturated, taurine begins to "overflow" into the second compartment. This is accompanied by a glutamic acid overflow which however is rapidly transformed to glutamine. Thus, taurine and glutamic acid levels remain steady but whatever small fluctuations of taurine occur in the glial compartment are accompanied by parallel fluctuations of glutamine (from 24). The "overflow" may represent an increased flow through valve a (glial uptake and glutamine synthesis rate) combined with a steady rate of outflow (glutamine and taurine release) at valve b.

With the assumption that the above presented data can be interpreted in the manner outlined, it should be possible to deduce from this type of analysis the changes which might occur when cerebral tissue is altered. In Fig. 3, results are presented from amino acid analysis of the undercut cat cortex. In such a hyperexcitable cortex, glutamic acid tends to decrease while taurine and glutamine values tend to increase (from 10). Glutamic acid retention is impaired as evidenced by an enhanced efflux of glutamic acid and decreased tissue content (6,10). The diminished glutamic acid storage capacity would suggest that the TAU-GLU compartment (neuronal) has decreased in size by undercutting of the cortex. If so, one would anticipate an absence of taurine-glutamic acid correlation, i.e. a saturation of the neuronal compartment (Fig. 2), and positively correlating fluctuations of the taurine and glutamine levels in the glial compartment. In effect, in the undercut cat cortex conditions have changed from one depicted in Fig. 1 to one shown in Fig. 2. The data analyses moreover would allow a presumption that, in spite of the somewhat higher total taurine content, less of this amino acid is now being stored in neuronal elements and more in glial structures (24).

Another instance of cerebral damage was obtained by abolishing the optic input to the tectum of a marine fish (from 31). This particular species exhibits very high cerebral levels of taurine. One would anticipate therefore that this animal, like the mouse (Fig. 2), should demonstrate rather steady levels of glutamic acid and taurine and combined correlations of the glutamine and taurine content. As shown in Fig. 4, this appears indeed to be the case, provided the retina remains dark-adapted or, to a lesser extent, when subjected to natural light conditions (Fig. 4A, right); the differences between dark adaptation (●) and normal physiological activation (□)* have been discussed elsewhere (30,31). Following optic nerve denervation 8 months earlier, the tectum demonstrates a loss of taurine and glutamine (Figs. 4A vs 4B, right) but glutamic acid levels remain virtually unaltered (no retention defect). Furthermore, the cortex of such fish shows permanent biochemical changes which suggest that especially the glial taurine-glutamine compartment has decreased in size relative to an unaffected taurine-glutamic acid compartment (Fig. 1). One interpretation would be that optic nerve severance results in a situation where normal release of glutamic acid and taurine is no longer matched by uptake of these two amino acids into metabolically altered glia (valve a impaired). Steady state fluctuations of taurine in this situation thus essentially reflect changes in the size of the remaining pool, present in neuronal structures.

Finally, previous investigations have already indicated that the compartmentalized metabolism of glutamic acid is reflected differently in various brain regions and that pathophysiological findings are expressed biochemically in accordance with such normal regional differences of tissue cytoarchitecture. Moreover these studies also demonstrated that all metabolic compartments implicated in the metabolism of glutamic acid may be affected, including the one associated with GABA and aspartic acid (ASP) metabolism (24,30). For example, when a cerebral cortex is stimulated or becomes spontaneously hyperexcitable, strong positive correlations between taurine and (GABA + ASP) content are seen (Fig. 5). This phenomenon thus also measures an indice of cerebral excitation. In a normal or unstimulated cerebral cortex, no such correlations are detectable. In contrast, within the cerebellum strong inverse correlations between taurine and (GABA + ASP) are noted under ordinary physiological conditions (Fig. 6, left). Cerebellar development in mice is strongly affected by conditions of neonatal malnutrition. That type of situation would therefore provide another opportunity to test whether the method of analysis as outlined above, could provide information of an anatomical or biochemical nature not easily detectable otherwise.

A comparison of TAU-GLU-GLN relationships between the cerebellum of a normal adult animal (40 days) and that of an adult animal subjected to malnutrition during the 1-16 days

*In ref. 31 the significance level and regression line drawn is in error.

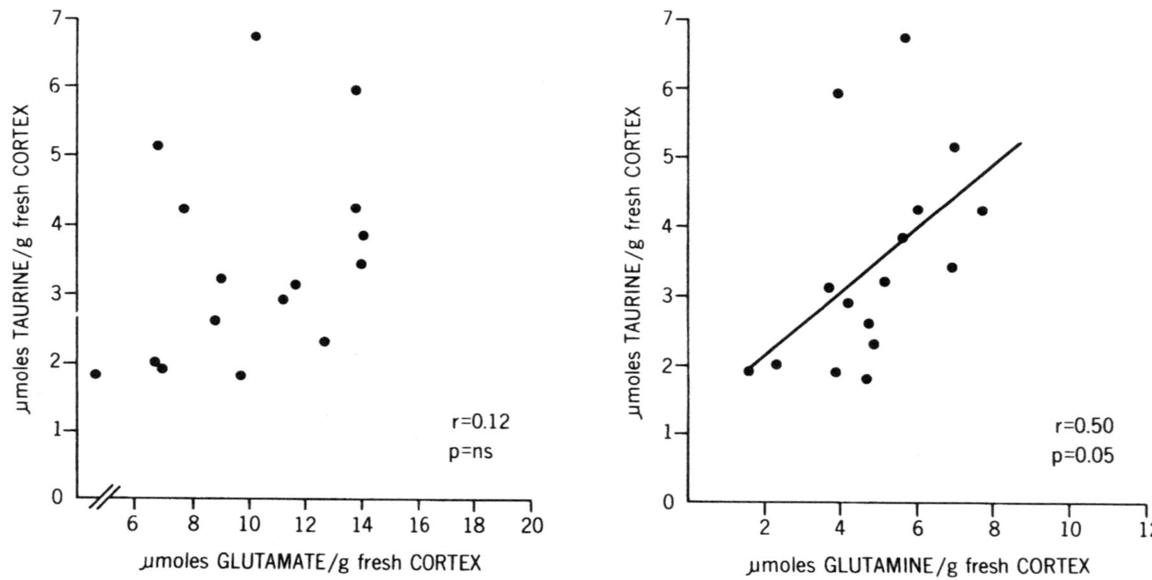

Fig. 3. Following undercutting of the cortex, glutamic acid content tends to decrease (50% vs 20% < 10 µmoles/g) and glutamine values tend to increase (60% vs 30% > 4 µmoles/g). This suggests that the GLU compartment has decreased in size so that in such a hyperirritable cortex glutamic acid has a greater tendency to overflow from that compartment, to be subsequently taken up by surrounding structures where it is converted to glutamine. The "saturation" of the glutamic acid storage capacity is accompanied by a similar change in the taurine retention capacity which will therefore also overflow into the glutamine containing glial structures (see Fig. 2). The findings indicate damage to some neuronal elements containing both glutamic acid and taurine and no change or little alteration of glial structures (from 10).

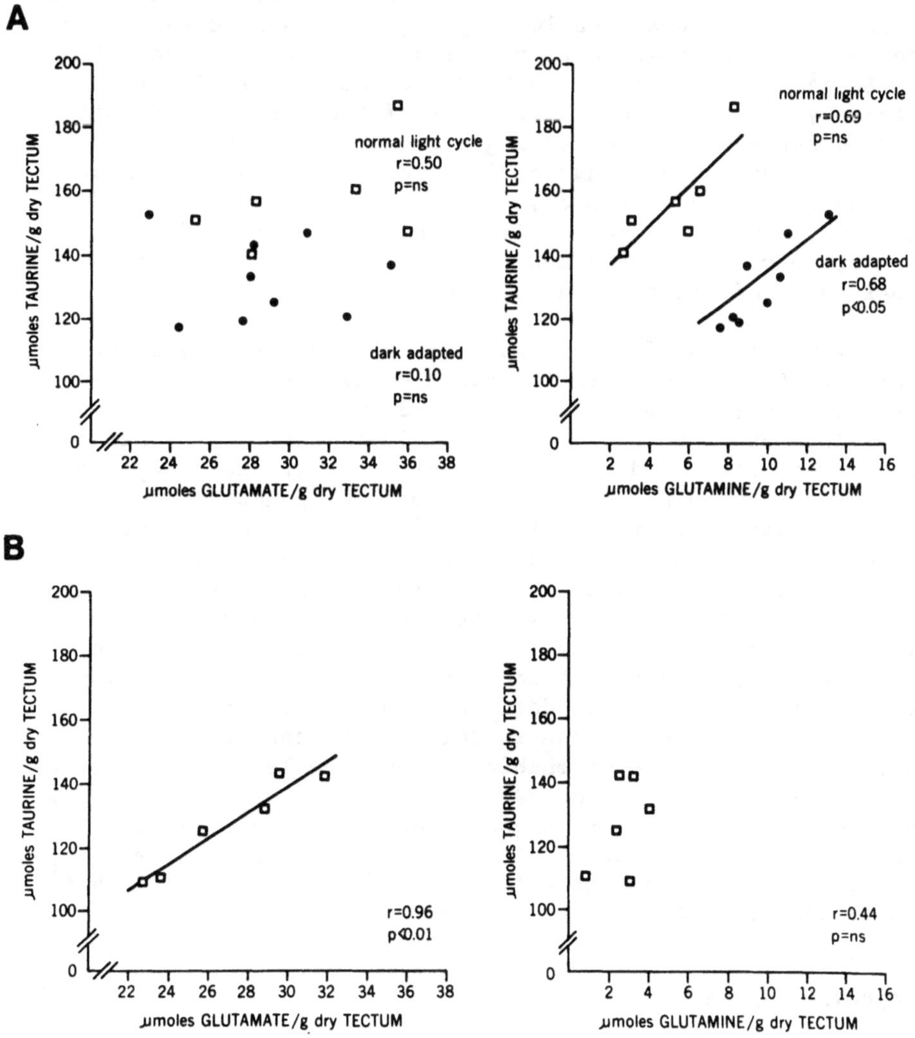

Fig. 4. Analogous to the situation in the mouse cortex, the tectum of a marine fish contains large amounts of taurine. The range of normal fluctuations in taurine and glutamic acid is much smaller than that observed in a normal cat cortex (saturation of TAU-GLU compartment, Fig. 2). This is seen in a tectum at rest or, to a lesser extent, under normal physiological conditions (but not in a stimulated tectum). Cytoarchitectural damage 8 months earlier decreases taurine and glutamine content in comparison to the appropriate control group (□, 4A); glutamate levels remain virtually unchanged. The findings suggest alterations of structures which store both taurine and glutamine but not glutamate and taurine, i.e. altered glial properties (see 27,31).

neonatal period, revealed no easily discernable biochemical differences. However, when the TAU-(GABA + ASP) relationship was examined (i.e. the third compartment not shown), an obvious alteration became discernable (Fig. 6, left vs middle). In the underdeveloped cerebellum taurine levels no longer fluctuate with (GABA + ASP) levels. Instead the taurine levels are remarkably stable at the lower limit of the normal range of values observed. When another group of malnourished mice, placed as usual on a standard diet from day 16 on, were also provided during this recovery period with 0.02% taurine (w/v) in the drinking water, normal relationships between (GABA + ASP) and taurine were restored and taurine levels returned to a normal range of values (Fig. 6, right).

Several lines of investigations are indicated by these findings. First, whatever the class of cerebellar structures, containing taurine, which was affected during a critical period of development as a consequence of malnutrition, these structures either recovered as a result of a nutritional supply of taurine, or, by means of this additional exogenous source of that amino acid they were able to maintain a normal regulation of the taurine content in the cerebellum. Secondly, the fact that this regulation usually occurs in reciprocal relationship to the (GABA + ASP) content (nerve terminal?), and that the relationship can be exactly restored, provides strong evidence that under certain circumstances and in certain brain regions where apparently a chronic deficiency of taurine has been created, dietary taurine can restore normal conditions. Finally, an inverse relationship between the levels of taurine and (GABA + ASP) in the cerebellum gives rise to a speculation that these amino acids are sequestered in separate nervous tissue elements. Release (retention) by these structures may be mutually exclusive in that activation of one may entail inhibition or non-activation of the other.

CONCLUDING REMARKS

The present article is intended to demonstrate that chronic cytoarchitectural alterations of nervous tissue will bring about specific biochemical changes. The detection of such changes, aside from being dependent on the species in which they occur, also seems to vary according to the brain region implicated, the nature of the anatomic manipulation inflicted, and the diet of an animal. This much can be reasonably concluded from the data shown here without raising much controversy; such variable factors should be born in mind when data from different laboratories are to be compared. While the underlying functional importance of taurine in excitable tissue may always be the same, the biochemical expression of that function may vary according to the species and type of tissue examined, as well as according to the nutritional requirement of the subjects studied and the diet they receive at that particular time.

More controversial may be the proposed method of data analysis employed to speculate regarding the identity of the structural elements most affected following chronic alterations in the physiological state or in the cytoarchitecture of a particular brain region. The method is based on the fact that changes in taurine content of nervous tissue for the most part will denote altered retention mechanisms in one or more cytoarchitectural structures of the tissue. These variations in taurine retention can be associated with closely linked changes in the content of glutamic acid and/or its immediate metabolites. The most important metabolic pool of each of these glutamic acid cycle constituents in turn can be assigned to a specific structural element of nervous tissue. Combined changes in the content of taurine and one of the constituents, as well as their altered relationships relative to the levels of the other constituents of the glutamic acid cycle, therefore not only appear to denote the anatomic structure most affected but, in addition, also seem to explain some of the functional disturbances in excitability which are observed.

To what extent this type of analysis will be of value in correlating anatomical information with biochemical changes must await experiments in which the cytoarchitectural alterations are more precisely defined (5,20). Nevertheless, in those studies where physiological and/or anatomical conditions were well defined, in combination with animo acid analysis, the method appears to be upheld (10,24,27,28,32). It would seem at this time that the alterations observed in the relationships between taurine and the compartmentalized metabolism of glutamic may provide a very sensitive indicator of nervous tissue excitation which is either induced or familial, with or without accompaniment of permanent cytoarchitectural changes.

Data shown are from studies supported by the Medical Research Council of Canada, the Université de Montréal (Québec), and by the Instituto Venezolano de Investigaciones Cientificas (to B.D.D.).

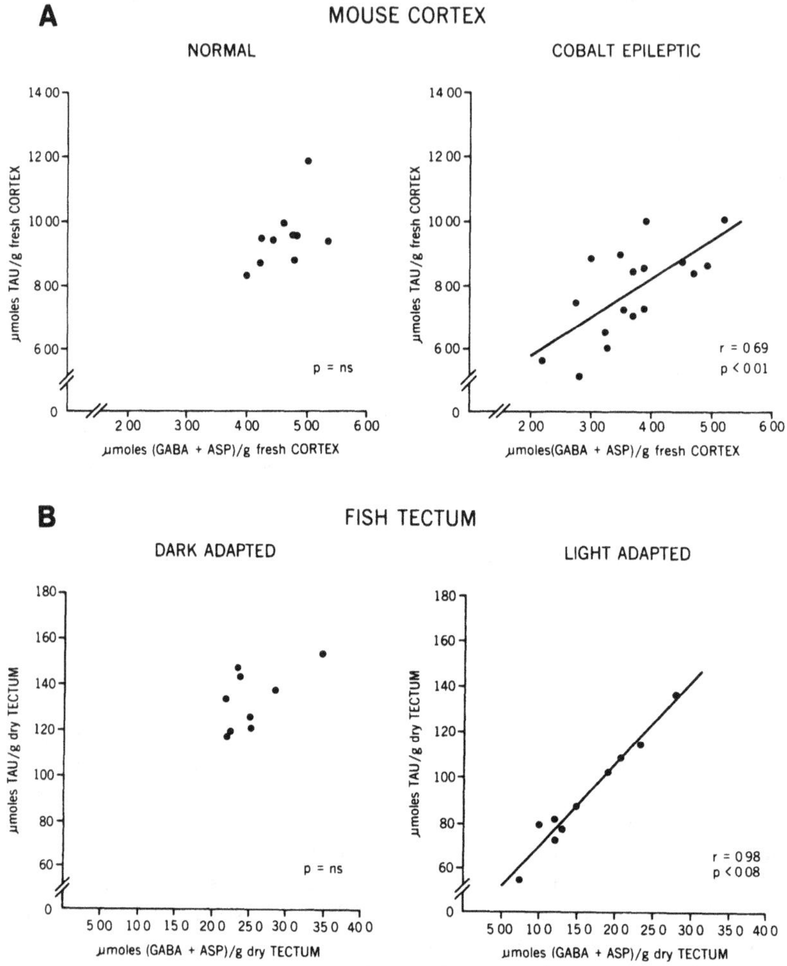

Fig. 5. In cerebral tissue exhibiting either hyperirritability or subjected to prolonged activation, a strong positive correlation between taurine and (GABA + ASP) can be noted which is not seen in a cortex at "rest". An analogous phenomenon is observed in the retina. These changes can thus serve to deduce whether the tissue had been subjected to stimulation prior to removal. High K^+ depolarization will also produce a similar effect. One interpretation would be that under non-excitable conditions taurine and the (GABA + ASP) content are regulated independently; as the tissue is stimulated these substances are released simultaneously and may become sequestered in a single compartment. Alternatively, a fraction of cerebral (GABA + ASP) and taurine is unaffected during (prolonged) stimulation and loss of these amino acids from other compartments now "unmasks" this unaffected pool (from 24,30).

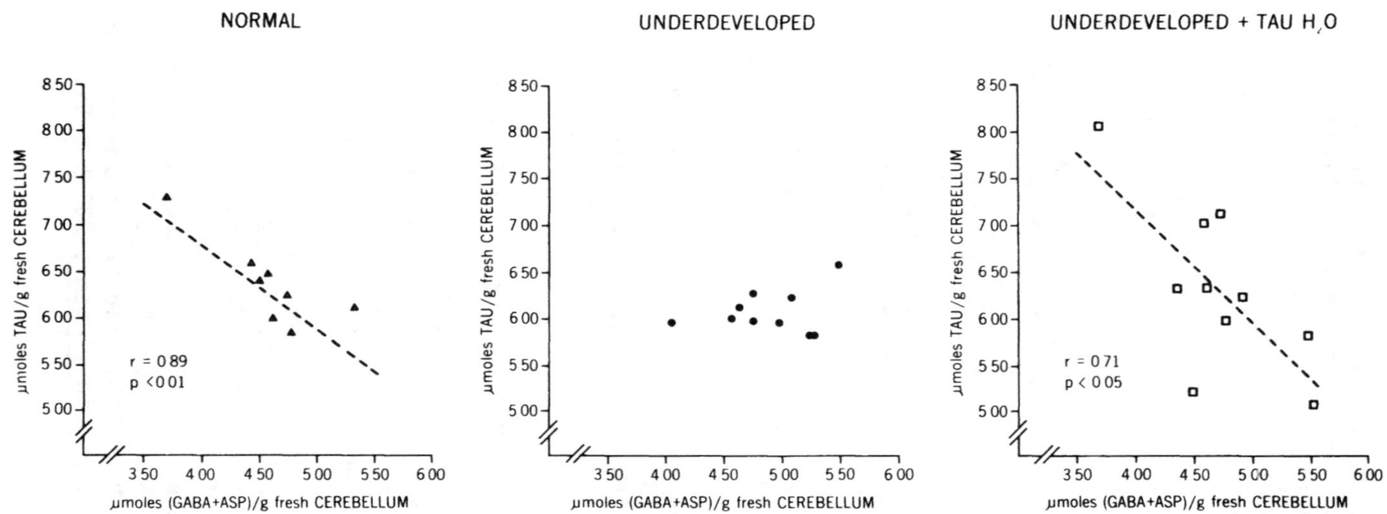

Fig. 6. As opposed to a stimulated or epileptic mouse cortex where TAU and (GABA + ASP) levels fluctuate together, in a normal cerebellum taurine levels tend to decrease as (GABA + ASP) levels increase, and vice versa. Despite having been fed a normal diet from day 16 to day 40, animals which suffered from malnutrition and, thus, cerebellar underdevelopment during the first 16 days of neonatal life, exhibit at adulthood a diminished cerebellar taurine content; cerebellar (GABA + ASP) appears normal. When in malnourished animals the ordinary diet was supplemented by 0.02% taurine (w/v) in the water supply during the 16-40 day recovery period, normal taurine and TAU-(GABA + ASP) relationships are restored. This suggests that a critical period of neonatal malnutrition subsequently leaves animals with a chronic taurine deficiency, either because cerebellar taurine retention mechanisms were damaged or the nutritional requirement for taurine had increased. If the inverse relationship between taurine and (GABA + ASP) levels were to reflect two excitable elements whose activation is mutually exclusive, then the findings may indicate that only one of these structures was selectively affected during a period of malnutrition. Such structures may respectively represent for example stellate cells or basket cells (taurine) and Purkinje cells (GABA).

REFERENCES

1. Armstrong, M.D., 1973, Decreased taurine excretion in relation to childbirth, lactation and progestin-estrogen therapy, Clin. Chim. Acta, 46:253-256.
2. Barbeau, A., and Huxtable, R., eds., Taurine and Neurological Disorders; various chapters, Raven Press, New York (1978).
3. Berl, S., Clarke, D.D., and Schneider, D., eds., in: "Metabolic compartmentation and neurotransmission: Relation to brain structure and function," Plenum Press, New York (1975).
4. Berl, S., Lajtha, A., and Waelsch, H., 1961, Amino acid and protein metabolism - VI Cerebral compartments of glutamic acid metabolism, J. Neurochem. 7:186-197.
5. Cremer, J.E., Changes within metabolic compartments related to the functional state and the action of drugs on the whole brain, in: "Metabolic compartmentation in the brain," R. Balasz and J.E. Cremer, eds., MacMillan Press, New York (1973), pp.81-93.
6. Dodd, P.R., Bradford, H.F., Abdul-Ghani, A.S., Cox, D.W.G., and ContinhoNetto, J., 1980, Release of amino acids from chronic epileptic and subepileptic foci in vivo, Brain Res., 193:505-517.
7. Hellstrom, K., and Schuberth, J., 1970, The effect of thyroid hormones on the urinary excretion of taurine in man, Acta Med. Scand. 187:61-65.
8. Jacobsen, J.G., and Smith, L.H., 1968, Biochemistry and physiology of taurine and taurine derivatives, Physiol. Rev. 48:424-511.
9. Koyama, I., 1972, Amino acid content in the cobalt-induced epileptogenic and non-epileptogenic cat's cortex, Can. J. Physiol. Pharmacol. 50:740-752.
10. Koyama, I., and Jasper, H.H., 1977, Amino acid content of chronic undercut cortex of the cat in relation to electrical after discharge: Comparison with cobalt epileptogenic lesions, Can. J. Physiol. Pharmacol. 55:523-536.
11. Kvamme, E, Regulation of glutaminase and its possible implication for GABA metabolism, in: "GABA - Biochemistry and CNS function," P. Mandel and F.V. De Feudis, eds., Plenum Press, New York (1979), pp.111-138.
12. Laidlaw, J., and Richens, A., eds., A textbook of epilepsy, Churchill Livingstone, New York, (1976) pp.66-108; 314-354.
13. Piez, K.A., and Eagle, H., 1958), The free amino acid pool of cultured human cells, J. Biol. Chem., 237:533-545.

14. Raghavan, K.G., and Nadkarni, G.B., 1970, Formation and excretion of taurine in X-irradiated rats, Int. J. Radiat. Biol., 18:41-49.
15. Rassin, D.K., and Gaull, G.E., Taurine and other sulphur containing amino acids: Their function in the central nervous system, in: "Amino acid as chemical transmitter," F. Fonnum, ed., Plenum Press, New York, (1978), pp.571-597.
16. Rylance, H.J., 1969, Hypertaurinuria in rheumatoid arthritis, Ann. Rheum. Dis., 28:41-44.
17. Rylance, H.J., and Myhal, D.R., 1971, Taurine excretion and the influence of drugs, Clin. Chim. Acta, 35:159-164.
18. Schousboe, A., and Divac, I., 1979, Differences in glutamate uptake in astrocytes cultured from different brain regions, Brain Res., 177:407-409.
19. Sturman, J.A., 1977, Taurine in nutrition, Comp. Ther., 3:59-65.
20. Tursky, T., Ruscak, M., and Lassanova, 1979, Effect of experimentally elicited astroglia proliferation on two compartments of glutamate metabolism in rat brain cortex slices, Physiol. Bohemoslov., 28:43-49.
21. van Gelder, N.M., 1972, Antagonism by taurine of cobalt-induced epilepsy in cat and mouse, Brain Res., 47:157-165.
22. van Gelder, N.M., Rectification of abnormal glutamic acid levels by taurine, in: "Taurine", R. Huxtble and A. Barbeau, eds., Raven Press, New York (1976), pp.293-302.
23. van Gelder, N.M., Glutamic acid and epilepsy: the action of taurine, in: "Taurine and neurological disorders," A. Barbeau and R. Huxtable, eds., Raven Press, New York, (1978a), pp.45-54.
24. van Gelder, N.M., ed., (1978b), Taurine, the compartmentalized metabolism of glutamic acid and the epilepsies, Can J. Physiol. Pharmacol., 56:362-374.
25. van Gelder, N.M., The biochemistry of hyperexcitability: Normal versus epileptic state, in: "Advances in epileptology: 11th Epilepsy International Symposium", R. Canger, F. Angeleri and J.K. Penry, eds., Raven Press, New York, (1980), pp. 285-288.
26. van Gelder, N.M., 1980, Glutamic acid metabolism and epilepsy. Neurosciences 6, (Suppl. 1):163-177.
27. van Gelder, N.M., Glutamic acid in nervous tissue and changes of taurine content: Its implication in the treatment of epilepsy, in: "Amino acid transmitters," P. Mandel and F.V. De Feudis, eds., Raven Press, New York, in press.
28. van Gelder, N.M., Aslam-Janjua, N., Metrakos, K., MacGibbon, B., and Metrakos, J.D., 1980, Serum amino acids in 3/sec spike-wave epilepsy, Neurochem. Res., 5:659-671.

29. van Gelder, N.M., and Courtois, A., 1972, Close correlation between changing content of specific amino acids in epileptogenic cortex of cats, and severity of epilepsy, Brain Res., 43:477-484.
30. van Gelder, N.M., and Drujan, B.D., 1978, Interrelated changes of amino acids in the retina and optic tectum of a marine fish with alterations of illuminating conditions, Brain Res., 159:137-148.
31. van Gelder, N.M., and Drujan, B.D., 1980, Alterations in the compartmentalized metabolism of glutamic acid with changed cerebral conditions, Brain Res., 200:443-455.
32. van Gelder, N.M., Edmonds, H.L., Hegreberg, G.A., Chatburn, C.C., Clemmons, R.M., and Sylvester, D.M., 1980, Amino acid changes in a genetic strain of epileptic beagle dogs, J. Neurochem., 35:1087-1091.
33. van Gelder, N.M., Koyama, I., and Jasper, H., 1977, Taurine treatment of spontaneous chronic epilepsy in a cat, Epilepsia, 18:45-54.
34. van Gelder, N.M., Sherwin, A.L., and Rasmussen, T., 1972, Amino acid content of epileptogenic human brain: focal versus surrounding regions, Brain Res., 40:385-393.
35. van Gelder, N.M., Sherwin, A.L., Sacks, C., and Andermann, F., 1975, Biochemical observations following administration of taurine to patients with epilepsy, Brain Res., 94:297-306.
36. Williamson, M.B., and Passman, J.M., 1960, Excretion of taurine during healing of experimental wounds, Clin. Chem., 6:140-147.

TAURINE, CYSTEINESULFINIC ACID DECARBOXYLASE AND GLUTAMIC ACID IN BRAIN

David K. Rassin

Division of Developmental Nutrition and Metabolism
Department of Pediatrics
University of Texas Medical Branch
Galveston, Texas

Despite the intensive studies of many investigators the role of taurine in mammals, other than as a component of taurocholate, is still unknown (2,3,11). The difficulty in assigning a function to taurine results from the large number of organs, actions and conditions with which the compound has been associated, such as neurotransmission (18,29), epilepsy (46,47), development (6,42), nutrition (30), retinal function (23), feline blindness (10,36), mongolism (8), adrenal function (21), cardiac function (9) and others. In addition, specific pharmacologic agonists and antagonists have not been identified, and pathologic changes, other than feline blindness, have not been associated with taurine deficiency or excess. Thus, conclusions cannot be made as yet concerning the function of taurine, and discussions of the importance of this compound still depend upon extrapolation from the descriptive data that are currently available.

One major problem in defining the place that taurine occupies in biological function is the difficulty in defining the source, exogenous or biosynthetic, of this compound and the factors that control the amount of taurine observed in various tissues and physiological fluids. It is the intent of this brief review to discuss the importance of cysteinesulfinic acid decarboxylase (CSAD) with respect to taurine content of brain and to include some information concerning the interaction of taurine with other amino acids, particularly glutamate.

In attempts to demonstrate that taurine fulfills the criteria necessary for identification as a neurotransmitter, it has been well established that the compound is present in both nerve terminals or synaptosomes (1,5,27,32,35,38) and in synaptic vesicles (5,27,32,35). Another criterion is that the enzymes that catalyze synthesis of the neurotransmitter be present at the nerve ending and this has proven to be a somewhat more controversial point. CSAD activity has been demonstrated in nerve terminals (1,24,28). However, there is still some question as to whether this activity is substrate specific for cysteinesulfinic acid (CSA) or merely a secondary activity of glutamic acid decarboxylase (GAD). If the decarboxylation of CSA is only secondary to that of glutamic acid it is not clear what the requisite mechanism is for supplying taurine to the nerve ending.

Two preliminary reports indicate that it is possible to separate two forms of CSAD from brain (26,39). One form of the CSAD coseparates with the activity of GAD (39). The other form of brain CSAD coseparates with the hepatic form of the enzyme and has little GAD activity (26). However, it is not clear to what extent the glutamic acid-insensitive form of CSAD is active _in vivo_ and where each form of the enzyme is located within the neuron. In addition, the changes in distribution and amounts with development of total GAD and CSAD are almost identical (13,33). This observation is difficult to reconcile with the finding that the products of the two enzyme activities, GABA and taurine, have opposite patterns during development. Taurine is present in high concentrations in fetal brain and declines during neonatal life to reach adult concentrations at about the time of weaning (41). In contrast, GABA is virtually absent in fetal brain and increases to adult concentrations during early neonatal development (33,34). GAD and CSAD have patterns of development very similar to that of GABA (13,33,34). This last finding is consistent with the concept that there is only one enzyme activity whose prime substrate is glutamic acid. However, there is some precedence for an enzyme activity developing inversely to its substrate. Histamine and histidine decarboxylase develop in a pattern very similar to that of taurine and CSAD respectively (19).

Another fact in support of two different enzymes, or perhaps, two different forms of the same enzyme, is the difference between the capacity of liver and brain to decarboxylate glutamic acid and CSA. Little or no GAD activity can be detected in the liver of those species that have been studied whereas considerable CSAD activity has been identified in the liver of a number of species. In fact, it has been suggested that hepatic CSAD may be a measure of the capacity of an animal to synthesize taurine (6). Species such as the rat have considerable hepatic CSAD activity and are relatively impervious to taurine deprivation. On the other hand, the cat and man have very low _in vitro_ hepatic CSAD activity and both these species may be liable to detrimental effects on dietary deprivation of taurine (Table 1). The large range of activity of CSAD in liver of various species is in contrast to the much smaller range of activity observed in brain (Table 2). In addition, the activity of CSAD and of GAD is lowest in man as compared to a number of other species (Table 2, reference 37). It is of interest that in those laboratories where both CSAD and GAD have been measured in brain

Table 1

Cysteinesulfinic Acid Decarboxylase
Activity in Liver

Species		Activity nmole/mg Protein/h	(n)	Reference
Rat:	Adult	557	(4)	28
	Adult	468	(5)	7
	Adult	527	(12)	40
	Adult	512	(3)	12
Dog:	Adult	790	(7)	40
	Adult	492	(4)	12
	84 Day	232	(3)	16
	35 Day	78	(5)	16
	21 Day	45	(4)	16
	5 Day	5	(4)	16
Cat:	Adult	4.5	(5)	7
	Adult	7.0	(2)	12
	Adult	6.6	(6)	40
Rhesus Monkey:	Adult	5.0	(6)	7
	Adult	8.2	(6)	40
	Fetus	3.5	(4)	7
Man:	Adult	0.3	(4)	7
	Adult	0.3	(6)	40
	Adult	0.2	(3)	12
	Fetus	0.3	(7)	7

at the same time, the activity of GAD is usually higher (Table 3). The last observation implies, perhaps, a preference for glutamic acid as a substrate by a single enzyme, although when the interaction between substrates is measured a non-competitive type of relationship is observed (12). Such determinations have only been performed in crude brain preparations, leaving open the possibility that purified preparations might behave differently.

Adding to the difficulty in evaluating the importance of CSAD to the maintenance of taurine in brain is the lack of information

Table 2

Cysteinesulfinic Acid Decarboxylase

Activity in Brain

Species		Activity nmole/mg Protein/h	(n)	Reference
Rat:	Adult	96.0	(6)	31
	Adult	62.8	(12)	40
	Adult	64.0	(3)	12
	Adult	138.3	(4)	1
	6 Day Old	14.1	(7)	1
	17 Day Old	58.3	(6)	1
Dog:	Adult	11.2	(3)	*
	Adult	54	(2)	12
	Adult	90.2	(7)	40
Cat:	Adult	55.4	(6)	14
	Adult	76	(1)	12
	Adult	54.2	(6)	40
Rhesus Monkey:	Adult	65.8	(6)	40
	Adult	20.8	(5)	34
	Neonate	17.1	(10)	34
	Fetus	3.9	(6)	34
Man:	Adult	5.7	(6)	40
	Adult	6.0	(2)	12
	Fetus	0.1	(3)	*

* Unpublished observations

concerning the amounts of possible precursors, CSA or cysteic acid, present in the brain. The presence of CSA in rat brain has been documented and an estimate of amount, a maximum of 1 mg per 200 mg of dry weight brain, has been made (4). Cysteic acid has been reported to be present in rat brain: 0.11 μmole/g of brain (17), <0.35 μmole/g (20), 0.01-0.03 μmole/g (43); and in chick brain 0.1 to 0.2 μmole/g (15). These amounts of cysteic acid and CSA are very low

Table 3

Ratio of Brain Cysteinesulfinic Acid Decarboxylase to Glutamic Acid Decarboxylase

Species		Ratio	Reference
Rat:	Adult	0.65	12
	Adult	0.65	1
	17 Day Old	0.52	1
	6 Day Old	0.45	1
Dog:	Adult	0.89	12
Cat:	Adult	1.06	12
Rhesus Monkey:	Adult	0.50	34
	Neonate	0.47	34
	Fetus	0.37	34

These data were calculated by dividing the activity of CSAD by that of GAD as reported in each reference identified in the table. These calculations were only performed when both enzymes were measured in the same animals.

compared with the amount of glutamic acid usually present in brain. The ratio of glutamic acid to cysteic acid where both were measured in the same samples was 44.8 in 8 day old chick brain and 117.3 in adult chick brain (15). In the light of these relative amounts of the two substrates for the decarboxylase enzymes it is difficult to understand how anything other than glutamic acid is decarboxylated in brain, unless much higher substrate specificity or much stricter metabolic compartmentation exists than has been demonstrated to date. In fact, the low enzyme activity of the sulfur amino acid pathway after the production of cystathionine, particularly the near absence of cystathionase (28), tends to exclude the pathway that includes methionine and cysteine as potential precursors of taurine within the brain.

A second area that pertains to control of taurine content in brain is the potential interrelationship with other amino acids, especially glutamic acid. Van Gelder first noted the existence of a taurine-glutamic acid interrelationship and has proposed that taurine may participate in the mechanism by which tissues retain glutamic

Table 4

Taurine and Glutamic Acid

in Rhesus Monkey Brain

	Adult		Neonate		Fetus	
			nanomole/g wet weight (n)			
Taurine	1519	(5)	4149	(10)	4956	(6)
Glutamic Acid	11255	(5)	14061	(10)	7920	(6)

Adapted from reference 34

acid, itself a potential neurotransmitter (45). In addition, Van Gelder has suggested that this relationship between taurine and glutamic acid may be important in understanding the etiology of some forms of epilepsy (44). Glutamic acid is a proposed neurotransmitter with excitatory properties, as opposed to the inhibitory properties of taurine. As does GABA, glutamic acid increases during development in contrast to taurine.

The biochemical development of sulfur amino acid metabolism in rhesus monkey brain has been studied in some detail as a model for normal primate brain development. The variation in glutamic acid and taurine from fetal, through neonatal to adult life led us to examine the potential correlation between these two amino acids. In fact, adult glutamic acid is higher than fetal, but lower than neonatal concentrations (Table 4) (33,34). On the other hand, the taurine in adult brain is significantly lower than in either fetal or neonatal brain with no easily discernible difference between fetal and neonatal brain taurine (33,34).

Considerable variation in brain taurine and glutamic acid concentrations may be observed amongst individual rhesus monkeys. However, a fairly consistent pattern of change occurs with the two compounds in any individual animal. Within an single age group of monkeys, when glutamic acid is increased taurine is increased and when glutamic acid is decreased taurine is decreased relative to the other animals in the same group. These variable concentrations and potential corresponding changes then seem to provide a good model for studying the correlation between glutamic acid and taurine. However, when all the data were examined as one group, no such correlation was observed ($r=0.26$, $\rho > 0.05$).

When the three groups of animals, fetal, neonatal, and adult, were studied separately a significant correlation between glutamic acid and taurine was observed for the fetal ($r=0.97$, $\rho < 0.01$) and for the adult ($r=0.99$, $\rho < 0.01$) but not for the neonatal animals. The neonatal animals could be divided into two groups, a younger group (less than 20 days of age) and an older group (more than 85 days of age), and combined with the fetal and adult groups respectively. These combined groups did show significant taurine to glutamic acid correlations ($\rho < 0.01$). The neonatal group appears to be a transitional group, perhaps indicating the development of a new pool of glutamic acid which is not directly involved in the interrelationship but which, once fully developed, lies beneath the pool that correlates increasing the baseline, or X-axis intercept, value (see Fig. 1).

In attempting to locate these various pools of taurine and glutamic acid, we examined the interaction in subcellular preparations of the same brains. Taurine and glutamic acid levels were correlated in the synaptosomal or nerve ending fractions of both fetal and adult brain but in the soluble fractions only in fetal brain. This last finding suggests that the taurine-glutamic acid relationship is of particular importance in the nerve-ending and that, perhaps, in the fetal brain differentiation of these pools has not yet occurred, explaining the correlation in the fetal brain soluble pool but not the adult brain soluble pool.

To check the validity of the taurine-glutamic acid relationship further a number of other species were examined. These data were obtained from previous experiments under a number of different conditions. As for the monkey, including all of one species in a single group often failed to give a significant correlation. When experimental groups within a species were examined positive correlations were observed. For example, four day and thirty-five day old beagle pups showed significant taurine-glutamic acid correlations respectively but the combined data from the puppies did not. Groups of adult control cats, gerbils and rats also had significantly correlated taurine-glutamic acid brain concentrations. It was interesting to note that brain of seizure-prone gerbils and of taurine-deficient cats did not have a significant correlation. The gerbil was of particular interest because no differences between control and seizure-prone animals were observed with respect to their brain taurine and glutamic acid concentrations. Perhaps changes in the relationship of taurine to glutamic acid is a more sensitive biochemical index than just a comparison of the absolute concentrations of these compounds.

In addition to these glutamic acid-taurine interrelationships, it is interesting to note that intrastriatal injections of the glutamic acid agonist, kainic acid, also affect taurine. Significant reductions in the endogenous content and high-affinity uptake of

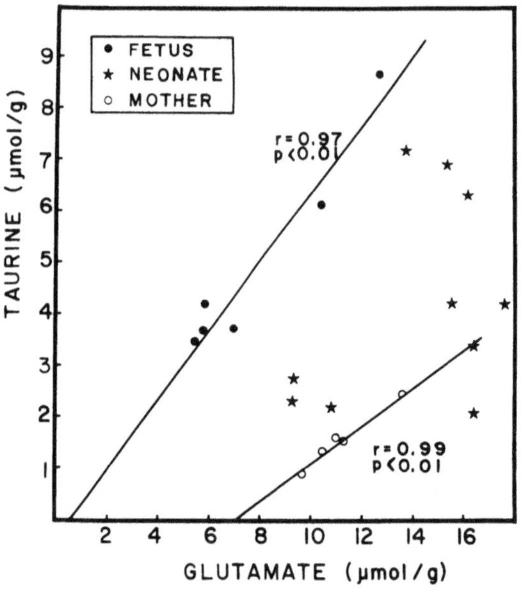

Fig. 1. Correlation coefficients were calculated by linear regression analysis of the concentrations of taurine and glutamic acid in whole homogenates prepared from rhesus monkey occipital lobe.

taurine have been observed following such injections, indicating that taurine is localized, at least partially, in kainic acid-sensitive neurones (22,25).

Thus, despite intensive investigation, it is not yet clear whether or not CSAD is a separate enzyme, or to what extent this enzyme controls the synthesis of taurine in the brain. Furthermore, it has been reported that CSAD is absent from axons (40). It is difficult to understand the meaning of this finding in the light of the considerable activity found in nerve terminals or synaptosomes. It seems unlikely that CSAD is synthesized within the nerve terminal so the amounts of enzyme being transported from the cell body at any one time may merely be below detectable limits. Immunocytochemical studies indicate that the same localization is true for GAD. It is present in nerve ending boutons but not in axons and cell somata (48). These latter authors suggest that the failure to detect GAD outside the nerve ending may either be due to the very low concentrations or to the synthesis of inactive forms of the enzyme at extraterminal sites (48). Similar studies need to be done comparing GAD and CSAD.

In addition to the structural similarities between taurine and GABA and the almost identical properties of their synthesizing enzymes, there may be a relationship between taurine and glutamic acid, the precursor of GABA. The possibility that normal brain function is dependent upon a carefully controlled ratio of taurine to glutamic acid is deserving of further investigation. It may be that the excitatory state of the brain is dependent upon the overall relationship between the amounts of the excitatory agent, glutamic acid, and the inhibitory agent, taurine. Such a control mechanism may eventually be found to go beyond just the regulation of such activity at the nerve ending and, thus, account for the large amounts of glutamic acid and taurine observed in brain.

ACKNOWLEDGMENTS

I am grateful for the longterm support and advice of Dr. Gerald E. Gaull. The secretarial skills of Kathy Horton are appreciated. Experimental results obtained by the author were carried out at the New York State Institute for Basic Research in Mental Retardation and supported by the New York State Office of Mental Retardation and Developmental Disabilities.

REFERENCES

1. Agrawal, H.C., Davison, A.N., and Kaczmarek, L.K., 1971, Subcellular distribution of taurine and cysteinesulphinate decarboxylase in developing rat brain, Biochem. J., 122:759-763.
2. Barbeau, A., and Huxtable, R.J., eds., "Taurine and Neurological Disorders, Raven Press, New York (1978).
3. Baskin, S., Kocsis, J.J., and Schaffer, S., "The Action of Taurine on Excitable Tissues," Spectrum Publications, Philadelphia, in press.
4. Bergeret, B., and Chatagner, F., 1954, Sur la presence d'acide cysteinesulfinique dans le cerveau du rat normal, Biochim. Biophys. Acta, 14:297.
5. DeBelleroche, J.S., and Bradford, H.F., 1973, Amino acids in synaptic vesicles from mammalian cerebral cortex: A reappraisal, J. Neurochem., 21:441-451.
6. Gaull, G.E., and Rassin, D.K., Taurine and brain development: Human and animal correlates, in: "Neural Growth and Differentiation," E. Meisami, and M.A.B. Brazier, eds., Raven Press, New York (1979).
7. Gaull, G.E., Rassin, D.K., Raiha, N.C.R., and Heinonen, K., 1977, Milk protein quantity and quality in low-birth-weight infants. 3. Effects on sulfur amino acids in plasma and urine, J. Pediatr., 90:348-355.
8. Goodman, H.O., King, J.S., and Thomas, J.J., 1964, Urinary excretion of beta-aminobutyric acid and taurine in mongolism, Nature, 204:650-652.
9. Grosso, D.A., and Bressler, R., 1976, Taurine and cardiac physiology, Biochem. Pharmacol., 25:2227-2232.

10. Hayes, K.C., Carey, R.E., and Schmidt, S.Y., 1975, Retinal degeneration associated with taurine deficiency in the cat, Science, 188:949-951.
11. Huxtable, R.J., and Barbeau, A., eds., "Taurine," Raven Press, New York (1976).
12. Jacobsen, J.G., Thomas, L.L., and Smith, L.H., Jr., 1964, Properties and distribution of mammalian L-cysteine sulfinate carboxylyases, Biochim. Biophys. Acta., 85:103-116.
13. Kaczmarek, L.K., Agrawal, H.C., and Davison, A.N., The biochemistry of taurine in developing rat brain, in: "Inherited Disorders of Sulfur Metabolism," N.A.J. Carson, and D.N. Raine, eds., Churchill Livingstone, London (1971), pp.63-69.
14. Knopf, K., Sturman, J.A., Armstrong, M., and Hayes, K.C., 1978, Taurine: An essential nutrient for the cat, J. Nutrition, 108: 773-778.
15. Levi, G., and Morisi, G., 1971, Free amino acids and related compounds in chick brain during development, Brain Res., 26: 131-140.
16. Malloy, M.H., Rassin, D.K., Gaull, G.E., and Heird, W.C., Development of taurine metabolism in beagle pups: Effects of taurine-free total parenteral nutrition, Biol. of the Neonate, in press.
17. Mandel, P., and Mark, J., 1965, The influence of nitrogen deprivation on free amino acids in rat brain, J. Neurochem., 12: 987-992.
18. Mandel, P., and Pasantes-Morales, H., 1976, Taurine: A putative neurotransmitter, Adv. in Biochem. Psychopharmacol., 15:141-151.
19. Martres, M.P., Baudry, M., and Schwartz, J.C., 1975, Histamine synthesis in the developing rat brain: Evidence for a multiple compartmentation, Brain Res., 83:261-275.
20. Mussini, M., and Marcucci, F., Free amino acids in brain after treatment with psychotropic drugs, in: "Amino Acid Pools," J.T. Holden, ed., Elsevier (1962), Amsterdam, pp.486-492.
21. Nakagawa, K., and Kuriyama, K., 1975, Effect of taurine on alteration in adrenal functions induced by stress, J. Pharmacol. Japan, 25:737-746.
22. Nicklas, W.J., Duvoisin, R.C., and Berl, S., 1978, Amino acids in rat striatum lesioned with kainic acid, Trans. Am. Soc. Neurochem., 9:91.
23. Pasantes-Morales, H., Klethi, J., Urban, P.F., and Mandel, P., 1972, The physiological role of taurine in retina: Uptake and effect on electroretinogram (ERG), Physiol. Chem. and Phys., 4:339-348.
24. Pasantes-Morales, H., Mapes, C., Tapia, R., and Mandel, P., 1976, Properties of soluble and particulate cysteine sulfinate decarboxylase of the adult and the developing rat brain, Brain Res., 107:579-589.
25. Placheta, P., Singer, E., Schonbeck, G., Heckl, K., and Karobath, M., 1979, Reduction of endogenous level, uptake and release of taurine after intrastriatal kainic acid injection, Neuropharmacology, 18:399-402.

26. Ransom, D.H., Oertel, W.H., Weise, V.D., Schmeckel, D.E., Krutzsch, H., and Kopin, I.J., 1980, Cysteinesulfinic acid decarboxylase isoenzyme in rat brain-Comparison with liver cysteinesulfinic acid decarboxylase, Soc. Neurosci. Abstr., 6:56.
27. Rassin, D.K., 1972, Amino acids as putative transmitters: Failure to bind to synaptic vesicles of guinea pig cerebral cortex, J. Neurochem., 19:139-148.
28. Rassin, D.K., and Gaull, G.E., 1975, Subcellular distribution of enzymes of transmethylation and transsulphuration in rat brain, J. Neurochem., 24:969-978.
29. Rassin, D.K., and Gaull, G.E., Taurine and other sulfur containing amino acids: Their function in the central nervous system, in: "Amino Acids as Chemical Transmitters," F. Fonnum, ed., Plenum Publishing Corporation, New York (1978), pp.571-597.
30. Rassin, D.K., and Gaull, G.E., Taurine: Significance in Human Nutrition, in: "The Action of Taurine on Excitable Tissues," S. Baskin, S. Schaffer, and J. Kocsis, eds., Spectrum, Philadelphia, in press.
31. Rassin, D.K., and Sturman, J.A., 1975, Cysteinesulfinic acid decarboxylase in rat brain. Effect of vitamin B_6 on soluble and particulate components, Life Sci., 16:875-882.
32. Rassin, D.K., Sturman, J.A., and Gaull, G.E., 1977, Taurine in developing rat brain: Subcellular distribution and association with synaptic vesicles of (^{35}S) taurine in maternal, fetal, and neonatal rat brain, J. Neurochem., 28:41-50.
33. Rassin, D.K., Sturman, J.A., and Gaull, G.E., 1979, Source of taurine and GABA in cerebrum of developing rhesus monkey, Trans. Am. Soc. Neurochem., 10:144.
34. Rassin, D.K., Sturman, J.A., and Gaull, G.E., Sulfur amino acid metabolism in the developing rhesus monkey brain: Subcellular studies of taurine, cysteinesulfinic acid decarboxylase, GABA, and glutamic acid decarboxylase, J. Neurochem., in press, 1981.
35. Rassin, D.K., Sturman, J.A., Hayes, K.C., and Gaull, G.E., 1978, Taurine deficiency in the kitten: Subcellular distribution of taurine and (^{35}S) taurine, Neurochem. Res., 3:401-410.
36. Schmidt, S.Y., Berson, E.L., and Hayes, K.C., 1976, Retinal degeneration in cats fed casein. 1. Taurine deficiency, Invest. Ophthalmol., 15:47-52.
37. Saito, K., Immunochemical studies of GAD and GABA-T., in: "GABA in Nervous System Function," E. Roberts, T.N. Chase, and D.B. Tower, eds., Raven Press, New York (1976), pp.103-111.
38. Sieghart, W., and Karobath, M., 1974, Evidence for specific synaptosomal localization of exogenous accumulated taurine, J. Neurochem., 23:911-915.
39. Spears, R.M., and Martin, D.L., 1980, Brain regional distribution of glutamate-insensitive cysteinesulfinate decarboxylase activity, Soc. Neurosci. Abstr., 6:443.
40. Sturman, J.A., Cysteinesulfinic acid decarboxylase activity in the mammalian nervous system: Absence from axons, J. Neurochem., in press, 1981.

41. Sturman, J.A., and Gaull, G.E., 1975, Taurine in the brain and liver of the developing human and monkey, J. Neurochem., 25: 831-835.
42. Sturman, J.A., Rassin, D.K., and Gaull, G.E., 1977, Taurine in development, Life Sci., 21:1-22.
43. Tyfield, L.A., and Holton, J.B., 1973, Alterations in sulphur-containing amino acids in rat brain extracts on storage, Clin. Chim. Acta., 45:415-421.
44. Van Gelder, N.M., 1978, Taurine, the compartmentalized metabolism of glutamic acid, and the epilepsies, Canad. J. Physiol. Pharmacol., 56:362-374.
45. Van Gelder, N.M., and Courtois, A., 1972, Close correlation between changing content of specific amino acids in epileptogenic cortex of cats and severity of epilepsy, Brain Res., 43:477-484.
46. Van Gelder, N.M., Sherwin, A.L., and Rasmussen, T., 1972, Amino acid content of epileptogenic human brain: Focal *versus* surrounding regions, Brain Res., 40:385-393.
47. Van Gelder, N.M., Sherwin, A.L., Sacks, C., and Andermann, F., 1975, Biochemical observations following administration of taurine to patients with epilepsy, Brain Res., 94:297-306.
48. Wood, J.G., McLaughlin, B.J., and Vaughn, J.E., Immunocytochemical localization of GAD in electron microscopic preparations of rodent CNS, in: "GABA in Nervous System Function," E. Roberts, T.N. Chase, and D.B. Tower, eds., Raven Press, New York (1976), pp.133-148.

PART VI

THE NEUROPHARMACOLOGY OF TAURINE

Ryan J. Huxtable

Department of Pharmacology
Health Sciences Center
University of Arizona
Tucson, Arizona

The neuropharmacology of taurine is a confusing and complex area. Despite this, many of the issues have become more clearly defined since the meeting in 1977. Perhaps the clearest message arising from the following papers is that these authors do not believe that taurine is a neurotransmitter.

In the first chapter in this section, Pasantes-Morales summarizes the evidence against a transmitter function for taurine. These includes the findings that depolarizing conditions do not release taurine under conditions where GABA is released, that no specific receptors for taurine can be found, and that the neuro-depressant actions of taurine may be antagonized either by glycine antagonists such as strychnine or by GABA antagonist such as bicuculline. No specific taurine antagonists have been found. Although none of these objections to taurine being a neurotransmitter are unanswerable, the increasing preponderance of evidence militates against it.

Pasantes-Morales supports an involvement of taurine in the contractile process, in structures as diverse as synaptosomes, retina, platelets and heart. In support of this, she shows that taurine release, although not affected by agents normally releasing neurotransmitters, is depressed by agents that disrupt contractile proteins. Furthermore, this proposed involvement of taurine in the contractile process may be closely related to its effects on calcium flux. A modulation of contractility would accord well with what is known of taurine in the heart. I have proposed that the effects of adrenergic stimulation of the heart, the so-called flight or flight response, is mediated by taurine, and that the effects of taurine and adrenergic agonists on calcium movements are but two aspects of the same phenomenon (1). When it is considered that contractile responses are involved in phenomena as diverse as excitation-contraction coupling, excitation-secretion coupling, and neurotransmitter release, modulation of contractility by taurine would unify and simplify a vast range of observations concerning this substance in both central nervous system and peripheral structures.

The next two chapters in this section discuss taurine binding sites, and both conclude that the evidence points to taurine not being a neurotransmitter. López-Colomé concludes that the majority of binding sites for taurine in the brain are in fact transport sites, based on the assumption that binding to post-synaptic receptors is sodium independent, whereas binding to transport sites is sodium dependent. She further finds that taurine binding is not especially concentrated into any one membrane fraction. Furthermore, the rather even distribution of binding sites for taurine throughout the brain, this not correlating with endogenous concentrations of taurine, raises the possibility that these sites include both modulatory and transport sites.

Whereas López-Colomé concentrated on taurine binding in the absence of sodium, Segawa and his coworkers have studied taurine binding in the presence of sodium. Binding was proportional to sodium concentration, and was unaffected by bicuculline. They conclude that taurine is binding to transport sites in glia and synaptic membranes. This latter binding they feel is in the presynaptic region. This could, therefore, be a modulatory site.

The fourth and fifth papers in this section delve into antagonistic actions of taurine on two peptides: angiotensin, and an enkephalin analogue. Izumi and his colleagues find that taurine suppresses akinesia and analgesia induced by the enkephalin analogue, and reverses the enkephalin-induced decrease in dopamine release. This is a further addition to the literature of interactions of taurine and opiate pharmacology in the central nervous system.

Furukawa and coworkers find that taurine antagonizes the central cardiovascular actions of angiotensin II. Apart from the direct neuropharmacological interest of this work, these and other studies indicate that the cardiac actions of taurine may not be limited to direct effects. The pharmacological studies suggest that taurine may have physiological effects on the heart via central, and not solely local, actions.

Some of the earliest neuropharmacological studies on taurine were concerned with thermoregulation. Chapters by Sgaragli and Lipton continue these studies. This action of taurine again raises the question as to the involvement of GABA and GABA receptors in this response. In light of the importance of sodium:calcium ratios in temperature regulation, one might also wonder about the involvement of calcium in the hypothermic response to taurine. This area is of clinical interest because of the insights it may yield on temperature control and fever regulation.

One neurological disorder in which there are tantalizing indications of disturbances in taurine regulation is Friedreich's ataxia. The work of Barbeau has been of seminal importance in this area, and he is to be credited with increasing our knowledge of Friedreich's ataxia by more within the last five years then we had learned over the previous century. It is apposite, therefore, that the last paper is a review by him of taurine and Friedreich's ataxia. Nutrition, cardiomyopathy, epilepsy, and cerebellar degeneration - these are some of the areas of public health and public concern in which taurine is increasingly being shown to be involved. The study of this fascinating substance is not only a worthy scientific problem in its own right, but is resulting in increased insights into the intricate balance of biochemistry that fuels our existence, and will result in an increased ability to moderate some of the ills and woes to which we are prey.

REFERENCE

1. Huxtable, R.J., and Chubb, J., 1977, Adrenergic stimulation of taurine transport by the heart, Science, 198:409-411.

THE ROLE OF TAURINE IN NERVOUS TISSUE: ITS EFFECTS ON IONIC FLUXES

H. Pasantes-Morales, N.E. Arzate and C. Cruz

Departamento de Neurociencias
Centro de Investigaciones en Fisiología Celular
Universidad Nacional Autónoma de México

The recognition of taurine and GABA, two neuroactive amino acids, as normal constituents of nervous tissue, occurred almost at the same time (1,43). Their structural similarity and the resemblance of their effects on neuronal activity had suggested similar roles for them in nervous function. However, whereas during the last few years an impressive amount of evidence has accumulated supporting a role for GABA as an inhibitory neurotransmitter (44), the elucidation of the role of taurine in nervous function has been particularly difficult. Some experimental evidence exists supporting a role for taurine as a neurotransmitter. It has been demonstrated that taurine, when applied via microiontophoresis, excerts a depressant effect on neuronal activity in several regions of the central nervous system (10-12). However, the unsuccessful search for specific blockers of this action of taurine, together with the antagonism shown by strychnine and bicuculline on inhibitory effects of taurine (10-12), have raised some doubts about the specificity of taurine as an inhibitory neurotransmitter itself. Since strychnine and bicuculline are likely to be specific antagonists of GABA and glycine receptors, the effect of taurine on postsynaptic neurons blocked by these drugs could be mediated through an interaction with GABA and glycine receptors. Studies *in vitro*, using biochemical techniques to characterize GABA and glycine postsynaptic receptors in several regions of the CNS, have clearly shown a significant interaction of taurine with GABA and glycine binding to their specific receptors (15,51). Furthermore, the identification of taurine receptors by this technique, has proved unsuccessful (López-Colomé, this volume).

Neurotransmitters, including amino acids, are released from nerve terminals in response to experimental conditions which simulate nerve excitation. The release of taurine coupled with changes in neuronal activity has been observed in a variety of nervous preparations, *in vivo* and *in vitro*. Electrical stimulation of brain slices and spinal cord slices has been shown to enhance the release of taurine (7,20,36). In contrast, depolarization by high concentrations of potassium did not increase taurine efflux from slices of six different regions of the CNS (27). One major objection to the use of tissue slice preparations for the study of transmitter release is the uncertainty as to the cellular localization of the site of release and the validity of this criterion as support for a neurotransmitter role is, therefore, conditional on the demonstration that the stimulated efflux occurs from the nerve terminals. Sieghart and Heckl (46) showed an increase in taurine release in response to depolarizing concentrations of potassium from a crude synaptosome fraction obtained from rat cerebral cortex. Experiments in our laboratory were carried out to extend this study to other depolarizing conditions, and for this purpose, synaptosomes labeled simultaneously with ^{3}H-GABA and ^{35}S-taurine were exposed to ouabain and veratrine. These agents failed to evoke an increased release of taurine, whereas high potassium concentration increased taurine efflux by 60% over the resting efflux value. In contrast, the release of GABA elicited by the three depolarizing conditions was increased more than 6 times (600%) over the prestimulation value (37). Similar results were observed using a crude synaptosomal fraction obtained from the chick retina (28). Since the same synaptosomal preparation was used for both GABA and taurine experiments, the failure of depolarizing agents to elicit taurine release cannot be attributed to technical deficiencies or variation in the biological preparation, but rather to a difference in the mechanism responsible for the release of the two amino acids.

It is generally accepted that under physiological conditions neurotransmitter release is triggered by the increased availability of calcium at the presynaptic terminal and, in keeping with this notion, the stimulated release of all neurotransmitters so far studied is a calcium-dependent process. In the case of taurine release calcium dependence is controversial; in some preparations, under certain stumulating conditions, taurine release is reduced in calcium free media (21,22,35), in others it is calcium-independent (28) or even, as shown by Sieghart (46), increased in the absence of calcium. In our experiments, the slight increase in taurine efflux elicited by potassium is higher in the absence than in the presence of calcium, in agreement with the observation of Sieghart et al. (46). All these results suggest that the release of taurine cannot be directly related to synaptic transmission on the basis of a neurotransmitter function. However, taurine could be released from other cell compartments or from glial cells and may thus be implicated in nervous function. In this respect, it should be pointed out that a specific mechanism for taurine transport, exhibiting high affinity constants, has been found in nervous tissue slices as well as in synaptosomes (16,26).

TAURINE AND CONTRACTILE PROCESSES IN NERVOUS TISSUE

Unlike GABA which is only found in the nervous system, taurine is present in many other animal tissues. Outside the CNS, taurine is found in particularly high concentrations in heart, striated muscle and secretory glands (9,19). Obviously, in muscle and secretory organs taurine is not playing a neurotransmitter role, and a different function should be envisaged. At this point it is pertinent to question whether taurine is involved in a single function in all tissues or if it plays a different role in each tissue. Taurine may be involved in a common mechanism mediating contractile, nervous and secretory functions. The high amounts of taurine present in muscle (23), as well as its pharmacological effects on cardiac function (14), have suggested its involvement in contractile mechanisms. However, contractile processes are not restricted to muscle; they participate in the basic function of several non-muscle cells, and this could provide the basis for a common role for taurine in different tissues. There is increasing evidence emphasizing the importance of contractile phenomena in secretory processes, in exocrine as well as in endocrine glands (8,29). Contraction also plays a key role in cell motion and in the maintenance of cell shape (39). The release of neurotransmitters also appears to involve a mechanico-chemical transduction comparable to the contraction of actomyosin (3). It is relevant to our speculation concerning a role in contractile mechanisms in non-muscle cells that the highest concentrations of taurine have been found in cells such as lymphocytes and platelets (19) in which the importance of contractile events is well documented. In retina, in which the distribution of the amino acid through the different retinal layers is now well known (33), taurine is found preferentially localized at the contractile elements of the retina. The highest taurine concentration in all species so far studied is found in the inner segment of photoreceptors and in the outer nuclear layer (33). In birds and fishes, the inner segment of photoreceptors is a true contractile element which elongates and contracts in response to light adaptation (50). In mammals, fibrillar structures presumably of a contractile nature, run through the photoreceptor, forming a tight envelope around the nucleus (48). A contractile protein, actin, has also been found in the outer segments of photoreceptors. In nervous tissue, the organelles exhibiting the highest concentration of taurine are probably synaptic vesicles (13), the presence of contractile proteins in these intrasynaptosomal structures being well documented (40). Therefore, one might speculate that the release of taurine under conditions which evoke neurotransmitter efflux is related to an interaction of taurine with contractile processes in the mechanism of transmitter release, rather than to an action as a true neurotransmitter. In order to test this possibility we studied the effect of drugs known to interfere with secretory processes involving contractile mechanisms; namely, colchicine, cytochalasin B and vinblastin, on the release of taurine from the whole retina elicited by light or by depolarizing agents. Redburn and Cotman (42) and Nicklas and Berl (32) have previously shown that the release of the neurotransmitter amino acids, GABA or glutamic acid, is unaffected by agents disrupting contractile proteins, while, in contrast, liberation of catecholamines is considerably reduced by these drugs. This difference is presumably due to the different intrasynaptosomal localization of neurotransmitter amino acids and catecholamines or acetylcholine. Whereas the latter are found stored in synaptic vesicles, transmitter amino acids appear to be localized in the synaptoplasm, and not related to synaptic vesicles. According to these observations, the release of taurine, if it were acting as a neurotransmitter, should be unaffected by colchicine and cytochalasin B. Table I shows the effect of these drugs on the release of taurine from the whole retina stimulated by a high potassium concentration, glutamate or veratrine. The presence of the alkaloids markedly reduced the release of taurine whereas that of GABA was unaffected. In agreement with the observations of Nicklas and Berl (32), the efflux of dopamine was practically abolished by colchincine and cytochalasin B (38). Fig. 1 shows the effect of these drugs on the light-stimulated release of ^{35}S-taurine from the chick retina. Also, in these experiments, the efflux of taurine was markedly reduced by colchicine in a dose-dependent manner, being inhibited by more than 40% in the presence of 10^{-4} M colchicine (Fig. 1) and by more than 90% when it is present at a concentration of 10^{-3} M (38). Cytochalasin B, at concentration of 10^{-5} M practically abolished the light stimulated release of taurine (Fig. 1). These observations indicate that the release of taurine evoked by light, as well as by depolarizing agents, is sensitive to the action of drugs disrupting contractile proteins, in contrast to that observed for transmitter amino acids. These results suggest an involvement of taur-

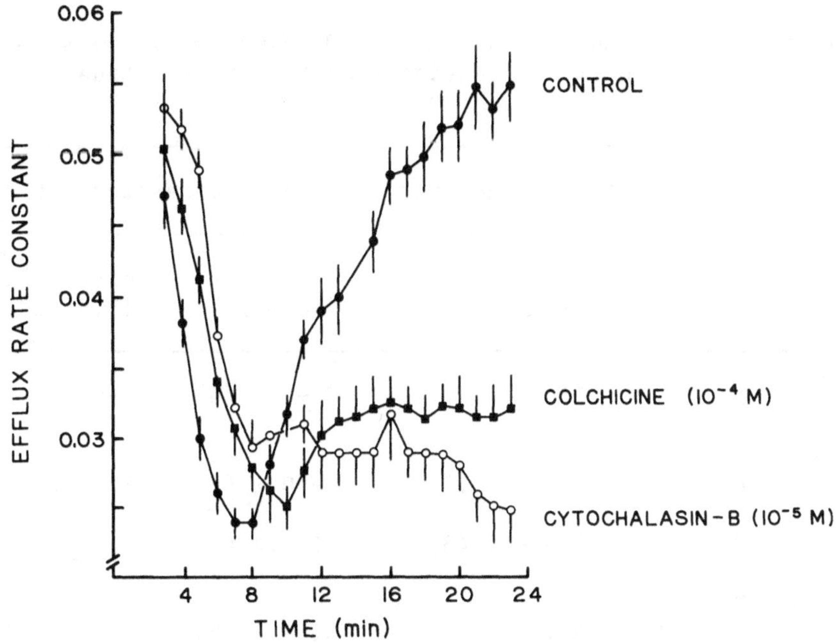

Fig. 1. The effect of colchicine (10^{-4} M) and cytochalasin B (10^{-5} M) on the light-stimulated release of ^{3}H-taurine from the chick retina. Retinas of dark adapted chicks were loaded with ^{3}H-taurine, washed and superfused with warmed, Krebs-bicarbonate medium, at a rate of 1.5 ml/min. Fractions were collected every minute; during the 8th collection period, 15 flashes were applied, and superfusion continued for 20 min. Cytochalasin B was present during loading and superfusion periods. Colchicine was present during a preincubation period of 20 min, as well as during the uptake and throughout all the prefusion period. Results are means ± SEM of 5 separate experiments. Control (●); Colchicine (■); Cytochalasin B (O).

ine in contractile processes occurring in retina at the synaptic level as well as at the photoreceptor level. These observations suggest a general role of taurine in contractile function in muscle and non-muscle tissues.

Turning now to the mechanism through which taurine might be implicated in contractile function, we have explored the possibility of a calcium mediated action, in view of the critical role played by intracellular calcium levels in contraction, secretion and neurotransmitter function.

TABLE I

The effect of colchicine and cytochalasin B on the stimulated release of (^{35}S)-taurine from the chick retina

Drug		Percentage increase over the resting efflux		
		Glutamate (10 mM)	KCl (70 mM)	Veratridine (0.01 mM)
Control		140 ± 7 (8)	68 ± 11 (5)	96 ± 8 (5)
Colchicine	0.5 mM	4 ± 0.2 (5)	8.3 ± 0.1 (6)	3 ± 0.1 (5)
	0.1 mM	103 ± 9 (3)	36 ± 4 (3)	—
Cytochalasin B	10^{-4} M	3.6 ± 0.6 (6)	32 ± 3 (5)	37 ± 3 (5)
	10^{-5} M	101 ± 11 (3)	52 (2)	37 ± 23 (5)
	10^{-6} M	112 ± 14 (3)	—	—

Retinas of dark adapted chicks were incubated in 1 ml of Krebs bicarbonate medium containing ^{35}S-taurine (0.5 μCi/ml, 15 μM) for 10 min, at 37°C. After incubation, loaded retinas were washed and perfused in glass chambers of 0.25 ml volume. Superfusion was carried out with warmed medium, pumped at a rate of 0.6 ml/min. Superfusate was collected as fractions over 1-min periods. Colchicine and cytochalasin B, at the indicated concentrations were present during the uptake and throughout the superfusion period. The stimulatory agents were added after the basal line of spontaneous release was attained, usually at the ninth perfusion fraction. Results are expressed as mean percentage increases over the resting efflux. Values are the means ± SEM of the number of experiments indicated in parenthesis.

TAURINE AND CALCIUM FLUXES

Previous observations of Dolara et al. (14) have shown that some actions of taurine in heart, namely its positive inotropic effect as well as its effects on the action potential of Purkinje fibers, are consistent with a taurine induced modification of intracellular calcium levels. The decrease of contractile force produced by perfusing hearts with calcium-free medium is counteracted when taurine is added to the perfusion buffer prior to the perfusion with calcium free medium. In those experiments, taurine was found to produce a modification in the kinetics of calcium efflux (14), resulting in a decreased loss of calcium from the heart, thereby leading to a larger pool of calcium available for contraction. These observation strongly suggested that the cardiac actions of taurine were calcium related.

According to our original hypothesis of a common role for taurine in contractile, secretory and nervous tissue, we explored the possibility that taurine actions in nervous tissue were also calcium related. For this purpose, the effect of taurine on calcium transport was examined in various nervous preparations. ^{45}Calcium accumulation by retinal subcellular fractions isolated from the chick retina is decreased when taurine is present in the incubation medium (34). This action of taurine is detected in a crude synaptosomal fraction containing pinched-off "conventional" nerve endings, as well as in isolated rod outer segments (34). Similar results were obtained by Huxtable et al. (17) using a sarcolemma preparation. In preliminary experiments we observed the same effect of taurine on ^{45}Ca accumulation by rat brain synaptosomes (35). Therefore, this action of taurine appears to be exerted in different biological preparations, providing some support for the hypothesis of a common mechanism involving taurine in the function of different tissues. Insight into the mechanism of action of taurine on calcium fluxes requires the characterization of the calcium transport mechanism which is affected by taurine. We attempted to do this in two preparations, rat brain synaptosomes as well as in rod outer segments, in order to look for differences or similarities between the two systems as well as to investigate eventual differences in the taurine effects. Calcium transport by brain synaptosomes appears to occur through at least three different mechanisms: a carrier mediated calcium transport which involves a sodium-calcium or calcium-calcium exchange; a net uptake of calcium through voltage sensitive channels, and a passive calcium diffusion (4). In rod outer segments, a calcium-sodium exchange has also been identified, together with a net ATP-dependent calcium uptake (45).

^{45}Ca Accumulation at High Concentrations of $CaCl_2$

Properties of the calcium accumulation process by rat brain synaptosomes and frog rod outer segments incubated in a Krebs bicarbonate medium containing 2.5 mM $CaCl_2$ are summarized in Table 2. Under these conditions, ^{45}Ca accumulation is unaffected by ATP, ouabain or depolarizing concentrations of KCL. It is significantly reduced by incubation at 4°C, but it is insensitive to oligomycin or dinitrophenol. These results indicate that calcium transport by synaptosomes, at this external calcium concentration, is not occurring through the voltage sensitive calcium channels described in synaptosomes. Reversal of sodium gradient increases calcium accumulation, a result suggesting that the sodium-calcium exchange mechanism might be involved. Omission of phosphate from the incubation medium markedly reduces ^{45}Ca accumulation by both fractions. Phosphate could not be replaced by other anions such as acetate or oxalate. To our knowledge, this effect of phosphate on calcium transport by nerve endings has not been described before. A phosphate-dependent calcium uptake has been observed in kidney and heart mitochondria in which calcium transport is highly stimulated by phosphate only at external calcium concentrations higher than 2.5 mM, whereas it is unaffected by phosphate at lower calcium concentrations (5,49). Also, a phosphate dependent calcium transport has been described in brain mitochondria (31). It has been suggested that phosphate is required for calcium binding sites on the inside of the inner mitochondrial membrane to accept calcium from the carrier site (31). Acetate and other anions cannot react with these membrane binding sites and are thus unable to remove calcium from the carrier. Although ^{45}Ca accumulation by isolated nerve endings is different from that by mitochondria, as is shown by its insensitivity to mitochondrial poisons, these two fractions appear to share the phosphate requirement.

Effects of Taurine at High Concentrations of Calcium

Taurine added to the incubation medium produces a dose-dependent decrease in calcium accumulation by both synaptosomes and rod outer segments (Fig. 2). This effect of taurine appears to be exerted on the phosphate-dependent calcium accumulation, since it is not observed in a phosphate-free medium and increases with increasing phosphate concentrations (Fig. 3). This effect of taurine seems to be specific; GABA, glycine or leucine do not modify calcium accumulation. The decreased ^{45}Ca accumulation induced by taurine is observed at pH's between 6.5 to 8.3, with the maximum effect at pH 7.8. The observation that taurine effects are produced on the phosphate supported calcium transport raises the possibility that its primary action is on phosphate transport and only secondarily on calcium accumulation. Fig. 4 shows the effect of taurine on the accumulation of ^{32}P by synaptosomes. A decrease in ^{32}P accumulation was observed at all the phosphate concentrations used, being maximal at 3 mM phosphate concentrations. The mechanism of phosphate transport by synaptosomes has not been described so far. In cardiac muscle, ^{32}P uptake is markedly dependent on the extracellular sodium concentration and appears to proceed via a reversible electroneutral cotransport system linked to a transmembrane sodium gradients (30). This was not observed in synaptosomes, suggesting that the transport process is different from that present in sarcolemma. ^{32}P accumulation is markedly decreased under conditions which reduce calcium transport, such as ruthenium red addition or in a calcium free medium, but it is unchanged at high Mg concentrations. These results suggest that calcium-phosphate co-transport is occurring.

^{45}Calcium Accumulation at Low Calcium Concentration

The properties of ^{45}Ca accumulation by synaptosomes and rod outer segments, incubated in a medium containing a low (50-100 μM) calcium concentration, differ from those observed at high calcium levels. Under these conditions, calcium transport is phosphate independent but it is markedly activated by ATP. In both preparations, increasing the ATP concentration, from 0.5 to 2.0 mM, leads to a concomitant enhancement of calcium accumulation. This process appears to show a specific requirement for ATP; other nucleotides, GTP, cAMP, and the non-hydrolizable analog β-γ-methylene ATP, do not replace ATP in its stimulatory effect on calcium accumulation (Table 3). This process is unaffected by the mitochondrial poisons oligomycin and dinitrophenol but is decreased by ruthenium red. These results raise the possibility that an ATPase is involved in calcium accumulation under these conditions. In this respect, the presence of a calcium-dependent ATPase in a synaptosomal membrane preparation has been recently described (41). Also, an ATP-dependent, oligomycin unsensitive calcium transport has been described in heart mitochondria (47), which shows some similarities with that observed in our preparation. Since a similar process for ^{45}Ca accumulation has been observed in the rat P_2 fraction, a crude mitochondrial fraction containing synaptosomes and mitochondria, the possibility that mitochondrial calcium uptake is also involved cannot be ruled out.

TABLE 2

Properties of ^{45}Ca accumulation by rat brain synaptosomes and by frog rod outer segments incubated in a medium containing 2.5 mM $CaCl_2$

SAMPLE CONDITION		^{45}Ca accumulation (nmole/mg protein) Synaptosomes	Rod outer segments
Control		1.11 ± 0.04 (9)	3.24 ± 0.32 (4)
4°C		0.22 ± 0.01 (4)	1.18 ± 0.13 (4)
ATP	(1 mM)	1.17 ± 0.16 (6)	3.45 ± 0.16 (12)
KCl	(70 mM)	1.00 ± 0.09 (4)	-
Ouabain	(10.4 mM)	0.99 ± 0.07 (4)	3.53 ± 0.29 (3)
Ruthenium red	(20 μM)	0.96 ± 0.08 (4)	1.69 ± 0.33 (4)
Oligomycin	(7 μg/ml)	1.09 ± 0.09 (4)	-
Low-sodium medium		1.88 ± 0.09 (3)	-
Phosphate-free medium		0.45 ± 0.03 (6)	1.63 ± 0.07 (4)

Synaptosomes were obtained from rat brain as previously described (35), and rod outer segments by gentle brushing of frog retinas. Synaptosomes (0.7-1.0 mg protein) were resuspended in 0.32 M glucose and incubated at 37°C for 5 min in 1 ml (final volume) of Krebs-bicarbonate medium (NaCl, 118 mM; KCl 4.7 mM; KH_2PO_4, 1.2 mM; $MgSO_4$, 1.17 mM, $CaCl_2$, 2.5 mM and $NaHCO_3$ 25 mM), pH 7.6. The incubation mixture contained 0.4 μCi of $^{45}CaCl_2$ and the modifications or additions indicated at each experiment. For the low sodium medium condition, synaptosomes were preincubated for 15 min in a Ca-free bicarbonate medium (118 mM NaCl) containing 10^{-3} M ouabain and then incubated for 5 min in the presence of $^{45}CaCl_2$ and 2.5 mM unlabeled $CaCl_2$ in a medium where NaCl was substituted of equimolar concentrations of choline chloride. Rod outer segments (0.5 - 0.7 mg protein) were incubated under the same condition except that the incubation temperature was 25°C. The reaction was terminated by centrifugation of 0.3 ml aliquots of the incubation mixture in a Beckman Microfuge and radioactivity incorporated was measured in washed, solubilized pellets after the addition of Tritosol. Results are means ± SEM of the number of experiments indicated in parenthesis.

Effects of Taurine at Low Concentrations of Calcium

When the concentration of calcium in the Krebs-bicarbonate medium is decreased to 50-100 μM, the inhibitory effect of taurine on ^{45}Ca accumulation is no longer observed and a stimulatory effect is observed instead. This is evident in both rat brain synaptosomes and isolated rod outer segments, in which 25 mM taurine increases ^{45}Ca accumulation by more than 80% (Fig. 5). The stimulatory effect of taurine appears to be exerted mainly on the ATP-stimulated ^{45}Ca accumulation. In synaptosomes, this effect is inhibited when phosphate is present in the incubation medium, whereas in rod outer segments, phosphate does not affect the activatory action of taurine. The effect of taurine is more evident when bicaronate buffer is used to maintain the pH, but it is also observed in a Krebs-Tris medium. When HEPES replaces Tris or bicarbonate, taurine has no effects on calcium accumulation. A stimulatory effect of taurine on calcium binding has also been described by Chovan et al. (6) in a preparation of sarcolemma, at low external calcium concentration. Also, a slight stimulatory effect of taurine on ^{45}Ca accumulation by rat brain mitochondria, has been observed by Kuriyama et al. (24).

Evidence presented in this discussion may be interpreted as supporting the involvement of taurine in the regulation of calcium fluxes in synaptic terminals. Identical results have been obtained in frog rod outer segments and the observations of Huxtable et al. (17) and of Chovan et al. (6) on sarcolemma suggest that it might be an extended property of taurine. Much, however, remains to be done before the mechanism responsible for the observed actions of taurine can be elucidated. From the results of the present work it is suggested that, at high calcium concentrations, taurine inhibits a phosphate dependent calcium transport or binding. Its effects might involve a dissociating effect on the formation of phosphate-calcium complexes which may be a previous step in the binding to carrier sites. In a cell free preparation, taurine shows a destabilizing effect on calcium-phosphate crystals (Unpublished observation). On the basis of the same effect, namely a dissociation of calcium phosphate, in experiments carried out at low calcium concentrations and in a phosphate-free medium, the only source of endogenous phosphate would be that produced by the hydrolysis of ATP. Taurine, by destabilizing calcium-phosphate complexes, would make more phosphate available for calcium transport since, as has been mentioned above, this anion appears to be necessary for the separation of calcium from its carrier sites.

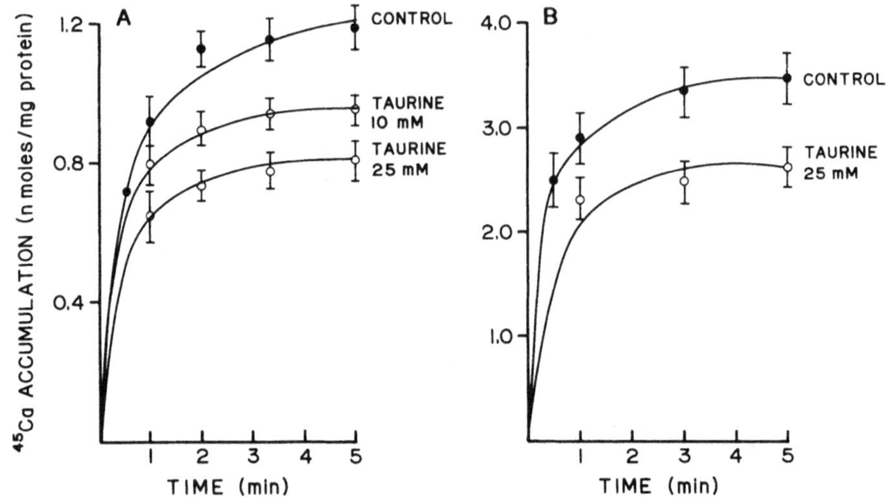

Fig. 2. The effect of taurine (25 mM and 10 mM) on ^{45}Ca accumulation by rat brain synaptosomes (A) and frog rod outer segments (B). Fractions were incubated in a Krebs-bicarbonate medium (see Table 2) containing 2.5 mM $CaCl_2$ and 0.4 μCi of ^{45}Ca. At the indicated times, aliquots were withdrawn and centrifuged in a Microfuge. The pellet was washed, solubilized and the radioactivity accumulated was measured by liquid scintillation spectrometry. Results are means ± SEM of 5-6 separate experiments. Control (●); Taurine (O).

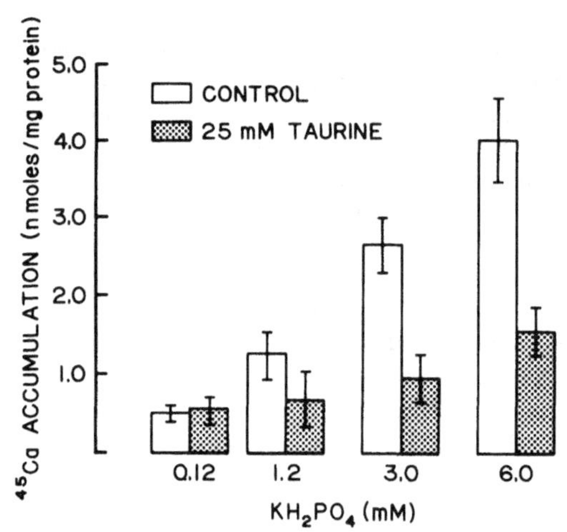

Fig. 3. The effect of taurine (25 mM) on ^{45}Ca accumulation by rat brain synaptosomes in the presence of varying concentrations of KH_2PO_4. Synaptosomes were incubated during 5 min in a Krebs-bicarbonate medium (1 ml final volume) containing 0.4 µCi of ^{45}Ca and the indicated concentrations of phosphate. The radioactivity accumulated was measured as described in Fig. 2. Values are the means ± SEM of 6-8 experiments except for points with the highest concentrations of KH_2PO_4 which are means of two separate experiments.

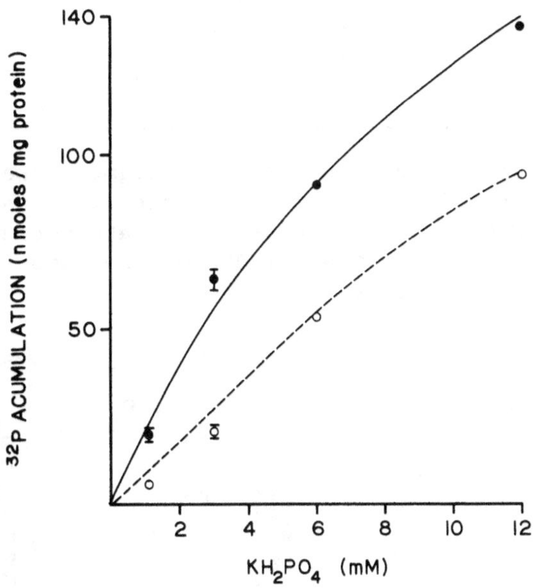

Fig. 4. The effect of taurine (25 mM) on the accumulation of ^{32}P by rat brain synaptosomes at different concentrations of KH_2PO_4. Synaptosomes were incubated for 5 min, at 25°C, in a Krebs-bicarbonate medium containing 1.0 µCi of ^{32}P and the concentrations of KH_2PO_4 indicated in the Figure. The radioactivity accumulated was measured as described in Fig. 2. Results are the means ± SEM of 5-8 experiments. Control (●); Taurine (0).

TABLE 3

Properties of ^{45}Ca accumulation by rat brain synaptosomes and by frog rod outer segments incubated in a medium containing 50-100 µM $CaCl_2$

Sample Condition	^{45}Ca accumulation	(nmole/mg protein)
Control	0.52 ± 0.03 (8)	0.83 ± 0.09 (8)
ATP 0.5. mM	0.74 ± 0.06 (6)	2.48 ± 0.17 (4)
1.0 mM	1.23 ± 0.09 (10)	3.58 ± 0.23 (8)
2.0 mM	2.19 ± 0.13 (6)	4.24 ± 0.37 (4)
GTP 1.0 mM	0.56 ± 0.07 (4)	—
CAMP 1.0 mM	0.54 ± 0.05 (4)	—
β-γ-methylene ATP (1.0 mM)	0.51 ± 0.04 (4)	—
Phosphate-free medium	0.49 ± 0.03 (4)	—

Rat brain synaptosomes and frog rod outer segments were obtained and incubated as described in Table 2, except that the concentration of $CaCl_2$ in the Krebs-bicarbonate medium was 50-100 µM. Results are the means ± SEM of the number of experiments indicated in parenthesis.

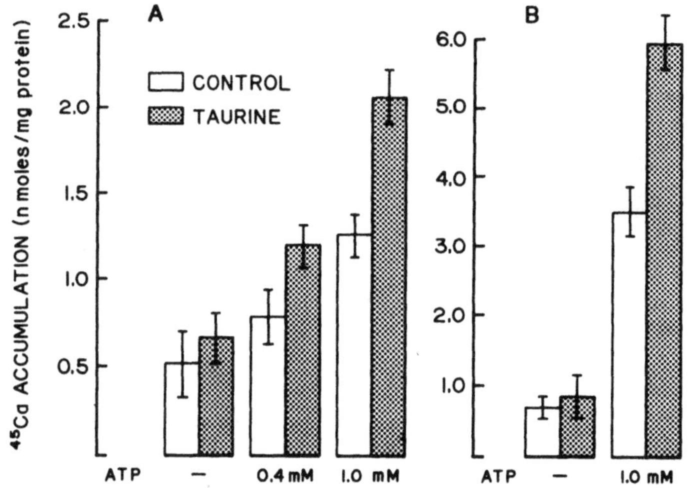

Fig. 5. The effect of taurine (25 mM) on ^{45}Ca accumulation by a crude synaptosomal fraction (P_2) (A) and frog rod outer segments (B) in a medium containing low $CaCl_2$ concentration (50-100 µM) and of varying concentrations of ATP. Incubation of fractions and measurements of ^{45}Ca accumulation were carried out as described in Fig. 2. Results are the means ± SEM of 4-8 separate experiments.

The question now rises as to whether the effects of taurine on calcium accumulation described in the present work have a physiological implication and also whether they could be responsible for the pharmacological effects of taurine in the CNS. Taurine shows a protective effect against a wide variety of convulsive states, with very different etiologies (2), suggesting that its action is mediated through a general mechanism regulating nervous excitability. Izumi et al. (18) have shown that the anticonvulsant effect of taurine against metrazol requires the presences of free calcium, thus relating its protective effect to a calcium-mediated mechanism. Taurine also protects against seizures produced by 4-aminopyridine, which probably causes convulsions by increasing calcium entry into the nerve terminal (25), and subsequently triggering a massive neurotransmitter release. These observations further support the notion that the anticonvulsant effect of taurine is related to calcium fluxes at the nerve terminal.

There is evidence that the gating of sodium and potassium channels in nerve membranes is influenced by fixed negative charges in the surface of the membrane. Normally, the divalent cations calcium and magnesium reduce the electrostatic effects of fixed charges on the outside of the membrane either by forming a double ionic layer near the charged sites or by binding to the membrane in the vicinity of the charged sites. Solutions with low divalent cation concentrations are thought to increase excitability by reducing the availability of cations which can fix negative charges on the outside surface of the nerve membrane. Taking these statements into account, we can speculate on whether the widespread effects of taurine as on anticonvulsant are due to a specific action on nervous membranes, by blocking calcium entry into the cell and then increasing its concentration in the vicinity of the membranes. The effects of taurine on calcium transport might also provide a background for its proposed role as a neuromodulator. The central role of calcium in mediating the release of neurotransmitters is widely recognized. Calcium entering into the nerve terminal triggers the release of neurotransmitters and is subsequently sequestered by intraterminal organelles, again lowering calcium levels to resting values. Taurine, by increasing calcium uptake into intraterminal mitochondria or other organelles, would contribute to the lowering of calcium levels and therefore to the termination of the process of neurotransmitter release.

ACKNOWLEDGMENTS

This work was partially supported by Grant No. 2 R01 EY 02540-03 from National Eye Institute. We are grateful to Araceli Ordoñez, QFB for technical assistance.

REFERENCES

1. Awapara, J., Landua, A.J., and Fuerst, R., 1950, Distribution of free amino acids and related substances in organs of the rat, Biochim. Biophys. Acta, 5:457-462.
2. Barbeau, A., Inowe, N., Tsukada, Y., and Butterworth, R.F., 1976, The neuropharmacology of taurine, Life Sci., 17:669-678.
3. Berl, S., and Nicklas, W.S., Contractile proteins in relation to transmitter release, in: "Metabolic Compartmentation and Neurotransmission," S. Berl, D.D. Clarke and D. Schneider, eds., Plenum Press, New York, (1975), pp. 247-272.
4. Blaustein, M.P., Ratzlaff, R.W., and Kendrick, N.K., The regulation if intracellular calcium in presynaptic nerve terminals, in: "Calcium Transport and Cell Function," A. Scarpa and E. Carafoli, eds., Ann. N.Y. Acad. Sci. Vol. 307, (1978) pp. 195-212.
5. Brierley, E., Murer, E., and Bachmann, E., 1964, Studies on ion transport. III, Accumulation of calcium and inorganic phosphate by heart mitochondria, Arch. Biochem. Biophys., 105:89-102.
6. Chovan, J.P., Kulakowski, E.C., Benson, B.W., and Schaffer, S.W., 1979, Taurine enhancement of calcium binding to rat heart sarcolemma, Biochem. Biophys. Acta., 551:129-136.
7. Collins, C.G.S., and Topiwala, S.H., 1979, The release of ^{14}C-taurine from slices of rat cerebral cortex and spinal cord evoked by electrical stimulation and high potassium ion concentration, Br. J. Pharmacol., 50:451-452.
8. Cooks, P.H., and Poisner, A.M., The role of cytoskeleton in adreno medullary secretion, in:"The Cytoskeleton in Normal and Pathological Processes," G. Gabbiani, ed., (1979), S. Karger, Basel, pp. 137-146.
9. Crabai, F., Sitzia, A., and Pepeu, G., 1974, Taurine concentration in the neurohypophysis of different animal species, J. Neurochem., 23:1091-1092.
10. Curtis, D.R., Duggan, A.W., Felix, D., Johnsten, G.A.R., and McLennan, H., 1971, Antagonism between bicuculline and GABA in the cat brain, Brain Res., 33:57-73.

11. Curtis, D.R., Hosli, L., and Johnston, G.A.R., 1968, A pharmacological study of the depression of spinal neurons by glycine and related amino acids, Exp. Brain Res., 6:1-18.
12. Curtis, D.R., and Tebecis, A.K., 1972, Bicuculline and thalamic inhibition, Exp. Brain Res., 16:210-218.
13. DeBelleroche, J.J., and Bradford, H.F., 1973, Amino acids in synaptic vesicles from mammalian cerebral cortex: a reappraisal, J. Neurochem., 21:441-451.
14. Dolara, P., Agresti, A., Giotti, A., and Pasquini, G., 1973, Effect of taurine on calcium kinetics of guinea pig heart, Eur. J. Pharmacol., 24:352-358.
15. Greenlee, D.V., Van Ness, P.C., and Olsen, R.W., 1978, Gamma-aminobutyric and binding in mammalian brain: receptorlike specificity of sodium-independent sites, J. Neurochem., 31:933-938.
16. Hruska, R.E., Huxtable, R.J., and Yamamura, H.I., High-affinity, temperature-sensitive, and sodium-dependent transport of taurine in rat brain, in: "Taurine in Neurological Disorders," A Barbeau and R.J. Huxtable, eds., Raven Press, New York, (1976), pp. 109-117.
17. Huxtable, R.J., Laird, H.E., and Lippincott, S.E., 1979, The transport of taurine in the heart and the rapid depletion of tissue taurine content by guanidinoethyl sulfonate, J. Pharmacol. Exp. Ther., 211:465-471.
18. Izumi, K., Igisu, H. and Fukuda, T., 1975, Effects of edetate in seizure supressing actions of taurine and GABA, Brain Res., 88:576-579.
19. Jacobsen, J.G., and Smith, L.H., 1968, biochemistry and physiology of taurine and taurine derivatives, Physiol. Rev., 48:424-511.
20. Jasper, H.H., and Koyama, I., 1969, Rate of release of amino acids from the cerebral cortex in the cat as affected by brain stem and thalamic stimulation, Can. J. Physiol. Pharmacol., 47:889-905.
21. Kaczmarek, L.K., and Davison, A.N., 1972, Uptake and release of taurine from rat brain slices, J. Neurochem., 19:2355-2362.
22. Kennedy, A.J., and Neal, M.J., 1978, The effect of light and potassium depolarization on the release of endogenous amino acids from the isolated retina, Exp. Eye Res., 26:71-75.
23. Kocsis, J.J., Kostos, U.J., and Baskin, S.I., Taurine levels in the heart tissues of various species, in: "Taurine", R. Huxtable and A. Barbeau, eds., Raven Press, New York, (1976), pp. 145-153.

24. Kuriyama, K., Muramatsu, M., Nakagawa, K., and Kakita, K., Modulatory role of taurine on release of neurotransmitters and calcium transport in excitable tissues, in: "Taurine in Neurological Disorders," A. Barbeau and R. Huxtable, eds., Raven Press, New York, (1978), pp. 201-216.
25. Lemeignan, M., 1972, Analysis of the action of 4-aminopyridine on the cat lumbar spinal cord. I. Modification of the afferent volley, the monesynaptic discharge amplitude and the polysynaptic evoked responses, Neuropharmacology, 11:551-558.
26. Lombardini, J.B., Regional and subcellular studies on taurine in the rat central nervous system, in: "Taurine," R. Huxtable and A. Barbeau, eds., Raven Press, New York, (1976), pp. 311-326.
27. López-Colomé, A.M., Salceda, R., Tapia, R., and Pasantes-Morales, H., 1978, K^+-stimulated release of labeled GABA, glycine and taurine in slices of several regions of rat central nervous system, Neuroscience, 3:1069-1074.
28. López-Colomé, A.M., Salceda, R., and Pasantes-Morales, H., 1978, Potassium stimulated release of GABA, glycine and taurine from the chick retina, Neurochem. Res., 3:431-441.
29. Malaisse, W.J., and Orci, L., The role of cytoskeleton in pancreatic β-cell function, in: "The Cytoskeleton in Normal and Pathologic Processes," G. Gabbiani, ed., Karger, Basel, (1979), pp. 112-136.
30. Medina, G., and Illingworth, J., 1980, Some factors affecting phosphate transport in perfused rat heart preparation, Biochem. J., 188:297-310.
31. Meela, L. and Wrobel-Kuhl, K., Special characteristics of brain mitochondrial calcium accumulation, in: Calcium Transport and Cell Function, A. Scarpa and E. Carafoli, eds., Ann. N.Y. Acad. Sci., Vol. 307 (1978), pp. 242-245.
32. Nicklas, W.J., and Berl, S., 1973, Effect of vinblastine and colchicine on uptake and release of putative neurotransmitter by synaptosomes and on brain actomyosin-line protein, J. Neurochem., 20:109-121.
33. Orr, H.T., Cohen, A.J., and Lowry, O.H., 1976, The distribution of taurine in the vertebrate retina, J. Neurochem., 26:609-612.
34. Pasantes-Morales, H., Ademe, R.M., and López-Colomé, A.M., 1979, The effect of taurine on $^{45}Ca^{++}$ transport by retinal subcellular fractions, Brain Res., 172:131-138.
35. Pasantes-Morales, H., and Gamboa, A., 1980, Effect of taurine on ^{45}Ca accumulation in rat brain synaptosomes, J. Neurochem., 34:244-246.

36. Pasantes-Morales, H., Klethi, J., Urban, P.T., and Mandel, P., 1974, The effect of electrical stimulation, light and amino acids on the efflux of ^{35}S-taurine from the retina of the domestic fowl, Exp. Brain Res., 19:131-141.
37. Pasantes-Morales, H., and Moran, J., Taurine as a modulator; its action on calcium fluxes and in neurotransmitter release, in: "Regulatory Mechanisms on Synaptic Transmission," R. Tapia and C. Cotman, eds., Plenum Press, In press.
38. Pasantes-Morales, H., Salceda, R., and López-Colomé, A.M., 1980, The effect of colchicine and cytochalasin B on the release of taurine from the chick retina, J. Neurochem., 34:172-177.
39. Pollack, R.E., and Kopelovich, L., The cytoskeleton in cultured cells: coordinate in vitro regulation of cell growth and shape, in: "The Cytoskeleton in Normal and Pathologic Processes, G. Gabbiani, eds., S. Karger, Basel, (1979), pp. 207-230.
40. Puszkin, S., and Schock, W., The role of cytoskeleton in neuron activity, in: "The Cytoskeleton in Normal and Pathological Processes," G. Gabbiani, ed., S. Karger, Basel, (1979), pp. 87-111.
41. Rahamimoff, H., and Abramovitz, E., 1978, Ca transport and ATPasa of synaptosomal vesicles from rat brain, FEBS Letters, 92:163-167.
42. Redburn, D.A., and Cotman, C.W., 1974, Calcium-dependent release of ^{14}C-GABA from vinblastine and colchicine treated synaptosomes, Brain Res., 73:550-557.
43. Roberts, E., and Frankel, S., 1950, γ-Aminobutyric acid in brain. Its formation from glutamic acid, J. Biol. Chem., 187:55-63.
44. Roberts, E., Chase, T.N., and Tower, D.B., GABA in Nervous System Function, Raven Press, (1976).
45. Schnetkamp, P.P.M., Daemen, F.J.M., and Bonting, S.L., 1977, Biochemical aspects of the visual process. XXXVI. Calcium accumulation in cattle rod outer segments: evidence for a calcium-sodium exchange carrier in the rod sac membrane, Biochim. Biophys. Acta, 468:259-270.
46. Sieghart, W., and Heckl, K., 1976, Potassium-evoked release of taurine from synaptosomal fractions of cerebral cortex, Brain Res., 116:538-543.
47. Sordahl, L.A., and Asimakis, G.K., Role of adenine nucleotides in calcium retention in heart mitochondria, in: "Calcium Transport and Cell Function," A. Scarpa and E. Carafoli, eds., Ann. N.Y. Acad. Sci., Vol. 307, (1978), pp. 238-241.

48. Uga, S., Nakao, F., Mimura, M., and Ikui, K., 1970, Some findings in the fine structure of the human photoreceptor cells, J. Electron Microscopy, (Tokyo) 19:71-84.
49. Vasington, F.D., and Morphy, J.V., 1962, Ca^{++} uptake by rat kidney mitochondria and its dependence on respiration and phosphorylation, J. Biol. Chem., 237:2670-2677.
50. Warren, R.H., and Burnside, B., 1979, Microtubules in cone-myoid elongation in the toleost retina, J. Cell Biol., 78:247-259.
51. Young, A.B., and Snyder, S., 1973, Strychnine binding associated with glycine receptors of the central nervous system, Proc. Nat. Acad. Sci., 10:2832-2836.

TAURINE RECEPTORS IN CNS MEMBRANES: BINDING STUDIES

A. M. López-Colomé

Departamento de Neurociencias
Centro de Investigaciones en Fisiología Celular
Universidad Nacional Autónoma de México
Apartado Postal 70-600
México 20, D.F.
México

Taurine is present in high concentration in a wide variety of animal tissues (26), especially in excitable and nervous tissue (34,40,45). Recently, great interest has developed in clarifying the function of taurine in the central nervous system. It has been demonstrated that taurine, which is present in the brain in high concentrations (40), has a hyperpolarizing effect on neurons from the spinal cord and cerebral cortex when iontophoretically applied (6-8). It also has been shown to depress reversibly the b wave of the electroretinogram (47,62). A high-affinity, sodium-dependent uptake system has been demonstrated for taurine accumulation into brain slices and into a specific synaptosomal subfraction (24,28,29,57). It has been reported that taurine is released from cerebral cortex in response to electrical stimulation (32) and from the retina upon the influence of light (48,55). Furthermore the release of taurine following depolarizing stimuli has also been demonstrated in rat brain and spinal cord slices as well as in mouse brain synaptosomes (3,4,27,32,36). These findings support the hypothesis that taurine might act as an inhibitory transmitter or a modulator in the central nervous system (9,39).

However, certain properties ascribed to a neurotransmitter substance have not yet been demonstrated for taurine. The stimulation evoked release of neurotransmitter candidates from nerve endings appears to be a calcium-dependent process (54), and although taurine is released upon stimulation, the calcium-independence of this phenomenon (59) does not support the hypothesis of taurine being a neurotransmitter.

Alternatively, a role for taurine as an osmotic effector (61), a membrane stabilizer (31) or as a regulatory of ionic fluxes has also been proposed (25,63).

The presence of specific post-synaptic receptors has been considered as evidence for a substance performing a neurotransmitter function. In the case of taurine, no such specific post-synaptic receptors nor the characteristics of its interaction to different kinds of binding sites in the membrane have been thoroughtly described. In recent years, a technique has been successfully developed for measuring amino acid receptors in nerve cell membranes (17,60). This technique has proved itself for GABA and glycine in the CNS. The correlation of the physiological effects of these substances with the action of specific antagonists of their post-synaptic effects seem to confirm the accuracy of the technique for measuring physiological post-synaptic receptors (17,18,53,56,66). According to these authors, uptake receptors, which require sodium for binding the compound to be transported, are better seen in freshly prepared membranes. Post-synaptic receptors, which are sparse compared to the uptake receptors and which do not need sodium to bind transmitter, become more evident in the absence of sodium due to the elimination of binding to the uptake sites. In addition, freezing and thawing of the membranes as well as treatment with neutral detergents such as Triton-X-100 seem to enhance post-synaptic GABA and glycine receptor binding, possibly due to the opening of membrane vesicles which could have resealed after osmotic shock. Basically this technique consists of binding the compound under study radiolabeled to high specific activity, or better, its labeled antagonist, to all possible receptor sites in the membrane. This gives the total binding. The radiolabeled compound is displaced by an excess of the physiological ligand or its agonists or antagonists, and the radiolabled compound which remains bound in the presence of this excess is considered unspecifically bound. The difference between total and unspecific binding is considered to be binding to specific receptors.

In an effort to provide further evidence for taurine being a neurotransmitter in the CNS or not, we have studied the characteristics of taurine binding to membranes obtained from different regions of vertebrate central nervous system.

BINDING TO THE RETINA

In the retina, the high content of taurine, and its synthesizing enzyme (50), as well as the depressant effect of taurine on several types of retinal neurons (49), and the fact that taurine is released from this organ following its natural stimulus, light, suggest an important physiological role for this amino acid, possibly as an inhibitory neurotransmitter.

Taurine was found to bind to membranes from the whole chick retina (37) as well as to membranes from the crude nuclear fraction (P_1), which contains, in addition to nuclei, photoreceptor outer segments, packed mitochondria, and the synaptic terminals from the photoreceptors. Taurine was also found to bind to crude synaptosomal fraction (P_2) which contains synaptosomes from the inner plexiform layer, membrane fragments and very few mitochondria (42).

Binding in the Absence of Na^+

In conditions in which post-synaptic receptors should be more apparent - frozen and thawed membranes in the absence of sodium - taurine binding is extremely low (0.014 ± 0.004 pmole/mg protein) compared to GABA (0.49 ± 0.16 pmole/mg protein) or glycine which have been considered retinal neurotransmitters (43). In addition, preincubation with detergent (Triton X-100 0.05%) almost abolishes taurine binding whereas GABA binding is greatly enhanced. GABA binding in this preparation was displaced by its specific antagonist bicuculline and glycine and taurine were displaced by strychnine to the same extent.

When binding to membranes from subcellular fractions was studied under the same conditions, both GABA and glycine binding were mainly concentrated in the P_2 fraction while taurine specific binding was very low and evenly distributed in P_1 and P_2 fractions (Table I). As in whole retinal membranes 20% of the bound taurine was displaced by 0.1 mM strychnine.

Binding in the Presence of Na+

When taurine binding was determined in freshly prepared membranes in the presence of 118 mM NaCl, conditions reported to favor the measurement of uptake receptors, taurine binding at 15 μM was 22.9 pmole/mg protein. This binding system showed two components with Kd's of 0.68 μM and 9.32 μM respectively. Maximum capacities were 4.4 and 38.2 pmole/mg protein when measured in a concentration range from 0.025 to 15 μM taurine. Binding showed an absolute requirement for sodium and was concentrated in the P_2 fraction although it was also considerable in the P_1 fraction (Table 2).

Taurine binding in these conditions was found to be specific, for of the compounds tested as displacers, only β-alanine displaced ^{3}H-taurine to the same extent as taurine itself. As in the frozen fraction strychnine effectively displaced ^{3}H-taurine binding (Fig. 1).

BINDING TO RAT CEREBRAL CORTEX AND OTHER CNS REGIONS

In order to elucidate if specific, high-affinity receptors for taurine are present with similar characteristics to the retina, binding to membranes from other regions of the CNS was examined. Cerebral cortex, hypothalamus, olfactory bulb and cerebellum were chosen because of their high taurine content and the possibility of this amino acid playing a neurotransmitter role in these regions (1,5,20,44). To eliminate the possibility of taurine receptors being unspecifically distributed in all CNS membranes, binding to membranes from dorsal root ganglia, in which no synapses are present, was also measured. To compare the characteristics of the binding process with those of an accepted neurotransmitter, GABA receptors were always measured under the same conditions.

Binding in the Absence of Na^+

^{3}H-GABA and ^{3}H-taurine binding in the absence of sodium, were measured in frozen and thawed membranes from cerebral cortex, cerebellum, olfactory bulb, hypothalamus and dorsal-root ganglia. In all cases, GABA specific binding was 10 to 100 times higher than taurine specific binding under the same conditions. Cerebellum showed the highest values for GABA specific binding, although it was also considerable in the cortex. ^{3}H-Taurine binding was low and similar in all regions. In dorsal root ganglion membranes, specific binding of either amino acid was almost absent (Fig. 2). Values obtained for ^{3}H-GABA binding were of the same order of those previously reported (17). Under the same conditions, practically no receptors for taurine were detected.

If binding to frozen and thawed membranes in a Na$^+$-free medium corresponds to post-synaptic receptors, it should be concentrated in the synaptosomal as compared to the whole homogenate membranes. This possibility was explored P_2 and synaptosomal fractions from cerebral cortex. ^{3}H-GABA specific binding was 5 times greater in synaptosomal membranes; taurine binding, total or specific, was very low and similar in all membrane fractions (Table 3). Treatment of the membranes with Triton enhanced GABA binding and diminished taurine binding.

It has been reported that glycine binding to its synaptic receptor is inhibited by Triton. In an effort to eliminate the possibility of our conditions being inappropriate for detecting taurine receptors, different Triton concentrations were tried, but none of them induced the unmasking of new receptor sites. In some cases, sonication rather than detergents have proven better for dispersing the membranes (58), but in the present case, it did not enhance significantly taurine binding (Table 4).

Recently, endogenous inhibitors of GABA and glutamate binding have been reported (58,65), which can be eliminated with extensive washing of the membranes and incubation at 37°C previous to the binding assay. Up to four washings of the membranes from cerebral cortex had no effect on the binding values reported in Table 3, nor did preincubation at 37°C for 30 min (Table 4).

118 mM NaCl was added to the assay medium to explore the existence of a Na$^+$-dependent process. However, taurine specific binding to frozen membranes was not enhanced (0.003 ± 0.000 pmole/mg protein).

Binding in the Presence of Na$^+$

Binding to freshly prepared membranes in the presence of 118 mM NaCl is supposed to occur to uptake receptors in addition to post-synaptic sites. We have studied taurine receptors under these conditions as compared to GABA receptors in an attempt to clarify if they correspond to an uptake mechanism or could be the mediators of the physiological synaptic action of taurine.

TABLE 1. Specific binding of taurine, GABA and glycine to frozen chick retinal membranes

	pmole/mg protein		
Membrane preparation	^{3}H-taurine	^{3}H GABA	^{3}H glycine
Whole retina	(9) 0.014±0.004	(3) 0.49±0.16	(6) 0.030±0.006
P_1 fraction	(2) 0.03	(4) 0.58±0.08	(4) 0.020±0.006
P_2 fraction	(2) 0.03	(4) 1.30±0.20	(4) 0.115±0.004

Membranes were obtained from whole retina or from primary subcellular fractions isolated as previously described (López-Colomé et al., 1978). The whole retinas, crude nuclear (P_1) or crude synaptosomal (P_2) fractions were osmotically shocked and the membranes were obtained as described by Enna and Snyder (1975). Binding was determined in a Na^+-free Krebs-Tris buffer, without glucose. ^{3}H-GABA and ^{3}H-glycine concentration was 8 nM; ^{3}H-taurine concentration was 25 nM. Specific binding was obtained subtracting the amount of radioactivity not displaced by 1 mM unlabeled amino acids from the total labeled amino acid bound. The results are the means ± SEM of the number of separate experiments indicated in parentheses.

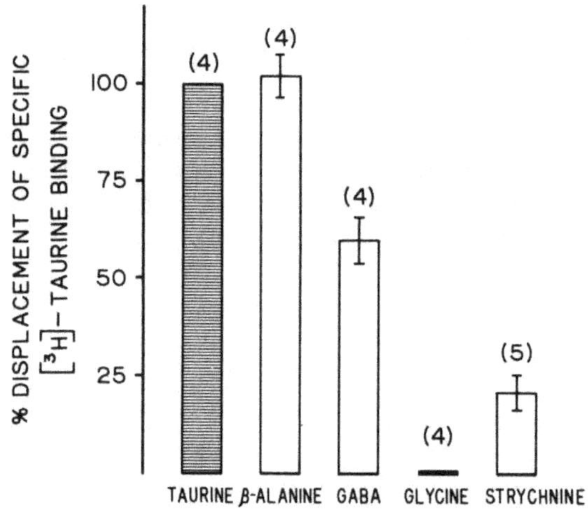

Fig. 1. Displacement of ^{3}H-taurine Na^+-dependent binding by other amino acids and strychnine. Membranes from whole retina (see Table 1) were incubated for the binding assay in the presence of 25 nM ^{3}H-taurine alone or in the presence of 1 mM unlabeled taurine, glycine, GABA or β-alanine. Strychnine concentration was 100 nM. Binding assays were performed with freshly prepared membranes in the presence of 118 mM NaCl. The number of separate experiments is indicated in parentheses.

TABLE 2. Effect of NaCl on the binding of taurine to retinal membranes

Condition	pmole/mg protein Fresh Tissue	Frozen Tissue
118 mM NaCl	(3) 22.8 ± 1.4	(3) 0.020 ± 0.004
-NaCl + 118 mM Choline chloride	(3) 9.2 ± 1.6	—
-NaCl	(3) 1.6 + 0.8	(3) 0.010 + 0.004

Freshly prepared or frozen and thawed membranes from whole retina were incubated in a Krebs-Tris medium and binding assay was performed as described by Enna and Snyder (1975). Values are the means ± SEM of the number of experiments indicated in parentheses.

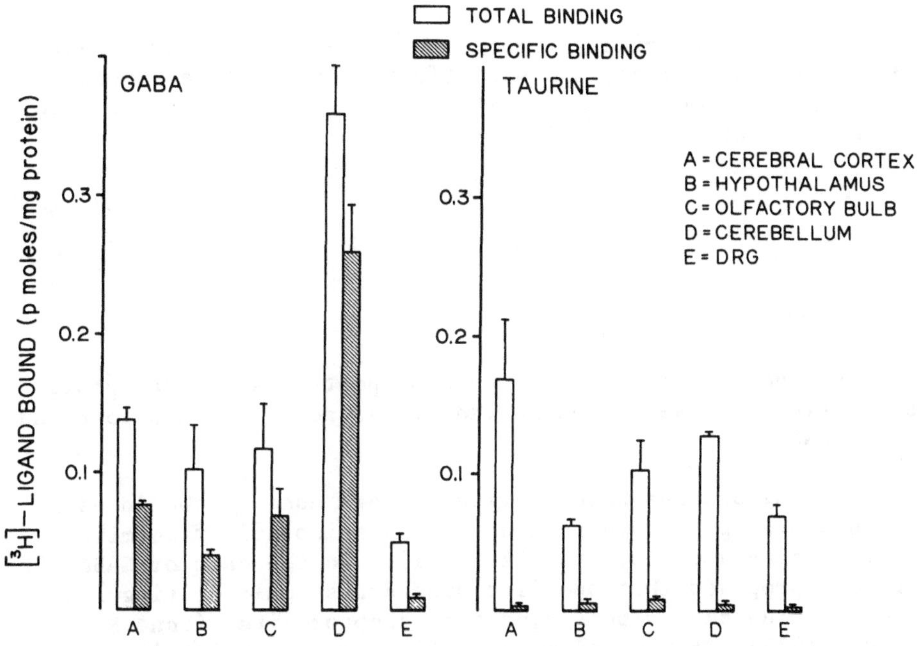

Fig. 2. Na^+-independent binding of ^3H-GABA and ^3H-taurine to membranes from central nervous system regions. Binding was measured in Krebs-Tris medium without NaCl. ^3H-GABA concentration was 8 nM and ^3H-taurine 25 nM. 1 mM unlabeled GABA or taurine were used as displacers. Values are the means of 3 to 10 experiments ± SEM. DRG=Dorsal root ganglion.

Taurine binding to freshly prepared membranes in the presence of sodium ranged from 0.02 to 0.04 pmole/mg protein in all regions studied; in membranes from dorsal root ganglia, only 0.009 pmole/mg protein were specifically bound (Fig. 3). Although taurine binding was about 10 times higher than in frozen membranes in the absence of sodium, these values are still low when compared to ^{3}H-GABA binding under the same condition, or even in the absence of sodium (see Figs. 3 and 4). As with frozen membranes, under these conditions, ^{3}H-GABA binding to cerebral cortex membranes was concentrated in the P_2 and synaptosomal membranes whereas ^{3}H-taurine binding was not (Table 3).

Kinetic analysis of taurine binding in the concentration range 0.15 - 12 μM showed it to be saturable with a single component for both whole cerebral cortex and synaptosomal membranes. A Kd of 4.06 μM and Bmax of 5.36 pmole/mg protein were calculated for whole cortex membranes, and for synaptosomal membranes, a Kd of 18.9 μM and Bmax of 6.14 pmole/mg protein (Fig. 4).

Binding of taurine showed an absolute sodium requirement under these conditions, and was effectively displaced (more than 100% when compared to taurine itself) by strychnine and bicuculline at concentrations as low as 20 μM.

DISCUSSION

We have explored the presence of post-synaptic receptors for taurine to assess a neurotransmitter role for this compound in the CNS.

Binding studies have previously been used in the investigation of receptor interactions of the amino acid transmitters GABA, glycine and glutamate (2,15,41). In the case of GABA, it has been proposed that specific procedures could distinguish binding to uptake or post-synaptic receptor sites (Enna & Snyder, 1975). For glycine, specific post-synaptic sites have been measured using the specific antagonist ^{3}H-strychnine (60). For glycine a single high-affinity binding system has been detected, whereas GABA binding systems show two components. The distribution of binding correlates with the inhibitory potency of the compounds in physiological preparations, thus indicating an interaction with post-synaptic receptors.

TABLE 3. Subcellular Distribution of Taurine and GABA Binding to Membranes from Rat Cerebral Cortex

Preparation	pmole/mg protein GABA Fresh		Frozen	
	Total	Specific	Total	Specific
Whole cerebral cortex	0.302±0.027 (7)	0.177±0.018 (7)	0.138±.009 (3)	0.076±0.001 (5)
Crude synapto-somal fraction	0.269±0.036 (5)	0.211±0.031 (5)	0.292±.066 (4)	0.166 (3)
Purified Synaptosomes	1.28 ±0.30 (6)	0.91 ±0.2 (6)	0.54 ±0.30 (5)	0.38 ±0.20 (5)
	pmole/mg protein TAURINE Fresh		**Frozen**	
	Total	Specific	Total	Specific
Whole cerebral Cortex	0.185±0.020 (6)	0.040±0.006 (6)	0.169±0.044 (6)	0.003±0.001 (6)
Crude synapto-somal fraction	0.05 ±0.08 (4)	0.013±0.002 (5)	0.116±0.023 (4)	0.006±0.002 (5)
Purified Synaptosomes	0.238±0.037 (3)	0.027±0.003 (3)	0.078±0.012 (4)	0.013±0.008 (3)

Crude synaptosomal fraction was obtained as described previously (DeRobertis et al., 1962). Purified synaptosomes were isolated as described by Hajos (1975). Whole cortex or previously isolated subcellular fractions were osmotically shocked, and membranes were pelleted by centrifugation at 45000 x g during 20 min. The pellet was washed once, and the binding assay was carried out as described in Table 1, after resuspending the pellet in Krebs-Tris buffer in the presence (fresh) or absence (frozen) of 118 mM NaCl. For determinations in the absence of Na^+, membranes were frozen for 48 hr and incubated with 0.05% Triton-X-100 prior to the binding assay. Data are means ± SEM of the number of experiments indicated in parentheses.

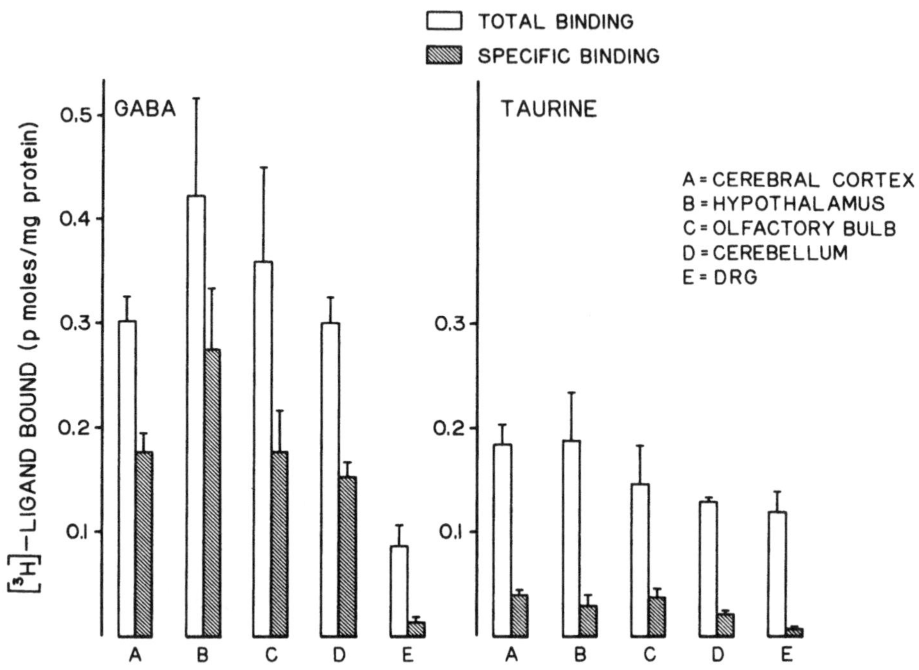

Fig. 3. Na^+-dependent binding of ^{3}H-GABA and ^{3}H-taurine to membranes from central nervous system regions. Membranes were prepared as described in Table 1 and binding conditions were as in Fig. 2 except for the presence, in this case, of 118 mM NaCl in the medium. Data are expressed as the means of 3 to 7 experiments ± SEM.

Except for the high-affinity binding component found in the retina (Kd 0.68 μM), the Kd values reported herein are close to those established for the high affinity uptake system for taurine in brain slices (7,8,35) and cortical synaptosomes (32,57). This binding process is antagonized by strychnine and bicuculline to about the same extent as has been reported for taurine transport processes (32). However, inhibition of transport requires high concentrations of these agents (100 μM), while the concentrations found to inhibit ^{3}H-taurine binding under our conditions (20 μM) are close to those reported to displace neurotransmitter binding to post-synaptic receptors (12,60).

The sodium dependence of these binding processes as well as their subcellular distribution suggest that neither of them are Membranes from whole cortex were prepared according to Enna and Snyder (1975). Binding assays were conducted in the absence of

TABLE 4 - Effect of different Triton concentrations, extensive washing and sonication on taurine binding to frozen membranes from rat cerebral cortex

Condition	^{3}H-Taurine binding (pmole/mg protein) Specific	Non Specific
Triton X100		
0.012	(3) 0.004 ± 0.001	(3) 0.124 ± 0.012
0.025	(3) 0.011 ± 0.001	(3) 0.094 ± 0.006
0.050	(6) 0.003 ± 0.001	(6) 0.166 ± 0.044
0.1	(3) 0.005 ± 0.001	(3) 0.100 ± 0.010
4 washes	(3) 0.003 ± 0.000	(3) 0.185 ± 1.010
3 min. Sonication	(3) 0.016 ± 0.003	(3) 0.212 ± 0.037

NaCl, at 4°C as described in Table I. Membranes were incubated 30 min. at 37°C prior to the binding assay with 0.05% Triton except when the effects of different concentrations was tested. Washes or sonication were performed before freezing the membranes. Results are the means ± SEM of the number of Separate experiments indicated in parentheses.

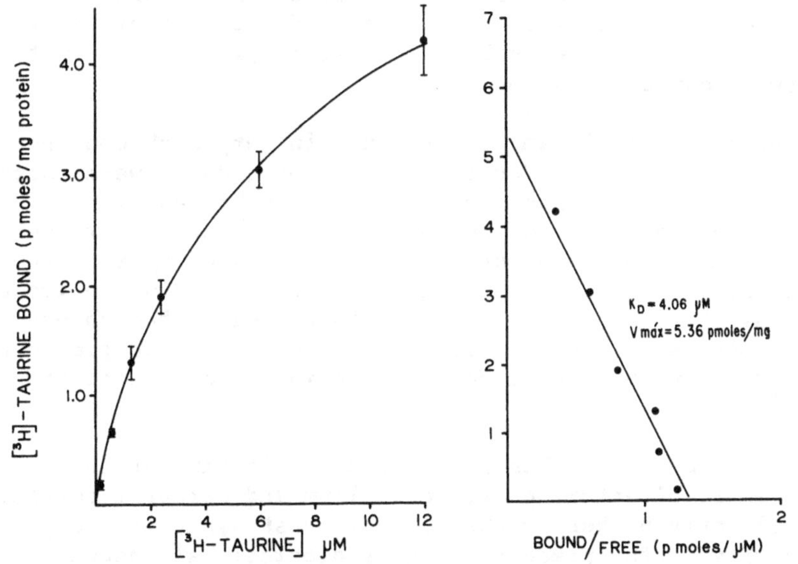

Fig. 4. Saturation of specific Na^+-dependent ^{3}H-taurine binding sites in membranes from rat cerebral cortex. Freshly prepared membranes were incubated in the presence of increasing concentrations of ^{3}H-taurine (0.15 - 12 μM) in the presence of 118 nM NaCl. Assays were performed as described (Table 1). Linear plot (left) and Scatchard plot (right) revealed a single component. Data are the means ± SEM of 4 separate experiments performed in duplicate.

related to post-synaptic taurine receptors in the membranes, for it has been assumed that interaction with such sites is a Na^+-independent process and should concentrate in synaptic membranes (17). In addition, Enna and Snyder (1975) have reported that GABA binding to post-synaptic sites is enhanced in frozen membranes treated with detergent prior to the binding assay whereas taurine binding is decreased in frozen membranes and almost abolished by incubation with Triton.

With regard to the high-affinity binding system in the retina, the Kd value is much higher than the one reported for GABA in the retina (18) which is in the low nM range, but very similar to the Kd for glutamate binding to cerebellar membranes as reported by Roberts et al. (19); however, this binding component is Na^+-dependent, while the glutamate Kd was estimated in the absence of sodium.

Kumpulainen et al. (30) and Lahdesmaki et al. (33) have found kinetic constants of 67 μM and 1.5 μM for ^{35}S-taurine binding to chicken and calf brain synaptic membranes respectively. However, taking into account the low specific activity of commercially available ^{35}S-taurine, which was used in those studies, as well as the contribution of unspecific binding to their calculations, it seems doubtful that these constants represent true values for taurine interaction with its post-synaptic receptors.

Taurine, like glutamate, is found in very high concentration in CNS tissue. For this reason procedures were carried out as to eliminate contamination of the membranes with endogenous amino acid, but neither extensive washing of the membranes, nor sonication or incubation at 37°C prior to the assay enhanced the almost negligible taurine binding to frozen membranes. Hence these values seem to be due to the absence of specific post-synaptic receptor sites, and not to be presence of endogenous taurine or of a specific inhibitor of the binding process.

The Na^+-dependent binding of taurine is similar in hypothalamus, olfactory bulb, cerebellum and cerebral cortex, but is slightly higher in the retina. Post-synaptic receptors are normally in the presence of physiological Na^+ concentrations (100 μM). Accordingly, binding to post-synaptic sites should be measured in the presence of this ion (11-13). However, Na^+-dependent of taurine in cerebral cortex membranes does not seem to represent this kind of interaction for the reason mentioned above regarding kinetic constants and distribution of binding. The possibility of this interaction being unspecifically present in all CNS membranes is eliminated by the finding that binding is practically absent from dorsal root ganglia membranes. In the case of the retina, although binding is concentrated in the crude synaptosomal fraction, this does not necessarily mean an interaction with post-synaptic receptors, for the fraction contains mitochondria and membrane small fragments in addition to synaptosomes and contamination with glial membranes or gliosomes cannot be ruled out (23).

The overall values of taurine binding at saturating concentration (12 μM) are similar in cerebral cortex and retina; specific binding varies in these two regions, being much higher in the retina. This, together with the presence of a higher affinity binding system, could suggest a more specific role for this amino acid in the retina.

The characteristics of taurine interactions with CNS membranes reported here do not correspond to those classically assigned to post-synaptic receptors. High-affinity synaptic-antagonist binding has proved to be a sensitive method for detecting post-synaptic receptors. As no such antagonists for taurine exist, the possibility of the existence of specific post-synaptic receptors which are present in small restricted cell populations exceeding the sensitivity of the method cannot be discarded at the moment.

Alternatively, the specific Na^+-dependent taurine binding could represent binding to a modulator site, for the similarity in the number of binding sites, regardless of the difference in the content of taurine in the different areas studied, seems to indicate that this binding is not to uptake sites only. Another possible explanation for the low and nearly uniform distribution of Na^+-dependent taurine binding could be a long-lasting interaction with the nerve cell membrane possibly related to the modulation of calcium fluxes (38,51,52,66), which in turn play an outstanding role in synaptic transmission.

Displacement of taurine by low concentrations of bicuculline and strychnine could also indicate that taurine interacts with GABA or glycine post-synaptic receptors under certain conditions (presence of NaCl) and would explain the depressant effect of taurine in physiological preparations, in which its action is antagonized by strychnine and bicuculline (21). The finding that taurine interacts at high concentrations (100 mM) with GABA and benzodiazepine receptors (64) also suggests a modulator role for taurine, perhaps through its interaction with other neurotransmitter receptors.

Although, the evidence presented here is not conclusive, it certainly points to the possibility of taurine having a more extended role than that of a classical neurotransmitter in nervous tissue.

REFERENCES

1. Austin, L., Recasens, M., Mathur, R.L., and Mandel, P., 1978, The distribution of taurine, cysteinyl sulphinilic acid decarboxylase and cysteinyl sulphinilic acid α-ketogluturate transaminase in the rat olfactory bulb and olfactory nucleus, Neurosci. Letters, 9:59-63.
2. Biziere, K., Thompson, H., and Coyle, J.T., 1980, Characterization of specific, high affinity binding sites for L- ^{3}H - glutamic acid in rat brain membranes, Brain Res., 183:421-433.
3. Clark, R.M., and Collins, G.G.S., 1976, The release of endogenous amino acids from the rat visual cortex, J. Physiol., 262:383-400.
4. Collins, G.G.S., and Topiwala, S.H., 1974, The release of ^{14}C - taurine from slices of rat cerebral cortex and spinal cord evoked by electrical stimulation and high potassium ion concentrations, Br. J. Pharmacol., 50:451-452.
5. Collins, G.G.S., 1974, The rate of synthesis, uptake and disappearance of ^{14}C-taurine in eight areas of the rat central nervous system, Brain Res., 76:447-459.
6. Curtis, D.R., and Watkins, J.C., 1960, The excitation and depression of spinal neurons by structurally related amino acids, J. Neurochem., 6:117-141.
7. Curtis, D.R., Hosli, L., and Johnston, G.A.R., 1968, A pharmacological study of the depression of spinal neurons by glycine and related amono acids, Exp. Brain Res., 6:1-18.
8. Curtis, D.R. and Watkins, J.C., 1973, The pharmacology of amino acids related to γ-aminobutyric acid, Pharmac. Rev., 17:347-391.
9. Davison, A.N., and Kaczmarek, L.K., 1971), Taurine a possible neurotransmitter Nature, 234:107-108.
10. De Belleroche, J.S., and Bradford, H.F., 1973, Amino acids in synaptic vesicles from mammalian cerebral cortex: A reappraisal, J. Neurochem., 21:441-451.
11. De Feudis, F.V., and Black, W.C., 1973, A comparison of the binding of ^{14}C - γ-aminobutyric acid and ^{3}H - acetylcholine to particulate fractions of 4 regions of rat brain in the presence of 40 mM NaCl, Brain Res., 49:218-222.

12. De Feudis, F.V., 1978, Can the binding of GABA, glycine and β-alanine to synaptic receptors be determined in the presence of a physiological concentration of Na^+?, Experentia, 34:1314-1315.
13. De Feudis, F.V., Maitre, M., Ossola, L., Elkouby, A., and Mandel, P., 1979, Bicuculline-sensitive GABA binding to synaptosome enriched fraction of rat cerebral cortex in the presence of a physiological concentration of sodium, Gen. Pharmac., 10:193-194.
14. DeRobertis, E., Pelegrino de Iraldi, A., Rodríguez de Lores Arnaiz, G., and Salganicoff, L., 1962, Cholinergic and non-cholinergic nerve endings in rat brain, I. Isolation and subcellular distribution of acetyl choline and acetylcholinesterase, J. Neurochem., 9:23-35.
15. DeRobertis, E., and Fiszer de Plazas, S., 1976, Isolation of hydrophobic proteins binding amino acids. Selectivity of the binding of L- ^{14}C - glutamate in cerebral cortex, J. Neurochem., 26:1237-1243.
16. Dolara, P., Ledda, F., Mugelli, A., Mantelli, L., Zilletti, F., Franconi, F., and Giotti, A., Effect of taurine on calcium, inotropism, and electrical activity of the heart. in: "Taurine and Neurological Disorders," A. Barbeau, and R. Huxtable, eds., Raven Press, New York (1978).
17. Enna, S.J., and Snyder, S.H., 1975, Properties of γ-aminobutyric acid (GABA) receptor binding in rat brain synaptic membrane fractions, Brain Res., 100:81-97.
18. Enna, S.J., and Snyder, S.H., 1976, Gamma-aminobutyric acid (GABA) receptor binding in mammalian retina, Brain Res., 115:174-179.
19. Foster, A.C., and Roberts, P.J., 1978, High affinity L-^3H-glutamate binding to postsynaptic receptor sites on rat cerebellar membranes, J. Neurochem., 31:1467-1477.
20. Guidotti, A., Badiani, G., and Pepeu, G., 1972, Taurine distribution in cat brain, J. Neurochem., 19:431-435.
21. Haas, H.L., and Hosli, L., 1973, The depression of brain stem neurones by taurine and its interaction with strychnine and bicuculline, Brain Res., 52:399-402.
22. Hajos, F., 1975, An improved method for the preparation of synaptosomal fractions in high purity, Brain Res., 93:485.
23. Henn, A.F., Anderson, D.J., and Rustad, D.G., 1976, Glial contamination of synaptosomal fractions, Brain Res., 101:341-344.
24. Hruska, R.E., Padjen, A., Bressler, R., and Yamamura, H.I., 1978, Taurine: Sodium-dependent, high-affinity transport into rat brain synaptosomes, Mol. Pharmacol., 14:77-85.

25. Izumi, K., Ngu, T.T., and Barbeau, A., Metabolic modulation in the central nervous system by taurine, in: "Taurine in Neurological Disorders," A. Barbeau and R. Huxtable, eds., Raven Press, New York (1978), pp.137-160.
26. Jacobsen, J.G., and Smith, L.H., 1968, Biochemistry and physiology of taurine and taurine derivatives, Physiol. Rev., 48:424-511.
27. Jasper, H.H., and Koyame, I., 1969, Rate of Release of amino acids from the cerebral cortex in the cats as affected by brainstem and thalamic stimulation, Can. J. Physiol. Pharmacol., 47:889-905.
28. Kaczmarek, L.K., and Davison, A.N., 1972, Uptake and release of taurine from rat brain slices, J. Neurochem., 19:2355-2362.
29. Kontro, P., and Oja, S.S., 1978, Sodium dependence of taurine in rat brain synaptosomes, Neuroscience, 3:761-765.
30. Kumpulainen, E., Jokisalo, V.J., and Lahdesmaki, P., 1978, Physiocochemical characterization of taurine binding with synaptic membranes in the chicken brain, Intern. J. Neuroscience, 8:123-128.
31. Kuriyama, K., and Nakagawa, K., Role of taurine in adrenal gland: A preventive effect of stress induced release catecholamines from chromaffin granules, in: "Taurine," R. Huxtable and A. Barbeau, eds., Raven Press, New York (1976), pp.335-345.
32. Lahdesmaki, P., Pasula, M., and Oja, S.S., 1975, Effect of electrical stimulation and chlorpromazine on the uptake and release of taurine, γ-aminobutyric acid and glutamic acid in mouse brain synaptosomes, J. Neurochem., 25:675-680.
33. Lahdesmaki, P., Kumpulainen, E., Raasakka, O., and Kyrki, P., 1977, Interaction of taurine GABA and glutamic acid with synaptic membranes, J. Neurochem., 29:819-826.
34. Lombardini, J.B., Regional and Subcellular studies on taurine in the rat central nervous system, in: "Taurine," R. Huxtable and A. Barbeau, eds., Raven Press, New York, (1976), pp.311-326.
35. López-Colomé, A.M., Salceda, R., and Pasantes-Morales, H., 1978, Potassium stimulated release of GABA, glycine and taurine from the chick retina, Neurochem. Res., 3:1069-1074.
36. López-Colomé, A.M., Salceda, R., Tapia, R., and Pasantes-Morales, H., 1978, K^+-stimulated release of labeled γ-aminobutyrate, glycine and taurine in slices of several regions of rat central nervous system, Neuroscience, 3:1069-1074.
37. López-Colomé, A.M., and Pasantes-Morales, H., 1980, Taurine interactions with chick retinal membranes, J. Neurochem., 34:1047-1052.

38. López-Colomé, A.M., and Pasantes-Morales, H., 1980, Effect of taurine on ^{45}Ca-transport in frog retinal rod outer segments, Exp. Eye Res., in press.
39. Mandel, P., Pasantes-Morales, H., and Urban, P.F., 1976, Taurine a putative transmitter in retina, in: "Transmitters in the Visual Process," S.L. Bonting, ed., Pergamon Press, Oxford, New York, (1976), pp.89-105.
40. Mandel, P., and Pasantes-Morales, H., Taurine in the Nervous System, in: "Reviews of Neuroscience. Vol. 3", Ehrenpreis and I. Kopin, eds., Raven Press, New York (1978), pp.157-193.
41. Michaelis, E.K., Michaelis, M.L., and Boyarski, L.L., 1974, High-affinity glutamate binding to brain synaptic membranes, Biochim. biophys. Acta., 367:338-348.
42. Neal, M.J., and Atterwill, C.K., 1974, Isolation of photoreceptor and conventional nerve terminals by subcellular fractionation of rabbit retina, Nature, London, 251:331-333.
43. Neal, M.J., 1976, Amino acid transmitter substances in the vertebrate retina, Gen. Pharmacol., 7:321-323.
44. Okamoto, K., and Sakai, Y., 1980, Localization of sensitive sites to taurine. γ-aminobutyric acid, glycine and β-alanine in the molecular layer of guinea-pig cerebellar slices, Br., J. Pharmac., 69:407-413.
45. Orr, H.T., Cohen, A.I., and Lowry, O.H., 1976, The distribution of taurine in the vertebrate retina, J. Neurochem., 26:609-611.
46. Pasantes-Morales, H., Klethi, J., Ledig, M., and Mandel, P., 1972, Free amino acids in chicken and rat retina, Brain Res., 41:494-497.
47. Pasantes-Morales, H., Urban, P.F., Klethi, J., and Mandel, P., 1972, Etude de l'effet de la taurine sur l'electroretinogramme de la retine en perfusion, C.R. Acad. Sci. D (Paris), 275:699.
48. Pasantes-Morales, H., Urban, P.F., Klethi, J., and Mandel, P., 1973, Light stimulated release of ^{35}S-taurine from chick retina, Brain Res., 51:375-378.
49. Pasantes-Morales, H., Bonaventure, N., Wioland, N., and Mandel, P., 1973, Effect of intravitreal injections of taurine and GABA on chicken ERG, Int. J. Neurosci., 5:235-241.
50. Pasantes-Morales, H., López-Colomé, A.M., Salceda, R., and Mandel, P., 1976, Cysteine sulfinate decarboxylase in chick and rat retina during development, J. Neurochem., 27:1103-1106.
51. Pasantes-Morales, H., Ademe, R.M., and López-Colomé, A.M., 1979, Taurine effects on $^{45}Ca^{2+}$ transport in retinal subcellular fractions, Brain Res., 172:131-138.

52. Pasantes-Morales, H., and Gamboa, A., 1980, Effect of taurine on $^{45}Ca^{2+}$ accumulation in rat brain synaptosomes, J. Neurochem., 34:244-246.
53. Peck, E.J., Schaeffer, J.M., and Clark, J.H., 1973, γ-aminobutyric acid, bicuculline and postsynaptic binding sites, Biochem. Biophys. Res. Commun., 52:394-400.
54. Rubin, R.P., 1970, The role of calcium in the release of neurotransmitter substances and hormones, Pharmacol. Rev., 22:389-428.
55. Salceda, R., López-Colomé, A.M., and Pasantes-Morales, H., 1977, Light-stimulated release of ^{35}S-taurine from frog retinal rod outer segments, Brain Res., 135:186-191.
56. Sano, K., and Roberts, E., 1963, Binding of γ-aminobutyric acid by mouse brain preparations, Biochem. Pharmacol., 12:489-502.
57. Schmid, R., Sieghart, W., and Karobath, M., 1975, Taurine uptake in synaptosomal fractions of rat cerebral cortex, J. Neurochem., 25:5-9.
58. Sharif, N.A., and Roberts, P.J., 1980, Problems associated with the binding of L-glutamic acid to synaptic membranes: Methodological aspects, J. Neurochem., 34:779-784.
59. Sieghart, W., and Karobath, M., 1974, Evidence for specific synaptosomal localization of exogenous accumulated taurine, J. Neurochem., 23:911-915.
60. Snyder, S.H., 1975, The glycine synaptic receptor in the mammalian nervous system, Br., J. Pharmac., 53:473-484.
61. Thurston, J.H., Hawhart, R.E., and Dirgo, J.A., 1980, Taurine: A role in osmotic regulation of mammalian brain and possible clinical signficance, Life Sci., 26:1561-1568.
62. Urban, P.F., Dreyfus, H., and Mandel, P., 1976, Influence of various amino acids on the bioelectrical response to light stimulation of a superfused frog retina, Life Sci., 18:473-480.
63. Welty, J.D., McBroom, J.J., Appelt, A.W., Peterson, M.B., and Read, W.O., Effect of taurine on heart and brain electrolyte imbalances, in: "Taurine," R. Huxtable, and A. Barbeau, eds., Raven Press, New York (1976), pp.155-163.
64. Williams, M., Risley, E.A., and Totaro, J.A., 1980, Interaction of taurine and β-alanine with central nervous system neurotransmitter receptors, Life Sci., 26:557-560.
65. Yoneda, Y., and Kuriyama, K., 1980, Presence of a low molecular weight endogenous inhibitor on ^{3}H-muscimol binding in synaptic membranes, Nature, 285:670-673.
66. Zukin, S.R., Young, A.B., and Snyder, S.H., 1974, Gamma-aminobutyric acid binding to receptor sites in the rat central nervous system, Proc. Natl. Acad, Sci., U.S.A., 71:4802-4807.

SPECIFIC BINDING OF TAURINE IN CENTRAL NERVOUS SYSTEM

Tomio Segawa, Atsuko Inoue, Takehiro Ochi, Yoshihiro Nakata and Yasuyuki Nomura

Department of Pharmacology
Institute of Pharmaceutical Sciences
Hiroshima University School of Medicine
Kasumi 1-2-3, Hiroshima 734, Japan

Taurine (2-aminoethanesulfonic acid) is present in the central nervous system (CNS) in high concentration and satisfies some of the criteria essential for identification as a neurotransmitter. These are high affinity uptake of taurine into synaptosomes or brain slices (6,18), the release of taurine from brain slices and synaptosomes by electrical stimulation or high concentrations of potassium ion (3,4,6,9,10) and specific binding of taurine to brain synaptic membranes (7,8,11,19). Furthermore, taurine has been shown to be bound to (19) and taken up by glial cells, suggesting that glial cells are involved in the functional role of taurine in CNS.

In this paper, in order to clarify the physiological role of taurine in CNS, we have investigated in more detail the characteristics of taurine binding in CNS and present evidence that the binding sites may involve postsynaptic receptor sites of neurons as well as uptake sites for the taurine accumulating system of glia.

MATERIALS AND METHODS

Male Wistar rats (150-200g) were used. Crude synaptic membranes, glial membranes and whole membranes were prepared by the method of Simon et al. (20), Nagata et al. (13) and Campochiaro et al. (2), respectively. The obtained membranes were suspended in 10 ml of distilled water and maintained at room temperature for 20 min to dissociate any taurine bound to the membranes. After centrifugation at 48,000 g for 10 min the pellets were suspended in Krebs-Ringer solution (pH 7.5) containing 16.2 mM Tris-HCl buffer to yield a final concentration of 0.4-1.0 mg protein/ml. To 0.8 ml of the suspension 0.1 ml of distilled water or non-radioactive taurine in distilled water (final concentration of 10 mM) was added and the mixture was incubated at 0°C for 5 min, after which 0.1 ml of [^{3}H] taurine (final concentration of 0.011-1.1 μM for crude synaptic membranes, 0.011-2.055 μM for glial membranes and 0.011-1.65 μM for whole membranes) was added and further incubation was carried out at 25°C for 10 min. Bound [^{3}H]taurine was separated from free by centrifuging the mixture at 48,000g for 10 min. The supernatant fluid was decanted, and the pellet was resuspended in 2 ml of modified Krebs-Ringer solution and recentrifuged at 48,000g for 10 min. Bound radioactivity in the pellet was determined in a model 3320 Packard Tri-Carb liquid scintillation spectrometer. Specifically bound [^{3}H]taurine is defined as the amount displaced by the unlabelled taurine. Protein was measured by the method of Lowry et al. (12). Assay of carbonic anhydrase was performed by the method of Okuma et al. (15).

The influence of temperature on specific binding of [^{3}H]taurine was investigated by changing medium temperature from 0°C to 65°C. Medium pH was changed from 8.5 to 6.0 by adding 1N HCl solution. For the experiment in which the influence of freezing and thawing on specific binding was examined, crude synaptic or glial membranes were maintained at -20°C for 18-137 hr after which 2 ml of distilled water was added and the mixture was incubated at 25°C for 20 min. After incubation, the medium was centrifuged at 4°C for 10 min. Thereafter the obtained pellets were resuspended in Krebs-Ringer solution (pH 7.5) and were homogenized with a Polytron for 1 min. The influence of ions on specific binding was examined by changing ion concentrations of NaCl (0-200 mM), KCl (0-50 mM) and $CaCl_2$ (0-20 mM). In experiments using GABA, β-alanine, taurine, hypotaurine, strychnine and bicuculline, crude synaptic membranes were suspended in Krebs-Ringer solution (pH 7.5), the appropriate amount of drug was added and the suspension was incubated at 0°C for 5 min. Following this incubation, 55 nM [^{3}H]taurine was added, and incubated at 0°C for 10 min.

The drugs used were: [1, 2-^{3}H]taurine (9 Ci/mmole, The Radiochemical Centre, Amersham), taurine (Taisho Pharmaceutical Co., Ltd.), GABA, β-alanine (Nakarai Chemical Co., Ltd.), hypotaurine (Calbiochem-Boehringer), Strychnine sulfate (Merck), Bicuculline (Sigma Chemical Co.).

RESULTS

Binding of [^{3}H]Taurine to Crude Synaptic, Glial and Whole Membranes

As shown in Table 1 a Scatchard analysis of the data for specific [^{3}H]taurine binding to crude synaptic, glial and whole membranes resulted in two distinct types of binding sites: one with high affinity and the other with low affinity. The Kd value for high affinity binding to glial membranes was the lowest, approximately one fourth of that for the binding to crude synaptic membranes, while Bmax value for high affinity binding to whole membranes was the highest, followed by the value for the binding to glial and crude synaptic membranes.

Subcellular Distribution of Specific [^{3}H]Taurine Binding

Table 2 shows subcellular distribution of specific taurine binding in rat cerebral cortex. In this experiment, each fraction from whole brain was subjected to the binding assay. The distribution in the primary fractions was even and was not highly localized in crude mitochrondrial P2 fraction. In P2 subfractions, the largest amount of taurine binding was observed in fraction P_2A which consists mainly of myelin. Therefore, taurine binding was not highly localized in the synaptic region.

Regional Distribution of Specific [^{3}H]Taurine Binding to Whole Membranes

The specific binding of [^{3}H]taurine to whole membranes was high in pons-medulla, spinal cord, diencephalon and mesencephalon and low in striatum and hippocampus. This regional distribution of the binding did not correlate with that of the specific activity of carbonic anhydrase, a marker of astroglial cells. This distribution also did not correlate with that of endogenous amounts of taurine (data are not shown).

Influence of Medium pH and Temperature on Specific Binding

Specific [^{3}H]taurine binding to three membranes was maximum at pH 7.5 (Fig. 1). However, a considerable percentage of maximum binding was also observed at pH 6.0 in glial membranes and at pH 8.0 in synaptic membranes. Binding was maximal at 250, and was not markedly affected on lowering the temperature to 0°C (Fig. 2). At temperatures exceeding 37°C there were steep decreases in the amounts of specifically bound [^{3}H]taurine in all membranes (Fig. 2).

Table 1. Scatchard analysis of specific [^{3}H]taurine binding to rat brain membranes

	Synaptic membranes	Glial membranes	Whole membranes
High affinity			
K_d (nM)	160	38	83
Bmax (pmole/mg)	0.21	0.29	0.34
Low affinity			
K_d (μM)	1.3	2.1	3.7
Bmax (pmole/mg)	1.2	7.1	9.0

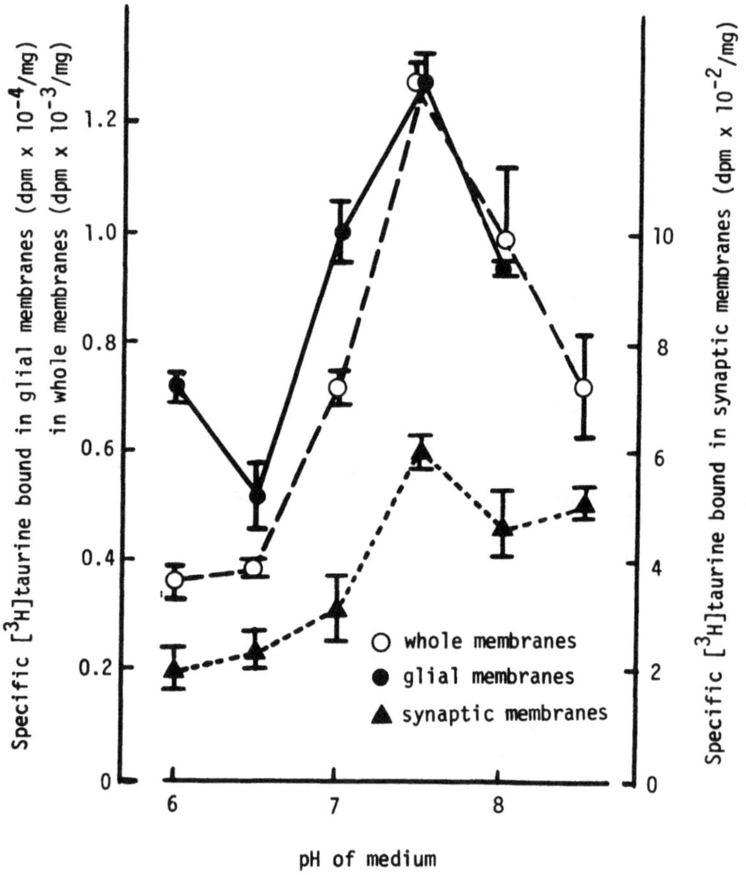

Fig. 1. Influence of medium pH on specific (^{3}H)taurine binding to rat brain membranes. Medium pH was changed by adding 1N HCl solution.

Influences of KCl and $CaCl_2$ on Specific [^{3}H]Taurine Binding

Although KCl failed to affect [^{3}H]taurine binding to crude synaptic and whole membranes, [^{3}H]taurine binding to glial membranes was influenced by KCl. Thus, compared to the binding at 4.8 mM KCl, the binding at 0 or 50 mM KCl was significantly lower (Fig. 3). At concentrations up to 50 mM, $CaCl_2$ had no effect on specific [^{3}H]taurine binding (Fig. 4).

Influence of NaCl on Specific [^{3}H]Taurine Binding

The specific binding of [^{3}H]taurine to the three types of membranes was linearly proportional to the concentration of NaCl over the range of 0-200 mM (Fig. 5). Replacement of NaCl, KCl, $CaCl_2$ and $MgCl_2$ with NaBr, KBr, $CaBr_2$ and $MgBr_2$ respectively, did not affect significantly specific [^{3}H]taurine binding (data are not shown).

Influence of Freezing and Thawing on NaCl-Dependent Specific [^{3}H]Taurine Binding

In crude synaptic membranes which had been previously frozen at -20°C for 18-137 hr, at 118 mM NaCl specific [^{3}H]taurine binding was significantly lower than that in fresh membranes, while the difference between these two values for fresh and frozen membranes was not statistically significant at lower concentrations of NaCl (Fig. 6).

In glial membranes which had been previously frozen at -20°C for 18-137 hr, binding was significantly lower than that to fresh membranes at 50, 118 and 150 mM NaCl (Fig. 7).

Influence of Drugs on [^{3}H]Taurine Binding to Crude Synaptic Membranes

Several drugs known to compete with [^{3}H]taurine uptake to synaptosomal fractions were tested for their ability to inhibit [^{3}H]taurine binding (Fig. 8), and IC_{50} values for displacement of [^{3}H]taurine binding were calculated for each drug by log-percent analysis of the data presented in Fig. 8 (Table 3). Taurine and hypotaurine were most potent followed by strychnine and β-alanine, while GABA was a weak displacer. Bicuculline was ineffective in displacing specific [^{3}H]taurine binding even at a concentration of 1 mM.

Hruska et al. (5) demonstrated the existence of temperature-sensitive and sodium-dependent transport mechanism for taurine into rat brain synaptosomal fractions. This uptake

Table 2. Subcellular distribution of specific [^{3}H]taurine binding in rat cerebral cortex

Fraction	Specific [^{3}H]taurine bound (dpm x 10^{-2}/mg protein)
Homogenate	701 ± 24
P_1	709 ± 13
P_2	477 ± 8
P_3	681 ± 18
P_2A	1413 ± 72
P_2B	113 ± 26
P_2C	30 ± 10

Values are the means ± SEM from three experiments.

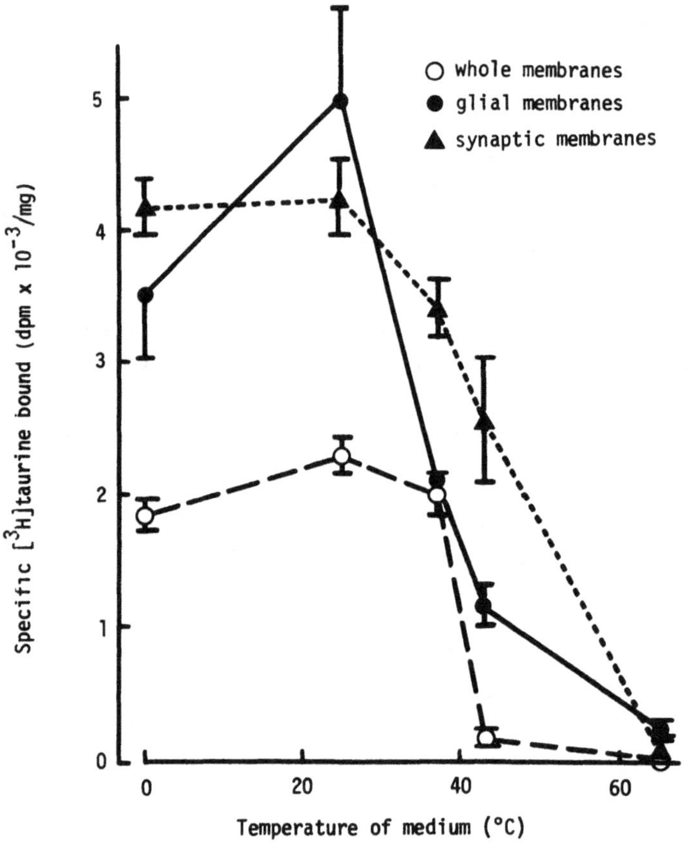

Fig. 2. Influence of medium temperature on specific (^{3}H)taurine binding to rat brain membranes.

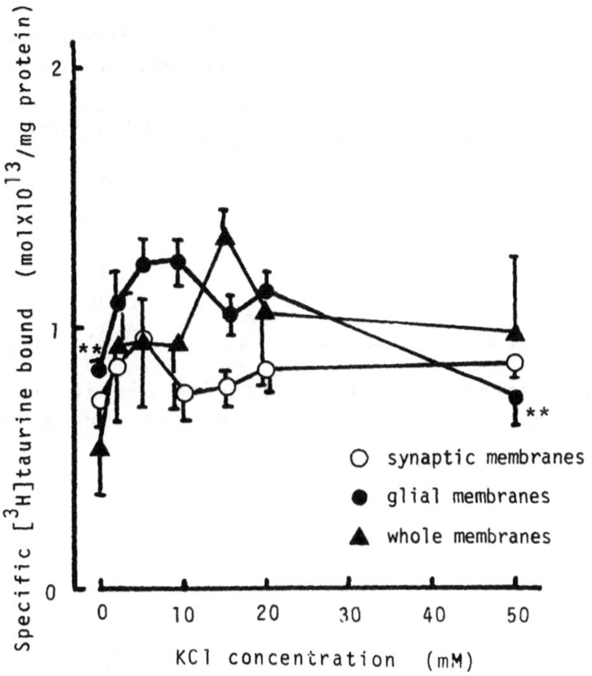

Fig. 3. Influence of KCl on specific (^{3}H)taurine binding to rat brain membranes. **P<.05 compared to value at 4.8 mM KCl.

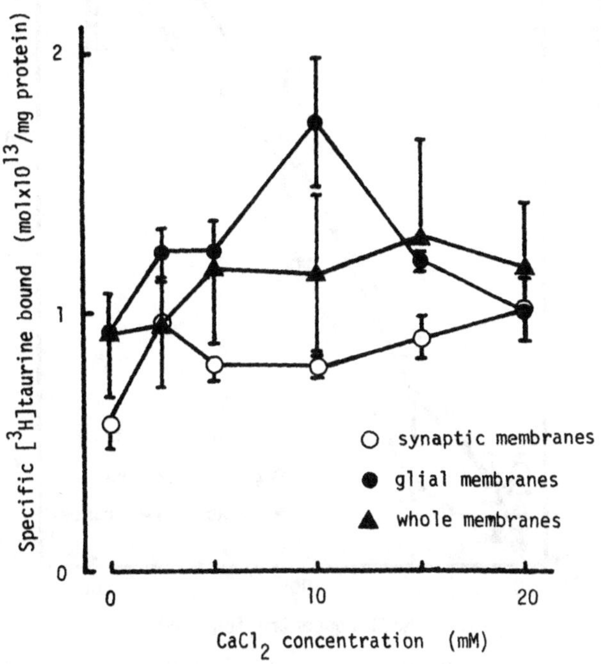

Fig. 4. Influence of $CaCl_2$ on specific (^{3}H)taurine binding to rat brain membranes.

Table 3. Influence of drugs on specific [^{3}H]taurine binding to rat brain crude synaptic membranes.

Drugs	IC_{50} (μM)
GABA	660
β-alanine	74
hypotaurine	9.4
strychnine	69
bicuculline	1000
taurine	8.7

IC_{50} values for displacement of [^{3}H]taurine binding were calculated by log-percent analysis of the data presented in Fig. 8.

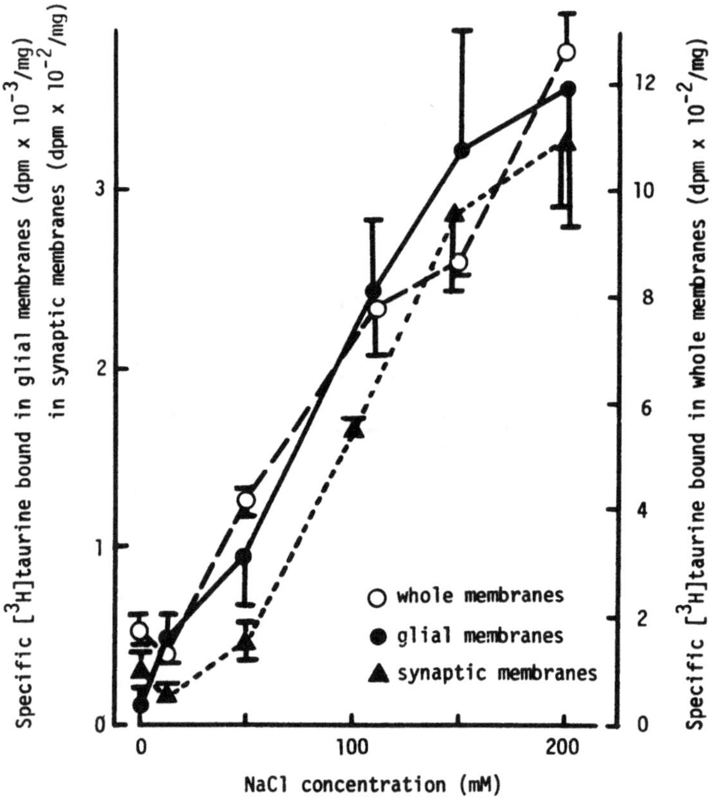

Fig. 5. Influence of NaCl on specific (^{3}H)taurine binding to rat brain membranes.

mechanism for [^{3}H]taurine was effectively inhibited by 100 μM taurine, hypotaurine, β-alanine, GABA, bicuculline or strychnine in this order. In Fig. 9, the IC_{50} values we find for displacement of [^{3}H]taurine binding are plotted against the percent inhibitions of [^{3}H]taurine uptake reported by Hruska et al. (5). No correlation between these two values for these drugs was observed.

DISCUSSION

In the present experiments we demonstrate the existence of two distinct types of binding sites for [^{3}H]taurine in whole, synaptic and glial membranes. Lahdesmaki et al. (11) observed only one type of binding site for taurine in synaptic membranes from calf brain cortex. Furthermore, their binding constant of 1.5 μM is about 10 times our value for high affinity binding. The reason for these differences are unclear. Probably the difference in taurine concentration and animal species contribute to the difference in results.

The existence of high affinity binding site for taurine in glial membranes suggests the importance of glial cells in a functional role for taurine in the CNS.

Regional distribution of specific [^{3}H]taurine binding in whole membranes correlated neither with that of the specific activity of carbonic anhydrase, a marker of astroglial cells, nor with that of the endogenous amounts of taurine.

The pH optimum of specific [^{3}H]taurine binding in three membranes was 7.5, suggesting that part of this binding to synaptic membranes may be the interaction with pharmacologically relevant receptors because neurotransmitter receptors, as surface membrane proteins, operate normally under conditions of physiological pH.

Specific [^{3}H]taurine binding was higher at 25°C than at 37°C. Probably taurine dissociates more from rapidly binding sites at 37°C than at 25°C.

Glial cells have an ability to take up K^+. This uptake phenomenon may function physiologically to remove K^+ from extracellular space and control the excitability of neurons. In preliminary experiments, we demonstrated that taurine could increase K^+ transport into glial cells (unpublished). Our findings that the maximum specific [^{3}H]taurine binding in glial membranes was observed at physiological concentrations of

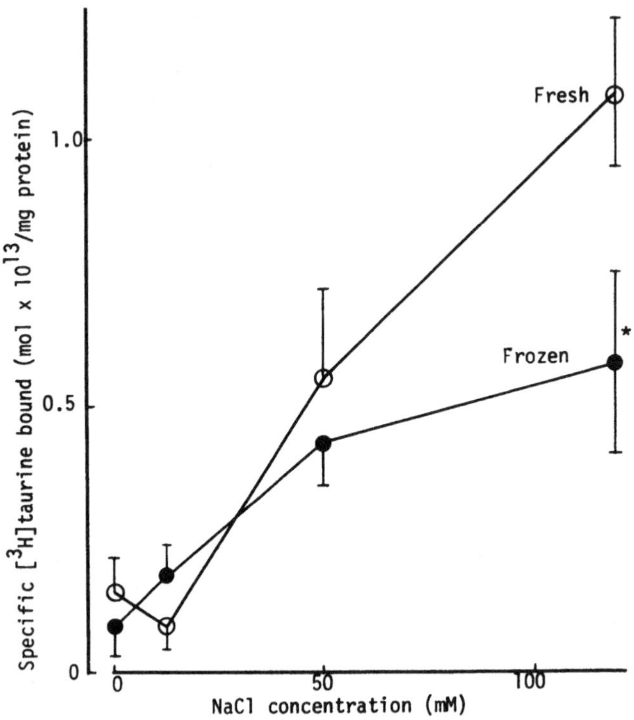

Fig. 6. 'Influence of freezing and thawing on NaCl-dependent specific (^{3}H)taurine binding to rat brain crude synaptic membranes.

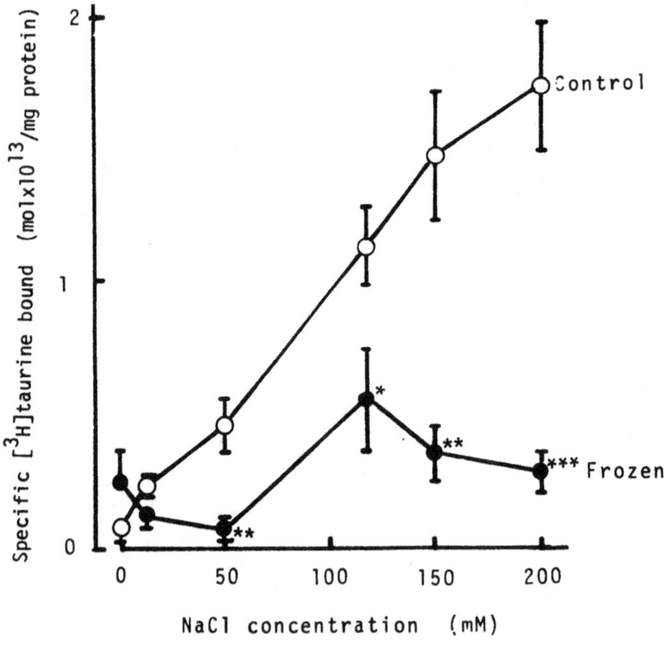

Fig. 7. Influence of freezing and thawing on NaCl-dependent specific (^{3}H)taurine binding to rat brain glial membranes.

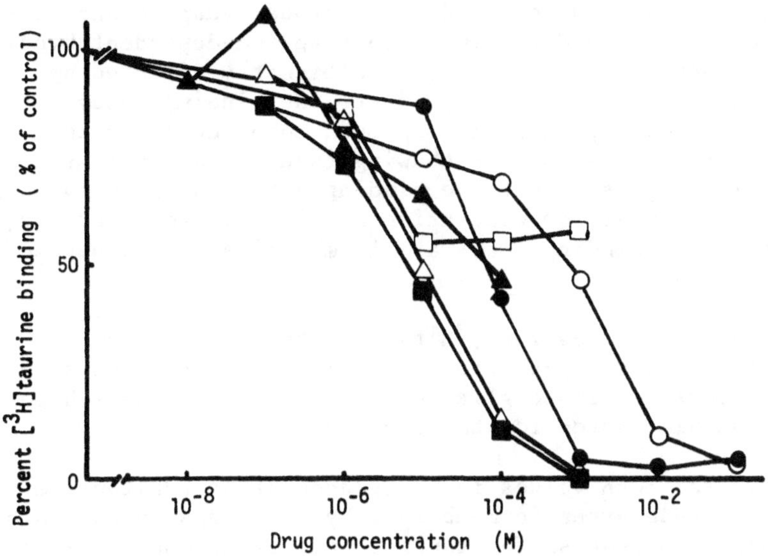

Fig. 8. Influence of drugs on specific (^{3}H)taurine binding to rat brain crude synaptic membranes. ○—○ GABA, ●—● β-alanine, △—△ hypotaurine, ▲—▲ strychnine, □—□ bicuculline, ■—■ taurine.

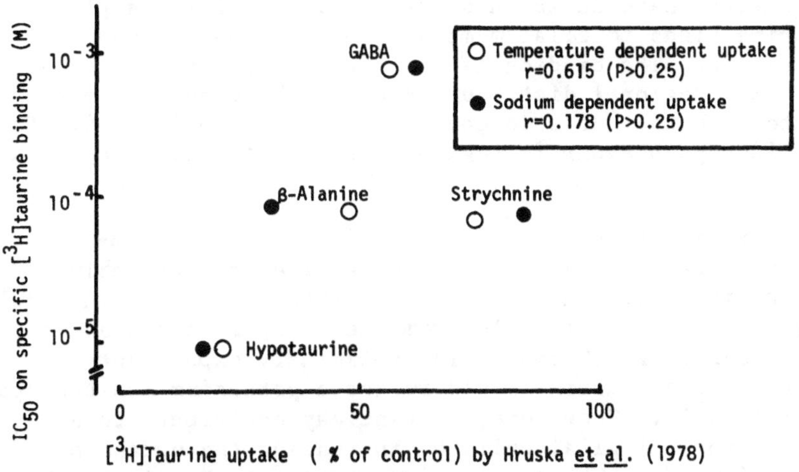

Fig. 9. Schematic representation of taurine binding sites in rat central nervous system.

K^+ suggests that taurine interacts with K^+ at the transport sites in glial membranes and may control the excitability of neurons.

Specific [^{3}H]taurine binding to crude synaptic and glial membranes was sodium-dependent. This sodium-dependent binding to crude synaptic membranes was significantly affected by storing the membranes frozen prior to the binding assay. Because the presynaptic region is much more sensitive to degeneration by freezing and thawing than postsynaptic region, this finding suggests that the binding sites in crude synaptic membranes at low Na^+ concentration are largely postsynaptic, but part of the binding sites at 118 mM Na^+ are presynaptic in glial membrane.

The binding sites for [^{3}H]taurine that are sodium-dependent and freezing-thawing sensitive may represent transport sites for taurine, since glial cells seem to possess a high affinity uptake system for taurine (14).

Since the IC_{50} values for displacement of [^{3}H]taurine binding to crude synaptic membranes by the drugs which inhibit the transport mechanism for taurine into synaptosomes did not correlate with the percent inhibition of [^{3}H]taurine uptake, sodium-dependent taurine binding to crude synaptic membranes may not involve taurine transport sites in presynaptic membranes.

Our results are rather preliminary and we need further detailed experiments before we reach a final conclusion, but the specific binding for taurine in CNS seems different from those of transmitter candidates in several points. For example: 1. Taurine binds to glial membranes as well as neuronal membranes. 2. The binding is not highly localized in synaptic region. 3. Regional distribution of the binding does not correlate with that of endogenous amounts of taurine. 4. Part of the binding in synaptic region occurs in the presynaptic area.

With regard to the binding to glial cells, sodium-dependent, potassium-sensitive binding sites are probably transport sites for taurine. Through these sites taurine could be taken up into glial cells from the synaptic cleft as is the case with GABA. Furthermore, in preliminary experiments we demonstrated that taurine could increase potassium ion transport into glial cells. Therefore, taurine may contribute to a functional role of glial cells by increasing the uptake of potassium ion and other neuroactive substances by binding to glial membranes and as a consequence, controlling excitability of neurons. If this is the case, in CNS, taurine may act as a neurotransmitter as well as a neuromodulator.

REFERENCES

1. Borg, J., Balcar, V.J., and Mandel, P., 1979, Characterization of taurine uptake by neuronal and glial cells in culture, J. Neurochem., 32:1801-1805.
2. Campochiaro, P., Schwarcz, R., and Coyle, J.T., 1977, GABA receptor binding in rat striatum: Localization and effects of denervation, Brain Res., 136:501-511.
3. Collins, G.C.S., and Topiwala, S.H., 1974, The release of [^{14}C]-taurine from slices of rat cerebral cortex and spinal cord evoked by electrical stimulation and high potassium ion concentrations, Br. J. Pharmac., 50:451-452.
4. Davison, A.N., and Kaczmarek, L.K., 1971, Taurine-a possible neurotransmitter? Nature, 234:107-108.
5. Hruska, R.E., Hustable, R.J., and Yamamura, H.I., High affinity, temperature-sensitive and sodium-dependent transport of taurine in rat brain, in: "Taurine and Neurological Disorders," A. Barbeau and R.J. Huxtable, eds., Raven Press, New York, (1978), pp.109-135.
6. Kaczmarek, L.K., and Davison, A.N., 1972, Uptake and release of taurine from rat brain slices, J. Neurochem., 19:2355-2362.
7. Kumpulainen, E., Jokisalo, V.J., and Lahdesmaki, P., 1978, Physiological characterization of taurine binding with synaptic membranes in the chicken brain, Intern. J. Neuroscience, 8:123-128.
8. Kumpulainen, E., Olinuora, M., and Lahdesmaki, P., 1979, Ability of calf brain synaptic membranes to bind [^{35}S] taurine to their Triton X-100 and chloroform extracts, Neurosci. Letter, 11:215-218.
9. Lahdesmaki, P., and Oja, S.S., 1972., Effect of electrical stimulation on the influx and efflux of taurine in brain slices of newborn and adult rats, Exp. Brain Res., 15:430-438.
10. Lahdesmaki, P., Pasula, M., and Oja, S.S., 1975, Effect of electrical stimulation and chlorpromazine on the uptake and release of taurine, γ-aminobutyric acid and glutamic acid in mouse brain synaptosomes, J. Neurochem., 25:675-680.
11. Lahdesmaki, P., Kumpulainen, E., Paasakka, O., and Kyrki, P., 1977, Interaction of taurine, GABA and glutamic acid with synaptic membranes, J. Neurochem., 29:819-826.
12. Lowry, O.H., Rosebrough, H.T., Farr, A.L., and Randall, R.J., 1951, Protein measurement with the Folin phenol reagent, J. Biol. Chem., 193:265-275.

13. Nagata, Y., Mikoshiba, K., and Tsukada, Y., 1974, Neuronal cell body enriched and glial cell enriched fractions from young and adult rat brain: Preparation and morphological and biochemical properties, J. Neurochem., 22:493-503.
14. Oja, S.S., and Kontro, P., Neurotransmitter actions of taurine in the central nervous system. in: "Taurine and Neurological Disorders," A. Barbeau and R.J. Huxtable, eds., Raven Press, New York (1978), pp.181-200.
15. Okuma, Y., Nomura, Y., and Segawa, T., 1979. Development of regional activities of 2',3'-cyclic nucleotide 3'-phosphohydrase and carbonic anhydrase in the rat brain: Influences of 6-hydroxydopa, Dyn., 2:391-396.
16. Riddall, D.R., Leach, M.J., and Davison, A.N., 1976, Neurotransmitter uptake into slices of rat cerebral cortex in vitro: effect of slice size, J. Neurochem., 27:835-839.
17. Schrier, B.K., and Thompson, E.J., 1974, On the role of glial cells in the mammalian nervous system, J. Biol. Chem., 249:1769-1780.
18. Segawa, T., Nakata, Y., and Nakamoto, N., 1979, Uptake and specific binding of taurine in synapse of the central nervous system, Sulfur-containing amino acids, 1:19-29.
19. Segawa, T., Nomura, Y., Nakata, Y., and Inoue, A., 1979, Specific binding of taurine in the central nervous system. Sulfur-containing amino acids, 2:47-54.
20. Simon, J.R., Contrera, J.F., and Kuhar, M.J., 1976. Binding of [^{3}H]kainic acid, an analogue of L-glutamate, to brain membranes, J. Neurochem., 26:141-147.
21. Wheler, G.H.T., Bradford, H.F., Davison, A.N., and Thompson, E.J., 1979, Uptake and release of taurine from cerebral cortex slices and their subcellular compartments, J. Neurochem., 33:331-337.

CENTRAL NEUROPHARMACOLOGY OF D-ALA^2-MET-ENKEPHALINAMIDE AND ITS INTERACTIONS WITH TAURINE IN RATS

Kanji Izumi[1], Eisuke Munekata[2], Hiroaki Yamamoto[3], Motoaki Yoshida[1], Takao Nakanishi[1], Pierre Wong[4] and André Barbeau[4]

[1]Department of Neurology
Institute of Clinical Medicine
[2]Institute of Applied Biochemistry
[3]Institute of Community Medicine
University of Tsukuba
Ibaraki 305, Japan
[4]Department of Neurobiology
Clinical Research Institute of Montreal
Quebec H2W 1R7, Canada

Methionine enkephalin, the amino acid sequence 61-65 (Tyr-Gly-Gly-Phe-Met) of β-lipotropin, and the related peptide leucine enkephalin (Tyr-Gly-Gly-Phe-Leu) are endogenous ligands for the opiate receptor in the brain (12). These peptides administered icv in rats (2) and mice (6) possess short-lasting and naloxone-sensitive analgesic properties. D-Ala^2-Met-enkephalinamide (DAME, Tyr-D-Ala-Gly-Phe-Met-$CONH_2$) is an enkephalin analogue and is pharmacologically similar to morphine and β-endorphin (β-LPH_{61-91}); DAME is more potent in the induction of analgesia than methionine enkephalin (24) and elicits the same catalepsy in rats (29) as β-endorphin (4,17,19, 22).

Taurine is an ubiquitous sulfur containing amino acid and is proposed to be a neurotransmitter or a modulator in the mammalian CNS (1,13). Recently, it has been shown that taurine can counteract certain actions of morphine; the increment in plasma growth hormone values in rats induced by morphine administration was completely blocked by the icv injection of taurine (11). On the basis of this report, we decided to investigate whether or not taurine acutely or chronically administered icv interferes with other symptoms such as akinesia, catalepsy or analgesia in rats induced by an acute injection of morphine-like peptides into the left lateral ventricle. The effects of GABA or L-leucine at the same molar concentration as that of taurine solution were also investigated in the present study, because these amino acids allow control of factors such as osmolarity and injection volume better than a buffer-saline solution. In addition, studies on effects of DAME upon dopamine release from the corpus striatum were also carried out in the presence or absence of taurine *in vitro*, since akinesia or catalepsy in animals caused by morphine-like peptides has been found to be closely related to dopaminergic involvement (17,21,26).

Akinesia Testing

Immobility (4,7) with muscle rigidity in animals induced by morphine-like peptides has often been described as catatonia (5,27, 29) or catalepsy (9,19,22). Catatonia is a well-recognized psychiatric condition, characterized by negativistic reactions, phases of stupor or excitement and implusive behavior, which is seen in patients with schizophrenia. A similar state of immobility and muscle rigidity is also seen as a cardinal symptom in patients with Parkinson's disease; the immobile state is described as akinesia in this extrapyramidal disorder, although akinesia is not well defined. Akinesia and muscle rigidity in Parkinson's disease are relieved by L-DOPA treatment, whereas catatonia or catalepsy is not. We feel that the term "akinesia" better describes the state of immobility induced by morphine-like peptides in rats than "catatonia" because this immobile state with muscle rigidity is completely reversed by L-DOPA (22), d-amphetamine (a dopamine releasor) (18) or apomorphine (a dopamine receptor agonist) (17). Throughout the present studies, we use the term "akinesia" for the state of immobility that is called catatonia or catalepsy by some others. Akinesia was evaluated by placing the forepaws of the rat on a horizontal bar 10 cm high. The duration of akinesia was individually recorded and the cut-off time was determined to be 300 sec at each point of estimation during a 120 min observation period. Akinesia was defined as positive when the animal sustained the position for at least 15 sec.

Analgesia Testing

Analgesia was assessed by a hot-plate procedure. The surface of the hot-plate was 40 cm^2 in size and was maintained at $54 \pm 0.1^{o}C$ (mean ± SD) by a thermoregulated water-circulating pump. The time between placing the rat on the hot-plate and the first appearance of any reaction such as licking the paws, rearing or rapid movements was counted for estimation of the duration of analgesia. The duration of analgesia was individually recorded and the cut-off time was determined to be 60 sec at each point of estimation during a 120 min observation period. Analgesia was defined as positive when the animal placed on the hot-place showed no response for at least 10 min.

Acute Effects of Amino Acids on Akinesia Induced by D-Ala^2-Met-Enkephalinamide (DAME)

Male Wistar rats (220-310 g) were used throughout the experiments investigating acute effects of amino acids on akinesia and analgesia caused by a ED_{99} of DAME (50 μg/10 μl). DAME was dissolved in 0.85% saline solution. Each amino acid was dissolved in 5 mM phosphate buffer in NaCl solution and adjusted to 280-290 mosM and pH 7.2-7.4.

Immediately after the icv administration of taurine and GABA at a dose of 95 mM/10 μl, all eighteen rats showed slightly decreased muscle tonus. Hypotonus lasted for several minutes and was normalized by 10 min after the injection of each amino acid. At a smaller dose (23.75 mM - 47.5 mM/10 μl) of these amino acids, none of the animals showed hypotonus. The injection of L-leucine (95 mM/10 μl) or 5 mM phosphate buffer in NaCl solution (10 μl) did not alter muscle tonus.

Effects of Taurine (Figs. 1,2; Table 1)

The duration of akinesia (274 ± 24 sec, mean ± SE, N=10) 10 min after the administration of DAME in rats which had been pretreated with 95 mM taurine was almost identical to that (280 ± 25 sec, N=20) in control animals with 5 mM phosphate buffer in NaCl solution (Fig. 1). However, subsequent observation demonstrated that these taurine-treated rats had a significant reduction in the duration of akinesia compared with that in control animals at later times: 187 ± 44 (N=10) versus 297 ± 3.0 sec (N=20) at 60 min after the injection of DAME (Fig. 1). Akinesia disappeared in seven out of 10 rats with taurine pretreatment at the 90 min estimation time. The remaining three rats showed a short duration of akinesia of 16 to 18 sec.

It should be emphasized that these effects of taurine on akinesia were not obtained in a uniform pattern. For example, 30 min following the administration of DAME, one of the 10 rats which had received 95 mM taurine already showed no akinesia and three had a markedly shorter duration of akinesia of 16 to 36 sec. The other six rats still had the maximum state of immobility. At 60 min, one of these six animals was completely released from akinesia and two had short duration of akinesia, each of 22 sec. The other three rats still had the prolonged duration of akinesia and needed more than 30 min for a reduction of the length of akinesia. In other words, some rats which had been injected with taurine had reduction in the duration of akinesia at a relatively early point, but others did so only at later times. Thus, the variation of the average duration of akinesia in rats pretreated with taurine, which is not represented on the figures, became rather great.

Nevertheless, the effects of taurine on decreasing the duration

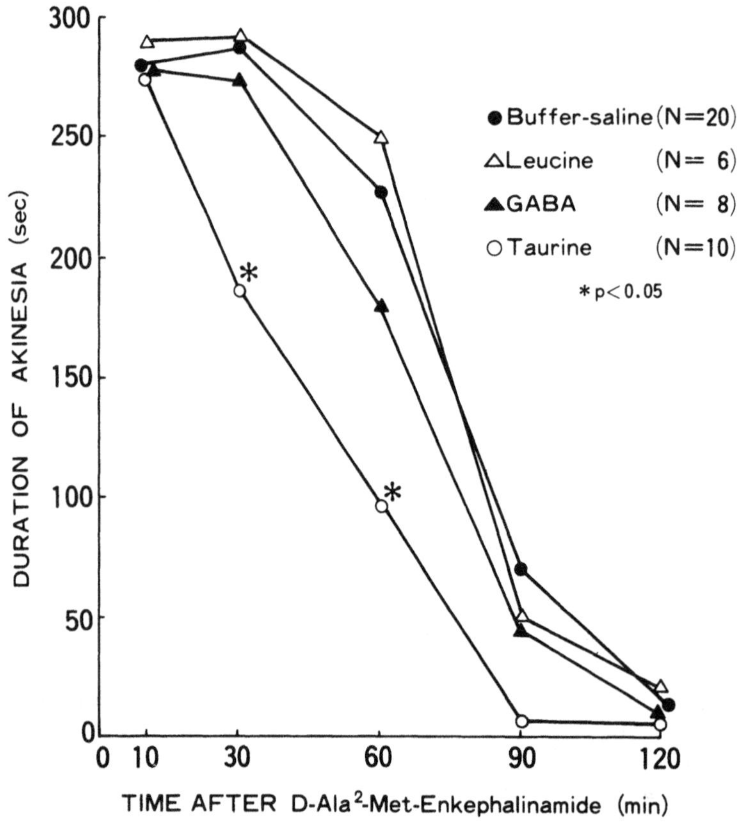

Fig. 1. Acute effects of amino acids on akinesia induced by D-Ala^2-Met-enkephalinamide (DAME). Test rats received each amino acid (95 mM/10 µl) 10 min prior to the administration of DAME (50 µg/10 µl). Control animals received 10 µl of 5 mM phosphate buffer in NaCl solution and 10 min later the same dose of DAME. All substances were administered intraventricularly. The maximum time of duration for the estimation of akinesia was determined to be 300 sec as the cut-off time.

of akinesia were observed in a dose-dependent manner, at least at estimation times between 30 min and 60 min following the injection of DAME (Fig. 2). To confirm these effects of taurine, an ED_{50} of DAME for akinesia was determined at 60 min following the injection of DAME in two groups of rats. As shown in Table 1, the ED_{50} of DAME was five times greater in the taurine-treated rats than that in the buffer-saline-treated animals. A calculated potency ratio (P.R.) was 5.0 and the factor P.R. was 2.94, indicating that the administration of taurine significantly decreased the occurrence of akinesia in rats evaluated at 60 min after the injection of DAME. At 30 min after the injection of DAME, an ED_{50} for akinesia could not be

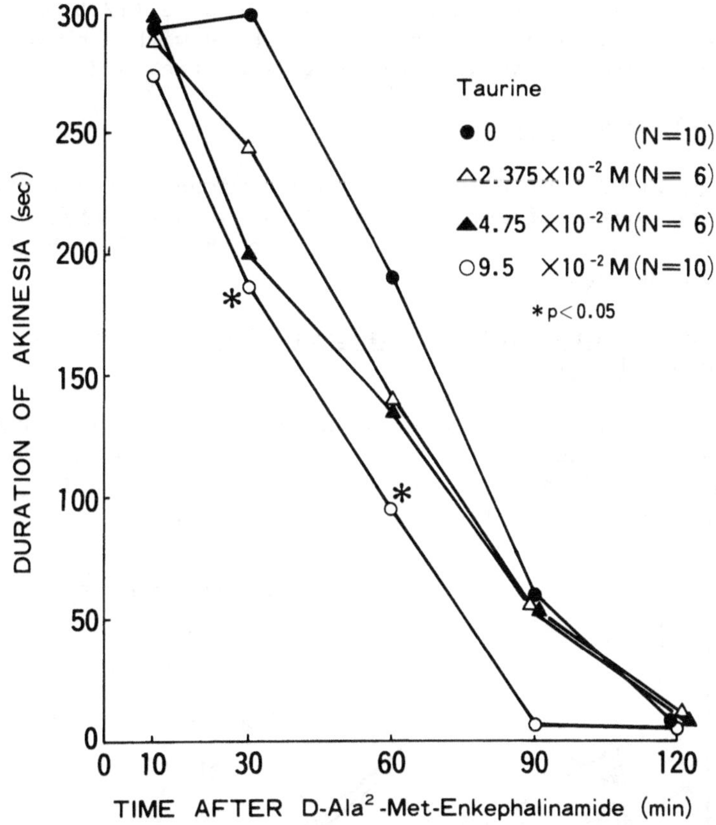

Fig. 2. Dose-dependent acute effects of taurine on akinesia induced by DAME. Different doses of taurine with a volume of 10 μl were injected intraventricularly 10 min prior to the intraventricular administration of DAME (50 μg/10 μl) in rats. The animals not receiving taurine received an identical volume of 5 mM phosphate buffer in NaCl solution.

obtained, since 100% occurrence of akinesia was observed even with a smaller dose (12.5 μg) of DAME in both test and control groups of rats.

Effects of GABA (Fig. 1; Table 1)

The duration of akinesia induced by DAME in rats was not modified by pretreatment with 95 mM GABA during a 120 min observation period (Fig. 1). An ED_{50} of DAME for akinesia in GABA-treated rats estimated at 60 min after the injection of the peptide analogue is not significantly different from that in control animals with 5 mM phosphate buffer in NaCl solution (Table 1). An ED_{50} of DAME for akinesia could not be obtained at 30 min, since the high occurrence of akinesia was observed even with a smaller dose (12.5 μg) of DAME; 83% rats pretreated with GABA and 67% of those with 5 mM phosphate buffer in NaCl solution showed akinesia after the DAME injection.

Effects of L-leucine (Fig. 1)

Although an ED_{50} of DAME was not determined in L-leucine-treated rats, the time course effects of this amino acid (95 mM/10 μl) on akinesia were found to be almost similar to those of 5 mM phosphate buffer in NaCl solution.

Acute Effects of Amino Acids on Analgesia Induced by D-Ala[2]-Met-Enkephalinamide (DAME)

Effect of each amino acid alone upon nociperceptive sensation was tested at one minute prior to the injection of DAME. None of amino acids used, administered icv 9 min before evaluation, caused either analgesia or hyperalgesia in rats (Fig. 3).

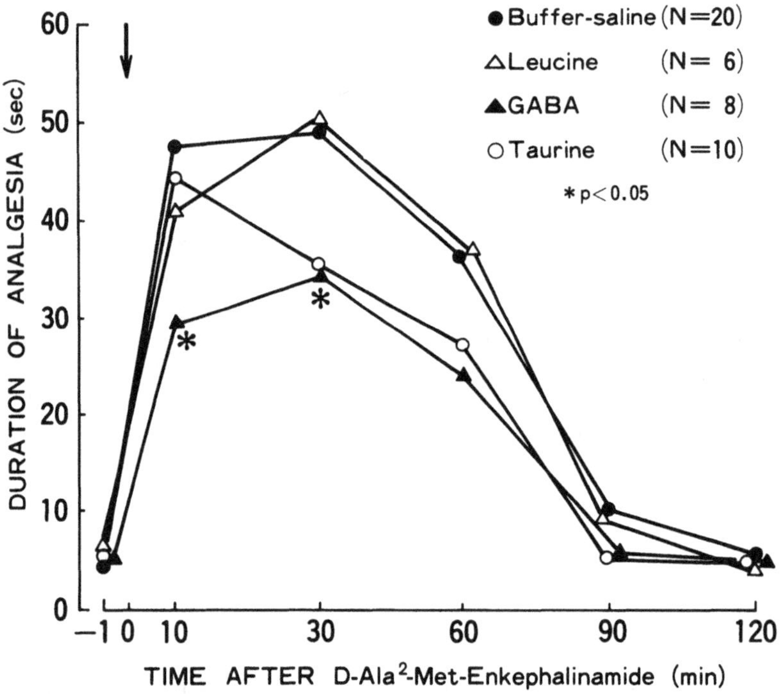

Fig. 3. Acute effects of amino acids on analgesia induced by DAME. One minute before the DAME injection, all rats were used for the examination of analgesia at basal values. The maximum duration for the estimation of analgesia was determined to be 60 sec as the cut-off time. The other experimental conditions were the same as those described in the legend for Fig. 1. The arrow indicates the time of injection of DAME.

Table 1

Changes in ED_{50} of D-Ala^2-Met-enkephalinamide for akinesia and analgesia by pretreatment with taurine or GABA in rats

Treatment	ED_{50} for akinesia at 60 min	ED_{50} for analgesia at 30 min	ED_{50} for analgesia at 60 min
Buffer-saline	8.91 (4.92-16.1)	6.76 (3.89-11.8)	17.8 (12.3-25.8)
Taurine	44.7 (18.2-109)*	12.5 (5.69-27.5)	29.5 (18.2-47.8)
GABA	11.3 (4.72-27.1)	20.9 (7.89-55.4)*	26.9 (17.8-55.4)

* $p<0.05$ Values are μg with 95% confidence units given in parentheses. Test substances were 95 mM/10 μl.

Effects of Taurine (Figs. 3,4; Table 1)

The duration of analgesia (44 ± 7 sec, mean ± SE, N=10) 10 min after the administration of DAME in rats which had been pretreated with 9.5×10^{-2}M taurine was almost identical to that (48 ± 6 sec, N=10) in control animals with 5 mM phosphate buffer in NaCl solution. Subsequently, a tendency to decrease the duration of analgesia was obtained at 30 and 60 min after the injection of DAME in taurine-treated rats. This tendency at these times was observed in a dose-dependent manner for this sulfur containing amino acid (Fig. 4). Individual case analysis revealed that there seemed to be a similar pattern for analgesic response by taurine pretreatment to that for akinesia. For example, 30 min following the injection of DAME, four out of 10 rats which had received 95 mM taurine showed a markedly short duration of analgesia, of 11 to 13 sec. The other six rats had durations of 25 to 60 sec. At 60 min, two rats which had had a duration of 25 or 48 sec at the 30 min estimation time, showed normal

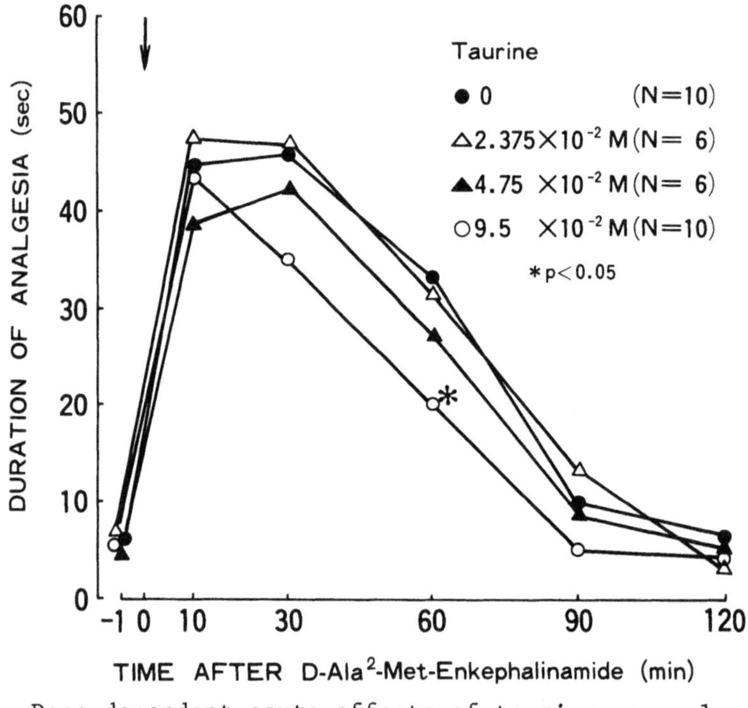

Fig. 4. Dose-dependent acute effects of taurine on analgesia induced by DAME. Different doses of taurine with a volume of 10 μl were injected intraventricularly 10 min prior to the intraventricular administration of DAME (50 μg/10 μl) in rats. The animals not receiving taurine received an identical volume of 5 mM phosphate buffer in NaCl solution. The arrow indicates the time of injection of DAME.

response against pain stimulation. The prolonged duration (60 sec) of analgesia in the remaining four animals was not altered by taurine pretreatment during a 120 min observation period.

An ED_{50} for analgesia determined at 60 min following the injection of DAME was slightly greater in taurine-treated rats than that in buffer-saline-treated animals (Table 1). A calculated potency ratio (P.R.) was 1.66 and the factor P.R. was 1.84, indicating that the two ED_{50}s are not significantly different. At 30 min, an ED_{50} for DAME was also determined and was found that the difference in the ED_{50} of enkephalinamide for analgesia in taurine-treated and buffer-saline-treated animals did not reach a significant level (Table 1).

Effects of GABA (Figs. 3,5; Table 1)

The duration of analgesia 10 and 30 min after the administration of DAME in rats pretreated with 95 mM GABA was significantly shorter than that at the same times in control animals treated with 5 mM phosphate buffer in NaCl solutions; 30 ± 8 (N=8) versus 48 ± 6 sec (N=20) at 10 min and 34 ± 7 (N=8) versus 49 ± 7 sec (N=20) at 30 min after the injection of DAME (Fig. 3). The suppressing effects of GABA on analgesia were observed dose-dependently during the first 60 min after the injection of enkephalinamide (Fig. 5). An ED_{50} for analgesia determined at 30 min following the injection of DAME was three times greater in GABA-treated rats than that in control animals (Table 1). A calculated potency ratio (P.R.) was 3.09 and the factor P.R. was 3.07, indicating that the administration of GABA significantly decreased the occurrence of analgesia in rats evaluated at 30 min after the injection of DAME.

Effects of L-leucine (Fig. 3)

Although an ED_{50} of DAME was not determined in L-leucine-treated rats, the time-course effects of this amino acid on analgesia were similar to those of 5 mM phosphate buffer in NaCl solution.

Chronic Effects of Amino Acids on Akinesia and Analgesia Induced by D-Ala^2-Met-Enkephalinamide (DAME)

Effects of taurine and L-leucine at a dose of 95 mM/10 μl chronically administered into the lateral ventricle on akinesia and analgesia caused by a single injection of DAME (50 μg/10 μl) were also investigated in rats (male Wister: 225-395 g). The repeated administration of taurine reduced significantly the duration of akinesia during the time between 10 and 60 min following the injection of DAME (Fig. 6). All six rats showed akinesia during these times. At 90 min, the duration (61 ± 52 sec, mean ± SE, N=6) of akinesia in rats with taurine was not significantly different from that (88 ± 47, N=6) in animals with L-leucine or that (115 ± 64, N=6) in controls

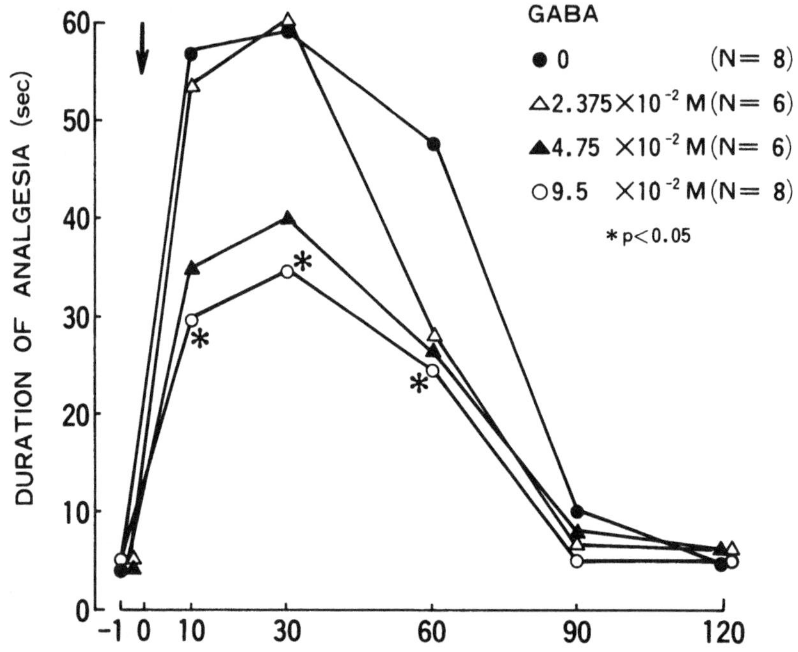

Fig. 5. Dose-dependent acute effects of GABA on analgesia induced by DAME. Different doses of GABA with a volume of 10 μl were injected intraventricularly 10 min prior to the intraventricular administration of DAME (50 μg/10 μl) in rats. The animals not receiving GABA received an identical volume of 5 mM phosphate buffer in NaCl solution. The arrow indicates the time of injection of DAME.

with 5 mM phosphate buffer in NaCl solution. However, at this estimation time, the number of rats with akinesia induced by DAME was significantly decreased with taurine treatment; only two out of six rats which had received daily taurine showed akinesia, while all six or five of six animals which had been treated with L-leucine or buffer-saline solution, respectively, had akinesia.

The duration of analgesia induced by DAME was also reduced in rats on repeated administration of taurine compared with that in control animals given buffer-saline solution, during the time between 30 and 60 min following the injection of DAME; 25 ± 12 (N=6) versus 44 ± 10 sec (N=6) at 30 min and 22 ± 12 (N=6) versus 34 ± 12 sec (N=6) at 60 min (Fig. 7). At the earliest point (10 min) or at 90 min observation time, the duration of analgesia in rats with taurine was not significantly different from that in controls with buffer-saline solution or that in animals with L-leucine. The number of rats with analgesia was not decreased with taurine- or L-leucine-treatment during the 120 min observation time following the DAME

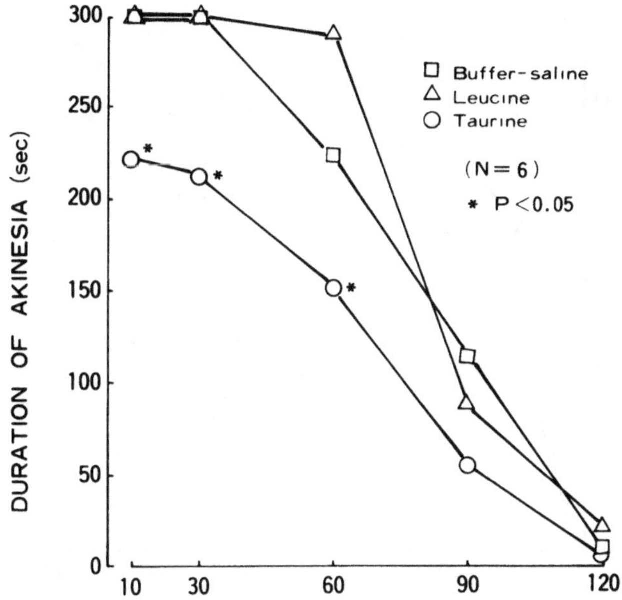

TIME AFTER D-Ala²-Met-Enkephalinamide (min)

Fig. 6. Chronic effects of amino acids on akinesia induced by DAME. The amino acids at a dose of 95 mM/10 μl were injected intraventricularly two times a day at around 10:00 a.m. and 4:00 p.m. for three days except on the third day when one injection with an amino acid solution was conducted 10 min prior to the intraventricular administration of DAME (50 μg/10 μl). Control rats received 10 μl of 5 mM phosphate buffer in NaCl solution instead of amino acid.

injection.

Effects of D-Ala^2-Met-Enkephalinamide on Dopamine Release from the Striatum and its Modification by Taurine

Male Wistar rats (250-280 g) were decapitated and the corpus striatum was dissected on ice as quickly as possible. Striatal slices, approximately 1.0 x 1.0 x 1.0 mm, were prepared. Slices were incubated at 37°C for 20 min in 1.0 ml of oxygenated (95% O_2, 5% CO_2) Krebs-Ringer bicarbonate buffer (pH 7.4) containing ascorbic acid (0.06 mM) and ^{3}H-dopamine (10^{-7}M: specific activity; 22.8 Ci/mmole). The Krebs-Ringer buffer consisted of 118 mM NaCl, 4.7 mM KCl, 1.2 mM KH_2PO_4, 2.5 mM $CaCl_2$, 1.2 mM $MgSO_4$, 25 mM $NaHCO_3$ and 10 mM glucose. Following incubation, the slices were rinsed twice with warm buffer and transferred to 13 mm-Swinnex Millipore filter holders (2 slices/holder) for superfusion. The slices were superfused for 20 min with normal Krebs-Ringer buffer and then superfused

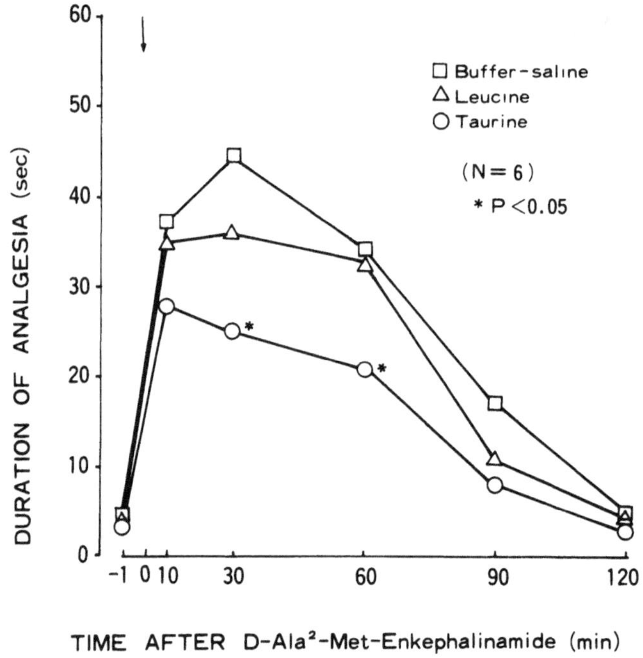

Fig. 7. Chronic effects of amino acids on analgesia induced by DAME. The experimental conditions were the same as those described in the legend for Fig. 6. The arrow indicates the time of injection of DAME.

with a Krebs-Ringer buffer solution containing 53 mM K^+ for 10 min, under the presence of DAME and/or taurine or the absence of both as a control. Finally the superfusing solution was switched back to normal Krebs-Ringer for another 10 min. The rate of superfusion was 0.175 ml/min and was maintained by a peristaltic pump. One-min fractions (0.175 ml) were collected directly into scintillation vials for counting radioactivity. At the end of the experiment the tissue slices were removed from the holder and their radioactivity also measured.

In a separate experiment, investigation by thin layer chromatography (TLC; BuOH:AcOH:H_2O, 4:1:1) was carried out for possible complex formation between taurine and DAME. It was found, however, that taurine and DAME at a concentration used in the present study do not form a complex.

Table 2 shows the summary of the results. The increased efflux of radioactivity induced by high K^+ from the striatal slices which had been preloaded with ^3H-dopamine was markedly reduced with superfusion of DAME. Almost complete inhibition of release was observed with high concentrations of DAME (9 μM). Naloxone, a specific opiate

Table 2

Effects of D-Ala^2-Met-enkephalinamide (DAME) and taurine on the potassium-evoked ^{3}H-dopamine release from rat corpus striatum

Treatment	n	Peak percentage of the total tissue radioactivity released by exposure to 53 mM K^+ (mean ± SE)
Control	4	3.0 ± 1.2
DAME (1 μM)	3	1.7 ± 0.1*
DAME (9 μM)	2	0.3 ± 0.1
DAME (9 μM) + Naloxone (9 μM)	3	2.8 ± 0.2
Naloxone (9 μM)	3	2.9 ± 0.3
Control	3	4.0 ± 1.5
Taurine (1 mM)	3	3.6 ± 1.3
Taurine (10 mM)	3	3.2 ± 1.4
Control	4	2.3 ± 0.2
DAME (5 μM)	3	0.9 ± 0.1*
DAME (5 μM) + Taurine (10 mM)	4	2.1 ± 0.1
Taurine (10 mM)	3	1.9 ± 0.1

n: number of experiments * $p<0.05$

Slices were incubated for 20 min with radiolabeled dopamine at concentration of 10^{-7}M and subsequently superfused with Krebs-Ringer bicarbonate buffer medium. Depolarization was effected with 53 mM K^+ for 10 min. The peak percentage of the total tissue radioactivity released by exposure to high K^+ was obtained by the following equation:

$$\text{peak percent release (\%)} = \frac{A - B}{C} \times 100,$$

where A = peak efflux stimulate by high K^+ (cpm), B = spontaneous averaged efflux in five collected fractions before stimulation (cpm), and C = total radioactivity uptaken into the slice (cpm).

antagonist, inhibited the decrease in efflux of activity, while on its own it had no effect on the K^+-induced release. Taurine did not significantly decrease the K^+-induced efflux of radioactivity, although the peak percentage release from the tissue tended to reduce in a concentration-dependent fashion. However, when striatal slices were superfused together with DAME and taurine, the decreased efflux by DAME alone (5 μM) was reversed almost to the control level.

DISCUSSION

L-Leucine was the only substance tested which did not modify akinesia and analgesia induced by DAME in rats. These findings imply that the actions of taurine and GABA observed in this study may be specific.

Effects of Taurine

Although suppressive effects of taurine on akinesia caused by DAME in rats were clearly observed, the effects on analgesia were not impressive. However, a tendency to decrease the duration of analgesia in a dose-dependent manner or increase the ED_{50} of DAME for analgesia was obtained with taurine treatment. Additional individual analysis revealed different suitable times for the suppressing effects of taurine in each rat upon both akinesia and analgesia; some animals which had been pretreated with taurine showed a decrease in the duration of analgesia as well as that of akinesia at 30 min after the administration of DAME, whereas others did so only at later times (60 or 90 min). This time difference in the effects of taurine in each rat might be due to a delicate difference in the intensity of akinesia and analgesia, which we could not estimate accurately in the behavioral procedure.

Considering these findings, the inhibitory action of taurine against analgesia observed in the present study may be significant, albeit mild. In fact, repeated administration of taurine decreased the duration of analgesia as well as that of akinesia caused by the acute injection of DAME. Thus it appears that taurine may exert similar actions upon both akinesia and analgesia elicited by the morphine-like peptide analogue. This conclusion may also be supported by the fact that the mode of onset of the taurine action upon analgesia is analogous to that upon akinesia; taurine pretreatment did not alter either the duration of akinesia or that of analgesia at the earliest point in time (10 min) after the DAME injection, but seemed to alter these symptoms at 30 min or more. For the chronic treatment with taurine, the duration of akinesia was also shortened slightly from the earliest time (10 min), but that of analgesia was not yet at this time; that was reduced significantly during the time between 30 and 60 min following the injection of DAME. These results imply that taurine acutely or chronically administered into the lateral ventricle may have an ability to promote a recovery from both akinesia and analgesia or suppress these behavior induced by morphine-like peptide.

It has been shown in several biochemical studies that methionine enkephalin and β-endorphin inhibit dopamine release from the striatum (21,26). The enkephalin analogue used in the present study also inhibited the potassium-stimulated overflow of ^{3}H-dopamine in the rat striatal slices. This inhibition of release appears to be a specific effect through the opiate receptors, since it is completely antagonized by naloxone. Thus DAME-induced akinesia that has been found to be reversed by apomorphine (unpublished data) may be largely due to the inhibition of dopamine release in the striatum by the morphine-like peptide analogue.

The relatively delayed effects of taurine suggest that its action may be indirect and may not be mediated by competition with DAME at opiate receptor sites. The counter-acting effects of taurine on akinesia induced by DAME appears to be related to its correcting action on decreased dopamine release in the striatum. Whether the correction of the inhibited dopamine release in the striatum may or may not be associated with the recovery from analgesia is controversial, since consistent pharmacological effects of dopamine or its related substances upon morphine analgesia have not been obtained (14). Thus dopamine in the striatum does not seem to be primarily involved for analgesia. Whichever neurotransmitter is essentially involved for nociperception, the underlying mechanisms of the counteraction of taurine on analgesia caused by DAME may be common to those on akinesia.

It has been demonstrated that β-endorphin has a selective depleting action on calcium content in synaptic membranes and synaptic vesicles, probably by inhibiting the binding of calcium to synaptic membranes (25). The calcium ion is indispensable for neurotransmitter release mechanisms. Taurine can also change intracellular Ca^{++} content in the CNS (15,20,23) and modify the release of norepinephrine and acetylcholine (20). In the present study, taurine tended to reduce the release of dopamine from the striatal slices. Recently Yamamoto et al. (in preparation) found that the decreased Ca^{++} uptake in synaptosomes observed with morphine was normalized by taurine. Therefore, one of the possibilities could be that the behavioral interaction observed after the administration of DAME and taurine in rats may be associated with Ca^{++} transport mechanisms in the CNS. Alternatively, the observed interaction between taurine and DAME might be due to a complex-formation with both substances. Hence the durations of akinesia and analgesia caused by DAME might be reduced by taurine treatment. Against this possibility is that the counteracting effects were not revealed at the earliest time (10 min) but were observed at the later (30-60 min) time after the DAME injection. In fact, a TLC study showed that taurine and DAME do not form a tight complex. Another possibility is that taurine might accelerate the excretion and/or metabolism of DAME from the CNS.

Effects of GABA

Unlike the delayed effects of taurine, GABA suppressed analgesia caused by DAME from the earliest time (10 min) following the injection of enkephalinamide. We do not know if GABA competes with DAME at opiate receptors. However, this is improbable, since GABA alone failed to cause analgesia in rats. The inhibitory action of GABA against DAME-induced analgesia, therefore, may be indirect. Perhaps by interacting with other neuronal systems, GABA, which is a potential neurotransmitter in the CNS, could modulate the analgesic actions of DAME.

Our results are consistent with those by Ho et al. (11) who demonstrated that elevating the GABA content of the brain by aminoxyacetic acid (an inhibitor for GABA transaminase) or inhibiting GABA uptake in the neurons with 2, 4-diaminobutyric acid, antagonized the analgesic action of morphine in mice. However, other conflicting data have also been previously reported; the administration of aminoxyacetic acid (30) or muscimol (a stimulant for GABA receptor) to mice and rats (3) potentiated morphine analgesia; decreasing the GABA content by semicarbazide (an inhibitor for GABA synthesizing enzyme) (30) or inhibiting GABA receptors by bicuculline (30) attenuated analgesia in mice. It is difficult to delineate at the present time the difference between our results, together with those of Ho et al. (11), and those conflicting data described above. Difference in strains of animals may be an important factor responsible for different analgesic response by morphine or morphine-like peptides (10) which in turn would cause opposite results with GABA treatments.

With respect to the effects of GABA on akinesia, it has been reported that the injection of muscimol to rats (3) or of aminoxyacetic acid to mice (28) enhanced "catalepsy" induced by morphine. Therefore, the administration of GABA was expected to prolong the duration of akinesia caused by DAME, although a subconvulsive dose of bicuculline or picrotoxin has been shown not to affect significantly morphine "catalepsy" in mice (28). In spite of this expectation, GABA did not alter the duration of akinesia in the present study. It might be possible that the estimation time for akinesia following the GABA injection at a dose of 95 mM/10 µl was too late to detect any efficacy of this neurotransmitter candidate, which is thought to be rapidly inactivated. However, this is unlikely for the following reasons: The first observation in the present study was carried out 20 min after the injection of GABA. Unlike akinesia, analgesia caused by DAME was fairly suppressed at the same time schedule and the same dose of GABA. Moreover, in our previous studies, this time lag following the intraventricular injection of GABA was found to be enough to protect rats from running seizures induced by ouabain, a cardiac glycoside (16).

When drugs that activate the GABA neuronal system in the brain are used, behavioral evaluation for akinesia should be performed carefully, since such drugs can suppress general motor activity and decrease muscle tonus. The intensity of akinesia evaluated could be overlaid by such symptoms and it might look as if akinesia was potentiated. We did not examine this possibility, because muscle tonus in rats pretreated with GABA was normal, at least at the time of the injection of DAME in the present study.

CONCLUSION

Effects of taurine, GABA and L-leucine on akinesia and analgesia induced by D-Ala^2-Met-enkephalinamide (DAME) were investigated in rats. Taurine administered acutely or chronically into the lateral ventricle promoted a recovery from akinesia and also analgesia. GABA suppressed only analgesia induced by DAME, while L-leucine did not either akinesia or analgesia. In a separate experiment, effects of DAME upon dopamine release from the rat striatum were examined in the presence or absence of taurine. DAME inhibited the high K^+-stimulated efflux of radioactivity in the striatal slices which had been preloaded with ^{3}H-dopamine. Taurine reversed this inhibitory effects of DAME on ^{3}H-dopamine release. Although the latter results appear to be related to the reversal of akinesia by taurine, underlying mechanisms which are common to those of recovery from both akinesia and analgesia caused by DAME would exist in the action of taurine. A possibility that Ca^{++} transport could be involved in such mechanisms is discussed.

ACKNOWLEDGMENTS

This study was supported in part by grants from the Ministry of Health and Welfare of Japanese Government and from Taisho Pharmaceutical Company.

REFERENCES

1. Barbeau, A., and Huxtable, R.J., eds., "Taurine and Neurological Disorders," Raven Press, New York (1978).
2. Belluzzi, J.D., Grant, N., Garsky, V., Sarantakis, D., Wise, C.D., and Stein, L., 1976, Analgesia induced in vivo by central administration of enkephalin in rat, Nature (Lond.), 260:625-626.
3. Biggio, G., Della Bella, D., Frigeni, V., and Guidotti, A., 1977, Potentiation of morphine analgesia by muscimol, Neuropharmacology, 16:149-150.
4. Bloom, F., Segal, D., Ling, N., and Guillemin, R., 1976, Endorphins: profound behavioral effects in rats suggest new etiological factors in mental illness, Science, 194:630-632.

5. Bradbury, A.F., Smyth, D.G., Snell, C.K., Deakin, J.F.W., and Wendlandt, S., 1977, Comparison of the analgesic properties of lipotropin C-fragment and stabilized enkephalins in the rat, Biochem. Biophys. Res. Comm., 74:748-754.
6. Büscher, H.H., Hill, R.C., Romer, D., Cardinaux, F., Closse, A., Hauser, D., and Pless, J., 1976, Evidence for analgesic activity of enkephalin in the mouse, Nature (Lond.), 261:423-425.
7. Chang, J.-K., Fong, B.T.W., Pert, A., and Pert, C.B., 1976, Opiate receptor affinities and behavioral effects of enkephalin: structure-activity relationship of ten synthetic peptide analogues, Life Sci., 18:1473-1482.
8. Collu, R., Charpenet, G., and Clermont, M.J., 1978, Antagonism by taurine of morphine induced growth hormone secretion, Canad. J. Neurol. Sci., 5:139-142.
9. Ferdberg, W. and Smyth, D.G., 1977, C-Fragment of lipotropin - an endogenous potent analgesic peptide, Birt. J. Pharmacol., 60:445-453.
10. Firgeni, V., Bruno, F., Carenzi, A., Racagni, G., and Santini, V., 1978, Analgesia and motor activity elicited by morphine and enkephaline in two inbred strains of mice, 1978, J. Pharm. Pharmacol., 30:310-311.
11. Ho, I.K., Loh, H.H., and Way, E.L., 1976, Pharmacological manipulation of gamma-aminobutyric acid (GABA) in morphine analgesia, tolerance and physical dependence, Life Sci., 18:1111-1124.
12. Hughes, J., Smith, T.W., Kosterlitz, H.W., Fothergill, L.A., Morgan, B.A., and Morris, H.R., 1975, Identification of two related pentapeptides from the brain with potent opiate agonist activity, Nature (Lond.), 258:577-579.
13. Huxtable, R., and Barbeau, A., eds., "Taurine," Raven Press, New York (1976).
14. Iwamoto, E.T., and Way, E.L., Opiate actions and catecholamines, in: "Neurochemical Mechanisms of Opiates and Endorphins," H.H. Loh and D.H. Ross, eds., Raven Press, New York (1979), pp. 357-407.
15. Izumi, K., Butterworth, R.F., and Barbeau, A., 1977, Effect of taurine on calcium binding to microsomes isolated from rat cerebral cortex, Life Sci., 20:943-950.
16. Izumi, K., Donaldson, J., Minnich, J.L., and Barbeau, A., 1973, Ouabain-induced seizures in rats: suppressive effects of taurine and γ-aminobutyric acid, Canad. J. Physiol. Pharmacol., 51: 885-889.
17. Izumi, K., Motomatsu, T., Chrétien, M., Butterworth, R.F., Lis, M., Seidah, N., and Barbeau, A., β-Endorphin induced akinesia in rats: effect of apomorphine and α-methyl-ρ-tyrosine and related modifications of dopamine turnover in the basal ganglia, Life Sci., 20:1149-1156.
18. Izumi, K., Viallet, C., Barbeau, A., Chrétien, M., and Seidah, N., 1979, Endorphin induced akinesia in rats: a pharmacological study, Clin. Neurol., (Jpn), 19:687-694.

19. Jacquet, Y.F., and Marks, N., 1976, The C-fragment of β-lipotropin: an endogenous neuroleptic or antipsychotogen? Science, 194:632-635.
20. Kuriyama, K., Muramatsu, M., Nakagawa, K., and Kakita, K., 1978, Modulating role of taurine on release of neurotransmitters and calcium transport in excitable tissues, in: "Taurine and Neurological Disorders," A. Barbeau and R.J. Huxtable, eds., Raven Press, New York (1978), pp. 201-216.
21. Loh, H.H., Brase, D.A., Sampath-Khanna, S., Mar, J.B., and Way, E.L., 1976, β-Endorphin in vitro inhibition of striatal dopamine release, Nature (Lond.), 264:567-568.
22. Motomatsu, T., Lis, M., Seidah, N., and Chrétien, M., 1977, Cataleptic effect of 61-91 beta-lipotropic hormone in rat, Canad. J. Neurol. Sci., 4:49-52.
23. Pasantes-Morales, H., and Gamboa, A., 1980, Effect of taurine on $^{45}Ca^{2+}$ accumulation in rat brain synaptosomes, J. Neurochem., 34:244-246.
24. Pert, C.B., Pert, A., Chang, J.-K., and Fong, B.T.W., 1976, (D-Ala^2)-Met-enkephalinamide: a potent, long-lasting synthetic pentapeptide analgesic, Science, 194:330-332.
25. Ross, D.H., and Cardenas, H.L., Nerve cell calcium as a messenger for opiate and endorphin actions, in: "Neurochemical Mechanisms of Opiates and Endorphins, H.H. Loh and D.H. Ross, eds., Raven Press, New York, (1979), pp. 301-336.
26. Subramanian, N., Mitznegg, P., Sprügel, W., Domschke, W., Domschke, S., Wünsch, E., and Demling, L., 1977, Influence of enkephalin on K^+-evoked efflux of putative neurotransmitters in rat brain, Naunyn-Schmiedeberg's Arch. Pharmacol., 299: 163-165.
27. Tseng, L.F., Loh, H.H. and Li, C.H., 1977, Human β-endorphin: development of tolerance and behavorial activity in rats, Biochem. Biophys. Res. Comm., 74:390-396.
28. Vanderwende, C., and Spoerlein, M.T., 1979, Morphine-induced catalepsy in mice. Modification by drugs acting on neurotransmitter system, Neuropharmacology, 18:633-637.
29. Wei, E.T., Tseng, L.F., Loh, H.H., and Li, C.H., 1977, Comparison of the behavorial effects of β-endorphin and enkephalin analogs, Life Sci., 21:321-328.
30. Yoneda, Y., Takashima, S., and Kuriyama, K., 1976, Possible involvement of GABA in morphine analgesia, Biochem. Pharmacol., 25:2669-2670.

CENTRAL EFFECTS OF TAURINE: ANTAGONISTIC EFFECTS ON CENTRAL ACTIONS OF ANGIOTENSIN

T. Furukawa, K. Yamada, K. Kushiku, N. Ono* and T. Tokunaga

Department of Pharmacology
School of Medicine and Pharmaceutical Sciences*
Fukuoka University
Fukuoka 814, Japan

Renin released from the juxtaglomerular apparatus of the kidney into the circulation produces angiotensin (AT) and this compound increases tubular absorption of sodium and water, resulting in an increase in circulating blood volume. In addition, AT evokes a net increase in sympathetic vasoconstrictor activity. The presynaptic effect of AT facilitates adrenergic transmitter release (16) and may also involve increased transmitter synthesis (3) and inhibition of adrenergic uptake (21). The postsynaptic effect appears to be due to an increase in sensitivity to norepinephrine (24). Furthermore, large doses of AT exert a constrictive action in the smooth muscle of blood vessels. Thus, the role of the peripheral renin-AT system in regulating the blood pressure is significant.

In the central nervous system, the renin-AT system and AT II receptors are unevenly distributed and are independent of the peripheral system (11,14). The AT I converting enzyme is present in relatively high concentrations in the striatum of the rat brain (27) and the caudate nucleus of the human brain (23), though relatively little is known of the presence of the enzyme in the central nervous system. Excitatory effects of AT II on neurons have been demonstrated in subfornical organs, the supraoptic nucleus and the area postrema and may participate in central actions of AT II. With regard to central actions of AT II, three distinct effects have been reported: drinking (1,25), antidiuretic hormone release (18) and pressor response (9). Endogenous substances which would antagonize AT II have not been described.

It has been reported that taurine decreases the blood pressure in SHRSP rats (Yamori et al., personal communication). However, in our previous studies on the peripheral effects of taurine on sympathetic activity and blood vessels, we did not obtain data which would validly explain the depressor response to taurine.

We have investigated the influence of taurine on the central actions of AT II and our findings are reported herein.

EFFECT OF TAURINE ON THE DRINKING BEHAVIOR ELICITED BY ANGIOTENSIN

Healthy male Wistar rats (300 - 350 g) were used. Water consumption was automatically recorded using a drinkometer according to the method previously described by Kuribara et al. (19), and measurement of drinking behavior was assessed from 10:00 a.m. to 4:00 p.m. All drugs were dissolved in saline and given intraventricularly in a volume of 10 μl through a guide cannula which had been implanted into the lateral ventricle, using the stereotaxic coordinates in the atlas of Pellegrino et al. (22). Taurine was injected 5 min before administration of AT II. In the water deprivation experiments, rats were deprived of water for 24 or 48 hr and taurine was administered 5 min before measurement of drinking behavior the following morning.

Rats which had been placed in the home cage during the day exhibited no particular drinking behavior nor did the rats given saline into the bilateral ventricles. However, AT II administered into the bilateral ventricles in a dose of 100 ng elicited a marked drinking behavior (Fig. 1), which began 1 - 2 min after administration, lasted about 30 min then all but ceased. This drinking behavior was strongly inhibited by previous intraventricular treatment with [Sar^1m Ile^8]-AT II, an AT II antagonist, in a dose of 2 μg (Fig. 1).

Other neuropeptides, such as substance P (10 μg), eledoisin (10 μg), eledoisin related peptide (2.5 μg), thyrotropin-releasing hormone (10 μg), α-melanocyte stimulating hormone (10 μg) and vasopressin (10 μg), did not result in the drinking behavior when these drugs were injected into the bilateral ventricle.

As shown in Fig. 2, the drinking behavior elicited by AT II, 100 ng, was depressed in a dose-dependent manner after intraventricular administration of taurine in doses of 400 and 800 μg. The behavior was similarly inhibited in a dose-dependent fashion when GABA was given by the same route in doses of 200, 400 and 800 μg (Fig. 3).

Carbachol injected into the ventricle at 300 ng also induced drinking behavior, but this behavior was not affected by intraventricular treatment with taurine, 400 μg (Fig. 4).

Rats deprived of water for 48 hr displayed a marked drinking behavior, as demonstrated in Fig. 5. Taurine injected bilaterally into the ventricle in a dose of 400 μg tended to inhibit this behavior, but the effect was not significant. GABA, 400 and 800 μg, administered by the same route exerted a similar effect to that of taurine.

EFFECT OF TAURINE ON PRESSOR AND POSITIVE CHRONOTROPIC ACTIONS OF ANGIOTENSIN II

Normotensive male Wistar rats weighing between 350 and 450 g were anesthetized with urethane (1.2 g/kg i.p.) and a 22G guide cannula of stainless steel was implanted into the right lateral ventricle and/or bilateral anterior or posterior hypothalamic nucleus, according to De Groot's stereotaxic coordinates (7).

Systemic blood pressure was measured in mm Hg using a pressure transducer by inserting a cannula into the right carotid artery. The heart rate was integrated with a cardio-tachometer from the electrocardiograph.

At II in a volume of 1 μl/injection/unilateral was injected into the bilateral posterior hypothalamic nucleus by means of 27G injection pipe through a pre-implanted guide cannula. The other drugs were given into the right lateral ventricle in a volume of 10 μl/injection, or intravenously into the right jugular vein through a catheter. At the end of each experiment, 'malachite green' dye was injected into the lateral ventricle and posterior hypothalamic nucleus. The brain was sectioned and the location of the drugs injected identified.

As shown in Figs. 6 and 7, taurine administered into the ventricle in doses of 20, 50, 100, 200 and 1000 μg induced dose-dependent decreases of blood pressure and heart rate, which appeared immediately after administration and lasted about 20

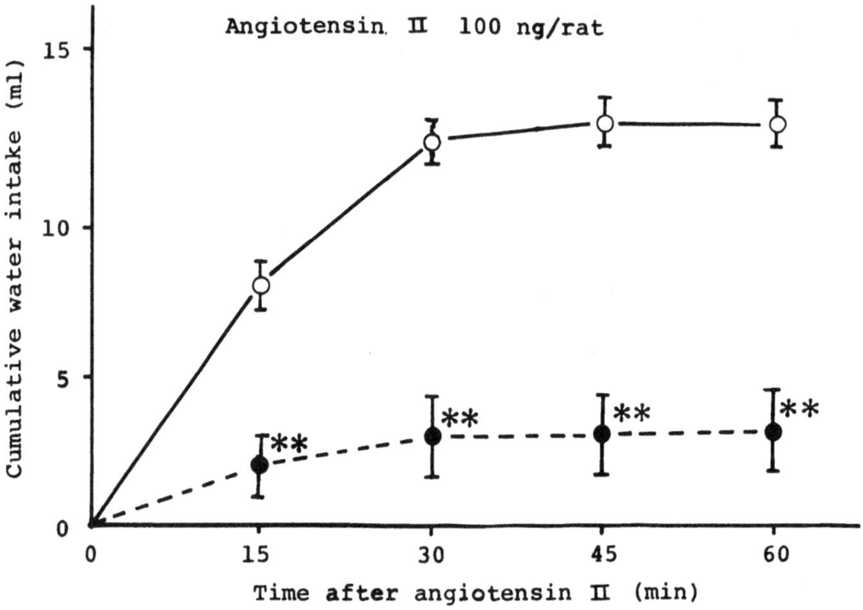

Fig. 1. Drinking behavior induced by angiotensin II and influence of (Sar^1, Ile^8)-angiotensin II on this behavior in rats. All agents were administered intraventricularly. ○—○ Saline (10 μl, n=8); ●---● (Sar^1, Ile^8)-angiotensin II (2 μg, n=8). Each point represents the mean ± SE. **P<0.01; significant difference from the saline injected group, as determined by the Student's t-test.

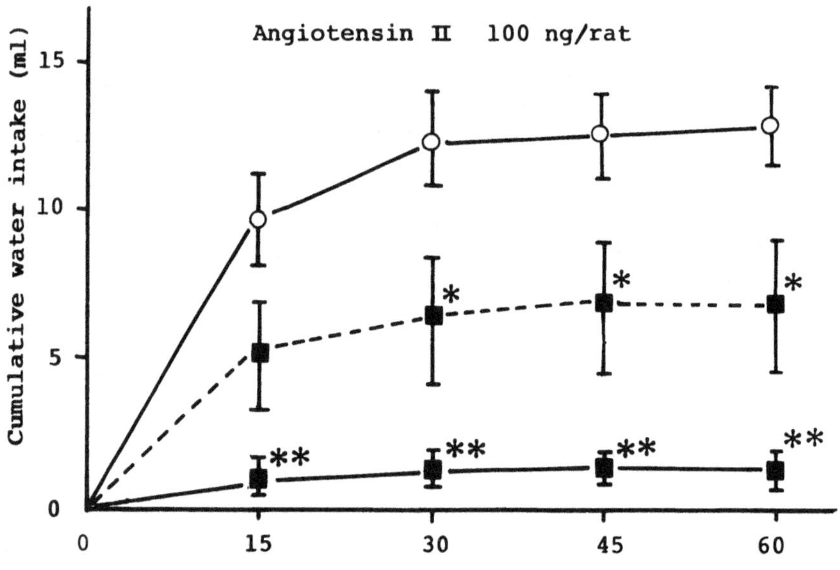

Fig. 2. Influence of taurine on the drinking behavior induced by angiotensin II in rats. ○—○ Saline (10 μl, n=13), ■---■ taurine (400 μg, n=12), ■—■ taurine (800 μg, n=9). *P<0.05, **P<0.01. See legend to Fig. 1.

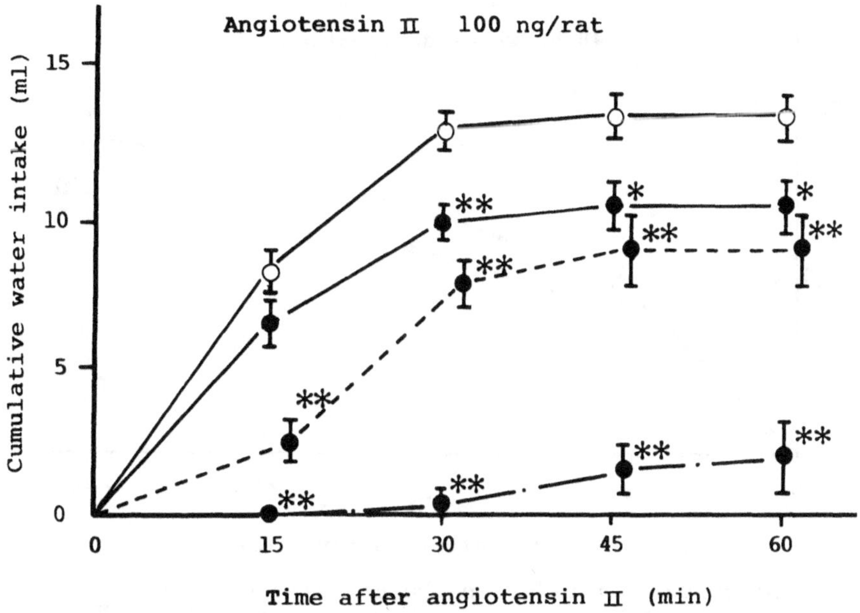

Fig. 3. Influence of GABA on the drinking behavior induced by angiotensin II in rats. O——O Saline (10 μl, n=8), ●——● GABA (200 μg, n=5), ●----● GABA (400 μg, n=5), ●—·—● GABA (800 μg, n=8). See legend to Fig. 1.

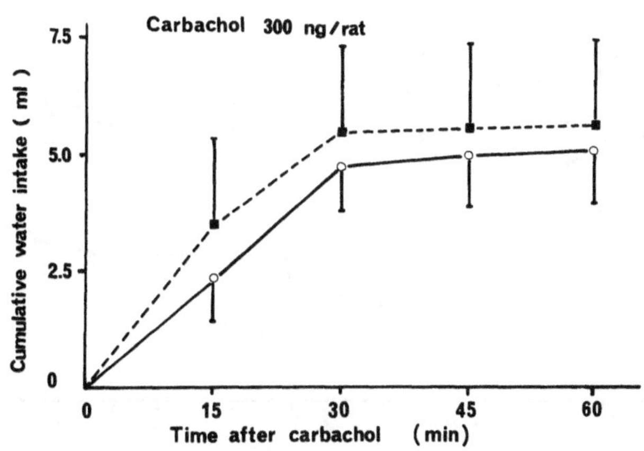

Fig. 4. The drinking behavior induced by carbachol and influence of taurine on this drinking in rats. O——O Saline (10 μl, n=10), ■---■ taurine (400 μg, n=10). See legend to Fig. 1.

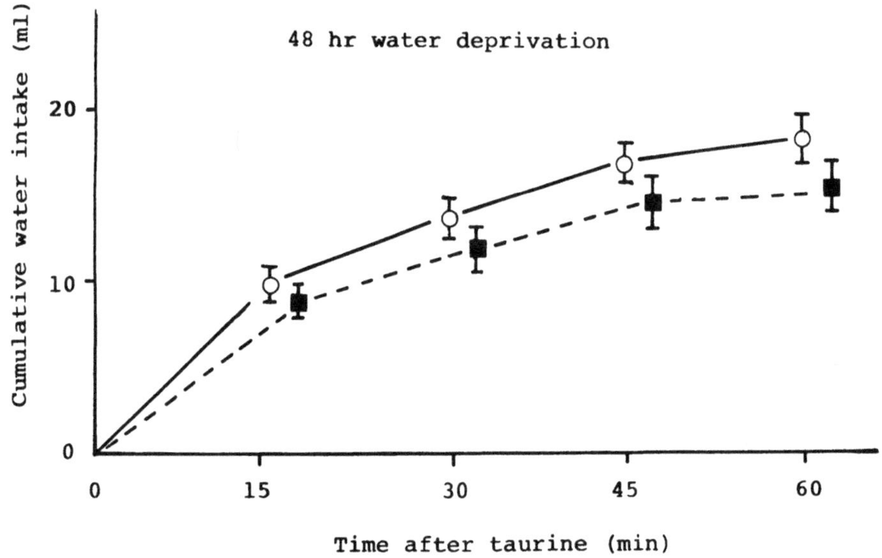

Fig. 5. Influence of taurine on the drinking behavior induced by water deprivation. ○——○ Saline (10 μl, n=10), ■----■ taurine (400 μg, n=10). See legend to Fig. 1.

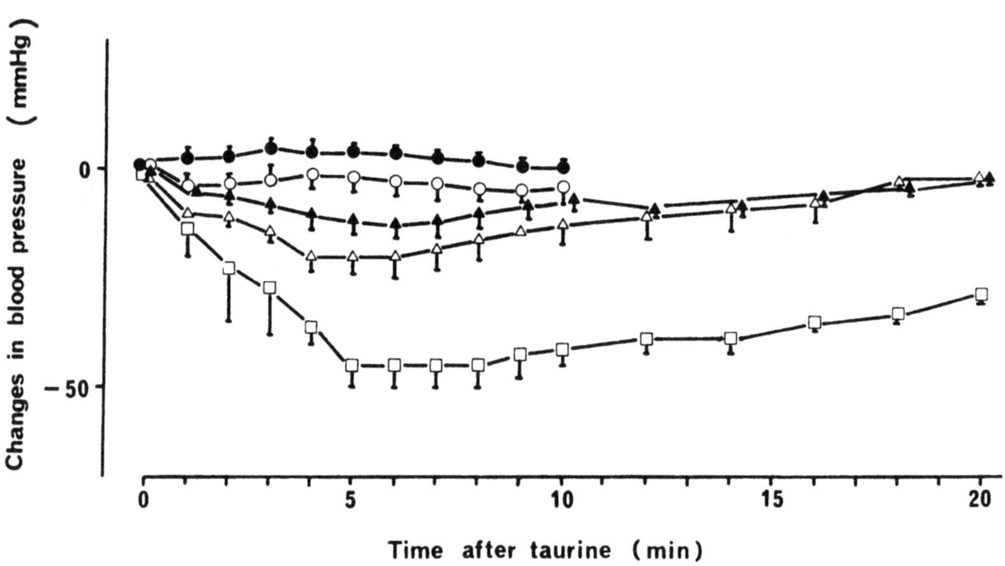

Fig. 6. Dose-response to taurine given icv on blood pressure in rats. ●——● Taurine (20 μg, n=7), ○——○ taurine (50 μg, n=12), ▲——▲ taurine (100 μg, n=12), △——△ taurine (200 μg, n=9), □——□ taurine (1000 μg, n=2).

min. However, when taurine was given intravenously in the same doses, no such actions were seen.

AT II administered into the ventricle in doses up to 100 μg tended to increase both blood pressure and heart rate, but these effects were not clear. When AT II was administered bilaterally into the anterior hypothalamus in doses up to 40 μg, neither blood pressure nor heart rate increased. AT II injected bilaterally into the posterior hypothalamus in a dose of 40 μg induced an increased in both blood pressure and heart rate, as seen in Fig. 8.

When both blood pressure and heart rate stabilized in about 20 - 30 min after intraventricular administration of taurine in doses of 50, 100 and 200 μg, AT II was injected bilaterally into the posterior hypothalamus in a dose of 40 μg. The pressor response to AT II was inhibited in a dose-dependent fashion by taurine, as shown in Fig. 8, though the inhibition by 50 μg was not significant. The positive chronotropic effect of AT II was also inhibited but this inhibition was weak and without statistical significance (Fig. 9).

INFLUENCE OF TAURINE ON GANGLIONIC STIMULATION EFFECT OF ANGIOTENSIN II

Mongrel dogs of either sex weighing between 8 - 16 kg were anesthetized with 30 mg/kg i.v. of pentobarbital sodium. The animals were respired with air using a Harvard respirator (24 strokes/min, 300 ml tidal volume). To eliminate reflex changes in autonomic activity, both vagus nerves at the cervical region and the spinal cord through atlanto-occipital foramen were severed. According to the method described by Fleisch et al. (13), after ligation of the right internal mammary and vertebral arteries, the right brachial artery was cannulated with a polyethylene catheter which was advanced proximally until the tip was approximately at the junction of the right internal mammary and subclavian artery and was tightly secured in place. Blood pressure was measured using pressure transducer and for the heart rate, a cardiotachograph was used. During the experiment, the blood pressure was maintained at about 100 mm Hg by constant i.v. infusion of dextran (6% in saline containing 5% glucose). The drugs were rapidly injected through the catheter in a volume of 0.1 - 0.2 ml. The compounds administered by this route reached the cardiac ganglia and had an effect on the peripheral function via the sympathetic nerves. After intra-arterial treatment with taurine or GABA, about 3 min was

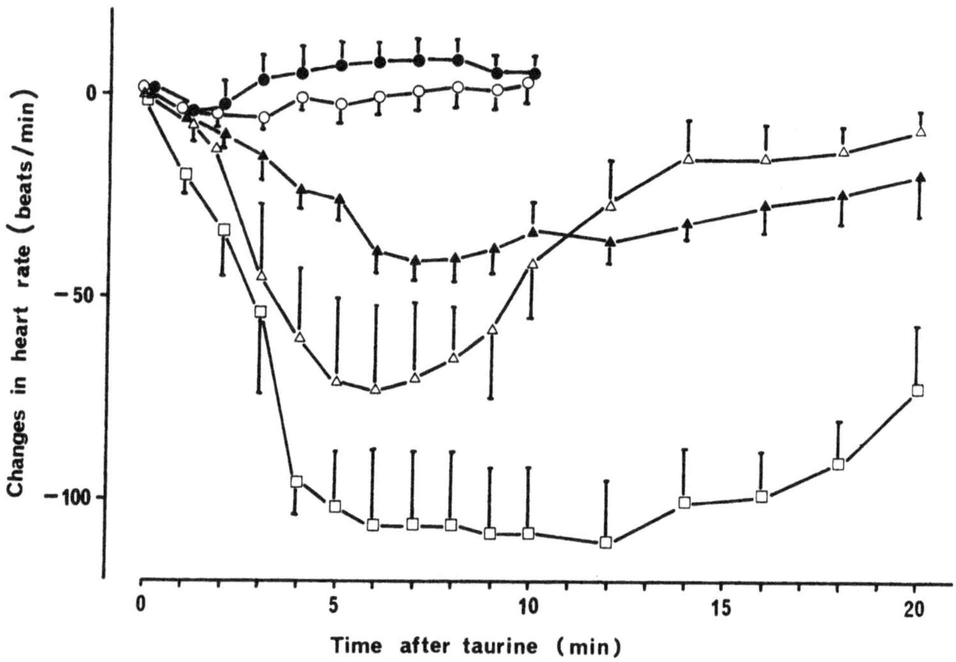

Fig. 7. Dose-response to taurine given icv on heart rate in rats. ●—● Taurine (20 μg, n=7), ○—○ taurine (50 μg, n=12), ▲—▲ taurine (100 μg, n=12), △—△ taurine (200 μg, n=9), □—□ taurine (1000 μg, n=2).

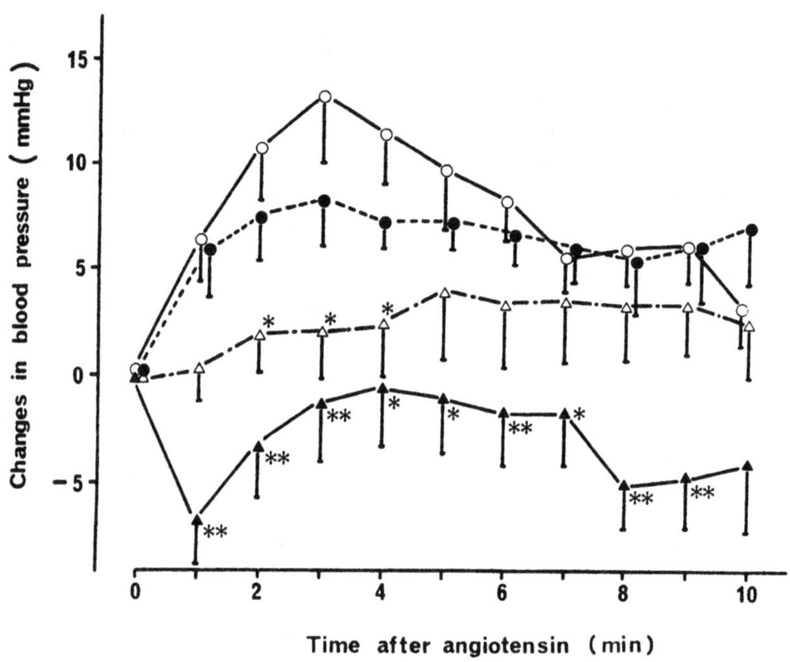

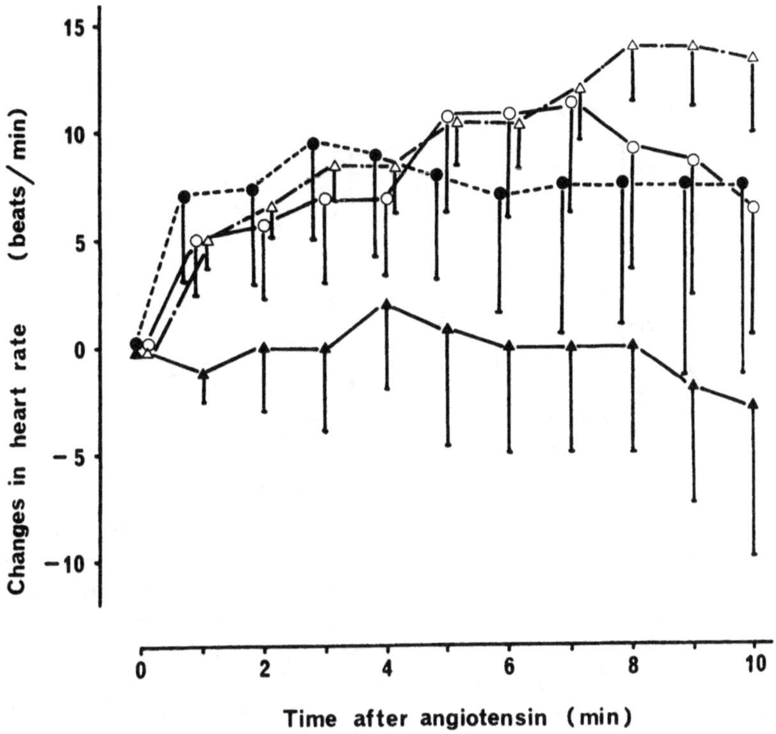

Fig. 9. Positive chronotropic response to angiotensin II and influence of taurine on the effect in rats. ○—○ Saline (10 μl, n=7), ●----● taurine (50 μg, n=8), △----△ taurine (100 μg, n=8), ▲—▲ taurine (200 μg, n=6). See legend to Figs. 1 and 8.

Fig. 8. Pressor response to angiotensin II and influence of taurine on the effect in rats. Angiotensin II (20 μg x 2) was administered bilaterally into the posterior hypothalamus. ○—○ Saline (10 μl, n=7), ●----● taurine (50 μg, n=8), △---△ taurine (100 μg, n=8), ▲—▲ taurine (200 μg, n=6). See legend to Fig. 1.

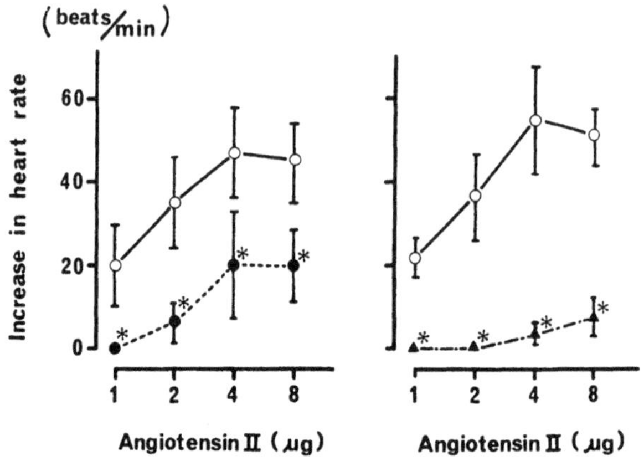

Fig. 10. Dose-response curve of angiotensin II for the positive chronotropic effects and influence of GABA on the effect in spinal dogs. All agents were administered through the right subclavian artery to reach the cardiac sympathetic ganglia. ○—○ Angiotensin II without pretreatment (n=5), ●----● angiotensin II after GABA (50 μg, n=5), ▲—·—▲ angiotensin after GABA (500 μg, n=5). *$P<0.05$; significant difference from the angiotensin II without pretreatment, as determined by the paired t-test.

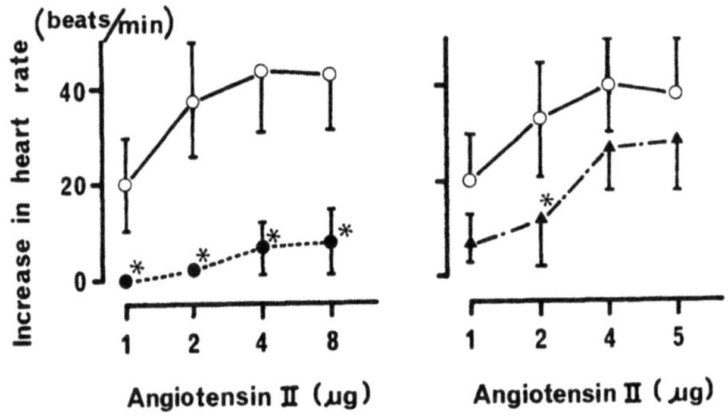

Fig. 11. Dose-response curve of angiotensin II for the positive chronotropic effects and changes by GABOB and taurine in spinal dog. ○—○ Angiotensin II without pretreatment (n=5), ●----● angiotensin II after GABOB (500 μg, n=5), ▲—·—▲ angiotensin II after taurine (20 mg, n=5). See legend to Fig. 10.

allowed to elapse before another dose-response for AT was determined.

AT II administered through the artery in doses ranging from 1 -8 μg produced a dose-dependent increase of heart rate with a latency of about 2 sec. The peak effect was reached within another 20 sec and the dose-dependent response lasted between 2 - 4 min. This effect of AT II was inhibited after treatment with saralasin, an AT II antagonist, in a dose of 100 μg.

These positive chronotropic responses to AT II were abolished after both right stellate and inferior cervical ganglionectomy but were unaffected after surigcal interruption of the preganglionic fibers. In addition, AT II given i.v. in a dose of 8 μg produced no signficant chronotropic action, although the blood pressure was increased. Accordingly, the positive chronotropic action of AT II is apparently due to its direct effect on the cardiac ganglia.

This dose-response curve for AT II was unaltered after successive i.a. administration of hexamethonium (10 mg/kg) and atropine (0.1 mg/kg), or after i.v. treatment with hemicholinium-3 (5 mg/kg). Consequently, the effects of AT II are due to direct stimulation of AT II receptors located at the ganglia and are not mediated through cholinergic activation.

This effect of AT II was inhibited after intraarterial administration of GABA (50 and 500 μg) or muscimol (50 and 100 μg) in a dose-dependent manner, the inhibition by 50 μg GABA being antagonized by an intraarterial administration of 2 mg picrotoxin.

GABOB administered intraarterially in a dose of 500 μg inhibited the ganglionic stimulatory effect of AT II, as did GABA, but taurine administered in a large dose of 20 mg only slightly inhibited the effect of AT II (Fig. 11). Glycine did not influence the effect of AT II, even in a dose of 20 mg.

DISCUSSION

Taurine induced a dose-dependent depressor and negative chronotropic action when administered into the ventricle (20 - 1000 μg), while it did not elicit such action when injected into the systemic circulation.

On the other hand, AT II administered intraventricularly produced a marked drinking behavior even at a relatively low dose. Both pressor and chronotropic actions were evident when AT II was injected into the posterior hypothalamus, in a larger dose, though this agent did not induce such action when injected into the anterior hypothalamus or into the ventricle. These results are compatible with the finding that electric stimulation of the posterior hypothalamus appears to enhance activity of the excitatory cardiovascular neurons while stimulation of anterior area inhibits the activity (17). Taurine and GABA, which are considered to be chemical transmitters or synaptic modulators in the central nervous system, blocked these central actions of AT II, in similar doses. However, both endogenous substances need larger doses than that of AT II to exert the blocking effect.

Taurine and GABA did not exert an inhibitory effect on drinking behavior induced after water deprivation. In the central actions of AT, prostaglandin (10), catecholamines (12) and acetylcholine (8,12) in the brain may also be involved. As these neuronal systems may work as compensatory mechanisms after water deprivation, taurine may not exert any inhibitory effect. In fact, saralasin, AT II antagonist, had no effect on the drinking behavior after water deprivation (15).

The sympathetic ganglia, in which the interneuron exists, are considered to be a simple model of brain synaptic transmission, albeit in the peripheral nervous sytem. AT II exerts a ganglionic stimulatory effect by acting directly on the AT receptors located at the ganglia. GABA inhibits this effect in doses similar to those administered into the central nervous system, whereas taurine exhibited no such inhibitory effects (inhibition occurred to a certain degree with large doses). GABA is present in the glial cell of the ganglia and may be under neuronal control (2,5,26). In addition, GABA can be released by stimuli, such as increased extracellular potassium concentrations (20), electrical stimulation of ganglia or preganglioni nerve trunk (4) and by certain GABA analogues which are substrates for the glial carrier process (5). Inhibition of the transport process enhanced the effect exogenous GABA on ganglion cell (6). It is thus likely that the GABA located at the ganglia may participate in ganglionic function, however, the presence and/or participation of taurine in the ganglia remains obscure.

SUMMARY

Taurine administered intraventricularly in doses of 20, 50, 100, 200 and 1000 μg elicited a dose-dependent decrease in the blood pressure and heart rate, but did not exert such actions when given intravenously in the same doses. On the other hand, AT II elicited a marked drinking behavior when administered intraventricularly in a dose of 100 ng, and a pressor and positive chronotropic actions when given bilaterally into the posterior hypothalamus in a dose of 40 μg. AT II, when administered directly into the cardiac ganglia through the artery in doses of 1 - 8 μg, produced a dose-dependent positive chronotropism. Taurine administered intraventricularly blocked the central actions of AT II, such as drinking behavior, pressor and positive chronotropic actions, but, when administered directly into the ganglia, had no effect on the ganglionic stimulation effect of AT II.

ACKNOWLEDGMENTS

We thank M. Ohara for critical reading of the manuscript.

REFERENCES

1. Anderson, B., Leksell, L.G., and Rundgren, M., 1975, Duration of central action of angiotensin II estimated by its interaction with CSF Na^+, Acta. Physiol. Scand., 93:472-476.
2. Bertilsson, L., Suria, A., and Costa, E., 1976, γ-aminobutyric acid in rat superior cervical ganglion, Nature, 260:540-541.
3. Boadle-Biber, M.C., Hughes, J., and Roth, R.H., 1970, Acceleration of noradrenaline biosynthesis in the guinea pig vas deferens by potassium, Br. J. Pharmacol., 40:702-720.
4. Bowery, N.G., and Brown, D.A., 1974, On the release of accumulated [^{3}H]-γ-aminobutyric acid (GABA) from isolated rat superior cervical ganglia, Br. J. Pharmacol., 52:436P-437P.
5. Bowery, N.G., Brown, D.A., Collins, G.G.S., Galvan, M., Marsh, S., and Yamini, G., 1976, Indirect effects of amino acids on sympathetic ganglion cells mediated through the release of γ-aminobutyric acid from glial cells, Br. J. Pharmacol., 57:73-91.
6. Brown, D.A., and Galvan, M., 1977, Influence of neuroglial transport on the action of γ-aminobutyric acid on mammalian ganglion cells, Br. J. Pharmacol., 59:373-378.
7. De Groot, J., 1959, The rat forebrain in stereotaxic coordinates, Trans. Roy. Neth. Acad. Sci., 52:1-40.

8. Elie, R., and Panisset, J.C., 1970, Effect of angiotensin and atropine on the spontaneous release of acetylcholine from cat cerebral cortex, Brain Res., 17:297-305.
9. Ferrario, C.M., Dickinson, C.J., and McCubbin, J.W., 1970, Central vasomotor stimulation by angiotensin, Clin. Sci., 39:239-245.
10. Feuerstein, G., Krausz, M., and Gutman, Y., 1978, Effect of indomethacin on water intake of the rat, Pharmacol. Biochem. Behav., 9:893-894.
11. Fischer-Ferraro, C., Nahmod, V.E., Goldstein, D.J., and Finkielman, S., 1971, Angiotensin and renin in rat and dog brain, J. Exp. Med., 133:353-361.
12. Fitzsimons, J.T., and Setler, Paulette, E., 1975, The relative importance of central nervous catecholaminergic and cholinergic mechanisms in drinking in response to angiotensin and other thirst stimuli, J. Physiol., 250:613-631.
13. Fleisch, J.H., Flacke, W., and Gillis, R.A., 1969, Nicotinic and muscarinic receptors in the cardiac sympathetic ganglia of the dog, J. Pharmacol. Exp. Ther., 168:106-115.
14. Ganten, D., Marquez-Jullio, A., Granger, P., Hayduk, K., Karsunky, K.P., Boucher, R., and Genest, J., 1971, Renin in dog brain, Am. J. Physiol., 221:1733-1737.
15. Hoffman, W.E., Ganten, U., Phillips, M.I., Schmid, P.G., Schelling, P., and Ganten, D., 1978, Inhibition of drinking in water-deprived rats by combined central angiotensin II and cholinergic receptor blockade, Am. J. Physiol., 234:F41-F47.
16. Hughes, J., and Roth, R.H., 1971, Evidence that angiotensin enhances transmitter release during sympathetic nerve stimulation, Br. J. Pharmacol., 41:239-255.
17. Isaac, L., 1980, Clonidine in the central nervous system: site and mechanism of hypotensive action, J. Cardiovasc. Pharmacol., 2:Suppl. 1, S5-S20.
18. Keil, L.C., Summy-Long, J., and Severs, W.B., 1975, Release of vasopressin by angiotensin II, Endocrinology, 96:1063-1065.
19. Kuribara, H., Hayashi, T., Alam, M.R., Tadokoro, S., and Miura, T., 1978, Automatic measurement of drinking in rats: Effects of hypophysectomy, Pharmacol. Biochem. Behav., 9:697-702.
20. Minchin, M.C.W., and Iversen, L.L., 1974, Release of [^{3}H] gamma-aminobutyric acid from glial cells in rat dorsal root ganglia, J. Neurochem., 23:533-540.

21. Peach, M.J., Bumpus, F.M., and Khairallah, P.A., 1969, Inhibition of norepinephrine uptake in hearts by angiotensin II and analogs, J. Pharmacol. Exp. Ther., 167:291-299.
22. Pellegrino, L.J., Pellegrino, A.S., and Cushmann, A.J., A stereotaxic atlas of the rat brain, Plenum Press, New York, (1979).
23. Poth, M.M., Heath, R.G., and Ward, M., 1975, Angiotensin-converting enzyme in human brain, J. Neurochem., 25:83-85.
24. Povolny, K.M., Jung, R.W., Kraft, E., and Zimmerman, B.G., 1977, Adrenergic potentiation by angiotensin II in isolated canine cutaneous blood vessels: Effect of bathing media and calcium, Blood Vessels, 14:105-115.
25. Simpson, J.B., and Routtenberg, A., 1973, Subfornical organ: site of drinking elicitation by angiotensin II, Science, 181:1172-1174.
26. Waniewski, R.A., and Suria, A., 1977, Alterations in γ-amino-butyric acid content in the rat superior cervical ganglion and pineal gland, Life Sci., 21:1129-1142.
27. Yang, H.Y.T., and Neff, N.H., 1972, Distribution and properties of angiotensin converting enzyme of rat brain, J. Neurochem., 19:2443-2445.

TAURINE AND THERMOREGULATION: BEHAVIORAL AND CELLULAR STUDIES

Vincenzo Carlà[1], Christopher G. Dacke[2], Neil Davidson[2], Alberto Giotti[1], Maurizio Magnani[1] and Giampietro Sgaragli[1]

[1]Istituto Interfacoltà di Farmacologia e Tossicologia
Viale G.B. Morgagni 65
50134 Firenze, Italy
[2]Department of Physiology
Marischal College
University of Aberdeen
United Kingdom

According to the generally accepted model of thermoregulation, the temperature of the body is detected by thermosensitive neurons which bring autonomic and behavioral mechanisms into play to counteract any deviation from the optimal state.

Until recently the neurochemistry of temperature regulation has been focused on classical neurotransmitters. However, the effect exerted by intracerebroventricular (icv) injection of taurine and 4-aminobutyric acid suggests that these endogenous amino acids play a role in temperature control mechanisms. Taurine given icv to the rat causes complete disruption of temperature regulation (21) and this effect has been shown to be at least in part mediated by central serotonergic systems (22). This effect of taurine on thermoregulation was later confirmed in rabbits (23,6) and in mice (7). Subsequently, it has been shown that taurine, administered icv to the conscious, unrestrained rabbit, causes a reduction in skeletal muscle tone and evidence has been produced that this effect is of central supraspinal origin (24).

TAURINE AND THERMOREGULATION

Fig. 1 shows the combined effects of icv injection of taurine in rabbits on rectal and ear skin temperatures and on the tone of the paravertebral longitudinal muscles. It is evident that the deep hypothermia following taurine injection is a result both of an increased heat dissipation subsequent to a reduced vasomotor tone and of a reduced heat production at muscular level. Similar effects are elicited by the icv injection of GABA.

These findings pose two important questions:
1) Where are the cells located that are responsible for the effects elicited by the icv injections of taurine and GABA?
2) Are these effects mediated by separate cellular taurinergic and GABAergic systems?

1. BEHAVIORAL STUDIES

Anatomical Identification of Taurine- and GABA-sensitive Cells.

When taurine and GABA are introduced into the brain, bypassing the blood-brain barrier, they may affect synaptic activity in much the same way as do the endogenous transmitters, but results obtained with this approach must be interpreted with caution. The artificial flooding of nervous centers, especially those facing the ventricles, with large amounts of these amino acids, given as highly concentrated solutions, could produce simultaneous and inappropriate inhibition of many different neuronal pathways and give results which are quite unrelated to the normal pattern of response. Recently we have shown (25) that among the neutral amino acids tested in the rabbit to define the specificity of taurine effects, 3-aminopropansulphonic acid (homotaurine) was more powerful than taurine, while GABA showed similar but less marked properties. In addition, 3-hydroxy-5-aminomethylisoxazole (muscimol), a conformationally restricted analogue of GABA (13), caused hypothermia and a reduction in skeletal muscle tone at doses three orders of magnitude smaller than that of taurine (see Table 1). The two compounds, however, as shown in Fig. 2, differed in their effects on the brain cortex electrocorticographic patterns. Taurine, in fact, and homotaurine were shown to induce a persistent electrocorticographic alertness in spite of the behavioral depression (Fig. 2, panel A). In contrast, animals injected with muscimol (Fig. 2, panel B) and with GABA exhibited a synchronization along with the behavioral depression; in addition, the arousal reaction to external stimuli was markedly depressed. These findings prompted us to localize, by using muscimol as a chemical probe, the group of cells responsible for the GABA-mediated effects. The following criteria have been used for the identification of

the muscimol target center:
1) The injection of muscimol into a restricted area of the brain must produce effects identical to those elicited by the intraventricular administration of a higher dose.
2) Moving away from the target center should produce an increase in the latency and a decrease in the strength of the effects.

In the experiments aimed at localizing the muscimol target center, injections of 8 nmole muscimol dissolved in 0.2 μl pyrogen-free saline were made into conscious, unrestrained rabbits via stereotaxically placed cannulas. The location of the chemically stimulated area was determined by injecting, at the end of each experiment, 0.2 μl of a saturated silver nitrate aqueous solution. After the animal had been killed with an overdose of ethylurethane, the brain was rapidly removed, appropriately cut, frozen and exposed to air until the silver nitrate was reduced to metallic silver. The brain specimen was then embedded in paraffin and 40 μ thick sections of the area marked by silver nitrate were collected. Sections were colored alternately with either thionine or luxol fast blue and cresyl violet and examined under light microscopy. The injections were localized in different levels within the diencephalon. All the injections were performed into the right hemisphere and the areas tested were: diagonal band of Broca, nucleus supraopticus diffusus, lateral hypothalamus, dorsal hypothalamus, periventricular nucleus of the hypothalamus and supramammillary area. The highest and most protracted effects both on colonic temperature and muscle tone, as well as those having the shortest latencies, occurred when muscimol was injected into the periventricular nucleus of the hypothalamus (Table 2). Pronounced effects were also elicited by injecting muscimol into the lower part of the third ventricle. By this route the effects of muscimol had even shorter latencies than by the periventricular nucleus; however, their intensities were much smaller. Hypothermia and decreased muscle tone were also evident when muscimol was injected into the medial preoptic area or dorsal hypothalamus, although their intensities were significantly smaller and their latencies much longer than those of the response elicited by injection into the periventricular nucleus. No effect was elicited from the diagonal band of Broca or lateral hypothalamus.

These data seem to indicate that in the rabbit the periventricular nucleus of the hypothalamus is made up of neurons which receive a GABAergic input and that it plays an important role in muscle tone control and thermoregulation. Experiments still in progress suggest that taurine-sensitive cells too are located in the proximity of the walls of the third

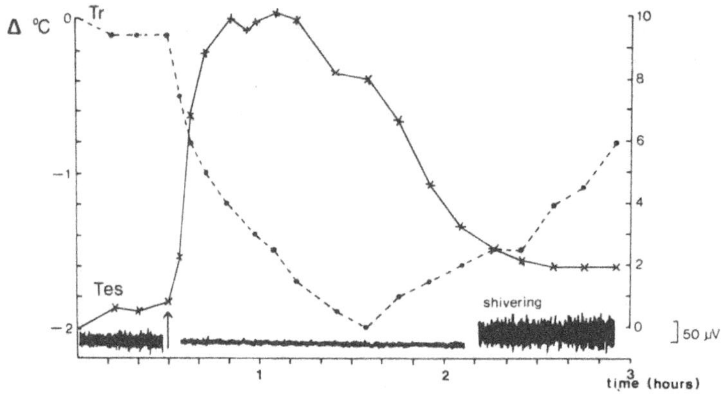

Fig. 1. Time course of the effects elicited by 15 μmole/kg taurine injected icv on rectal, ear skin temperature and muscle tone in one rabbit. The experiment was performed at an ambient temperature of 20°C Tr, rectal temperature; Tes, ear skin temperature.

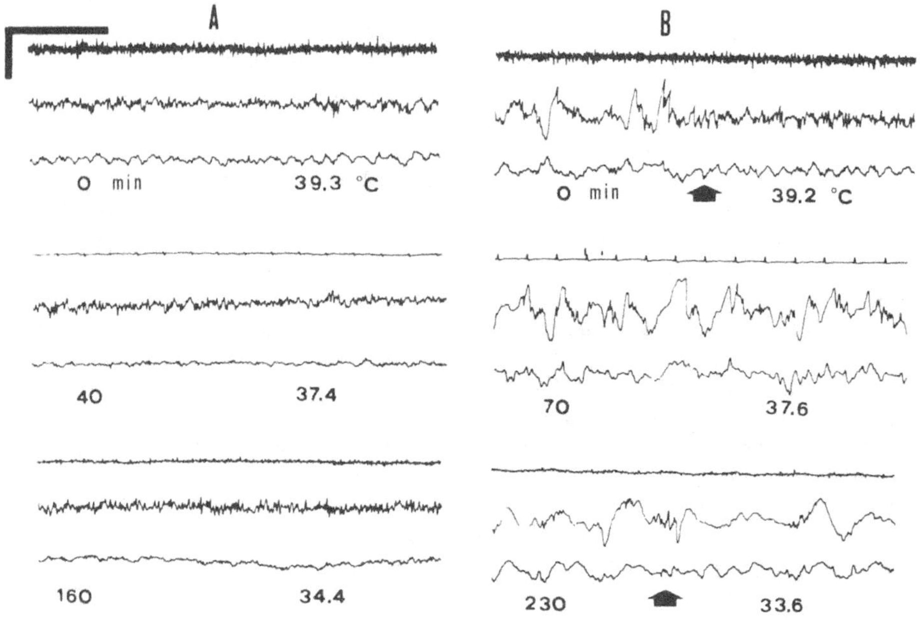

Fig. 2. Effects of taurine (A) and muscimol (B) on electromyogram (EMG) and electrocorticogram (ECoG) in rabbits. Taurine (20 μmole/kg) and muscimol (0.012 μmole/kg) were injected into the lateral ventricle of unrestrained, unanesthetized rabbits. Tracings indicate, from the top, EMG, ECoG from the anterior sensorimotor cortex and ECoG from the limbic parieto-occipital cortex. Times after injection and the corresponding rectal temperatures are shown. Calibrations: 1 sec and 50 μV. At the arrow light pressure was applied to the nose of the animal. (From Sgaragli et al., ref. 25, with permission.)

TABLE 1. EFFECTS OF COMPOUNDS INJECTED ICV ON RECTAL TEMPERATURE AND MUSCLE TONE.

Compound	dose (μmole/kg)	Maximum reduction in rectal temperature (oC)	Muscle tone (% of control)
Taurine	20	2.2 ± 0.3	2
Homotaurine	20	3.4 ± 0.6	0
GABA	80	2.6 ± 0.3	36
	0.02	1.4 ± 0.2	21
Muscimol	0.03	2.5 ± 0.2	0
	0.04	4.4 ± 0.2	0

Solutions prepared immediately before use with "pyrogen-free" water and adjusted to pH 7.4 were injected into unanesthetized, unrestrained animals. The experiments were performed at an ambient temperature of 20°C. Mean values ± SEM from 5 animals.

TABLE 2. EFFECTS OF MUSCIMOL INJECTED IN DIFFERENT AREAS OF BRAIN ON RECTAL TEMPERATURE AND MUSCLE TONE.

Injection site	Dose (nmole/kg)	n	Maximum variation in rectal temperature (°C)	Onset time of muscle tone effects (min after injection)
Lateral ventricle	20	5	1.4 ± 0.2	10 ± 3
Third ventricle	20	3	1.2 ± 0.4	4 ± 2
Periventricular nucleus of hypothalamus	8	4	2.9 ± 1.3	7 ± 1

Solutions prepared immediately before use with "pyrogen-free" water and adjusted to pH 7.4 were injected into unanesthetized, unrestrained animals. The experiments were performed at an ambient temperature of 20°C. Mean values ± SEM.. n, number of animals.

ventricle. As shown in Table 3, in fact, the placement of taurine into the third ventricle elicited more prompt and stronger effects than when the compound was injected into the lateral ventricles.

2. CELLULAR STUDIES

Existence of Separate Taurinergic and GABAergic Systems.

It is obviously important to determine whether or not the various actions of taurine injected into the third ventricle and, possibly, into the periventricular nuclear region, are mediated by separate cellular taurinergic and GABAergic systems. Although a high affinity uptake of ^{3}H-taurine into hypothalamic slices has been reported (12), the cellular location of this uptake remains unknown, and no autoradiographic studies appear to have been performed to examine at the cellular level the distribution of labelled GABA and taurine in the periventricular region of the hypothalamus. However, in another brain region, the cerebellar cortex, where GABA and taurine again share some similar characteristics such as high affinity transport (15), a Ca^{2+}-dependent release (4,5) and depression of neuronal firing rate, there is some evidence that not only are different neuronal populations involved, but that GABA is likely to function as a neurotransmitter while taurine is more likely to have some neuromodulatory role. Fig. 3 shows autoradiographic profiles of 1 μ sections of glutaraldehyde fixed rat cerebellar cortex after pre-incubation with 100 μCi/ml of either ^{3}H-taurine or ^{3}H GABA. Fig. 3a shows clearly the concentration after ^{3}H GABA pre-incubation, of silver grains in small cells in the molecular layer, likely to be inhibitory basket or stellate cells, an observation similar to one reported earlier by Shon and Iversen (26). Fig. 3b shows a completely different autoradiographic profile after pre-incubation with ^{3}H-taurine. In this figure silver grain deposition was found to occur systematically in Purkinje cell bodies and dendrites only.

DISCUSSION

Taurine and GABA and their respective agonists homotaurine and muscimol when introduced into the lateral ventricle of rabbit act as inhibitors of at least two of the autonomic thermoregulatory effectors (peripheral vasomotor tone and heat production). The possible important role exerted by taurine and GABA in thermoregulation is indicated in the neuronal representation in Fig. 4, which was recently proposed by Bligh (2). Our data strongly suggest that the group of GABA-sensitive cells, identified within the periventricular nucleus of the hypothalamus, is a strategic portion of the CNS interface of the thermoregulatory system. Experiments still in progress will define whether or not the periventricular nucleus also contains taurine-sensitive cells. The possibility that a very restricted area of the brain contains both GABA- and taurine-sensitive cells raises the question whether exogenously applied GABA or taurine exert similar actions by inhibiting each other's re-uptake systems. Although Kaczamarek and Adey (11) showed that ^{3}H-taurine efflux from cat cerebral cortex could be increased in the presence of 25 mM unlabelled GABA, this increased efflux is unlikely to be due to an inhibition of taurine re-uptake. Kaczamarek and Davison (10) have already shown that the high affinity taurine re-uptake system is relatively specific, not being significantly inhibited by the presence of other amino acids except for its carboxylic analogue β-alanine. On the other hand, it is noticeable that taurine can potentiate K^+-induced release of ^{3}H-GABA from superfused slices of rat brain cerebral cortex (14). Interestingly, the potentiation of stimulus-induced release of GABA by taurine does not appear to be due to inhibition of re-uptake of GABA.

Since electrophysiological (3) and receptor binding studies (27,17) have indicated that taurine does not seem to interact appreciably with GABA receptors in mammalian brain, the data reported here should reasonably suggest that exogenous taurine and GABA interact with different neuronal pathways which are dealing with thermoregulation. This view is supported by our observation of a the differential effect on the electrocorticogram induced by taurine-homotaurine and by GABA-muscimol.

It has been shown recently that unilateral intranigral injection of taurine induces intense contralateral circling (8), as do GABA and some GABA agonists (18,19,16). The finding, however, that strychnine inhibited taurine-induced but not GABA-induced circling (9) indicates that even in the substantia nigra the effects of taurine and GABA are not mediated by the same receptors.

In spite of the finding (20) that taurine is taken up by glial cells in vitro, which suggested that taurine could mediate the modulation of neuronal function by glial cells, our histoautoradiographical studies in the cerebellum did not show accumulation of labelled taurine into glial cells. On the contrary, there was an accumulation of labelled taurine into dendrites and cell bodies of Purkinje cells.

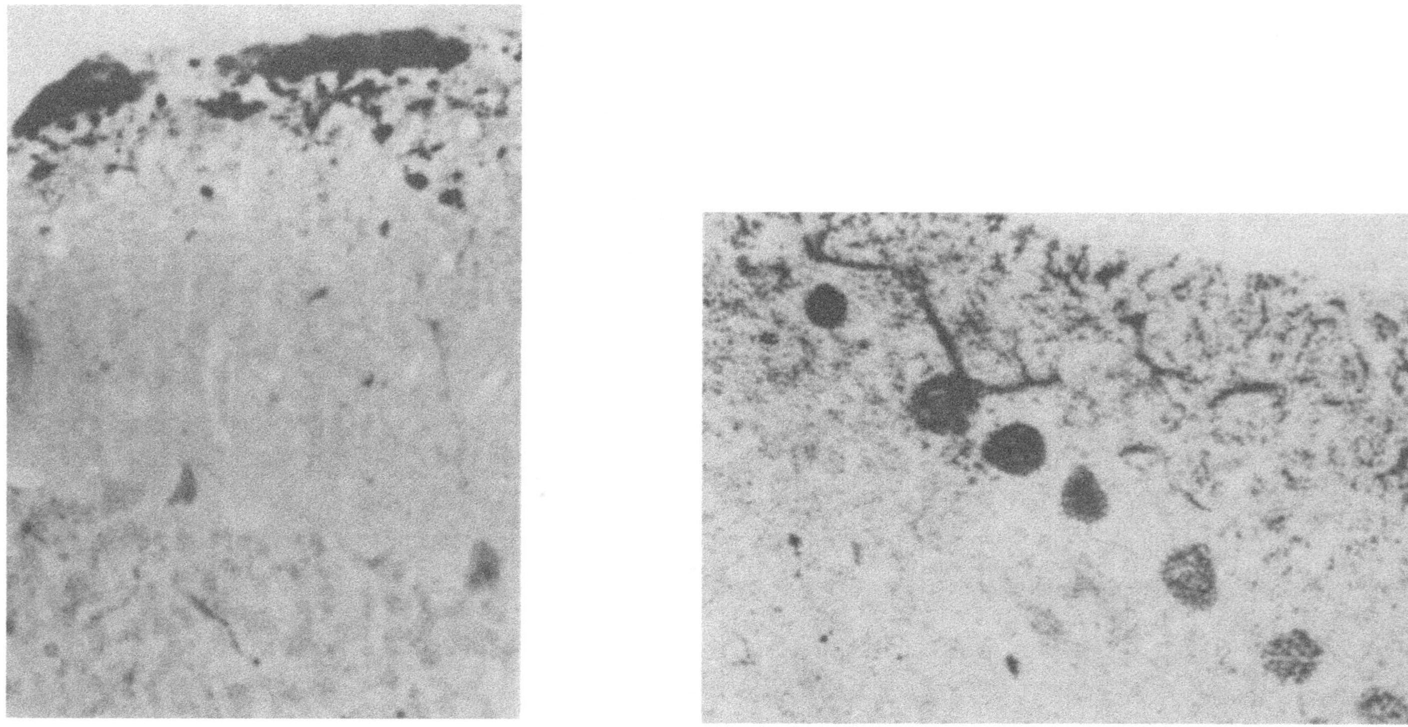

Fig. 3. Dark-field autoradiograms of rat cerebellar cortex (sagittal, 1 μ sections) after tissue loading with A: (^{3}H)GABA; or B: (^{3}H)taurine. Heavy silver grain accumulation can be seen in several cells in the superficial molecular layer in A, and in the single row of Purkinje cell bodies between the molecular and granular layer, as well as in their dendrites in the molecular layer, in B. (From Assumpcão et al., ref. 1, with permission).

TABLE 3. EFFECT OF TAURINE INJECTED IN DIFFERENT AREAS OF BRAIN ON RECTAL TEMPERATURE AND MUSCLE TONE.

Injection site	Dose (μmole/kg)	n	Maximum variation in rectal temperature (°C)	Onset time of muscle tone effects (min after injection)
Lateral ventricle	2	5	no effect	no effect
	10	5	0.5 ± 0.3	13 ± 3
Third ventricle	2	3	0.6 ± 0.1	9 ± 2

Solutions prepared immediately before use with "pyrogen-free" water and adjusted to pH 7.4 were injected into unanesthetized, unrestrained animals. The experiments were performed at an ambient temperature of 20°C. Mean values ± SEM.. n, number of animals.

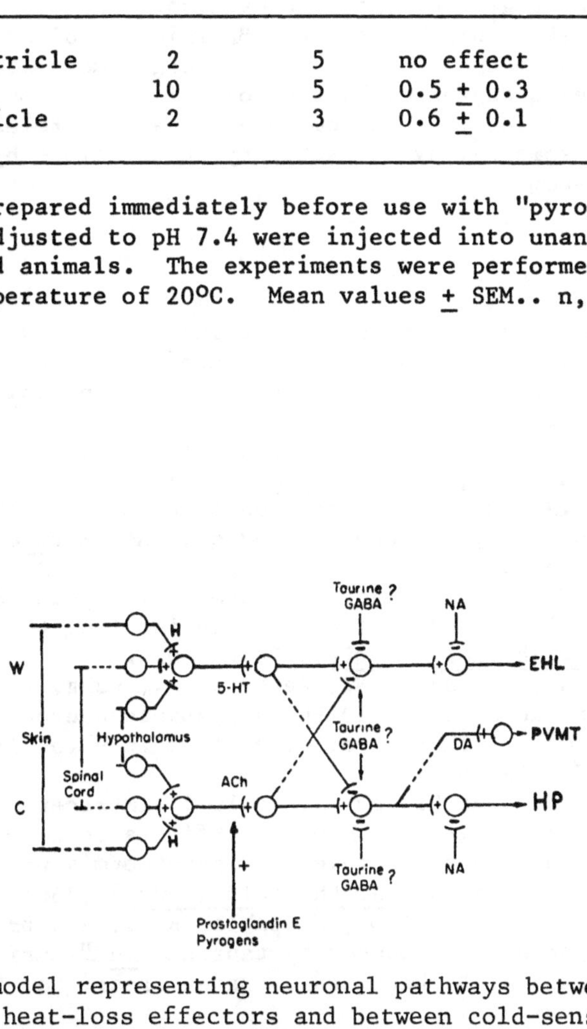

Fig. 4. A model representing neuronal pathways between warm-sensors and heat-loss effectors and between cold-sensors and heat-production effectors with crossed inhibitory pathways between them. The model also includes possible afferent pathways from central and extracentral thermosensors.
C, cold-sensors; W, warm-sensors; DA, dopamine; EHL, evaporative heat loss; H, histamine; NA, noradrenaline; PVMT, peripheral vasomotor tone of the ears; HP, heat production.

It has repeatedly been suggested that the role of taurine in the brain is that of a neuromodulator rather than a neurotransmitter. The post-synaptic localization of labelled taurine in cerebellar Purkinje cell bodies and dendrites supports this concept.

Experiments are now in progress to ascertain whether or not similar cellular localization of taurine and GABA exist in the periventricular region of the hypothalamus. One possible functional relationship between a GABAergic inhibitory transmitter system and a close-by taurinergic modulator system has already beern suggested by the work of Kaczamarek and Adey (11). Thus, the level of neuromodulator taurine release may be determined in some way by the activity in the close-by GABAergic inhibitory system.

REFERENCES

1. Assumpcão, J.A., Bernardi, N., Dacke, C.G., Davison, N., and Eichelberger, H.G., 1979, Purkinje cell uptake "in vivo" of (^{3}H)taurine in the rat cerebellum, J. Physiol (Lond.), 298:35-36P.
2. Bligh, J., 1979, The central neurology of mammalian thermoregulation, Neuroscience, 4:1213-1236.
3. Curtis, D.R., Hösli, L., and Johnston, G.A.R., 1967, Inhibition of spinal neurones by glycine, Nature, 215:1502-1503.
4. Davison, N., 1979a, High potassium, veratridine and electrically induced release of taurine from the cerebellar cortex, J. Physiol. (Paris), 75:673-676.
5. Davison, N., Bernardi, N., Fawcett, M., Wong, E., Assumpcão, J.A., and Dacke, C.G., 1979b, A possible neuronal release of (^{14}C) GABA from the rat cerebellum in vivo, Pflügers Arch., 379:149-155.
6. Harris, W.S., and Lipton, J.K., 1977, Intracerebroventricular taurine in rabbits: Effects on normal body temperature, endotoxin fever and hypothermia produced by PGE_1 and amphetamine, J. Physiol. (Lond.), 266:397-398.
7. Hruska, R.E., Thut, P.D., Huxtable, R.J., and Bressler, R., 1976, Hypothermia produced by taurine, in "Taurine," R. Huxtable and A. Barbeau, eds., Raven Press, New York, 347-356.
8. Kaakkola, S., and Käriänen, I., 1980, Controlateral circling behavior induced by intranigral injection of taurine in rats, Acta pharmacol. et toxicol., 46:293-298.

9. Kaakkola, S., and Käriänen, I., 1980, Circling behavior induced by intranigral injections of GABA and muscimol in rats. Psychopharmacology, 68:31-36.
10. Kaczamarek, L.K., and Davison, A.N., 1972, Uptake and release of taurine from rat brain slices, J. Neurochem., 19:2355-2362.
11. Kaczamarek, L.K., and Adey, W.R., 1974, Factors affecting the release of (^{14}C) taurine from rat brain: the electrical effects of taurine on normal and seizure prone cortex, Brain Res., 76:83-94.
12. Kerwin, R.W., and Pycock, C.J., 1979, Role of taurine as a possible transmitter in the thermoregulatory pathways of the rat, J. Pharm. Pharmacol., 31:466-470.
13. Krogsgaard-Larsen, P., Johnston, G.A.R., Curtís, D.R., Game, C.J.A., and McCulloch, R.M., 1975, Structure and biological activity of a series of conformationally restricted analogues of GABA, J. Neurochem., 25:803-809.
14. Leach, M.J., 1979, Effect of taurine on release of (^{3}H) GABA by depolarizing stimuli from superfused slices of rat brain cerebral cortex "in vitro", J. Pharmac. Pharmacol., 31:533-535.
15. Lombardini, J.B., 1978, "Taurine and neurological disorders," A. Barbeau and R. J. Huxtable, eds., Raven Press, New York, pp. 119-135.
16. Martin, G.E., Papp, N.L., and Bacino, C.B., 1978, Controlateral turning evoked by the intranigral microinjection of muscimol and other GABA agonists, Brain Res., 155:297-312.
17. Möhler, H., and Okada, T., 1977, GABA receptor binding with ^{3}H(+)bicuculline-methiodide in rat CNS, Nature, 267:65-67.
18. Olpe, H.R., Schellenberg, H., and Koella, W.P., 1977, Rotational behavior induced in rats by intranigral aplication of GABA-related drugs and GABA antagonist, Eur. J. Pharmacol., 45:291-294.
19. Scheel-Kruger, J., Arnt, J., and Magelund, G., 1977, Behavioral stimulation induced by muscimol and other GABA agonists injected into the substantia nigra, Neuroscience Letters, 4:351-356.
20. Schrier, B. K., and Thompson, F. J., 1974, On the role of glial cells in the mammalian nervous system. Uptake, excretion and metabolism of putative neurotransmitters by cultured glial cells, J. Biol. Chem., 249:1769-1780.
21. Sgaragli, G. P., and Pavàn, F., 1972, Effect of aminoacid compounds injected into cerebro-spinal fluid spaces on colonic temperature, arterial blood pressure and behavior of the rat, Neuropharmacology, 11:45-56.

22. Sgaragli, G. P., Pavàn, F. and Galli, A., 1975a, Is taurine-induced hypothermia in the rat mediated by 5-HT?, Naunyn Schmiedeberg's Arch. Pharmaco., 288:179-184.
23. Sgaragli, G.P., Carlà, V., Galli, A., Magnani, M., and Giotti, A., 1975b, Central effects of taurine and homotaurine, intraliquorally injected, in rat and rabbit: Probable role of central 5-HT systems. International Society for Neurochemistry - Satellite symposium on "New first and second messengers in nervous tissue," Brescia (Italy). August 28-30, abstracts.
24. Sgaragli, G.P., Magnani, M., Carlà, V., and Giotti, A., 1976, Muscle relaxation induced in the rabbit by intracerebroventricular taurine injection: A supraspinal effect, Naunyn Schmiedeberg's Arch. Pharmacol., 295:95-97.
25. Sgaragli, G. P., Carlà, V., Magnani, M., and Giotti, A., 1978, Homotaurine and muscimol mimic taurine and GABA effects on muscle tone and temperature regulation, Naunyn Schmiedeberg's Arch. Pharmacol., 305:155-158.
26. Shon, F. and Iversen, L.L., 1972, Selective accumulation of (^{3}H) GABA by stellate cells in rat cerebellar cortex, Brain Res., 42:503-507.
27. Zukin, S. R., Young, A.B., and Snyder, S.H., 1974, Gamma aminobutyric acid binding to receptor sites in the rat central nervous system, Proc. Natl. Acad. Sci. U.S.A., 71:4802-4807.

INFLUENCE OF CENTRALLY ADMINISTERED TAURINE ON THERMOREGULATION AND FEVER

J. M. Lipton
Departments of Physiology and Neurology
University of Texas
Health Science Center
Dallas, Texas

One of the most striking and consistent effects of central administration of taurine is hypothermia. First observed in rats in 1972 by Sgaragli and Pavan (22), this effect has since been found to occur in several other species (Table 1). The relatively consistent finding of hypothermia among various mammalian species given taurine centrally in thermoneutral and cold environments indicates that taurine has a rather uniform and compelling influence on temperature control of homeotherms. In characterizing the nature of the influence of a substance on body temperature one of the first steps is to administer it at different ambient temperatures (3). When this was done using rats (22) and rabbits (12, Fig. 1) taurine caused hypothermia in thermoneutral environments, greater hypothermia in the cold and had no significant effect on body temperature in the heat. Similar effects of taurine have recently been observed in young and aged squirrel monkeys (Clark and Lipton, unpublished). However, in these experiments there was a small decrease in body temperature when taurine was given in a 30°C environment. This difference in response from that of other species may reflect a higher thermoneutral zone in the squirrel monkey.

One other deviation has been reported from the general pattern of effects described above. In sheep (2) taurine caused decreases in body temperature in the cold, both increases and decreases in a 20°C environment, and rises in temperature in the heat (40°C). Such differences in temperature responses of the sheep, particularly in the hot environment, may reflect differences in the importance to this species of particular thermoeffector systems that are influenced by taurine, a point considered later in this report.

Despite these differences between the sheep and other species

Table 1

Influence of Central Taurine on Thermoregulation and Fever

SPECIES	ROUTE	DOSE	AMBIENT TEMPERATURE (°C)	EFFECT ON BODY TEMPERATURE (Δ°C)	REMARKS	INVESTIGATOR
Mouse 30-40 g	icv acute	125 μg	22-23	↓2		Hruska, et al. 1976
		250 μg			killed 4/6	
		500 μg			killed all	
	i.p.	∼1.1 g/kg	22-23	↓1.7		
		2.7 g/kg		↓2.7		
Rat 200-250 g	intracisternal	10 mg/kg	22-23	↓5.5	anesthetized rats	Sgaragli and Pavan, 1972
	icv	10-12.5 mg/kg	24	↓4.5		
			0-4	↓<6.0 in 30 min		
			37	↑not different from controls		
Rat 220	icv	0.1-3.0 mg		↓1.4-3.0 dose related		Garcia de Yebenes Prous et al., 1978
Rat 250-300 g	icv	5 μg	22	→		Kerwin and Pycock, 1979
		10 μg		↑1.8		
		20 μg		↑1.0		
	pre-optic/ anterior hypothalamus	2.5 μg	22	↑0.9		

		5 µg 10 µg		↑1.3 ↓0.9		
Rabbit	icv	2.5 mg/kg	20	↓	less hypothermia than homotaurine	Sgaragli 1978
Rabbit 2-3 kg	icv	2.5 mg/kg	19 ± 1.0	↓2.23		Sgaragli, Carla, Magnani and Giotti 1978
Rabbit 2-3 kg		5 mg/kg		↓profound longlasting		Sgaragli, Magnani, Carla and Giotti 1976
Rabbit 2-4 kg	icv	0.01-6.0 mg	23	↓dose-related		Harris and Lipton, 1977
		0.01-1.0 mg 0.01-6.0 mg	10 30	↓dose-related →		
	icv	2 mg	23	↓0.8-1.1	daily injections for five days	Harris and Lipton 1977
	icv	0.5 mg bolus + infusion	23	inhibited fever during infusion, caused prolonged fever when infusion was stopped		
	pre-optic/ anterior hypothalamus	0.01-1.0 mg	23	→		Harris and Lipton, 1977
	medulla	0.01-1.0 mg	23	→		

Table 1 (continued)

SPECIES	ROUTE	DOSE	AMBIENT TEMPERATURE (°C)	EFFECT OF BODY TEMPERATURE (Δ°C)	REMARKS	INVESTIGATOR
Rabbit 2-4 kg	icv	0.5 mg	23	reduced PGE hyperthermia		Harris and Lipton, 1977
		5.0 mg	23	prevented PGE hyperthermia caused hypothermia		
		0.5 and 5.0 mg	23	no consistent effect on amphetamine hyperthermia		
Rabbit 2-4 kg	icv	0.5 mg + infusion (0.1-0.2 mg/min for 1 hr)	23	Inhibited LP fever and PGE hyperthermia, prolonged fever after infusion stopped		Lipton and Ticknor, 1979
Sheep 25-35 kg	icv	50 μg/kg	0	↓	unshorn	Bligh, Silver, Smith and Bacon, 1979
		100 μg/kg		↓	shorn	
		50 μg/kg	10	↓	shorn	
		100 μg/kg		↓	shorn	
		50 μg/kg	20	↓	unshorn	
		100 μg/kg		→	unshorn	
Sheep		50 μg/kg	40	→	unshorn	
		100 μg/kg		↑	unshorn	
		125 μg/kg		↑	unshorn	
Squirrel Monkey	icv	0.5-4.0 mg	23	↓dose-related in animals under 9 years, greater decreases in animals over 9 years		Lipton and Clark, unpublished observations
		0.5-4.0 mg	15	↓dose-related in animals under 9 years		
		0.5 mg	15	↓near-lethal in animals over 9 years		
		0.5-4.0 mg	30	↓dose-related in both age groups		

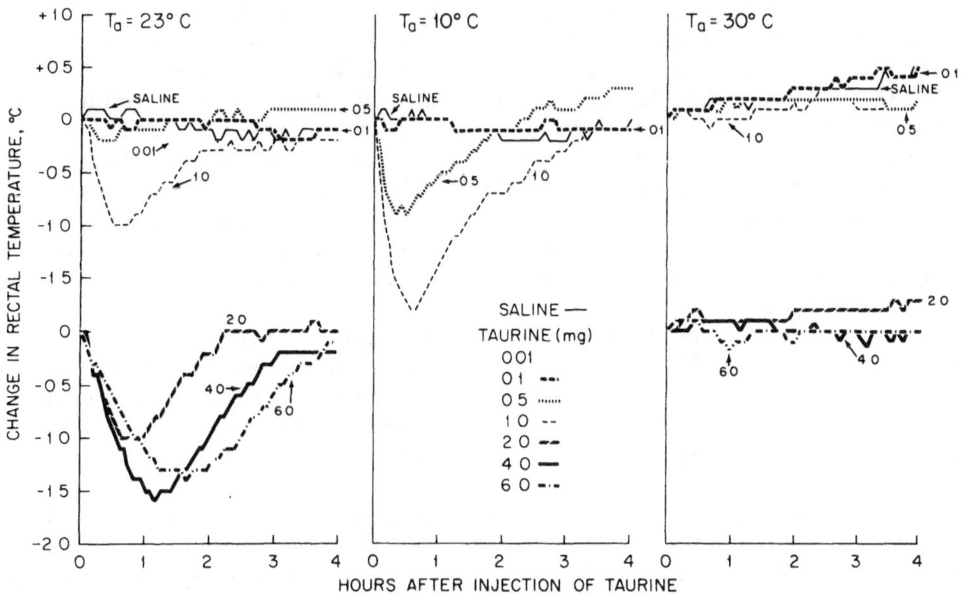

Fig. 1. Effect on rectal temperature of taurine injected into a lateral cerebral ventricle of rabbits exposed to neutral, cold and hot ambient temperatures (T_a). Each line represents the mean temperature response of six animals. Hypothermia developed after doses of 1.0 mg and greater in the 23°C environment and after 0.5 and 1.0 mg doses in the cold environment, but no changes were observed after taurine injections in the 30°C environment. (From 12).

and the finding in one experiment of hyperthermia after central taurine administration in the rat (15, Table 1), the bulk of the data collected so far suggests that this sulphonated amino acid causes hypothermia in neutral and cold environments.

Doses of taurine that have no effect on temperature of afebrile animals also inhibit induction of fever by Salmonella typhosa endotoxin (12) and leukocytic pyrogen (18). Such inhibition is associated with marked vasodilatation and behavioral quieting. The rise in body temperature produced by intracerebroventricular injection of PGE was also inhibited by subsequent injection and infusion of taurine (12,18). On the other hand, central administration of taurine had no consistent effect on hyperthermia induced by amphetamine (12). Either taurine does not effect central pathways essential to the response to amphetamine or, more likely, it does not reach peripheral sites of amphetamine action.

The mechanisms by which taurine produces hypothermia and inhibits fever have not been fully explained. However, it is likely that the well-known capacity of taurine to suppress neuronal firing (5,10) underlies these effects. Because taurine, like GABA and glycine, has strong inhibitory effects on neuronal activity and because it occurs naturally in synaptic vesicles (20), this amino acid is thought to be an inhibitory neurotransmitter (10) or neuromodulator (14). It has been proposed that the hypothermia induced by taurine is caused by an increase in the extracellular level of this amino acid in the hypothalamus (13). It is reasonable to expect that inhibition of neuronal activity in this part of the brain would explain the hypothermia, but taurine has no effect when injected directly into tissue in the pre-optic/anterior hypothalamus region, at sites where injection of pyrogen does cause fever, nor for that matter, into a part of the medullary reticular substance that is thought to be important to secondary temperature control (12). Sgaragli and Pavan (22) also proposed that taurine modifies the function of neurons involved in thermoregulation although they did not rule out the possibility that the effect on temperature was secondary to changes in respiration and blood circulation.

Whether the hypothermia and inhibition of fever can be traced to an influence on central temperature controls per se, or whether they represent secondary effects of actions on other physiological systems is pivotal to the understanding of the importance of taurine in thermoregulation. It is clear that taurine does not reduce the central setpoint of temperature control since the changes in body temperature it produces depend upon ambient temperature. The finding that taurine does not cause hypothermia when injected into sites of primary and secondary central temperature control suggests that its effects on body temperature may be indirect or that only effector outflow is altered. The influence of taurine on vasomotor and muscle tone can be construed to reflect a coordinated influence on heat loss and heat production mechanisms but, again, these effects may reflect nonspecific actions of taurine. Taurine causes vasodilatation in the febrile and afebrile rabbit (12) and lowers arterial blood pressure in the rat (22) presumably through changes in peripheral vasomotor tone. Peripheral vasomotor tone is maintained through active firing in neuronal pathways which derive from the floor of the fourth cerebral ventricle. Direct inhibitory actions on this pathway would produce vasodilatation much as observed in the experiments described above. Taurine also causes muscle relaxation (21) and would therefore tend to inhibit heat production derived from maintenance of normal muscle tone and through shivering. From these observations sufficient doses of taurine would be expected to lower body temperature in thermoneutral and cold environments independently of action on temperature controls per se, by promoting heat loss via vasodilatation and by inhibition of shivering.

Additional data which suggests that the effects of taurine on body temperature may be incidental comes from research on the sheep (2). In this species centrally injected taurine inhibited all three thermoregulatory effectors: evaporative heat loss by panting, heat conservation by peripheral vasoconstriction, and heat production by shivering. Such widespread effects might not be expected of a neurotransmitter with specific functions in central temperature control.

The finding that taurine reduces arousal may help to explain the effects of this amino acid on central temperature controls. Taurine in known to cause sedation (23) and behavioral quieting (12,18), to depress motor activity (1,7,22) and to cause muscle relaxation (21). Taurine reduces the amplitude of electrical activity of the brain, a finding which corroborates the observations of reduced arousal (21). Such reductions in arousal are known to affect body temperature control. For example, the temperature of children declines and is regulated at a lower level during afternoon naps (6). Similar phenomena have been observed in man sleeping in 60 and 80°F environments (16). Vasodilatation occurs and hypothalamic temperature of the Rhesus macaque falls during sleep (11). Although the reduced state of arousal undoubtedly contributes, it is unlikely that taurine affects thermoregulation solely through depression of CNS arousal mechanisms. The decreases in body temperature caused by taurine are much greater than those observed in sleep. Further, body temperature appears to be regulated at a lower level in sleep and there was no evidence of a simple reduction in set-point of temperature control after taurine.

From the evidence presented above it is not clear whether taurine affects central temperature controls specifically or if body temperature changes are secondary to multiple actions on physiological systems that are also used in thermoeffector activity. Taurine reduces arousal and it may thus affect the control of body temperature by altering the "tone" of central temperature controls. The final effect of taurine on body temperature is perhaps a result of a combination of these influences.

Many of the doses used in these experiments are large in comparison with the amounts of other putative neurotransmitters which alter temperature when given centrally. Effects produced by certain substances are often labelled "pharmacological" as opposed to "physiological" on the basis of the greater magnitude of doses required to produce effects. When no other variables are compared such as transport or enzymatic inactivation of the substances, their relative access to sites of action within CNS tissue, possible excitatory and inhibitory effects on the same system, etc., such a differentiation is arbitrary, particularly in the early phases of research on the substance. In the current understanding of central control of body temperature many of the variables of potential importance to the effects of central administration of a drug are unknown and perhaps are unknowable, at least at present. In the case of taurine, central transport processes may remove much of the amino acid before it can reach its site of action, and/or perhaps only small amounts penetrate the parenchyma to reach sites deep in the brain tissue. Until more is known about these variables the possibility that taurine has a role in normal thermoregulation, based on the data collected after central administration of the amino acid, cannot be ruled out. On the other hand, the large doses required to affect body temperature tends to move taurine down on the list of potential neurotransmitters essential to central temperature control.

Specificity of the Hypothermogenic Effects of Taurine

A number of amino acids cause body temperature to fall when injected into the brain (13). These results raise further questions about the specificity of the influence of taurine on temperature. The most comparable data come from experiments in which the effects of taurine and other substances have been tested together. Sgaragli and Pavan (22), for instance, found that L-alanine, GABA, glycine and L-serine also caused significant hypothermia when given intracisternally in 80 µmole/kg doses to rats whereas sulphonic acid, ethylurethane, L-lysine, L-histidine at 160 µmole/kg and 4-hydroxybutyric acid had no significant effects on temperature. Aminoethanesulfonic acid, the one-carbon analogue of taurine; 3-aminopropanesulfonic acid, the three carbon analogue; β-alanine, the carboxyl analogue; and aminoisobutyric acid, a transportable but non-metabolizable amino acid were injected icv in mice by Hruska et al (13). All caused hypothermia of greater, lesser, or in the case of β-alanine, the same magnitude as that produced by taurine. Sgaragli et al. (21) found that 3-aminopropanesulfonic acid was more potent in producing hypothermia than taurine whereas GABA was less potent. In experiments related to our initial observations on taurine (12) we have found (8,9) that L-serine, glycine and L-alanine share hypothermogenic qualities with taurine. In comparisons in the sheep (2) icv injections of glycine up to 1000 µmole/kg altered neither thermoeffector activity nor body temperature. In these experiments GABA was generally ineffective in altering body temperature but occasionally had some inhibitory effects on thermoregulatory effectors. On the other hand, small doses of muscimol, the GABA agonist, produced marked changes in thermoeffector activity. It is clear from the results cited above that amino acids related to taurine either structurally or in inhibitory effects on neuronal activity share some of its effects on body temperature.

Augmentation of Fever by Taurine

The events within the CNS that underlie the cessation of fever are poorly understood. Based on the general theory of the pathophysiology of fever, defervescence occurs when pathological processes responsible for release of endogenous pyrogen (EP) abate and the amount of EP reaching the brain, and influencing central temperature control, subsequently declines. In addition, however, something must happen within the brain to terminate the action of EP already present. Otherwise a fever, once initiated, would persist indefinitely. One possibility is that EP is inactivated within the brain by transport away from receptor sites, as in the case of certain neurotransmitters.

Data supporting this possibility were obtained in the course of experiments on taurine inhibition of fever described above. In these experiments rabbits were made febrile by iv injection of Salmonella typhosa endotoxin (12, Fig. 2). Taurine infused icv inhibited development of the fever. However, when the infusion was stopped body temperature rose and remained elevated for 8.75 - 17.5 hr, much longer than the usual 6 hr fever produced by endotoxin alone. Infusion of a similar amount of taurine at the same rate in the same animals when they were afebrile had no effect on normal temperature and did not cause fever when infusion ceased. Similar results were obtained when homologous leukocytic pyrogen (LP) was used to produce fever (Fig. 3). These results indicate that there is an essential relation between the pyrogen that reaches the brain and central taurine infusion in the production of the prolonged fevers. The taurine effect was specific in that it did not prolong hyperthermia produced by icv PGE or iv amphetamine. The long term fevers resulting from endotoxin/taurine administration resembled those caused by central administration of LP in which the LP bolus presumably causes

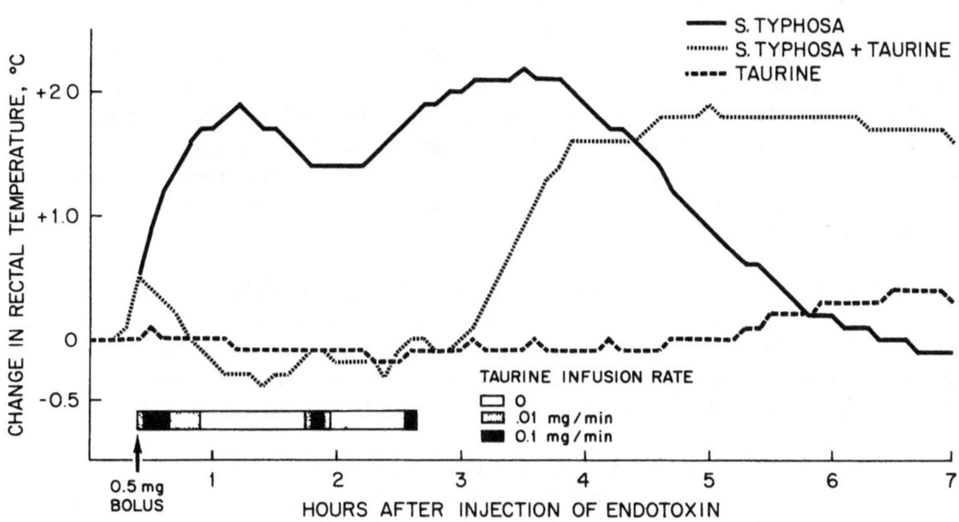

Fig. 2. Taurine infused into a lateral cerebral ventricle blocked the characteristic febrile response to peripheral endotoxin. Examples from a single rabbit showing: normal febrile response to iv Salmonella typhosa endotoxin (——), blockade of the febrile response by a bolus and controlled infusion of taurine in another experiment (.....) and the lack of effect of taurine control injections and infusions made in the same animal when it was afebrile (-----). Rate of infusion of taurine: 0, □; 0.01 mg/min ▨; 0.1 mg/min ■. (From 12).

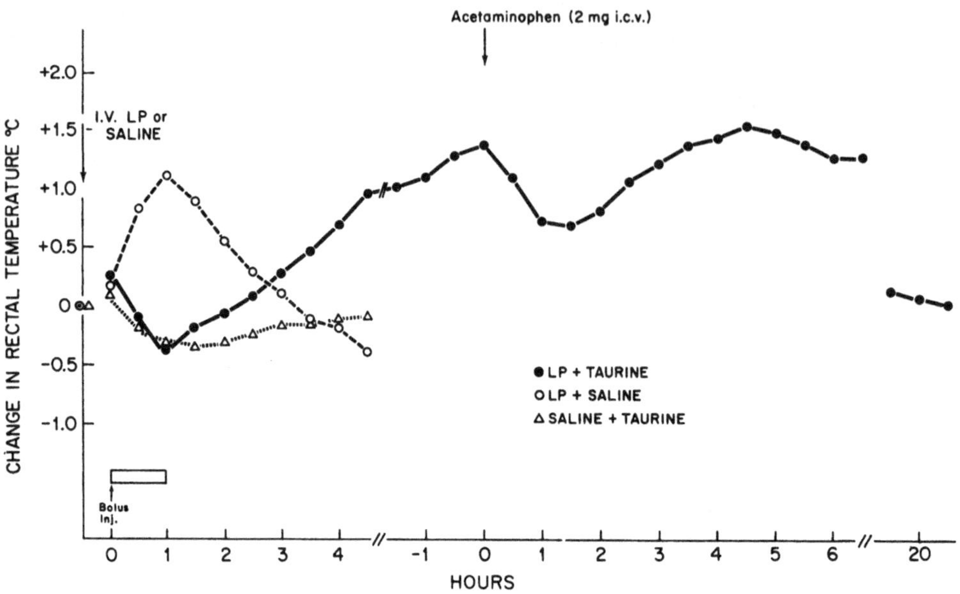

Fig. 3. Effects of bolus injection (0.5 mg) and 1 hr infusion (0.01-0.2 mg/min) of taurine on LP fever. Scores are means of observations on five animals. At a time when the febrile response is generally ending body temperature after taurine rose and remained elevated. A total dose of 2 mg acetaminophen given icv in three equal doses 10 min apart around the time indicated by the arrow reduced the prolonged fever for 2-3 hr but did not prevent a return of body temperature to elevated levels. The time between taurine infusion and acetaminophen injections was not the same in each animal (note break in line). (From 18).

fever until it is destroyed or removed. One possibility is that LP is inactivated by removal from receptor sites via the same transport processes as taurine and that taurine prolongs fever by saturating a common carrier essential to inactivation of the endogenous pyrogens, thus causing LP which reaches the brain to accumulate (17).

Probenecid, an inhibitor of facilitated transport, given icv to rabbits made febrile by iv LP caused dose-related augmentation of the fevers (4). This transport inhibitor also prolonged hyperthermia caused by icv PGE_2, an autacoid that has been implicated in central mediation of fever. However, it is unlikely that probenecid augmented LP fever by prolonging the action of PGE released by LP action since antipyretics which have no effect on PGE hyperthermia reduced fever induced by LP and prolonged by probenecid. The effects of other transport inhibitors were also determined: perchloric acid, an inhibitor of the iodide transport system; iodipamide which inhibits the L or liver-like system and para-amino hippuric acid, the classic inhibitor of the H transport system. Probenecid has low specificity and is known to influence both the H and L systems. Iodipamide given 10 min before LP augmented the ensuing fever in a dose-related fashion whereas PAH and perchloric acid had no influence. Iodipamide also augmented PGE_2 hyperthermia but again when given during LP-induced/iodipamide-augmented fever acetaminophen lowered the temperature. This rules out the possibility that the prolonged fever is due to prolongation of central PGE action. Augmentation of LP fever by iodipamide is less than that caused by probenecid, the "catch all" transport inhibitor. This indicates that transport systems other than the L system may be involved in the central inactivation of LP.

Specificity of Amino Acid Transport Systems in Prolonging Fever

If taurine slows or prevents termination of the effect of central LP by interfering with its transport, other substances which compete with taurine for transport in CNS tissue (14) might also affect fever. To test this idea hypotaurine and β-alanine, the sulfinic and carboxyl analogues of taurine, were given icv to rabbits after iv LP injections (18). Hypotaurine inhibited the development of fever, increased the fever peak and prolonged the febrile response (Fig. 4). β-Alanine consistently inhibited fever development, increased fever maximum in 60% of the experiments and prolonged fever in 80% (Fig. 5). The inhibition of fever development by taurine and these two analogues is presumed to result from effects described earlier in this paper. That all three amino acids prolong

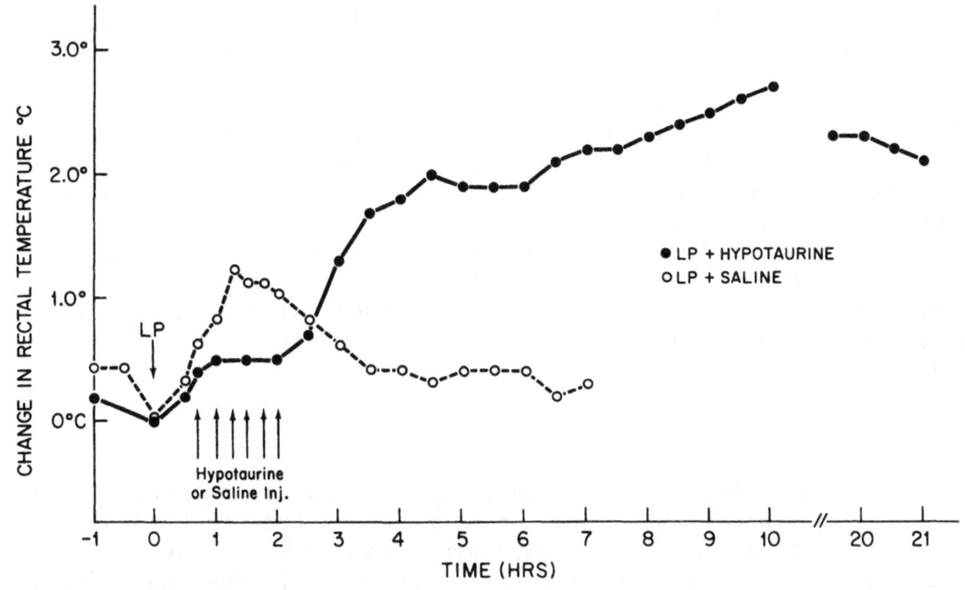

Fig. 4. Effects of repeated icv injections of hypotaurine (1 mg) on fever produced by iv LP in a single rabbit. (From 17).

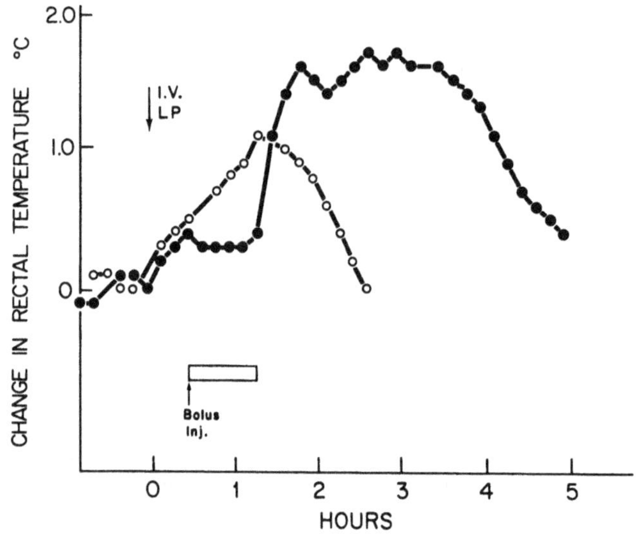

Fig. 5. Effects of bolus injection (0.5 mg) and infusion (0.1-0.2 mg/min) of β-alanine on fever produced by iv LP in a single animal. ●, LP + β-alanine; ○, LP + saline. (From 18).

fever suggests that the transport system they share is also involved in normal termination of the effects of LP or related central fever mediators.

The capacity of three other amino acids, L-serine, L-alanine and glycine to prolong fever was tested in related experiments. Single and multiple icv injections of serine and infusions sufficient to arrest LP induced fever for 1 hr, did not prolong or enhance fever (8, Fig. 6). In this case the LP + serine temperature curve paralleled the LP + saline defervescence curve to baseline. Thus while serine shares with taurine certain hypothermogenic and fever-inhibiting properties, in its failure to prolong fever it is unlike taurine. Single icv injections of glycine (1 mg) and L-alanine (0.5-2.0 mg) inhibited fever and did not prolong it (8, Glyn and Lipton, unpublished observations). However, the effects of these two amino acids on fever must be tested further using prolonged icv infusion before they can be adequately compared with those of taurine.

SUMMARY

The results cited above indicate that the most consistent effect of centrally administered taurine on body temperature is hypothermia. Taurine also inhibits rises in temperature caused by pyrogens and PGE which act on central temperature controls but it does not affect hyperthermia produced by peripheral actions of amphetamine. Taurine has well-known inhibitory effects on neurons, and it may be that

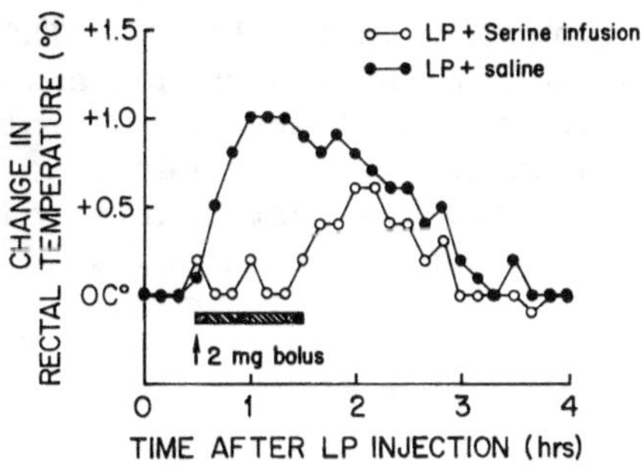

Fig. 6. Effect on LP fever of serine infusion in one rabbit. A bolus icv injection of serine was given after rectal temperature had risen 0.3°C, followed by an infusion of serine (0.2 mg/μl) at a rate which blocked the fever normally produced by LP. (From 8). 0.05 mg/min, ; 0.10 mg/min, ; 0.15 mg/min, ; 0.20 mg/min, .

such an action on neurons in central temperature controls may underlie the hypothermia and fever-inhibiting effects. However, several findings must be accounted for before the effects of taurine on temperature control are accepted as specific. For example, taurine does not appear to produce its effects by acting at neuronal sites associated with primary and secondary temperature controls. It is also not clear whether peripheral vasodilatation and inhibition of muscle tone are primary effects due to an action on central thermoregulatory pathways or secondary effects due to a nonspecific action of taurine. Effector organs that have other roles than control of body temperature, and other innervation, are used in thermoregulation and influences of taurine on the effector pathways could incidentally affect thermoregulation. The inhibitory effects of taurine on heat loss, heat conservation and heat production in the sheep (2) suggests a broad influence of taurine on central pathways. There is also a depression of arousal by taurine which may promote nonspecific effects on body temperature. Finally, rather large doses of taurine are required to affect temperature and the effects of this amino acid are shared by several analogues and other putative inhibitory neurotransmitters.

Whereas the data on hypothermia and fever-inhibiting effects of taurine may confound simple and direct interpretation in terms of central temperature control, prolongation of fever by centrally administered taurine has opened a new approach to the understanding of defevescence and the central inactivation of endogenous pyrogen. The evidence indicates that taurine interacts with central pyrogen, but not PGE, to prolong fever. The long-term fevers are similar to those produced by a central injection of LP and to those caused by peripherally injected pyrogens plus central administration of transport inhibitors. Taurine may prolong fever by saturating a transport carrier, thus inhibiting central inactivation of pyrogen by transport and causing it to be accumulated in the brain. This idea is supported by the finding that hypotaurine and β-alanine, analogues that compete with taurine for transport in CNS tissue, also prolong fever. Serine does not prolong fever which suggests further that the effect of taurine is specific. The discovery that taurine can prolong the effects of endogenous pyrogen opens new avenues to understanding of fever control.

ACKNOWLEDGMENT

Supported by grants from the National Institute of Neurological and Communications Diseases and Stroke (NS 10046) and from the National Institute on Aging (AG 00109).

REFERENCES

1. Baskin, S.I., Hinkamp, D.L., Marquis, W.J., and Tilson, H.A., 1974, Effects of taurine on psychomotor activity of the rat, Neuropharmacology , 13: 591-594.
2. Bligh, J., Silver, A., Smith, C.A., and Bacon, M.J., 1978, The effects on thermoregulation in the sheep of intracerebroventricular injection of taurine, glycine, GABA and muscimol, J. Thermal. Biol., 4: 9-14.
3. Clark, W.G., 1979, Changes in body temperature after administration of amino acids, peptides, dopamine, neuroleptics and related agents, Neurosci. Biobehav. Revs., 3: 179-231.
4. Crawford, I.L., Kennedy, J.I., Lipton, J.M., and Ojeda, 1979, Effects of central administration of probenecid on fevers produced by leukocytic pyrogen and PGE_2 in the rabbit, J. Physiol., 287: 519-533.
5. Curtis, D.R., and Watkins, J.C., 1960, The excitation and depression of spinal neurones by structurally related amino acids, J. Neurochem., 6: 117-141.
6. Day, R, 1941, Regulation of body temperature during sleep. Am. J. Dis. Child., 61: 734-746.
7. Garcia De Yebenes Prous, J., Carlsson, A., and Mena Gomez, M.A., 1978, The effect of taurine on motor behavior, body temperature and monoanine metabolism in rat brain, Naunyn-Schmiedebergs Arch. Pharmacol., 304: 95-99.
8. Glyn, J.R., and Lipton, J.M., in press, Central administration of serine causes hypothermia and inhibits fever in rabbits, Brain Res. Bull., in press.
9. Glyn, J.R., and Lipton, J.M., in press, Comparison of the effects of central administration of serine and glycine on body temperature of the rabbit, Brain Res. Bull., in press.
10. Haas, H.L., and Hösli, L., 1973, Depression of brain stem neurones by taurine and its interaction with strychnine and bicuculline, Brain Res., 52: 399-402.
11. Hammel, H.T., Jackson, D.C., Stolwijk, J.A.J., Hardy, J.D.,

and Stromme, S.B., 1963, Temperature regulation by hypotholamic proportional control with an adjustable set point, J. Appl. Physiol., 18: 1146-1154.

12. Harris, W.S., Jr., and Lipton, J.M., 1977, Intracerebroventricular taurine in rabbits: effects on normal body temperature, endotoxin fever and hyperthermia produced by PGE_1 and amphetamine, J. Physiol., 266: 397-410.
13. Hruska, R.E., Thut, P.D., Huxtable, R., and Bressler, R., 1976, Hypothermia produced by taurine, in: "Taurine," A. Barbeau and R. Huxtable, eds., Raven Press, New York, pp.347-356.
14. Kaczmarek, L.K., and Davison, A.N., 1972, Uptake and release of taurine from rat brain slices, J. Neurochem., 19: 2355-2362.
15. Kerwin, R.W., and Pycock, C.J., 1978, Role of taurine as a possible transmitter in the thermoregulatory pathways of the rat, J. Pharm. Pharmacol., 31: 466-470.
16. Kreider, M.B., Iampietro, P.F., Buskirk, E.R., and Bass, D.E., 1959, Effect of continuous cold exposure on nocturnal body temperature of man, J. Appl. Physiol., 14: 43-45.
17. Lipton, J.M., 1980, Central inactivation of endogenous pyrogens, in: "Fever," J.M. Lipton, ed., Raven Press, New York, pp.71-80.
18. Lipton, J.M., and Ticknor, C.B., 1979, Central effect of taurine and its analogues on fever caused by intravenous leukocytic pyrogen in the rabbit, J. Physiol., 287: 535-543.
19. Lipton, J.M., Whisenant, J.D., Gean, J.T., and Ticknor, C.B., 1979, Effects of fever of central administration of transport inhibitors, Brain Res. Bull., 4: 297-300.
20. Lombardini, J.B., 1976, Regional and subcellular studies on taurine in the rat central nervous system, in: "Taurine," A. Barbeau and R. Huxtable, eds., Raven Press, New York, pp.311-326.
21. Sgaragli, G.P., Magnani, M., Carla, V., and Giotti, A., 1976, Muscle relaxation induced in the rabbit by intracerebroventricular taurine injection, Naunyn-Schmiedebergs Arch. Exp. Path. Pharmacol., 295: 95-97.
22. Sgaragli, G.P., and Pavan, F, 1972, Effects of amino acid compounds injected into cerebrospinal fluid spaces, on colonic temperature, arterial blood pressure and behavior of the rat, Neuropharmacology, 11: 45-56.
23. Sugihara, H, Nagasawa, S., and Okabe, H., 1936, Experimentelle und klinesche Untersuchungen uber Taurin., Klin. Wschr., 15: 751-756.

TAURINE AND FRIEDREICH'S ATAXIA: AN UPDATE

A. Barbeau[1], S. Melancon[2], R. J. Huxtable[3] and B. Lemieux[4]

[1]Department of Neurobiology
Clinical Research Institute of Montreal
110 Pine Ave West, Montreal Quebec, H2W 1R7
[2]Centre de Recherche Pédiatrique
Hôpital Ste-Justine
Département de Pédiatrie et Neurologie
Université de Montréal
[3]Department of Pharmacology
Health Sciences Center
University of Arizona
Tucson, Arizona
[4]Centre Hospitalier
Universitoice de Sherbrooke, Quebec, Canada

Friedreich's ataxia is a recessively inherited spinocerebellar ataxia appearing in childhood and relentlessly progressing to invalidity or death in middle age at the latest. A number of signs and symptoms have been found to be present in 100% of patients and are obligatory for a diagnosis, while other signs are definitely of a progressive nature (19). Incoordination of limbs (ataxia) is linked to evidence of damage to the posterior columns of the spinal cord and of dorsal root ganglia (loss of vibration and position sense), and of the pyramidal tract (absence of knee and ankle jerks, trophic changes). Most of these symptoms first appear in the lower limbs. Also observed are skeletal deformities (pes cavus, kyphoscoliosis). Electrically, almost all cases of the typical disease have absent or considerably diminished sensory action potentials in the sural nerves, while motor conduction velocities are either normal or only slightly decreased. Retinitis pigmentosa is not usually seen in typical Friedreich's ataxia, but it is often present in association with a variety of other forms of hereditary ataxia. A final feature of Friedreich's ataxia that must be stressed is the constant presence of a cardiomyopathy as evidenced by abnormal echocardiograms, vectocardiograms or electrocardiograms, singly or in combination. This cardiomyopathy is usually hypertrophic in type and occasionally obstructive; it appears to be an integral part of Friedreich's ataxia, and not a phenomenon secondary to the disease process. The most important histological changes are myocardial fibrosis and foci of degeneration of the cardiac muscle cells. Granular deposits of calcium or iron salts can occasionally be found in these muscle cells.

Thus, Friedreich's ataxia is a complex multisystem disease, the many components of which must eventually be explained biochemically. At the start of the Quebec Cooperative Study on Friedreich's ataxia in 1975, very little was known on this subject (37), but successive publications from our group have unravelled some of the mysteries (4,5,7,8). The pathophysiology of most of the symptoms of the disease is now fairly clearly understood, and has been detailed elsewhere (8). One of the components of this puzzle is the role that the amino acid taurine may play in the development of some of the symptoms. In the 1978 Symposium (6) we had the occasion to review some of the initial findings concerning the possible function of taurine in Friedreich's ataxia while elsewhere we reviewed the neuropharmacology of taurine (2). In the present paper, we will attempt to update these findings.

(a) Taurine in Blood, Urine and CSF in Ataxia

Taurine concentrations have been determined with precise methods in a number of tissues and fluids from patients with the typical form of Friedreich's ataxia. Findings are summarized in Table 1. As can be seen there are a number of significant differences between the patients and the controls. While blood and CSF concentrations (30,31) are entirely within normal limits, it can be seen that the urinary excretion of taurine is doubled in Friedreich's patients. Not indicated on the Table is the fact that β-alanine excretion is also markedly increased (6.4 ± 11.5 versus 0.8 ± 1.3 μM/min/1.73m^2). Because of the wide variations in urinary taurine values it is difficult to conclude whether all patients, or the group as a whole, suffer from this taurine loss. To investigate this point in more detail we treated patients and controls with a uniform taurine load (250 mg) and studied timed urinary samples (18). As can be seen in Fig. 1 ataxic subjects excreted twice as much of the taurine load as the age-matched controls. Such a high urinary excretion has been observed in a number of physiological as well as pathological conditions. Thus, high urinary taurine concentrations are found after stress, corticosteroids and "idiopathic scoliosis" (29), as well as in camptodactyly with metal retardation (34), in myotonic dystrophy associated with generalized amino aciduria (9) and in familial cerebellar dyssynergia (21). It is evident that ataxics cannot retain taurine as well as normal subjects. In view of the normal intestinal absorption and serum concentrations previously demonstrated, this loss could be due to a defect in uptake or reuptake mechanisms or, more simply, to a decrease in the total muscle mass which normally holds more than 70% of the body pool of taurine.

In further studies (17,33) we were able to show that uptake of taurine into platelets (which are normally taurine-rich) is entirely normal in Friedreich's ataxia (Table 2) and that taurine uptake into fibroblasts is also within normal limits (Fig. 2). A localized renal transport defect for taurine and β-alanine, which utilize the same transport system (13), is still possible, but would require micropuncture studies to be proven. However, recent observations (20) made in epileptic patients, may throw a new light on this problem. High and low taurine excretors were found in the general population, presumably indicating alleles at the same locus. We are presently carrying out similar studies to see if ataxics would have a tendency to fall within the high excretor subgroup.

(b) Taurine in Tissues of Ataxics

There is strong evidence implicating taurine as a neurotransmitter in the cerebellum (32) and it is therefore of great interest that this amino acid has been found to be in low concentrations in certain animal models clinically manifested by ataxia. Thus, in thiamine-deficient encephalopathy (Table 3) there is a specific deficiency in taurine in the dorsal spinal cord with a similar trend in other parts of the CNS (11). In cerebellar ataxia produced by 3-acetyl pyridine in the rat, taurine concentrations are markedly and significantly reduced in the cerebellum and medulla oblongata, but not in other areas studied (12).

In two cases of Friedreich's ataxia, the concentration of various amino acids has recently been studied (26). Glutamate concentration, as well as the concentration of aspartate, are

Table 1

TAURINE IN FRIEDREICH'S ATAXIA

	CONTROL	FRIEDREICH
Plasma (μmole/100 ml)	4.39 ± 2.93	4.38 ± 3.36
Urine (μmole/min/1.73m^2)	47.2 ± 46.1	91.5 ± 67.0*
CSF (μmole/L)	8.9 ± 4.0	8.9 ± 4.2
Renal Clearance (ml/min/1.73m^2)	14.2 ± 3.8	58.9 ± 65.5*
Cerebellar Hemispheres (μmole/g wet weight)	2.2 ± 1.0	3.8 ± 0.6*
Anterior Vermis (μmole/g wet weight)	1.6 ± 0.7	2.6 ± 0.1*
Dentate Nucleus (μmole/g wet weight)	1.3 ± 0.7	3.5 ± 3.2

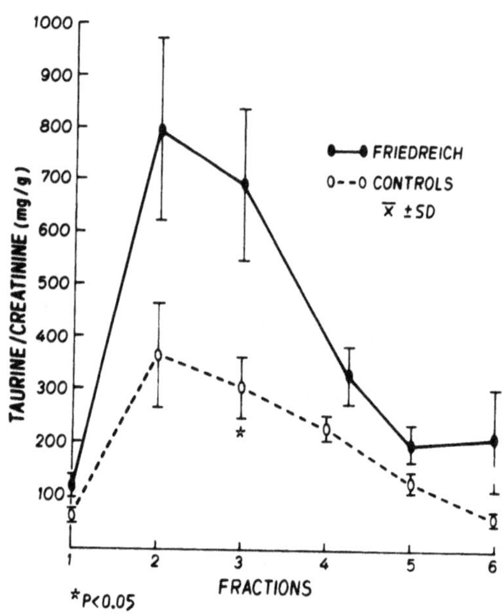

Fig. 1. Urinary excretion of taurine in patients with Friedreich's ataxia and in controls in timed urinary samples. Fraction numbers correspond to 2 hours collections.

Table 2

KINETICS OF TAURINE UPTAKE BY PLATELETS

	N	Vmax (mmole X 10^{-6}/15 min/10^{11}pl)	Km (M X 10^{-5})
Controls	20	0.44 ± 0.06	5.33 ± 0.74
Patients	20	0.48 ± 0.09	5.12 ± 0.79

The results (mean ± SE) were obtained using platelets from 20 ataxics and 20 controls. Platelets were incuabated with 2.4, 4.8, 24, 40 μM (^{14}C-1,2)taurine for 15 minutes. Km and Vmax were calculated according to the Michaelis-Menten equation used to describe saturable enzyme/substrate interactions.

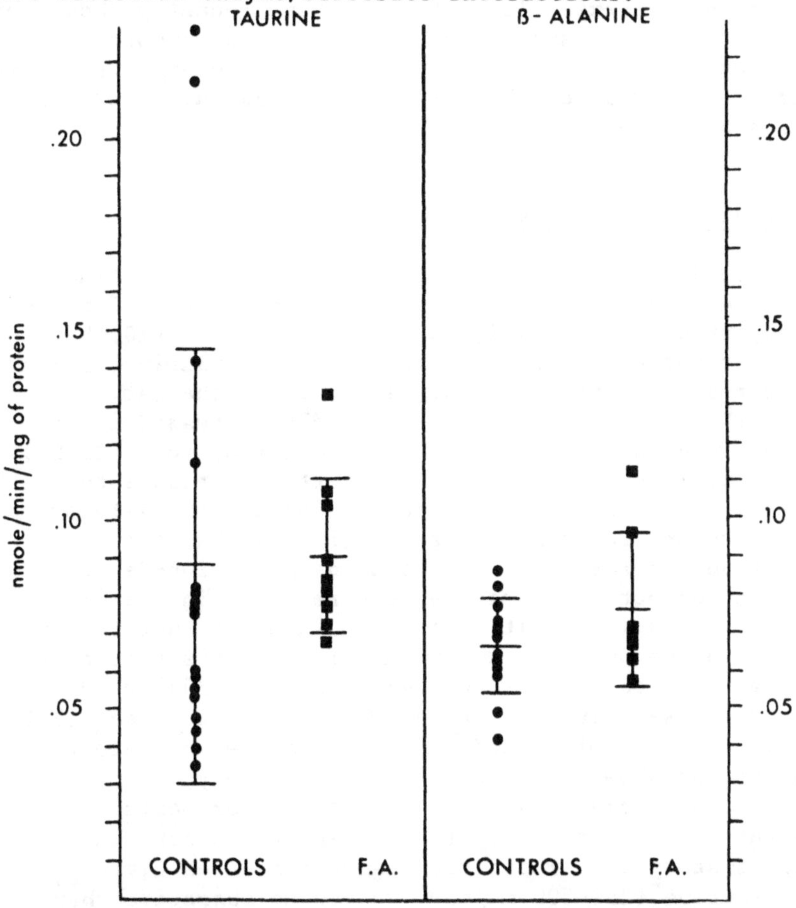

Fig. 2. Taurine and β-alanine uptake into fibroblasts of subjects with Friedreich's ataxia and normal age-matched controls (Reproduced with permission from Melancon et al., 1980).

decreased in the cerebellar hemispheres and posterior cerebellar vermis. GABA concentrations are significantly decreased in the same two areas, and also in the anterior cerebellar vermis. On the other hand, taurine was the only amino acid to be increased in all brain areas examined (Table 1). It is not yet known whether this corresponds to glial taurine and thus secondary to cicatricial gliosis, but is becoming evident, particularly from the studies of Van Gelder (38), that the concentration of taurine is often in an inverse relationship to that of glutamate. The fact that taurine can be construed as an inhibitory neurotransmitter (2) and that it probably modulates the central nervous system glutamate and aspartate levels through modifications in calcium binding or transport (28,3,27), may explain why large (pharmacological) doses of taurine can themselves cause ataxia (Table 4). We have recently shown that low aspartate and glutamate were related to the symptom "ataxia", because we were able to completely reverse it, in rats rendered ataxic by 3-acetyl pyridine, through intraventricular injections of the excitatory amino acids (15).

(c) Taurine and Retinitis

As stated above, retinitis pigmentosa is a rare finding in Friedreich's ataxia, but it is not uncommon in other forms of hereditary ataxias. As early as 1962, Brotherton (10) had shown that the rat retina contains large amounts of taurine, a finding soon confirmed in other species, particularly the cat. He also indicated that retinal taurine is markedly decreased in rat retinitis pigmentosa. Later Hayes and collaborators (22,23) found that taurine deficiency in the kitten is associated with retinal degeneration. Further studies located the site of disruption to the tapetum (39) and indicated that the electroretinogram responses are reduced in taurine-depleted cats. This subject has recently been reviewed in detail by Sturman and Hayes (36). It is thus of interest that taurine has been tried in the possible treatment of retinitis pigmentosa. The initial results reported by Reccia et al. (35) in 10 patients are disappointing. After one year large doses (1-2 grams/day) of this amino acid failed to induce any detectable improvement, as shown by clinical and laboratory investigations. Some subjective general and/or ocular improvements were reported by the patients. In our own unpublished study in 8 retinitis pigmentosa cases given low doses of taurine (250-500 mg/day) we saw no objective change in any of the visual fields, but all patients and their spouses reported improvement in night vision. This was confirmed by tests of walking in dark rooms.

Table 3

TAURINE IN THIAMINE DEFICIENCY

μmole/g wet wt.

REGION	CONTROL	THIAMINE DEF.
Cerebellum	3.3 ± 0.2	3.3 ± 0.3
Medulla	2.1 ± 0.1	2.2 ± 0.1
Midbrain	2.0 ± 0.2	1.7 ± 0.1
Cortex	4.6 ± 0.4	4.1 ± 0.2
Caudate	6.2 ± 0.4	5.7 ± 0.4
Cervical cord	1.9 ± 0.1	1.5 ± 0.2
Dorsal cord	1.2 ± 0.3	0.7 + 0.2*
Lumbar cord	0.8 ± 0.2	0.6 ± 0.1

Table 4

BEHAVIORAL MANIFESTATIONS FOLLOWING INTRAVENTRICULAR INJECTION OF TAURINE

BEHAVIOR	CONTROL (N=6) (artificial CSF)	TAURINE (N=6) (9.5×10^{-2}M + artificial CSF)
Decrease of locomotor activity	0	6
Loss of body balance	0	6
Ataxic gait	0	6
Circling movement	0	3
Sniffing	5	4
Licking	6	2
Grooming	6	2

(d) Taurine and Cardiomyopathy

Taurine is abundant in the mammalian heart (29) and has been claimed to have a calcium modulating effect in that organ (16). It has been shown that taurine concentrations increase in both human and experimental congestive heart failure (24,25). It is thus of great interest that decreased myocardial taurine levels and hypertaurinuria has recently been described in a kindred with mitral-valve prolapse and congestive cardiomyopathy (14). These findings are quite compatible with the experiments in hamsters with congenital cardiomyopathy, where it was shown that taurine feeding partially prevents the accumulation of calcium in the left ventricle (1).

SUMMARY

This short review demonstrates that taurine may have a role to play in the development of certain of the symptoms of Friedreich's ataxia, particularly ataxia and the cardiomyopathy. A possible deficiency in taurine exists in that disease, as demonstrated by large urinary losses. This lead deserves much more detailed studies.

ACKNOWLEDGMENTS

The studies reported in this paper were the result of Cooperative undertaking. The list of co-authors is to be found in the original papers cited. This work was made possible by grants from L'Association Canadienne de l'Ataxie de Friedreich and the Department of Health and Welfare (Ottawa).

REFERENCES

1. Azari, J., Brumbaugh, P., Barbeau, A., and Huxtable, R., 1980, Taurine decreases lesion severity in the heart of cardiomyopathic hamsters, Can. J. Neurol. Sci., 7:435-440.
2. Barbeau, A., Inoue, N., Tsukada, Y., and Butterworth, R.F., 1975, Minireview: the neuropharmacology of taurine, Life Science, 17:669-678.
3. Barbeau, A., Tsukada, Y., and Inoue, N., Neuropharmacologic and behavioral effects of taurine, in: "Taurine," R. Huxtable and A. Barbeau, eds., Raven Press, New York (1976), pp. 253-266.
4. Barbeau, A., 1976, Friedreich's ataxia 1976 - An Overview, Can. J. Neurol. Sci., 3:389-397.

5. Barbeau, A., 1978, Friedreich's ataxia 1978 - An Overview, Can J. Neurol. Sci., 5:161-165.
6. Barbeau, A., Taurine and Friedreich's ataxia, in: "Taurine and Neurological Disorders," A. Barbeau and R.J. Huxtable, eds., Raven Press, New York (1978), pp. 429-440.
7. Barbeau, A., 1979, Friedreich's ataxia 1979 - An Overview, Can J. Neurol. Sci., 6:311-319.
8. Barbeau, A., 1980, Friedreich's ataxia 1980 - An Overview of the physiopathology, Can J. Neurol. Sci., 7:455-468.
9. Blahd, W.H., Bloom, A., and Drell, W., 1955, Qualitative study of amino aciduria in muscular dystrophy and myotonia dystrophica, Proc. Soc. Exp. Biol. Med., 90:704-706.
10. Brotherton, J., 1969, Studies on the metabolism of the rat retina with special reference to retinitis pigmentosa. 2-amino acid content as shown by chromatography, Exp. Eye Res., 3:837-843.
11. Butterworth, R.F., Hamel, E., Landreville, F., and Barbeau, A., 1979, Amino acid changes in thiamine-deficient encephalopathy: some implications for the pathogenesis of Friedreich's ataxia, Can. J. Neurol. Sci., 6:217-222.
12. Butterworth, R.F., Hamel, E., Landreville, F., and Barbeau, A., 1978, Cerebellar ataxia produced by 3-acetyl pyridine in rat. Can. J. Neurol. Sci., 5:131-133.
13. Chesney, R.W., Jax, D.K., Scriver, C.R., and Mohyuddin, F., Taurine transport in mammalian kidney, in: "Taurine and Neurological Disorders," A. Barbeau and R.J. Huxtable, eds., Raven Press, New York (1978), pp. 73-93.
14. Darsee, J.R., and Heymsfield, S.B. 1981, Decreased myocardial taurine levels and hypertaurinuria in a kindred with mitral-valve prolapse and congestive cardiomyopathy, New Eng. J. Med., 304:129-135.
15. De Michele, G., Jolicoeur, F.B., Rondeau, D.B., Butterworth, R.F., and Barbeau, A., 1980, Effects of glutamate and aspartate on ataxic gait induced by 3-acetyl pryidine in rats, Can J. Neurol. Sci., 7:451-454.
16. Dolora, P., Agresti, A., Giotti, A., and Pasquini, G., 1973, Effect of taurine on calcium kinetics of Guinea pig heart, Eur. J. Pharmacol., 24:352-358.
17. Filla, A., Butterworth, R.F., Geoffroy, G., Lemieux, B., and Barbeau, A., 1978, Platelet taurine uptake in spinocerebellar degeneration, Can. J. Neurol. Sci., 5:119-123.
18. Filla, A., Butterworth, R.F., and Barbeau, A., 1979, Pilot studies on membranes and some transport mechanisms in Friedreich's ataxia, Can. J. Neurol. Sci., 6:285-289.
19. Geoffroy, G., Barbeau, A., Breton, G., Lemieux, B., Aube, M., Leger, C., and Bouchard, M.P., 1976, Clinical description and roentgenologic evaluation of patients with Friedreich's ataxia, Can. J. Neurol. Sci., 3:279-286.

20. Goodman, H.O., Connally, B.M., McLean, W., and Resnick, M., 1980, Taurine transport in epilepsy, Clin. Chem., 26:414-419.
21. Hall, C.D., Stowe, F.R., and Summer, G.K., 1974, Familial cerebellar dyssynergia and myoclonus epilepsy associated with a defect of amino acid metabolism, Neurology, 24:375.
22. Hayes, K.C., Carey, R.E., and Schmidt, S.Y., 1975, Retinal degeneration associated with taurine deficiency in the cat, Science, 188:949-951.
23. Hayes, K.C., Rabin, A.R., and Berson, E.L., 1975, An ultrastructural study of nutritionally - induced and reversed retinal degeneration in cats, Am. J. Pathol. 78:505-524.
24. Huxtable, R.J., and Bressler, R., 1974, Taurine concentration in congestive heart failure, Science, 198:409-411.
25. Huxtable, R.J., 1978, Cardiac pharmacology and cardiomyopathy in Friedreich's ataxia, Can. J. Neurol. Sci., 5:83-92.
26. Huxtable, R.J., Azari, J., Reisine, T., Johnson, P., Yamamura, H., and Barbeau, A., 1979, Regional distribution of amino acids in Friedreich's ataxia brains, Can. J. Neurol. Sci., 6:255-258.
27. Izumi, K., Butterworth, R.F., and Barbeau, A., 1977, Effect of taurine on calcium binding to microsomes isolated from rat cerebral cortex, Life Sciences, 20:943-950.
28. Izumi, K., Ngo, T.T., and Barbeau, A., Metabolic modulation in the central nervous system by taurine, in: "Taurine and Neurological Disorders," A. Barbeau and R.J. Huxtable, eds., Raven Press, New York, (1978), pp. 137-149.
29. Jacobsen, J.G., and Smith, L.L.H., 1968, Biochemistry and physiology of taurine and taurine derivatives, Physiol. Rev., 48:424-511.
30. Lemieux, B., Barbeau, A., Beroniade, V., Shapcott, D., Breton, G., Geoffroy, G., and Melancon, S., 1976, Amino Acid metabolism in Friedreich's ataxia, Can. J. Neurol. Sci., 3:373-378.
31. Lemieux, B., Giguere, R., Barbeau, A., Melancon, S., and Shapcott, D., 1978, Taurine in cerebrospinal fluid in Friedreich's ataxia, Can. J. Neurol. Sci., 5:125-129.
32. McBride, W.J. and Frederickson, R.C.A., Neurochemical and neurophysiological evidence for a role of taurine as an inhibitory neurotransmitter in the cerebellum of the rat, in: "Taurine and Neuological Disorders," A. Barbeau and R.J. Huxtable, eds., Raven Press, New York (1978), pp. 415-427.
33. Melancon, S.B., Grignon, B., Ledru, E., Geoffroy, G., Potier, M., Dallaire, L., and Vanasse, M., 1980, The Beta-amino acid transport system in Friedreich's ataxia, Can. J. Neurol. Sci., 7:441-446.

34. Nevin, N.C., Hurwitz, L.J., and Neill, D.W., 1964, Camptodactyly with mental deficiency, J. Med. Genet.. 3:265-268.

35. Reccia, R., Pignalosa, B., Grasso, A., and Campanella, G., 1980, Taurine treatment in retinitis pigmentosa, Acta Neurologica (Napoli), 18:132-136.

DISCUSSIONS ON TAURINE

Ryan J. Huxtable

Department of Pharmacology
Health Sciences Center
University of Arizona
Tucson, Arizona

Open discussions were an important part of the meeting. Six were held in a conversational style around a table, and each took as its topic an important current question in taurine research. Apart from this, the discussions were almost completely unstructured. The remainder of this volume comprises edited transcriptions of these discussions. References, figures and tables have been added as appropriate. As is to be expected, the conversations sometimes wander, and there are some redundancies from one discussion period to another. Despite this loose format, the editors hope that researchers working on taurine, and those outside the field, will find these transcripts interesting and informative.

When experts talk with their guard down, a clearer view of the area may emerge than is possible from the polished prose of stately reviews and scholarly research articles. These conversations may be mined for research suggestions, and they leave a strong impression as to where this group of interdisciplinary experts think the field is moving. For example, the growing importance of nutritional research on taurine is clearly evident. In the definition of the expanding circle of insight our ignorance is as important as our knowledge. Our deficiencies in knowledge are dragged blinking into the sunshine in these conversations.

New and unpublished results from a number of laboratories are discussed throughout the sessions, and work already published is perhaps sometimes explained in simpler language and with the intent clearer than is the case in the original publication. We hope this will be of value not only to students, but also to those of us fully engaged in research on taurine. It may even be that in some cases the clearer explication allows the work to be more critically examined, but that is to the benefit of all. If someone reading of an experiment in these transcripts is moved to exclaim, "Hey, I could have done that better and more convincingly" then these transcripts will fulfill one of the functions that the editors have in mind for them.

Not least, we hope some of the vividness and excitement of research appears in the following pages. This is an aspect of science that is usually carefully distilled out of scientific publications, leaving a seamless flow of logic from initiation to conclusion. Of course, we all know that in the laboratory, science does not flow so smoothly, and perhaps the direct speech style used in editing these tapes catches some of the alluring reality of science.

The topics discussed are indexed, and readers interested in particular aspects should be able to trace them using the index.

SESSION I

HOW ARE TISSUE TAURINE CONCENTRATIONS REGULATED?

Moderator: Paul Mandel

MANDEL: The first question is: where does taurine come from? Although we have already learned today that taurine can be either of exogenous or endogenous origin, it looks as if the derivation is not the same in different tissues, and one might expect different pools in a single tissue. What we have also learned today is that the biosynthesis of taurine will depend on its exogenous availability. What are the regulatory mechanisms linking exogenous availability with endogenous biosynthesis? If you maintain an animal on a protein-free diet, there is a decrease of taurine in tissues like liver and muscle where tissue disappears at the same time. There is no decrease of taurine in brain. Taurine behaves in the brain like an essential amino acid, and so there is no decrease of brain weight and no decrease of brain taurine. That means there are regulatory mechanisms whereby an animal can maintain taurine levels. Now who would like to comment?

STURMAN: I would like to show two slides regarding taurine biosynthesis. One of the interests that we had developed in our lab was using partial hepatectomy in rats to produce a rapidly growing tissue (Fig. 1). This is four hours after partial hepatectomy and this is also following an injection of ^{35}S cystine. If you use a fairly acidic elution buffer, which separates hypotaurine from taurine, you get quite an accumulation of labeled hypotaurine (Fig. 1, A). The second part of the slide just shows my observation it is indeed hypotaurine and is readily oxidized to taurine (Fig. 1, B). One of the problems with looking at rat livers is the great variability of taurine concentration. The figure shows that the range is very large. The concentration of taurine in normal rat liver, compared to that in liver remaining after partial hepatectomy is significantly different only over the first 48 hr (Fig. 2). At the time that taurine concentration in rat liver is significantly decreased, hypotaurine accumulates. It is difficult for me to see how a pathway not involving hypotaurine could be involved in taurine biosynthesis.

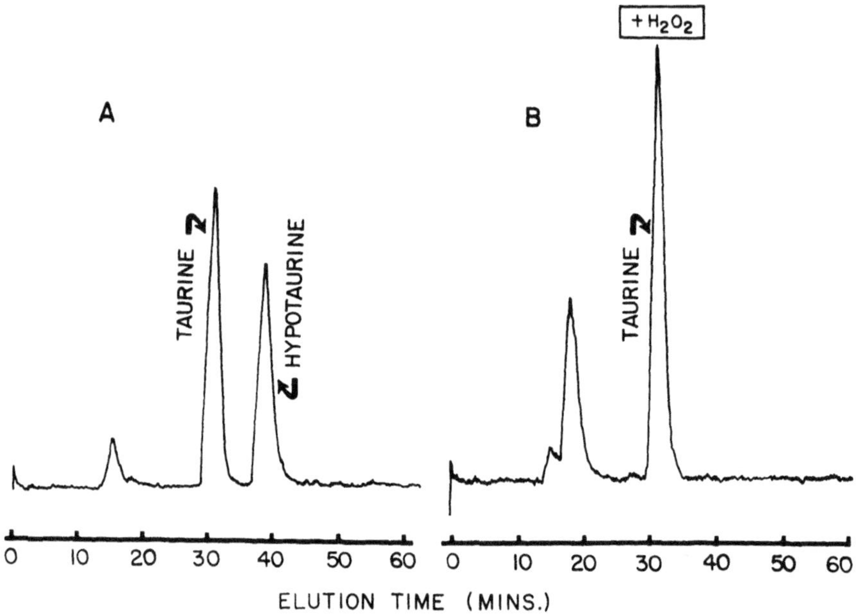

Fig. 1. STURMAN AND FELLMAN. Flow-cell chromatogram of (A) an extract of liver 4 hr after partial hepatectomy and ip injection of (^{35}S)cystine, and (B) the same extract after oxidation with a drop of hydrogen peroxide.

BASKIN: John, How do you know that the neon lights or other conditions are not converting hypotaurine to taurine nonenzymatically?

STURMAN: I'm sorry, what?

BASKIN: How do we know? I'm not sure of your methodology?

STURMAN: The important point is that in rat liver you don't ordinarily see hypotaurine. Here we have actually produced an accumulation of hypotaurine in an animal with altered metabolism. Figure 2 illustrates ^{35}S hypotaurine accumulating from ^{35}S cystine. We have shown the same thing from labeled CSA, and labeled methionine.

MANDEL: This is a lateral question. Let's go back to the main problem. Who would like to comment about this topic; exogenous and endogenous regulation?

HAYES: I would comment on a statement that you made on the conclusion from this morning, Ryan, that when the exogenous source, that is to say the dietary source, of taurine is diminished there is an enhancement of synthesis. I don't know that we have the data to say that. Maybe you do. But in our cat studies, we looked at

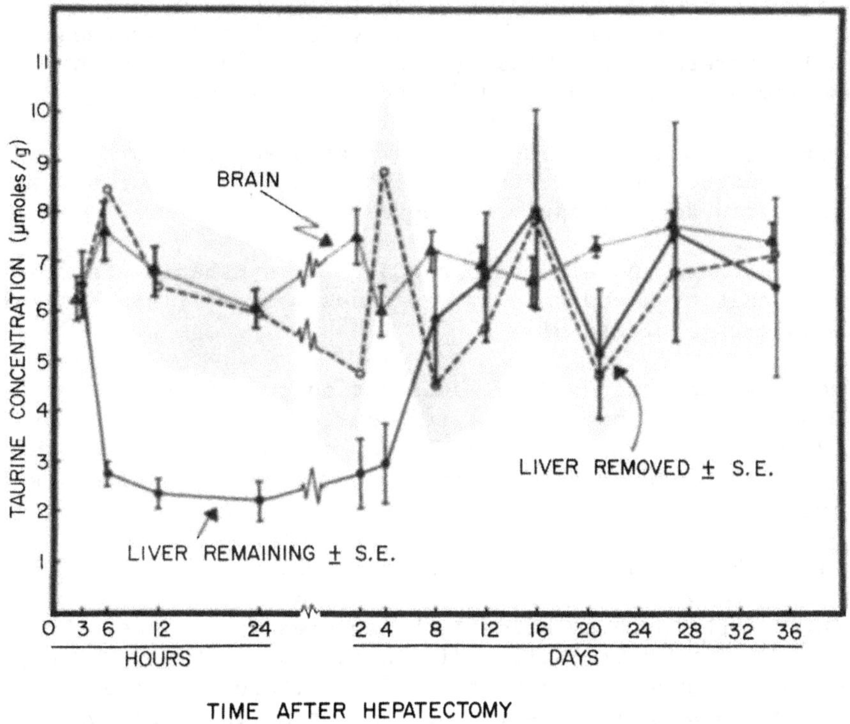

Fig. 2. STURMAN AND FELLMAN. Concentration of taurine in rat liver remaining after partial hepatectomy, in liver removed at operation, and in brain. The shaded area represents ± 1 SE about the mean value of the liver removed at operation. The values obtained for liver remaining after operation, and for brain, are illustrated ± SE. Each value represents the mean obtained from 8 rats.

it a little differently. When we removed the dietary source, there was no increase in synthesis based on CSAD activity. Therefore, you have to be careful, I think, in assessing the endogenous contribution. The apparent increase in synthesis may be due to reduced excretion, in other words, a prolonged half-life allowing it to be reutilized as opposed to a contribution from new synthesis or enhanced synthesis.

HUXTABLE: The evidence that CSA decarboxylase is involved in taurine biosynthesis is all negative. There is really no good positive evidence that it is and the opposing developmental characteristics whereby taurine concentrations decrease while CSAD activity increases is still a problem that has not been answered.

HAYES: In interpreting your data, let's look at it a different way then. What contribution of the build up in percent of endogenous taurine is due to the fact that the kidney doesn't excrete it anymore?

So you could have 100% from biosynthesis simply because synthesis goes on at the same rate and it keeps every damn bit it synthesizes, whereas if it excreted 50% of that you could you see play with numbers and I'm not sure your data did that this morning.

STURMAN: If the cat cannot respond with increased taurine biosynthesis, and indeed doesn't, the cat just loses it's taurine, so the generalization doesn't apply to the cat.

MANDEL: But in those animals which are able to synthesize taurine it looks like that when you have an exogenous source it synthesizes less, like as with cholesterol.

HAYES: The only indirect evidence does not support that concept, since CSAD activity did not change with taurine status in cats or monkeys (12,16). Does the body synthesize less or the same amount and retain more or less by controlling kidney excretion? That is the question I ask.

STURMAN: I think Dr. Rassin has an answer here.

RASSIN: Part of the problem here is the species difference. This is a slide of some work Dr. Malloy did in our lab in New York on puppies fed total parenteral nutrition (TPN) without taurine. The table shows the CSAD activities in these animals (Table 1), while the controls are animals nursed by their mothers. During early development, the TPN-fed, or taurine deficient animals, have very large increases in the activity of CSAD which you can see at 5, 21, and 35 days. At 84 days, it's interesting that the control CSAD activity has increased to about the same as, or maybe slightly more, than that of the deficient animals, so perhaps later on in development we may increase our capacity to synthesize taurine to try and compensate for the lack of taurine in the diet. Now I emphasize this is puppies; not cats, and not man. But there is a response of CSAD to a taurine deficient diet.

HAYES: But Ryan might say that those data may not mean much because they represent enzyme activity (CSAD). Ryan, correct me if I'm wrong, but the data you presented simply say that removal of dietary taurine leads to more of the body taurine pool comes from synthesis. Those are the data. One interpretation could be that, yes, you accelerate synthesis to balance the loss, or you simply cut down excretion and sequester the small amount synthesized for the tissues and reduce the rate or precent of taurine turnover.

HUXTABLE: I agree. Indeed, we have recently obtained information, which will appear in my chapter although I did not discuss it this morning, showing that synthesis of taurine is not increased in taurine-deprived animals. Instead, biosynthesized taurine is excreted more slowly.

TABLE 1

HEPATIC CYSTEINESULFINIC ACID DECARBOXYLASE ACTIVITY IN CONTROL AND TOTAL PARENTERALLY NOURISHED (TPN) PUPPIES

Age (Days)	Control	TPN	Period of TPN (Days)
	(nanomoles CO_2 produced/mg protein/hr ± SD)		
5	3.8 ± 0.7	18.2 ± 8	0 - 5
21	24.0 ± 7.0	143.6 ± 59	10 - 21
35	40.7 ± 13.0	112.3 ± 58	21 - 35
84	158.0 ± 11.0	-	-

The control puppies were suckled by their mother (beagle milk contains 190 μmoles of taurine/100 ml (23)). The TPN fed pups were suckled until the first day of the experimental period. The table is adapted from reference (20).

MANDEL: We measure taurine in urine in protein deprived animals. There is still an excretion of taurine, but this taurine derives from liver and muscle which are destroyed, so as part of the destruction of cells there is elimination of taurine.

STURMAN: In the experiments that I did on B6 deficiency I included a control group plus a control group supplemented with taurine. I would agree with you that the concentrations in tissues are not significantly changed. However, the elimination of the radiolabeled taurine that was in those tissues is speeded up, which is one point that Casey's seeking.

HAYES: Well, the other aspect of excretion, of course, involves bile acid metabolism. In the cat, for instance, taurine is recycled in the bile acids, and if you haven't got much taurine you hang on to it for dear life. Otherwise it is excreted in the feces. So to evaluate balance you must also collect the feces.

MANDEL: Bile salt production depends on what is eaten.

HAYES: True to an extent, but changing from a high fat to a low fat diet doesn't change the dynamics of bile acid metabolism that much. If the calorie intake is constant per day for the rat, it doesn't change bile production markedly.

IRVING: I want to return to the observation of the fall in the excretion of taurine as a result of dietary or metabolic depletion of taurine. Two hypotheses can be proposed: (1) That there is an active change in the renal reabsorption of taurine stimulated by the falling plasma levels, or (2) that the fall in excretion is actually a passive response which results from the existence of taurine stores within the body. When the taurine stores or binding sites are full, then excess dietary taurine spills out into the urine so you see an excretion of excess taurine in the well-fed state. When the animals are depleted of taurine then the stores fall and then there is no more spill over and apparent conservation.

MANDEL: My feeling would be that you have active reuptake of taurine in kidneys, and that this is a normal process which is regulated by the level of taurine in plasma.

GAULL: Unfortunately, that's contrary to the facts. Put this slide back on please. The plasma doesn't fall, the plasma is maintained.

MANDEL: The plasma is maintained because it has recovered.

GAULL: Of course it's recovered, but you can't postulate it falls if it doesn't fall.

VAN GELDER: But aren't there groups of people that are so-called high taurine excretors?

GAULL: I have seen Goodman's data and I don't consider it very good because there is no information about the dietary intake (11).

VAN GELDER: Well, there are a few patients, epileptic patients in whom taurine was injected. These patients definitely cannot retain steady plasma levels and they seem to excrete taurine at a much faster rate than if you do the same procedure with an unaffected population.

MANDEL: We have to go to the next problem - where does taurine go to? Who would answer such a question?

BASKIN: In respect of the central nervous system, I don't think taurine goes to all areas to the same degree. For example, some areas of the central nervous system are behind the blood brain barrier. I think it's maybe inappropriate to compare those areas behind the blood brain barrier in terms of where taurine goes with those areas that are not behind the blood brain barrier. In the heart a similar situation may exist. Different areas of the myocardium take up taurine to different extents. We can't say yet that there's pharmacokinetic non-uniformity. Dr. Huxtable's data suggest it but Ryan didn't make that point whether there were differences between spinal cord, colliculi, cerebellum or cerebrum in terms of the pharmacokinetic data.

CRASS: I wonder if somebody could summarize the subcellular distribution of taurine. Has anyone really ever looked at this intensively? We are talking about transient fluxes of taurine. We can be talking about depletion of a given organellar pool which could be critical to its function in that particular tissue, and yet perhaps not alter the size of other pools. Within the tissue or organ some of the biosynthetic regulatory phenomena that Ryan was talking about a few minutes ago may be quite localized.

STURMAN: We have done quite a few studies on subcellular distribution of taurine in the brain and it's sort of disappointing in one sense, because if you look at the taurine-depleted cat brain versus a normal cat brain there is no difference in the proportion of taurine in the different subcellular pools (24). The proportion is exactly the same. In another study, we did what we thought was a very sophisticated experiment. We delivered ^{35}S taurine to the brain by axonal transport from the retinal ganglion cells. At the same time, we delivered ^{3}H taurine by intracerebral injection. The subcellular distribution of those two isotopes was identical (14). We also looked at the distribution during development and that's the only time when there is any change in the proportion (22). I've done some experiments recently along the lines that Ryan showed today where I labeled female rats to constant specific activity with ^{35}S taurine given intraperitoneally before mating, and then I've done subcellular fractionation of the brains of progeny following a subsequent injection of ^{3}H taurine. For the first time I have found differential labeling of the P_2 fraction. So I think that in early gestation there is a portion of the taurine in the P_2 fraction - presumably synaptosomes but it could be mitochondria - that is not accessible to taurine injected into the animal after this (30).

MANDEL: We must expect that the problem will be rather difficult because with GABA we have two quantitatively very different pools. It is probably the same story with taurine. So far we have no approach for investigating specifically a functional pool of taurine that may constitute only 10% of the total taurine.

SGARAGLI: Is there any evidence that taurine is a substrate for oligopeptides? There is a group in Hungary that is working on a purified dipeptide.

BARBEAU: Yes, can taurine serve as a building block for peptides?

MANDEL: This is not an important aspect of taurine. If we looked, we could find all kinds of brain peptides. But it might be that there is somewhere a very small fraction of taurine peptides, and it is an open question whether these are the type of peptides in which GABA is involved for which we have no physiological explanation.

VAN GELDER: Could we ask Dr. Oja to comment on that?

OJA: I don't have experience of my own, but I have seen work done by an Hungarian group (J. Somogyi, personal communication) and Lähdesmäki's group in Finland (19). They have found a number of peptides which contained taurine, and I think that Dr. Lähdesmäki also demonstrated that brain tissue is able to synthesize these peptides starting from simple compounds, such as amino acids and taurine (18).

GAULL: It seems to me there were some Japanese workers, 8 or 10 years ago, who identified taurine-containing peptides in the brain, and that John Blass, when he was in McElwain's lab. 10 or 12 years ago, also identified some peptides in the brain, so I don't think that this is a new observation.

HUXTABLE: I remember one report by Reichelt and Kvamme in 1973 (26), but there was never any follow up to it. With this work of Feuer's in Hungary on γ-glutamyl taurine, has anyone seen any critical reports? The only material I've seen on the mammalian studies has been press-release type of material rather than scientific communication, although I know of one accessible report on amphibians (6).

GAULL: Oh! Yes, that's right!

HUXTABLE: He spoke about this at the A. N. Richardson Symposium in Philadelphia in 1979.

VAN GELDER: So do the transpeptidases enter into this role?

FELLMAN: Possibly, yes.

SCHAFFER: What is interesting about γ-glutamyl taurine is that it has radioprotective effects 100-1000 fold more potent on a concentration basis than taurine (3). So that means that the intracellular concentration of the peptide can be much lower than that of taurine.

HUXTABLE: But that's not very convincing because taurine is such a very poor radioprotective agent (2). Cysteamine is a much more potent radioprotective agent, as are simple derivatives of it, such as 2-aminoethylisothiourea.

BASKIN: We have injected γ-glutamyl taurine, a large dose, i.p. into rats. Initially, for 2 days, there is a reduction in serum T_3 and T_4. Subsequently, there is an increase in T_3 and T_4 over control (at 2 weeks).

FELLMAN: Have you observed γ-glutamyl taurine as he has?

BASKIN: He made many claims for that compound. I doubt if anyone has performed all the experiments that...

MANDEL: It will be difficult to come to an answer about γ-glutamyl taurine. We will keep it in mind that it may be something interesting and we will see in two years time if people have really found it interesting. We can discuss the next question, which seems to be a basic one. How are taurine transport sites regulated? We find a high affinity uptake of taurine in neuronal cells as well as glial cells, *in vitro*. I should say that we found also in our lab the same thing for choline which means that the presence of high affinity uptake doesn't mean that it is neurotransmitter. Now who would like to comment on this problem? Herminia?

PASANTES-MORALES: I would say that the high affinity uptake of taurine is probably not the most important mechanism in talking about the function of taurine. Probably the low affinity uptake of taurine is more important. We cannot disassociate this question concerning the origin of taurine from that of the functional role of taurine. Since I think that taurine is involved in a basic slow mechanism, I think that low affinity uptake is as least as important as high affinity uptake. But there is a difference concerning the synthesis and transport of taurine and GABA. For instance, in brain the decarboxylases acting on taurine and GABA are probably the same. Transport of taurine and GABA are definitely different; and so we have two different transport systems for the two amino acids. Probably the high affinity transport for taurine in brain is related to a synaptic function for taurine, not necessarily that of a neurotransmitter. For the rest of the cell, there may be a kind of a stabilizer effect of taurine on membranes: that taurine is moving in and out of the cell continuously in a sort of protection to membranes. I think that the low affinity uptake would be as important as the high affinity.

MANDEL: I think that we cannot discuss the problems in terms of what is important because in the living animal everything is important. Everything is necessary for some such function or another. We found high affinity uptake of taurine in cell populations without synapses. In addition, we shouldn't think that the low affinity uptake is something which is of minor importance, because, for instance, glutamine uptake by neurons to produce GABA is by a low affinity uptake. So that it is not a criterion. We have discussed the central nervous system; what about regulation in other tissues?

HUXTABLE: Taurine levels in the heart are regulated by transport, and not by synthesis within the heart, and that transport is stimulated by cyclic AMP (13). This is unique to β-amino acids, and this effect is only seen in tissues that contain β1 receptors, such as the heart and salivary gland of the rat, but it is not seen in organs that have β2 type adrenergic receptors. I believe Dave Klein has found similar interactions in pineal.

KLEIN: Just the opposite, release!

HUXTABLE: Cyclic AMP stimulation of release is rather hard to understand, because for uptake you're moving taurine against an incredible concentration gradient, so some type of active transport system is needed. Why you need this for release, I don't know because a simple change in membrane permeability would result in a large amount of taurine released.

VAN GELDER: Are you talking about release or secretion?

KLEIN: It looks like release, or some sort of an efflux. There is no evidence of calcium dependency.

FRANCONI: We studied the taurine uptake into cardiac sarcolemma vesicles isolated from guinea-pig ventricles. It was strictly sodium-dependent. However, the incubation of these membranes in a calcium, magnesium and potassium free medium produced a modification in the kinetic of taurine uptake (8).

CHESNEY: I want to comment about the renal accumulation of taurine because, of course, it is also sodium dependent as we can see in the slide (Fig. 3), where we are looking at brush border membrane vesicles. If we preincubate our vesicles in a mannitol-containing medium and then put them in a medium containing sodium we see accumulation of taurine into the vesicles. If we put them into a medium containing mannitol, we don't see this active accumulation, and if we initially incubate them with sodium and then put them into a sodium-containing medium we again don't see the accumulation. Therefore it appears that the renal handling, this is by the high affinity system, is sodium-dependent and we had shown at the last meeting in Tucson that we saw an effect of dibutyryl cyclic AMP on the enhancement of the accumulation by kidney cortex (5).

MANDEL: Is uptake also energy dependent?

CHESNEY: Yes, it's very much energy dependent, and is inhibited by hypoxia, anoxia or other metabolic inhibitors.

HUXTABLE: Do you know whether this action of dibutyryl cyclic AMP is an adrenergic one, and whether you are dealing with β1 or β2 receptors?

CHESNEY: I don't know that. That's an intriguing possibility.

SCHAFFER: This afternoon I will be discussing the transport of taurine into isolated reconstituted vesicles of heart sarcolemma. We find that this process is also sodium dependent, and has characteristics similar to the process described by Ryan Huxtable (13) in isolated perfused heart. The sarcolemma contains two taurine binding proteins; one that has a high affinity for taurine and the other one has a low affinity. We feel that the high affinity

a

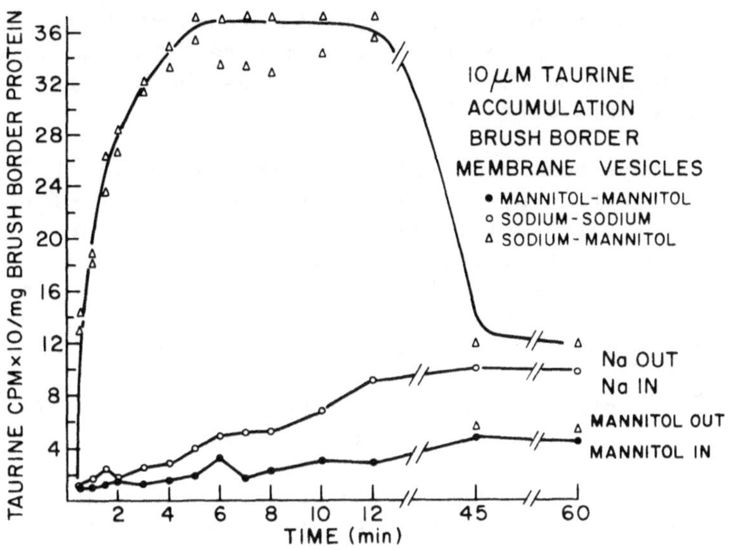

b

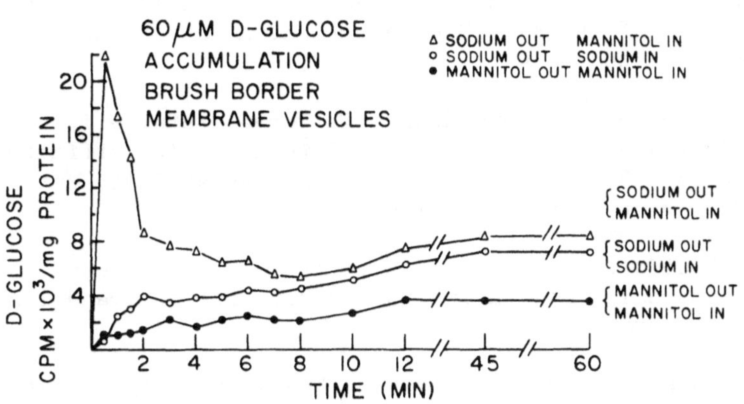

Fig. 3. CHESNEY. The accumulation of taurine (Fig. 3a) and glucose (Fig. 3b) by isolated renal tubule brush border membrane vesicles of rat. Note that both taurine and glucose uptake are dependent on external sodium and a sodium gradient going inwards. The abrupt overshoot found using glucose -- over within 90 sec -- is not seen for taurine uptake where increased uptake lasts for 6 to 12 min.

protein is related to taurine uptake while the low affinity protein is probably involved in altering calcium kinetics in the heart. This means that the so-called functional protein for taurine would have a low affinity, not a high affinity for taurine.

CHESNEY: I think that's probably true in all biologic systems.

Certainly in terms of the quantity that's transported, the low affinity system is going to transport far more. This is true for sugars, amino acids and other organic solutes, both in the gut and the kidney where active transfer occurs (27).

KLEIN: I've always been confused about the search for the high affinity taurine uptake system in many tissues because of the high concentrations of circulating taurine that are available. It seems to me that the high affinity system may in fact not really play an important role. The low affinity system has a tremedous capacity, and in fact its Km is such that it could probably handle most of the transport. Maybe I've been wrong.

MANDEL: We know nothing about the extracellular space. You don't know how much taurine you have in extracellular space.

CHESNEY: The Km for renal handling of taurine is somewhere in the range of 0.15 mM in the rat, and this is roughly what we find in the plasma level.

VAN GELDER: I don't agree with this division between high and low affinity. These Km's are derived from Michaelis-Menten kinetic analysis, and half the time Michaelis-Menten conditions were not obeyed when the studies were done!

MANDEL: No, no, no, we have done this very carefully in tissue culture, and there is no doubt that there is a high affinity system. The Km is about 20 μM, so it really is a high affinity uptake. It exists. Now, what is the function?

FELLMAN: Shouldn't we be cautious in the use of the word transport? Aren't these studies done using radioactive tracers, and isn't the word exchange more accurate?

SCHAFFER: This clearly depends upon the system. In our studies, we used isolated vesicles, in which no taurine was present inside the vesicles. Any accumulation must represent uptake.

FELLMAN: That's a critically important system. Bob Cutler at the last Tucson meeting showed that there was release of taurine as the taurine was taken up as radiolabeled taurine, and he raised the question of how accurate or how precise is the word transport or uptake.

MANDEL: All this can be calculated very carefully and doesn't make a big problem. I think that we are aware already as to what we have to do to know all about transport. So let's move on. Are rates of turnover important? As you know, the data in the literature including from our own laboratory indicate that the turnover rate is rather low. It cannot be compared, for instance, with norepinephrine or serotonin. However, we should keep in mind that perhaps 5% or even less of taurine can have a very high turnover rate, which we are unable to measure at present. Who would like to comment about the turnover rates, either in brain or in other tissues?

PASANTES-MORALES: I think that turnover rates of taurine are part of the function in each different tissue. Which is rather interesting because, as was found some years ago (28), excitable tissues have slower turnover rates. That could mean that in these tissues taurine is involved in a basic function peculiar to these tissues, and that the taurine is not very rapidly used, but it can be either slowly used or used several times for its unknown role.

MANDEL: But how can you distinguish whether you have two pools of taurine? Errors in turnover experiments are of about 10% and if you have a fraction which has a high turnover rate comprising about 5% of the total taurine you can never find it. It's impossible to make the calculation. We tried it. It's impossible. So we need, perhaps, to find a model in which we can exclude, or increase one of the pools.

PASANTES-MORALES: John Sturman has shown that taurine influx changes in the cats on a taurine-free diet (31). So he has shown a model in which taurine can...

MANDEL: No, but here probably taurine plays a role of cellular constituent. This is the pool which is quantitatively high, but we don't know if we have a small pool of taurine doing something else. We cannot show it! We can't exclude it, we cannot include it.

PASANTES-MORALES: But with catecholamines there have been shown to be multiple pools, in which newly synthesized or newly taken-up norepinephrine or dopamine are released preferentially (25). If you do experiments like these with taurine you can follow exactly the same line of investigation to solve the question.

MANDEL: That is because these agents do not have a cellular role, or only a slight one, compared with the neurotransmitter and messenger role. With GABA, if the quantity of GABA which is metabolized is compared with the quantity which is used as a neurotransmitter, it is less than 5%. If the GABA shunt is activated then it will be less than 1%, and in that situation the neurotransmitter pool can never be discovered.

HABER: Mr. Chairman, is this really an important question at this stage of our knowledge of taurine, where we really don't know what taurine pools are coupled to? Isn't it a little premature to get involved in discussions of taurine turnover? What are they going to tell us when we don't know what taurine is doing? It might be more fruitful to address ourselves to issues other than whether it's a high or low affinity system. It's probably not important what these transport rates may be coupled to. We perhaps ought to be focusing on some global issues which have not been emerging from this discussion, which I would like to see emerge and that is what might taurine really be doing? Are there such things as rate limiting enzymes of taurine biosynthesis? Is taurine synthesis important in the maintenance of intracellular pools? These are the questions I would like to see answered.

MANDEL: Good. *(Laughter)* You would like the answer to a different question. But it is still extremely important to know how the turnover of taurine occurs.

HABER: I hate to disagree with you, Mr. Chairman, but if we really don't have a feeling as to what is the importance of the biosynthetic pathways or degradative pathways to the maintenance of taurine levels, how can we intelligently discuss taurine turnover? This has been clearly demonstrated for acetylcholine, norepinephrine and GABA. We want to have some feeling for the importance of biosynthetic and degradative pathways.

VAN GELDER: Let's reverse the argument: Why is it that the tissues have to maintain such very steady levels of taurine? The issue is, why is it that this slowly metabolized amino acid is so carefully conserved in all the tissues where it is present?

KLEIN: You're restating the question.

MANDEL: This is one of the basic questions to which there is no answer.

RASSIN: Could I add fuel to this a little bit. When you look at these deficient cats why the hell aren't they globally affected with enormous pathology when you get rid of 90% of the taurine in the brain, 90% of the taurine in the heart - why is only the retina affected, if taurine is so important?

STURMAN: And the tapetum.

RASSIN: You're enormously affecting taurine pools with minimal obvious pathology if it's such an important compound.

STURMAN: On this point, I think we've got to pick up each little piece of information as we find it, and we know specifically in rats and we know in humans that the taurine you put in the diet can all be accounted for in the urine. We've gone up to 9 grams a day in some studies in man and it was all excreted in the urine, provided their taurine intake was adequate before (32). In rat studies that I did many years ago, I collected all the urine, all the feces, and found that the more taurine we put into the rat the more you get out in the urine (29). The tissues won't allow the concentrations to increase. The only exceptions I found were that plasma did go up a bit and that liver will go up. Liver is the only tissue which shunts taurine off somewhere else by conjugation.

MANDEL: I should stress that when you have equilibrium with input and output being the same it doesn't mean that it is of importance, because it happens for all amino acids. You have a quantity of nitrogen. You give it to a man or to an animal, and after four or five days you have again an equilibrium.

STURMAN: I'm saying that it is an important phenomenon.

MANDEL: It's not important.

STURMAN: I'm saying that it's important that those tissues will not allow any more taurine into them, as Nico was saying.

GAULL: Rephrased, the equilibrium is very carefully maintained.

CHESNEY: In non-mammalian species, such as the skate, taurine seems to serve as an osmoregulator. I wonder if anybody thinks that this has any pertinence to mammalian tissues (7). There are some data, they are very sketchy, that in young rats made hypernatremic, taurine may be important in regulating cell size and I wondered what other people think (4).

BASKIN: Fugelli in Norway has found in a number of fish that taurine is very important not only in red blood cells but in heart cells (10). He originally published that in certain brain cells taurine is important as an osmoregulator (9).

CHESNEY: That may be one of the reasons that we have these relatively stable levels of taurine.

MANDEL: Perhaps Dr. Gaull can give you an answer. In his lymphocytes which survived there is a variable level of taurine.

GAULL: They have essentially none. They survive but they don't do as well but these are very preliminary experiments. We have no idea as yet what happens to the morphology of the cell. We only know that they will be passed from generation to generation. They will survive, but of course they are in a medium where they may be obtaining other substances that are substituting for taurine in maintaining osmolarity.

HABER: Claude Baxter's work shows that clearly taurine is one of the amino acids, both in plasma and in brain, which responds in a very significant manner to changes in osmolarity. I think taurine may well have a major role to play in the maintenance of cellular osmolarity.

MANDEL: Let's go now to a very important question, at least in the brain. Is taurine stored in particular subcellular organs? Someone?

STURMAN: The only situation in which I've been able to locate a nonexchangeable, nonaccessible pool of taurine is in labeling from conception (30). By doing this, I've been able to find a pool that's not accessible or less accessible later in life to exogenously supplied taurine.

MANDEL: How about the data published on synaptosomal taurine? Are there some comments about that?

VAN GELDER: You will have to have a synaptosomal marker. This is, in a sense, why, if you look at figure 6 of my article (this book), I chose to compare GABA and aspartic acid, which I think is a synaptosomal marker. If you use that as a marker and then look at taurine you could conceivably, or I can make a case for it, although I'm not saying it is so, you can make distinctions between the pools in the various cells. Now whether in the end this turns out to be true or not, I don't know, but if, for example, there is an inverse correlation between taurine and GABA, you might start supposing that there are two structures in which excitation is mutually exclusive. If one is stimulated the other is not stimulated, and so you could play around with that sort of scheme. But I think that in all these studies you will have to have some type of marker for each individual cellular component.

MANDEL: There may be a discrepancy, which is that during the isolation of subcellular particles there is a loss of all kinds of compounds.

RASSIN: No, I don't think that's really true.

MANDEL: So it is difficult to give a quantitative evaluation.

RASSIN: I think that everybody who has done subcellular studies has found essentially the same thing. Even if you look at the areas in these fractions where the taurine would be released if it was being redistributed during the separation after *in vivo* administration of (^{35}S)taurine you don't find labeled taurine in those areas of the gradients, and if you measure taurine added to the homogenates you don't find it being accumulated in the synaptosomes during preparation when done at 4°C. I think the question is; are those pools of taurine that are isolated in synaptosomes contained in synaptic vesicles?

MANDEL: Are these vesicles taking up taurine?

RASSIN: But you don't want uptake during preparation of the fractions because that is how you get redistribution. The question is, is there redistribution during preparation? If the fractionation is done at 4°C, there isn't any redistribution. The people who get problems are people who don't do the fractionation at 4°C. The real synaptosomal preparers, the Whittakers and the DeRobertis' (GAULL: And the Rassins) will argue that if you don't have frostbite on your fingers during preparation you haven't done it correctly, and that is when you get into trouble with redistribution.

VAN GELDER: What happens the first time you homogenize?

MANDEL: Yes, already.

PASANTES-MORALES: Does anybody know about subcellular distribution in muscle? Is taurine related with some particular subcellular structure? Can you localize taurine by audioradiography?

GAULL: You have trouble with audioradiography with taurine because of its solubility and its small size.

STURMAN: Even that is exaggerated. We've looked at retina and you really can localize it, as has also been done with GABA.

BASKIN: I think we should hear from Dr. Oja. He's done an awful lot of work on synaptosomal fractions.

OJA: That's true. We have done very extensive studies collecting maybe 30,000 numbers showing the distribution of taurine in different cell fractions and in different areas in the bovine brain (17,21). The impression was that the distribution of taurine in the bovine brain in different areas was not very heterogenous but the distribution at subcellular levels was very heterogenous indeed. It appears that taurine perhaps is most highly enriched in these subcellular fractions which contain membranes. My impression was that taurine was associated or bound to those membranes, and therefore if there is some kind of physical binding to the membranes taurine is not lost during isolation to such an extent as other amino acids.

BARBEAU: You have not differentiated between one type of membrane and the other? Have you localized it to one form of membrane or to a series of membranes?

VAN GELDER: Like synaptosomal membranes?

BARBEAU: But there are other membranes also when you look at the synaptosome. You have all sorts of particles like mitochondria and they have their own membranes and so on. When you're talking about membranes in the synaptosomal fraction of a Whittaker type separation, you can't say this is synaptosomal membrane. You could be talking about mitochondrial membrane.

OJA: Yes, but taurine is enriched in all membrane fractions we have isolated.

KLEIN: Ryan, you were saying earlier that 85% of the taurine that is in the tissue is released on homogenization. Is that what everybody finds? Most of the taurine in cells is free, therefore?

HUXTABLE: That's not an accurate deduction necessarily. It only says that the energy of binding within the cell is less than the energy of disruption.

KLEIN: Yes, but it's not tightly bound, O.K.

PASANTES-MORALES: I am convinced of the importance of taurine in tissues, so I was very disappointed by the response obtained by people working on a taurine-free diet in which we see depletion of 90% of the taurine without any behavioral, neurological or physiological changes in the animal. Let's put the question to experts. Taurine depletion takes a long time to be achieved, so do you think that tissues have enough time to adapt to the depletion and that any role played by taurine could be adjusted or adapted to so the small amount that remains could be providing normal conditions? Have we more accurate models of depletion? I know, of course, of the work of Ryan with guanidinoethyl sulfonate, but in his case I always wonder if the transport inhibitor is not also an agonist doing the same thing as taurine.

STURMAN: We discussed this in England in 1978, when I first heard of your muricidal rats. The olfactory bulb is the one area of brain that resists taurine depletion, but it's interesting that the concentration of GABA goes up about the same amount that the taurine goes down (1,30). So there's a possibility that if it were not for this compensation we might end up with more aggressive cats.

MANDEL: But still you obviously can't say there is a difference of taurine in the brain after at least four weeks of a taurine deficient diet if the brains still have the normal amount of taurine.

GAULL: I'd like to make a point from a nutritional point of view. The science of nutrition grew up looking at acute changes in things like nitrogen balance. There's an increasing amount of evidence now that there are certain kinds of nutritional changes the manifestations of which one sees only over a very long time span, such as the changes that one might expect with cholesterol and arteries, if indeed those are conjoined in a way the American Heart Association wishes us to believe. But in any case, these things take place over a very long time span and we're used to thinking of the changes taking place over a very short time. All these observations on taurine are going to fall into place only when we find what the central cellular role is.

SCHAFFER: There are a couple points which should be made regarding Herminia's comment. First of all, in a tissue such as the heart, there's so much ATP present that one can deplete it substantially and still saturate the ATPases (15). Similarly it may be necessary to deplete taurine by more than 90% to see an effect. Secondly, perhaps we have to stress tissues such as the heart or brain following taurine depletion. In the unstressed situation no effect would be observed, while we would be able to detect something in the stressed animal.

STURMAN: We were lucky I think with Casey's original observation. Photoreceptor cell degeneration is a massive pathologic change. We shouldn't be looking for animals to come down with seizures or cardiac problems. Supposing there's a 10% memory loss because the membranes in the hippocampus are affected. Are any of us doing studies that will pick that up? No!

GAULL: Or taking Schaffer's point: There's an excellent example recently. When I was in school, people were having symposia about vitamin E the way we're having symposia about taurine. Now we know that it's a potent antioxident, and that sometimes you don't see the effects of vitamin E deficiency until, from a dietary point of view, you stress an infant with a high concentration of unsaturated fatty acids. Then, if you add extra iron onto that you will get hemolytic anemia. You won't see any of that unless you bring those other factors to bear. So Steve raised a very important point that we may be expecting, because of old nutritional models, much more dramatic changes than we are going to get until we find exactly how to aim the vectors.

MANDEL: So that what we can conclude that there is a relationship between the need and the availability of taurine. In cats, there's a great need; in other animals there is a lower need for taurine so you have to wait for a longer period of time before you see a depletion.

HAYES: To get back to Baskin's question this morning, if we want to use one species to get the answers faster, we should all use the cat.

VAN GELDER: I disagree. *(Laughter)*

MANDEL: No, first let's go to a very important question: what are outstanding or unanswered questions in taurine metabolism? There is one question obviously outstanding with regulation. When I think about regulation, is the problem, number one, a function of the availability of exogenous taurine, or number two, as a function of cysteine sulfinic acid transaminase, in other words, availability of precursor? Now who would like to comment on this topic?

RASSIN: You know, we do a lot of comparisons between taurine and GABA and their metabolic pathways, and I have a very hard time trying to conceive regulatory pathways with taurine when we know so little about how much substrate is available in the body. Can anybody here really give a number for how much cysteine sulfinic acid is actually in liver or in brain? These values are almost nonexistent in the literature, and yet when we talk about regulation of GABA, we know so much about glutamate and glutamate concentrations are so high compared to cysteine sulfinic acid. I don't think you can start talking about regulation until you know the substrate concentrations.

FELLMAN: Let's look at what we can learn from classical biochemistry. If you look at tyrosine metabolism, for example, there's substantially no steady-state concentration of parahydroxyphenylpyruvate, homogentisic acid and all the fragments that finally end up as CO_2 and water. I think it's fair to conclude that the rate limiting enzyme, which for tyrosine is tyrosine aminotransferase, is the one that we should be looking at for taurine biosynthesis. My view right now is that it is cysteine dioxygenase which is the overall rate limiting enzyme in taurine biosynthesis and we shouldn't expect to see any steady-state concentration of the intermediates. In some experiments that I participated in with John Sturman, we found that partially hepatectomized rats show an elevation in hypotaurine levels after four hours, and yet CSA decarboxylase, the enzyme that allegedly produces hypotaurine, was half the concentration that is normally present. In fact the levels were unrelated to the appearance of hypotaurine. So it seems to me that the thing we should be focusing on is the rate limiting enzyme, and that's cysteine dioxygenase.

STURMAN: We know almost nothing about the last to final step. Why is it we almost never see hypotaurine in tissues? Dr. Oja and Dr. Fellman have been looking for this step - the so-called hypotaurine oxidase - for a long time, and they haven't come up with an answer satisfactory to themselves.

MANDEL: Because it is extremely difficult to isolate such a small fraction of hypotaurine when there is such a huge amount of taurine. We are now working on this problem and it is a very very hard problem.

HUXTABLE: There are major problems with all pathways of taurine biosynthesis (Fig. 4). All pathways that can be seriously considered except one go through cysteine sulfinic acid, and the one that does not is the one from cysteine by some peculiar pathway to cysteamine and then by a well characterized enzyme to hypotaurine.

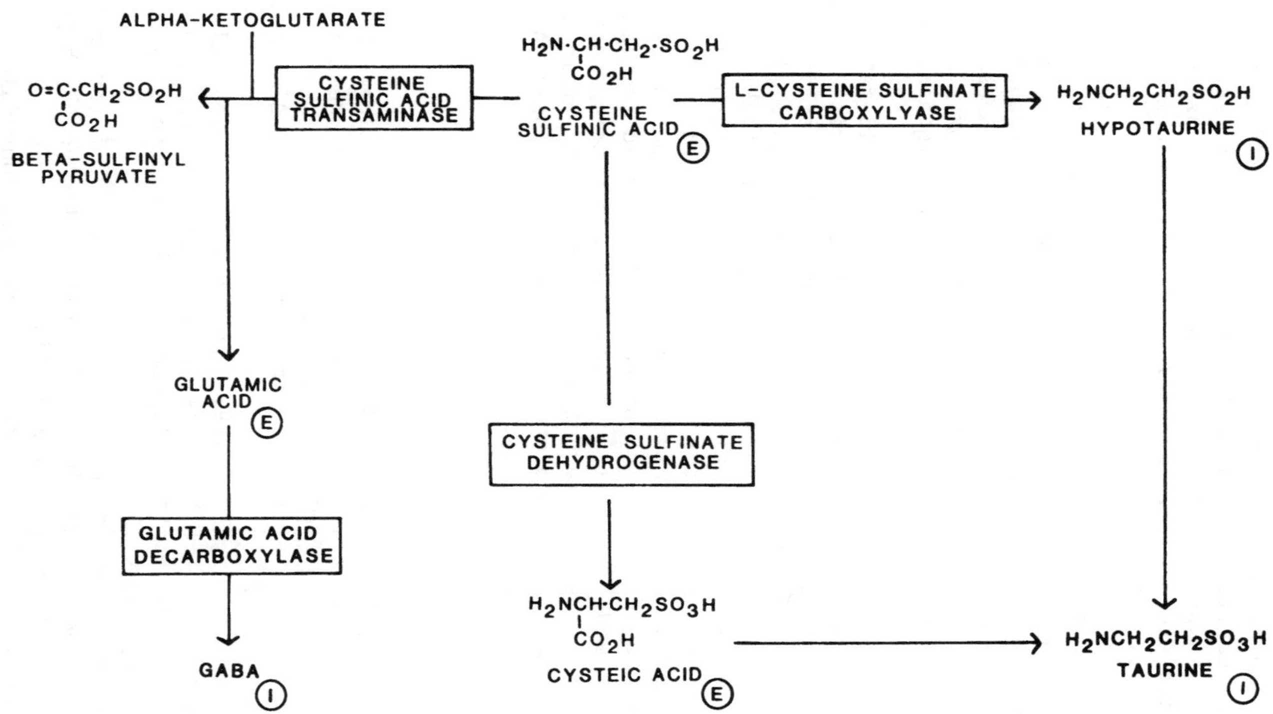

Fig. 4. HUXTABLE. Significance of cysteine sulfinic acid metabolism in brain excitability.

E = excitatory amino acid. I = Inhibitory amino acid.

The last step of hypotaurine to taurine is mysterious, as I think Dr. Oja and Dr. Fellman will discuss later in the meeting. Another pathway involves going through cysteine sulfinic acid to cysteic acid, via the poorly defined enzyme cysteine sulfinate dehydrogenase, and then on to taurine by another well characterized decarboxylase. Finally, CSA can be decarboxylated to hypotaurine. In general, there are mysteries about all of these paths but the enzymes that have been studied are present in general in both brain and in the liver, although for some species this may not be true. Another enzyme that acts on CSA is CSA transaminase. I would like to suggest that in the liver enzymatic pathways available for CSA fulfill basically a metabolic function, and that they are high capacity pathways in that the major route of sulfur metabolism in mammals goes through cysteine sulfinic acid via the transaminase (Fig. 4). So in the sense, in the liver, cysteine dioxygenase is unlikely to be a regulatory enzyme for CSA and the synthesis of taurine because its product fulfills such an important metabolic function for sulfur. I would also put it up for consideration that in the brain these same enzymes metabolizing CSA comprise a low capacity system that is involved in the regulation of excitability rather than having a metabolic role. I say this because all the substrates and products of these various enzyme reactions are neuroactive: CSA, cysteic acid and glutamic acid are excitatory; hypotaurine and taurine are inhibitors. Glutamic acid can go on to GABA which is inhibitory. So the enzymes acting on this one substance, cysteine sulfinic acid, are quite strategically placed for regulating excitability, and I think it's worthwhile considering that although these pathways are present in two organs, the brain and liver, that they may serve completely different functions: A metabolic versus a regulatory function.

REFERENCES

1. Anderson, P.A., Baker, D.H., Corbin, J.E., and Helper, L.C., 1979, Biochemical lesions associated with taurine deficiency in the cat, J. Anim. Sci., 49:1227-1234.
2. Z.M. Bacq: Chemical protection against ionizing radiation, C.C. Thomas, 1965.
3. Benkö, G., and Feuer, L., 1979, Investigation of the radiation protective effect of Litoralon (gamma-L-glutamyl taurine) in experimental animals, Sixth International Congress of Radiation Research, Tokyo, May 13-19.
4. Chan, P.H., and Fishman, R.A., 1979, Elevation of rat brain amino acids, ammonia and idiogenic osmoles induced by hyperosmolality, Brain Res., 161:293-301.
5. Chesney, R.W., Jax, D.K., Scriver, C.R., and Mohyuddin, F., Taurine transport in mammalian kidney, in: "Taurine in Neurological Disorders," R. Huxtable and A. Barbeau, eds., Raven Press, New York (1978), pp.73-94.

6. Feuer, L., Torok, L.J., Kapa, E., and Csaba, G., 1978, The effect of gamma-L-glutamyl-taurine (litoralon) on the amphibian metamorphosis, Comp. Biochem. Physiol., 61C:67-71.
7. Forster, R.P., and Goldstein, L., 1979, Amino acids and cell volume regulation, Yale J. Biol. Med., 52:497.
8. Franconi, F., Martini, F., Manghi, N., Galli, A., Bennardini, F., and Giotti, A., 1981, Uptake of ^{3}H-taurine into myocardial membranes, Biochem. Pharmacol., 30:77-80.
9. Fugelli, K., 1967, Regulation of cell volume in flounder (pleuronectes flesus) erythrocytes accompanying a decrease in plasma osmolarity, Comp. Biochem. Physiol., 22:253-260.
10. Fugelli, K., and Zacharussen, R.E., 1976, The distribution of taurine gamma-aminobutyric acid and inorganic ion between plasma and erythrocytes in flounder (platochthep flesus) at different plasma osmolarities, Comp. Biochem. Physiol., 55A:173-177.
11. Goodman, H.O., Connolly, B.M., McLean, W., and Resnick, M., 1980, Taurine transport in epilepsy, Clin. Chem., 26:414-419.
12. Hayes, K.C., Stephan, Z.F., and Sturman, J.A., 1980, Growth depression in taurine-depleted infant monkeys, J. Nutrition, 110:119-125.
13. Huxtable, R.J., Chubb, J., and Azari, J., 1980, Physiological and experimental regulation of taurine content in the heart, Fed. Proceedings, 39:2685-2690.
14. Ingoglia, N.A., Sturman, J.A., Rassin, D.K., and Lindquist, T.D., 1978, A comparison of the axonal transport of taurine and proteins in the goldfish visual system, J. Neurochem., 31: 161-170.
15. Katz, A.M., 1970, Contractile proteins of the heart, Physiol. Rev., 50:63-158.
16. Knopf, K., Sturman, J.A., Armstrong, M., and Hayes, K.C., 1978, Taurine: an essential nutrient for the cat, J. Nutrition, 108:773-778.
17. Kontro, P., Marnela, K.-M., and Oja, S.S., 1980, Free amino acids in the synaptosome and synaptic vesicle fractions of different bovine brain areas, Brain Res., 184:129-141.
18. Lähdesmäki, P., and Airaksinen, K., 1980, Demonstration of biosynthesis on N-acetylaspartyl peptides in mouse brain homogenates by autoradiography on thin-layer chromatography plates, Acta Univ. Oul. A 97, Biochem., 29:117-121.
19. Lähdesmäki, P., Airaksinen, K., Vartiainen, M., and Halonen, P., 1980, Characterization of two synaptosomal peptides in calf brain, Acta Chem. Scand., B34:343-348.
20. Malloy, M.H., Rassin, D.K., Gaull, G.E., and Heird, W.C., 1981, Development of taurine metabolism in beagle pups: Effects of taurine-free total parenteral nutrition, Biol. of the Neon., in press.
21. Marnela, K.-M., Kontro, P., Pitkänen, R.I. and Oja, S.S., 1980, Free amino acids in synaptic subfractions of bovine brain, Acta Univ. Oul. A 97, Biochem., 29:11-16.

22. Rassin, D.K., Sturman, J.A., and Gaull, G.E., 1977, Taurine in developing rat brain: Subcellular distribution and association with synaptic vesicles of (^{35}S)taurine in maternal, fetal and neonatal rat brain, J. Neurochem., 28:41-50.
23. Rassin, D.K., Sturman, J.A., and Gaull, G.E., 1978, Taurine and other free amino acids in milk of man and other mannals, Early Human Develop., 2:1-13.
24. Rassin, D.K., Sturman, J.A., Hayes, K.C., and Gaull, G.E., 1978, Taurine deficiency in the kitten: Subcellular distribution of taurine and (^{35}S)taurine in brain, Neurochem. Res., 3:401-410.
25. Raiteri, M., and Levi, G., Release mechanisms for catecholamines and serotonin in synaptosomes, in: "Reviews of Neuroscience," E. Ehrenpreis and I. Kopin, eds., Raven Press, New York, Vol. 3 pp.77-130.
26. Reichelt, K.L., and Kramme, E., 1973, Histamine-dependent formation of N-acetyl-aspartyl peptides in mouse brain, J. Neurochem., 21:849-859.
27. Scriver, C.R., Chesney, R.W., and McInnes, R.I., 1976, Genetic aspects of renal tubular transport: Diversity and topology of carriers, Kidney Int., 9:149-171.
28. Spaeth, D.G., and Schneider, D.L., 1967, Taurine synthesis, concentration and bile salt conjugation, Proc. Soc. Exp. Biol. Med., 147:855-859.
29. Sturman, J.A., 1973, Taurine pool sizes in the rat: Effects of vitamin B_6 deficiency and high-taurine diet, J. Nutr., 103: 1566-1580.
30. Sturman, J.A., 1981, Origin of taurine in developing rat brain, Developmental Brain Res., in press.
31. Sturman, J.A., and Hayes, K.C., The biology of taurine in nutrition and development, in: "Advances in Nutritional Research, H.H. Draper, ed., Plenum Publ. Corp., Vol. 3, pp.231-299.
32. Sturman, J.A., Hepner, G.W., Hofmann, A.F., and Thomas, P.J., 1975, Metabolism of (^{35}S)taurine in man, J. Nutr., 105:1206-1214.
33. Sturman, J.A., Rassin, D.R., Hayes, K.C., and Gaull, G.E., 1978, Taurine deficiency in the kitten: Exchange and turnover of ^{35}S taurine in brain, retina, and other tissues, J. Nutr., 108: 1462-1476.

SESSION II

ARE THE PHARMACOLOGICAL ACTIONS OF TAURINE RELATED TO ITS PHYSIOLOGICAL FUNCTIONS?

Moderator: Kinya Kuriyama

KURIYAMA: The first question is what is the relationship between taurine and calcium? I will show two slides on calcium and taurine in the heart. The first figure shows some data on sarcolemmal calcium binding (Fig. 1). As you can see here, we preincubated isolated sarcolemmal membrane with taurine and subjected it to the binding assay for calcium. We definitely get an increase in calcium binding. When we analyzed these results kinetically, some increases of affinity of calcium for binding sites were noted. We got similar results in synaptosomal membrane from the brain (Fig. 2). So let's start the discussion by asking what has happened between the taurine and calcium on these membranes. Does anyone have a comment on this type of interaction of taurine and calcium?

SCHAFFER: We have done a number of studies on this relationship in the heart. According to our evidence hypotaurine is more potent than taurine in increasing sarcolemmal calcium binding (7). Hypotaurine also interacts more strongly with the membrane than does taurine. Thus, we feel that taurine is causing these changes by interacting with the protein we referred to as component I (Schaffer et al., this book). It remains to be determined whether or not the increase in the amount of calcium bound to the membrane regulates calcium uptake by the cell.

HUXTABLE: The effect of taurine on cell calcium seems to work both ways though, in that our work on the genetically cardiomyopathic hamster, which is a calcium overload condition, has shown that taurine actually decreases the amount of calcium accumulated in the cell (2,3).

BASKIN: The problem is that we obtain a biphasic response, that at one concentration of taurine there is an increase in calcium binding, and at higher concentrations that appears to reverse and we get a decrease. I'm not really sure what that means.

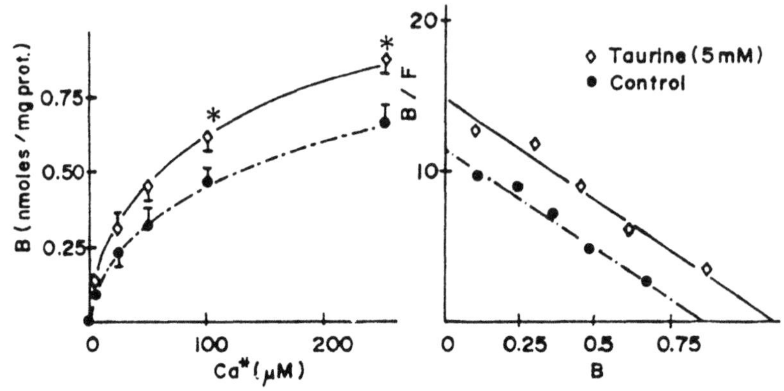

Effect of Taurine on Sarcolemmal ^{45}Ca Binding

Fig. 1. KURIYAMA. Left: Taurine was added at the beginning of preincubation to a final concentration of 5 mM in Na^+-medium. Incubation was initiated by the addition of various amounts of $^{45}CaCl_2$. The mean ± SEM obtained from 6-9 experiments was shown. * Significantly different from control value ($p<0.05$). Right: Scatchard plot of sarcolemmal ^{45}Ca binding in the presence or absence of 5 mM taurine. Ordinate; bound ^{45}Ca (B: nmoles/mg protein) Abscissae; bound ^{45}Ca/free ^{45}Ca (B/F: nmoles/mg protein/mM).

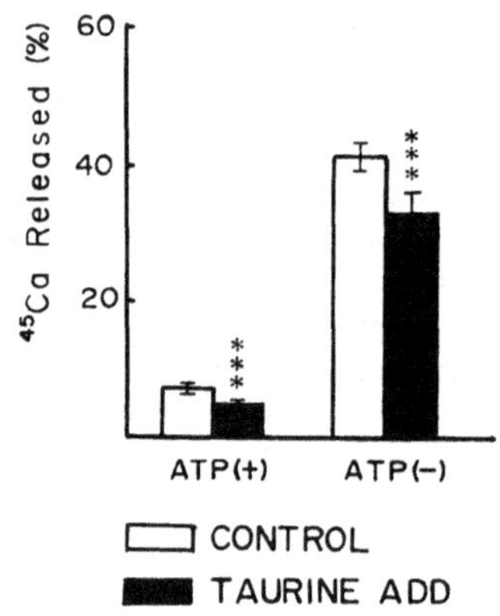

Effect of Taurine on ^{45}Ca Release from Crude Synaptosomal Fraction from Rat Brain.

Fig. 2. KURIYAMA. P_2 fraction from rat brain was preloaded by incubation with the standard medium containing 0.1 mM of $^{45}CaCl_2$ at 30°C for 8 min. The ^{45}Ca loaded P_2 fraction was incubated for 8 min at 30°C in the standard medium containing 0.01 mM of $CaCl_2$. Taurine (10 mM) was added in the latter incubation medium. Mean ± SEM obtained from 4 separate experiments is presented. *** $P<0.01$, Compared with control.

KURIYAMA: Despite these questions, I feel if we use sarcolemmal membrane we may get many different calcium effects compared to mitochondrial membrane.

BASKIN: Our preparation was that of Harigaya and Schwartz (11), which is not the cleanest preparation, but a pretty clean preparation of sarcolemma.

SCHAFFER: This is not the myopathic hamster, but the calcium paradox model, which is also a calcium overload model (Fig. 3). Addition of high concentrations of taurine to the perfusate, delays the onset of failure (14).

HUXTABLE: This is a perfused heart?

SCHAFFER: This is a perfused heart.

HUXTABLE: O.K. You're perfusing with a very high taurine concentration here, so the question is does this pharmacological effect have any relationship at all to any endogenous actions of taurine?

SCHAFFER: Well, the only thing I can say in this regard is that in this particular model there's an 80% loss of intracellular taurine. The protective effect of taurine may be due to the replenishment of cell taurine. We haven't studied this, but it's a key question.

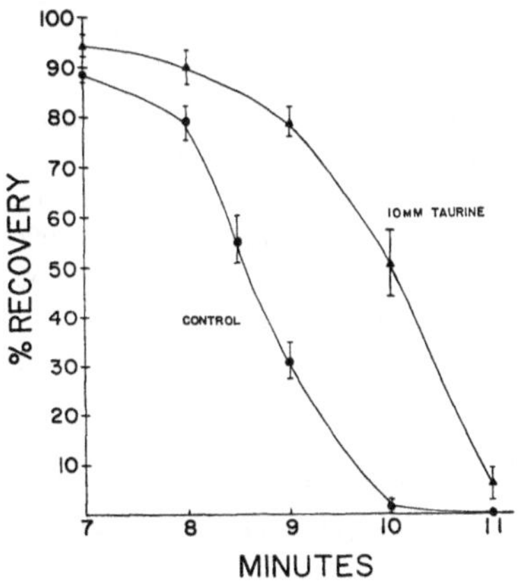

Fig. 3. SCHAFFER. Effect of taurine on the calcium paradox. Hearts were stabilized 15 min prior to perfusion with Krebs- Henseleit buffer lacking calcium. Following the designated period of calcium free perfusion, hearts were reperfused for 15 min with buffer containing 2.5 mM calcium. % recovery is a comparison of cardiac work during stabilization and after 15 min of reperfusion. Each time point represents the mean ± SEM of 4 to 10 hearts. Data taken from Kramer et al. (14).

HUXTABLE: One of the problems we're faced with in taurine is that we can measure so many pharmacological effects in the central nervous system, in the heart, and in other organs, but it's a difficulty relating these to any possible physiological functions.

SCHAFFER: I accept that problem. *(Laughter)* I agree with you I don't have an answer to that.

PASANTES-MORALES: Do we know how to detect physiological actions of taurine in the kind of experiments we are doing here?

KURIYAMA: Dr. Baskin, do you have something to say?

BASKIN: I just have a comment on the paradox model, but I think I'll refrain.

KURIYAMA: O.K.

SCHAFFER: I've a couple slides here to show the calcium effect, which is what you were talking about, Ryan. Incubation of normal membranes with taurine and calcium (0.1 - 0.4 mM) enhances sarcolemmal calcium binding (Fig. 4). On the other hand, in the calcium overload model, taurine prevents the increase in calcium binding (Fig. 5). That's somewhat analogous to what you're saying Ryan. Taurine can either enhance calcium binding, or in the failure model, reduce calcium binding.

FRANCONI: In my lab, we measured calcium content in the heart by atomic spectrophotometry and we saw that taurine (20 mM) increased calcium concentration. If we incubated the same ventricular strip with D600 we observed a decrease in calcium concentration and at the same time, of course, a decrease in contractility. When D600 and 20 mM taurine were administered together we found in the treated strips the same calcium content as in the control, and the cardiac contraction was also the same as control. These results suggest that taurine can interact with slow inward channels at sarcolemma level.

SCHAFFER: I agree. We've found the same thing with verapamil and taurine; taurine reverses the negative inotropic effect of verapamil (7).

HUXTABLE: And we find the same thing with low calcium concentrations in the perfusion of hearts with taurine. So it appears that, whereas, at normal calcium levels taurine does not have too much effect, if you raise the calcium levels or lower them, then taurine has a sort of homeostatic effect. It brings conditions towards the equilibrium state.

VAN GELDER: I would like to remind you that taurine does exactly the same thing to glutamate levels! *(Laughter)*

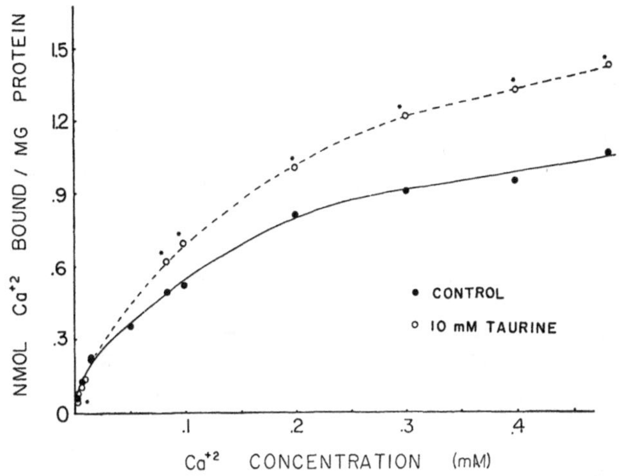

Fig. 4. SCHAFFER. Stimulation of Ca^{2+} binding to rat heart sarcolemma by taurine. Sarcolemma suspended in Tris maleate buffer, pH 7.4, containing 10 mM $MgCl_2$, 100 mM KCl and 5 mM NaCl were preincubated for 10 min in the presence and absence of 10 mM taurine. Following a 10 min incubation with $^{45}CaCl_2$ (1.0 - 500 μM), membrane calcium binding was determined by Millipore filtration procedure (Schaffer et al., this book). All values were corrected for nonspecific interactions.

KURIYAMA: We have discussed how taurine affects calcium binding - Next, we must discuss how does calcium affect taurine release? This afternoon Dr. Klein showed some very interesting data. In his case, calcium was not related to the efflux of taurine from the pineal. We found a similar situation in the optic nerve (Fig. 6). We used optic nerve preparation because it has no synapses, and consists of just axons and ganglion cells. But we still got calcium dependent release of taurine. So I'm wondering what kind of meaning we can give to this criterion of calcium dependent release?

STURMAN: Is it whole nerve though?

KURIYAMA: Yes, We have used intact optic nerve preparations.

HABER: A point I'd just like to bring up, is that there is a variety of glial preparation from which you can show a calcium dependent release. I think, perhaps, the optic nerve may not be a bad glial preparation too. However, the calcium dependent release sounds very different from the traditional neuronal kind of calcium dependence, and I think we begin to fall into the trap of thinking that everything that's calcium dependent has to be a kind of neuronal excitatory mechanism. Probably it is not and may well be very different - although calcium dependent.

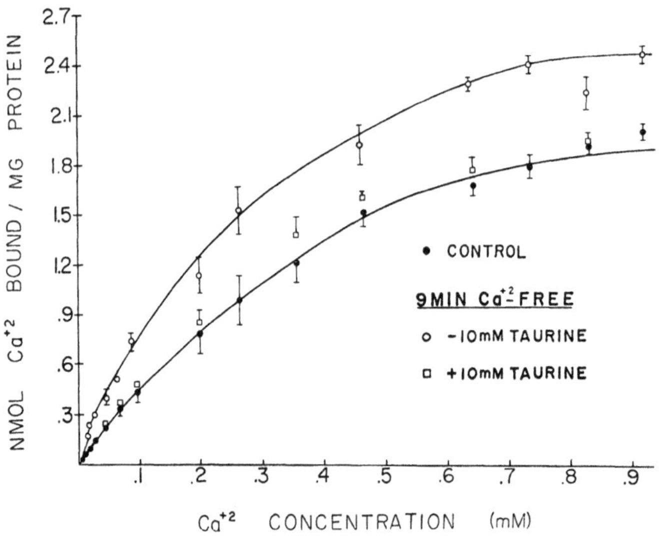

Fig. 5. SCHAFFER. Effect of taurine on sarcolemmal calcium binding of the failing myocardium. The calcium paradox model of heart failure was induced by subjecting rat hearts to 9 min of calcium free perfusion followed by 15 min of reperfusion with Krebs-Henseleit buffer supplemented with either 2.5 mM calcium (O) or 2.5 mM calcium and 10 mM taurine (□). Sarcolemma, suspended in Tris maleate buffer, pH 7.4, containing 5 mM KCl, 10 mM $MgCl_2$ and 100 mM NaCl, were prepared from these hearts, as well as normal hearts (●). Following a 10 min incubation with appropriate concentration of $^{45}CaCl_2$, membrane calcium binding determined by the Millipore filtration procedure (Schaffer et al., this book). All values represent the mean ± SEM of 3 preparations and were corrected for nonspecific interactions.

MANDEL: How do you make the distinction between the two kinds of release, what is your test?

HABER: Between what kind of release, Paul?

MANDEL: Between calcium dependent release from neurons, and calcium dependent release from glial cells?

HABER: I didn't make that distinction. Andres Hamberger and others made that distinction. You can distinguish that by using tetrodotoxin. In one case tetrodotoxin will have an effect, and in the other it will not. The distinction is not mine.

KLEIN: Of course, you could ascribe the potassium induced release to the calcium induced release of a transmitter, so the first question is can you get this effect with acetyl choline?

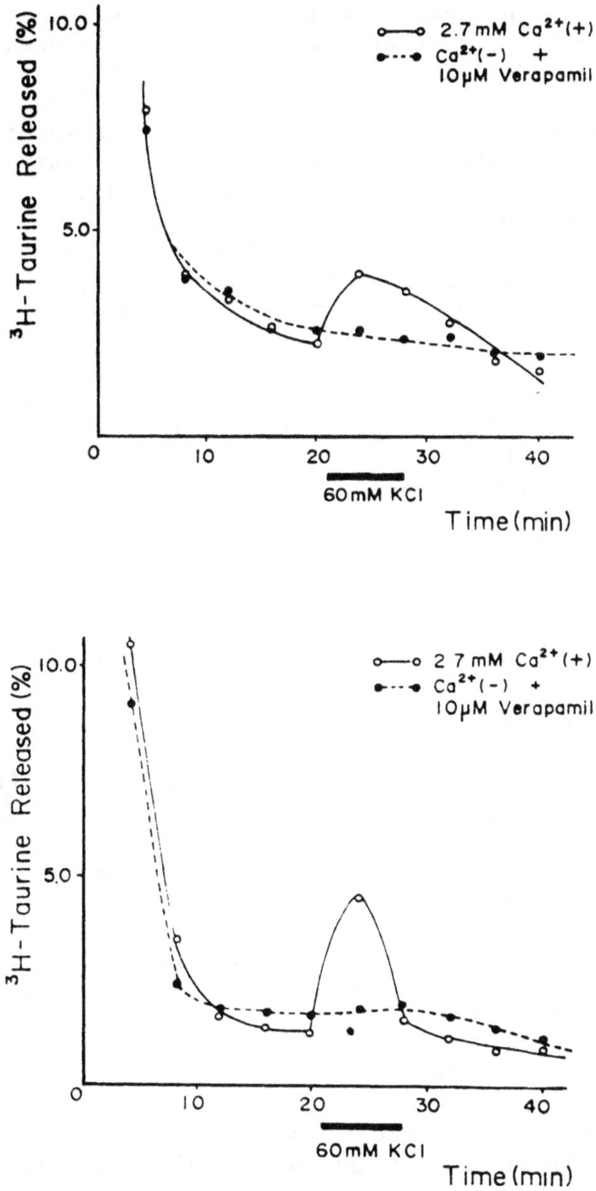

Fig. 6. KURIYAMA. Effect of high concentrations of KCl (60 mM) on (^{3}H)-taurine release from cerebral cortical slices (A) and optic nerve preparations (B). The mean values obtained from 4-5 separate experiments are shown.

KURIYAMA: I don't know, but I can say that optic nerves have no potassium-induced and calcium dependent release of GABA.

KLEIN: It doesn't do that?

KURIYAMA: No. Furthermore, I would like to make clear that in cerebral cortical slices, there is no potassium-induced release of GABA in the absence of calcium.

KLEIN: The pericervical ganglia which is of some interest, has taurine in it. Also the posterior pituitary has taurine in it. We saw that in certain tissues, potassium alone had a releasing effect, I'm trying to sort out in my mind which one it was, but that doesn't seem to be a uniform response because in one tissue that has a lot of taurine there was no effect of depolarization, so I would suspect that you may have a transmitter involved in the effect that you're seeing.

BASKIN: Since everyone else is putting in their own commercial, in this month's Circulation Research (4) we described the fact that verapamil or tetrodotoxin as well as no calcium, also reduces taurine fluxes in Purkinje fibers and I...

KLEIN: Purkinje fibers?

BASKIN: In the heart Purkinje fibers, I'm sorry. *(Laughter)*

BARBEAU: There's only one Purkinje; the heart really doesn't count.

BASKIN: Which is a model for the nerve.

HUXTABLE: Taurine uptake into heart, though, is not affected by calcium.

BASKIN: The primary effect of taurine in the heart is on the Purkinje fibers. Just kidding.

STURMAN: Kinya, how do you know that taurine is acting axonally? It seems to me you may well have loaded the glial cells around the outside and not the axon.

KURIYAMA: That's possible.

STURMAN: Suppose you load the cells by injecting the animal in the eye and let it axonally transport and then do that experiment? What happens?

KURIYAMA: That's really hard. We tried that. It's not so easy to do those types of experiments.

STURMAN: Why is it hard to analyze an experiment like that? I don't understand.

KURIYAMA: Because the essential aim of our study is that we want to compare the release of GABA and taurine from optic nerve preparations, and the optic nerve seems to have no axonal transport system for GABA.

STURMAN: What animal are you using?

KURIYAMA: We have used the rat.

STURMAN: You could use rabbit. It's very easy since the eye and nerve are much larger.

HABER: I'm just wondering in terms of looking at the involvement of calcium and taurine, if it may not be worthwhile to talk about some other systems in which calcium is involved other then release phenomena, which are complicated for a variety of reasons, and if it might not be worthwhile to talk about cell differentiation, process formation, cell adhesion and things like that which are calcium dependent phenomenon. Just for openers I'd like to throw out some unpublished, and not as yet repeated observations that I have done partly with the assistance of Herminia. Taurine (25 mM) seems to block the, if you want to call it that, the differentiation of glial cells, and the response to dibutyryl cyclic AMP. The cells do not seem to put up very big processes, if they do that at all. Unfortunately I did not bring any kind of pictures. Perhaps this might be one other paradigm. It would be profitable to study what taurine and calcium are doing, rather than just focusing on release phenomenon, which perhaps are irrelevant.

HAYES: There is an interesting correlation here that relates taurine with cell differentiation and calcium metabolism. Back when I was young and eager I worked on vitamin A deficiency. One aspect of that deficiency is an extraordinary excretion of calcium early on in the deficiency which continues. Urinary excretion of taurine also increases and cell differentiation is markedly altered. Furthermore, the oxidation of sulfur amino acids seems to be accelerated in vitamin A deficient situations, so that the taurine excretion may reflect an increased oxidation of sulfur amino acids. The muscle mass doesn't necessarily deplete, there's no evidence that you're losing taurine from tissue stores. I don't know if anybody here has worked in this area. I'd be interested to hear comments on what is the spin off for taurine excretion when the metabolism of sulfur amino acids or their oxidation is accelerated.

MANDEL: So it's a similar situation to X-irradiation. There is also a release of quantities of calcium.

GAULL: And taurine!

VAN GELDER: And wound healing. In experimental wounds during the healing process there's apparently a tremendous taurine sparing action also (13,19).

GAULL: In what animal and at what age does that occur in vitamin A deficiency, Casey?

HAYES: The original observation on calcium excretion was in calves (20), but there has since been similar data published in the vitamin A-deficient rat (21). The accelerated urinary excretion of taurine has been noted in rats (16). Apparently vitamin E deficiency resulting in muscle degeneration also leads to an increased urinary taurine excretion as well (unpublished).

RASSIN: Generalized malnutrition seems to be associated with increased taurine excretion, so maybe you're not differentiating between specific deficiencies, but only seeing a global response (5).

HAYES: Well, one of the things I have looked for in these reports is whether or not a wasting malnutrition situation existed. Early in vitamin A deficiency the animals are still growing so they are not wasting away. In protein-calorie malnutrition the situation may be different.

RASSIN: But even with a protein-calorie malnutrition, we don't know when the increase in taurine excretion starts, whether it starts before you can measure muscle wasting or afterwards.

HAYES: Yes, but I think in the studies that I've looked at I've tried to factor out muscle wasting *per se* that would release taurine.

HABER: I'm very ignorant, but has anyone done any kind of experiments on looking at the effect of taurine, perhaps, on calcium clearance in, say, a muscle preparation, where clearance of calcium is very slow. It seems to me that these are the kind of preparations we should start looking at.

BASKIN: Someone's done that experiment.

HABER: And what happened?

BASKIN: Ah! I don't know. You're going to have to ask them.

PASANTES-MORALES: I don't know if I hear you correctly but we've been doing studies on the effect of taurine on calcium transport, I'm not going to talk very much about this now otherwise I have hardly anything more to say tomorrow, *(Laughter)* but we have observed that taurine modifies calcium transport, depending mainly on the concentration of external calcium. We observed the same action of taurine in preparations as different as isolated synaptosomes, crude synaptosomal fractions containing mitochondria, and isolated rod outer segments. Briefly, our results show that taurine increases calcium uptake or binding at low calcium concentrations and decreases calcium binding or accumulation at high calcium concentrations. This has been also observed in sarcolemma preparations by Steve Schaffer and Ryan Huxtable, respectively. So it seems that the effect of taurine on calcium is observed in a wide variety of biological preparations. There is a general problem here of taurine/calcium relationships, not restricted to a particular tissue.

KURIYAMA: O.K., we have only 34 minutes left *(Laughter)* so let's go to the second question. What is the nature of the neuromodulatory action of taurine? It seems to be a big issue. Who wants to start?

PASANTES-MORALES: I think I'd like to...

MANDEL: And the definition of a neuromodulator perhaps?

PASANTES-MORALES: Well, a neuromodulator modifies nervous excitability, not directly related to the arrival of nerve impulses,...

MANDEL: Neuromodulatory of dopaminergic nerves?

PASANTES-MORALES: Yes.

MANDEL: It modulates?

PASANTES-MORALES: Yes, yes, sure, sure any...

MANDEL: So GABA will be a neuromodulator?

PASANTES-MORALES: Anything that modifies neurotransmitter release could be considered a neuromodulator. I like to have all my results in a well-ordered scheme, so I was worried to know if the effects we observed on calcium transport could be related to some of the pharmacological or physiological effects of taurine. Concerning the possible neuromodulator role of taurine, we have observed that taurine decreases potassium-stimulated GABA release from isolated synaptosomes. The figure shows a preparation of crude synaptosomal fraction. In the presence of depolarizing concentrations of potassium, the release of GABA is increased more than 10 times over the basal levels. In the presence of taurine, 25 mM, we have two effects: one is an increase in the spontaneous efflux of GABA and then a very marked decrease of potassium stimulated release of GABA. This is the same pattern we observed when we did the experiment in the absence of taurine and in the absence of calcium, so the presence of taurine mimicks the effect of a calcium-free medium on GABA release, both on the spontaneous release and on the potassium stimulated release. In a calcium-free medium we have both an increase in spontaneous release of GABA and a decrease in the potassium stimulated release of GABA. So it seems that the effect of taurine in reducing calcium transfer could simulate the effect of calcium-free media on the release of GABA. If taurine could do this *in vivo*, this could be related to its pharmacological actions, and it may alter the release of neurotransmitters. This could be an explanation for the neuromodulator role of taurine.

KURIYAMA: But Herminia we have trouble here because both GABA and taurine are inhibitory, so if you want to explain the inhibitory effects of taurine by the suppression of release of GABA, you may have a problem. Could you explain clearly on this point?

PASANTES-MORALES: No, this could be a very general effect of taurine. In calcium-free media, we reduce the release of every neurotransmitter since all of them are calcium related. Therefore, the neuromodulatory effects of taurine, or pharmacological effects of taurine, on reducing neurotransmitter release could be due to this ability of taurine to reduce the calcium available for the release of neurotransmitters.

SGARAGLI: Herminia, Leach (17) in Britain has recently shown that...

PASANTES-MORALES: The opposite! *(Laughter)*

SGARAGLI: No, no I didn't say that...

PASANTES-MORALES: No, but he did!

SGARAGLI: He has shown that only the potassium induced release is increased by taurine. The basal release is unaffected.

PASANTES-MORALES: Yes, I know this paper, and we were worried about it, of course. We are using a different preparation. They are using cerebral cortex slices and we are using isolated synaptosomes, the P_2 fraction. We did the same experiments using brain slices, and we could not observe any effect of taurine, no reduction on increase in the release of GABA, and no effect on potassium depolarization. I am not sure I can see the reasons for that. But what I could say is that this effect of taurine on reducing calcium uptake or binding to synaptosomes is not easily observed in brain slices.

KURIYAMA: What magnitude of stimulation on release can you get compared to the slices. It might be very small.

PASANTES-MORALES: Oh! No, no! It's much greater when we are using the P_2 fraction.

KURIYAMA: But usually if we destroy the plasma membrane, you know, potassium effect decreases tremendously.

PASANTES-MORALES: Yes, in brain slices the effect of potassium on increasing GABA release is much smaller than when you are using the P_2 fraction. Probably there's a problem of diffusion of taurine in slices, but we couldn't see that taurine increased the potassium-stimulated release of GABA. However, we didn't observe a decrease either.

MANDEL: What occurs at a lower concentration of taurine? Because at 25 mM it has a chelating effect. *(Laughter)* It is so! So what about 10 mM or 5 mM?

PASANTES-MORALES: Well, this is a dose dependent effect. At 10 mM it is decreased and at 5 mM the effect is not observed. If 25 mM taurine has a chelating effect, I would remind you that a physiological concentration is, for instance, in photoreceptors 40 - 50 mM, so that...

MANDEL: Taurine is like serotonin in platelets. It is an absolutely unusual situation! The retina has a tremendous concentration of taurine in order to keep calcium in the outer segment. It's a very peculiar situation that you cannot generalize. You cannot extrapolate from serotonin in platelets to serotonin in the brain.

PASANTES-MORALES: It is not a very unusual situation for taurine. Taurine is 25 mM in leukocytes, 25 mM or so in platelets, 25 - 45 mM in muscle, 20 mM in pineal, 100 mM in neurohypothesis and so on and...

MANDEL: And the brain?

PASANTES-MORALES: 8 mM in the rat, so if it is 8 mM in the whole brain, you could guess that at the point where it is really concentrated it could be much higher than 8 mM.

MANDEL: As acetylcholine too. You can have a tremendous concentration, but it doesn't mean that when you do an experiment to see a stimulation with acetylcholine that you have to use 5 mM acetylcholine.

PASANTES-MORALES: Yes, I agree that these effects are pharmacological. O.K. But probably taurine is also depleted in our preparation. When we prepare P_2 fractions, taurine is lost from synaptosomes, and probably all we are doing is replenishing all those pools of taurine, which is important, and that's why I'm very happy hearing about the effect of bicarbonate. I have showed that these effects of taurine on calcium transport, both increasing calcium binding at low calcium concentrations or decreasing of calcium binding at high calcium concentrations, are absolutely dependent on the presence of bicarbonate. So it seems bicarbonate appears to decrease taurine transport. We could conclude that these effects are due to the transport of taurine inside the cells and this makes our pharmacological results more closer to the physiological.

KURIYAMA: What is the effect of bicarbonate in the case of brain slices? If you use them instead of homogenate or P_2 fraction, do you get a similar effect?

PASANTES-MORALES: No, with slices we cannot observe the effect of taurine on calcium transport. Probably, slices have their full amount of taurine present in the cells, so you cannot see any effect.

CHESNEY: I presented this in 1977 (6), that in all uptake studies that we did in renal slices comparing Tris buffer and bicarbonate buffer we saw no difference. But using the tubules that are isolated where there is no presumed bathing medium around the tubules within the slice, we see a definite bicarbonate effect that I showed yesterday (Chesney, Fig. 3., this book).

KURIYAMA: In the brain, we saw a similar difference, and I'm wondering which preparation we should use for studies? Slices seem more reasonable to me.

RASSIN: One comment about that. The P_2 fractions probably contain more mitochondria in them that they do synaptosomes, just by their nature. Mitochondria transport calcium. How can you really differentiate between whether this is a synaptosomal or mitochondrial effect?

PASANTES-MORALES: We are using mitochondrial-containing P_2 fraction because it's much more rapid and easy to isolate, but when we did our controls we observed identical effects on GABA release when we compared synaptosomal fractions isolated by the method of Hajos (10) with the mitochondrial-containing fraction. I think the effects of taurine on calcium transport, not on GABA release of course because mitochondria cannot release GABA, but on calcium transport are also observed in pure mitochondria.

RASSIN: That should be easy to test, though, because you can isolate mitochondria from synaptosomes.

PASANTES-MORALES: Yes, for calcium transport studies I agreed that we have absolutely to separate mitochondria from synaptosomes, but probably not for GABA experiments since mitochondria are useless.

RASSIN: The question is; when is it better to use slices, P_2 fractions, or P_2B fractions. The three preparations probably should be compared.

KURIYAMA: That's right!

CHESNEY: Is there a bicarbonate effect when you use purified synaptosomes though?

PASANTES-MORALES: Yes, absolutely. We have observed the bicarbonate effect in all preparations we have studied. I would like to make a comment also on these effects of bicarbonate. It is well known from the papers of Altura and his group (1,18) that bicarbonate is necessary in order to have good responses in muscle. When you replace bicarbonate by Tris or another buffer, you have both a decreased calcium uptake and a decreased contractility. Also for the release of neurotransmitters, we have observed that replacement of bicarbonate by Tris, HEPES or other ions, there is a decreased GABA release. For instance, in this same preparation the stimulation of GABA release by high potassium concentration in a bicarbonate medium is about 10 times greater than basal, whereas when we use Tris or HEPES release decreases to 3 or 4 times the basal rate, so it seems that bicarbonate has a physiological role to play in excitable tissues.

HABER: I think perhaps I am making this comment because Nico told me I'm suppose to speak this afternoon, and meanwhile to shut up! One of the issues in terms of what kind of preparation one is using, P_2 or P_2B or slices, perhaps the real issue, is what kind of compartment one is releasing from. Certainly the compartmentalization is not identical in all the 3 preparations, and that may account for some of the disparity of the results being obtained. One may be getting release from different compartments.

IWATA: Maybe it's a very silly question, but as far as I heard, is bicarbonate also a modulator?

KURIYAMA: For taurine uptake I think.

BARBEAU: It may not be a neuromodulator!

IWATA: The same question to Dr. Mandel?

MANDEL: The pH can also be a modulator. *(Laughter)*

HABER: What are the effects of taurine on the membrane potential of excitable cells?

VAN GELDER: Inhibition by taurine can be blocked by strychnine and, in general, its action is very similar to that of GABA or glycine (8,9,12): It increases chloride permeability or conductance. This tends to drive the cell potential towards the negative chloride equilibrium potential.

HABER: Alright, that's a change in chloride conductance, but I'm wondering are there effects of taurine in terms of what I might call membrane stabilization which might reflect resistance to changes in membrane potential?

MANDEL: I have a question, what is membrane stabilization? *(Laughter)*

VAN GELDER: That's exactly what it is! If you keep your potential clamped at a certain level by allowing the movement of the chloride (or potassium) ion to compensate for any influx of sodium ions, then you have stabilized your membrane (15).

MANDEL: Yes, but when you say membrane stabilization you have to say which property of the membrane you stabilize.

HABER: If that's the case, Nico, one ought to be able to predict that taurine ought to have some very pronounced inhibitory effects on acetylcholine release in the neuromuscular junction. I don't believe it does that.

VAN GELDER: There are some reports, as a matter of fact, that it inhibits the spontaneous release of acetylcholine.

HABER: Well, I'm happy to hear that.

HUXTABLE: It inhibits the release of acetylcholine from the adrenals, too; both stimulated and basal release.

KURIYAMA: Any other comment?

PASANTES-MORALES: In this relationship to this point, do you think that these effects of taurine on modifying chloride transport are related to specific receptors, or could it be using receptors of other amino acids like GABA or glycine, which have a more widespread distribution in excitable membrane? The point is, are there receptors specific for taurine which could mediate the increase in chloride conductance?

VAN GELDER: I believe what you're saying is that chloride ionophore can be stimulated either by glycine receptors or by taurine receptors, not by beta or by GABA receptors. In any case, I think it was Snyder who first showed that if you used something like picrotoxin, which is supposed to close, or to prevent the opening of, the chloride ionophore, that picrotoxin will inhibit any action of glycine, any action of GABA and any action of taurine. So it is a specific ionophore effect, which can be mediated by different receptors of inhibitory transmitters.

PASANTES-MORALES: But do you suppose that this chloride ionophore is different, for instance, in spinal cord than in cortex?

VAN GELDER: No.

PASANTES-MORALES: Because in cortex, glycine is not able to modify chloride conduction, whereas it is able to do it in the spinal cord.

VAN GELDER: It is the coupling of a specific amino acid receptor to an ionophore, so it is the receptor for the amino acids which can be modified, but the ionophore itself stays the same.

PASANTES-MORALES: So the question is are there available this kind of coupled receptor for taurine?

MANDEL: No, you cannot have an effect if the substance doesn't bind to a receptor, or something which should receive the message. How can you have an answer without reception of the message? If a message is received, then taurine must have been bound to the molecule which will give the answer.

KURIYAMA: I feel that receptor binding and simple membrane binding are different.

MANDEL: I didn't say binding, I said when they have an answer, or response. When I am testing binding it is only binding that is measured, but when I am testing the answer I measure a block of the chloride ionophore, so I am testing the receptor site.

BASKIN: To try to answer Dr. Mandel's point or Herminia's point, what we did we gave SITS which is supposed to be, in quotes, "a specific chloride inhibitor" and we found that SITS could inhibit the fast component of taurine uptake in bovine cardiac Purkinje fibers.

HUXTABLE: What is SITS, could you tell us? *(Laughter)*

BASKIN: Rosenstein up in Toronto uses it as a what he calls a specific chloride and bicarbonate transport inhibitor.

KURIYAMA: Would you describe what kind of mechanism is involved in the action of the inhibitor?

HABER: I think it inhibits chloride-bicarbonate exchange.

VOICE: If that's the case, it doesn't help.

CHESNEY: It's an exchange inhibitor.

MANDEL: Let's wait. To explain something by something else what you don't know how it works is a little bit difficult.

BASKIN: That's called pharmacology!

VAN GELDER: Inhibitors of the chloride ionophore that I know of can either block GABA effects, in which case they can also stop the taurine effect, or they can block the glycine effect and they then can also block the taurine action, but there does not seem to exist a chloride ionophore inhibitor which seems only to act in conjunction with a "taurine receptor". It's either taurine and GABA, or taurine and glycine.

HABER: Nico, let's not fall in love with GABA, and its intimate link to hyperpolarization and chloride conductances, because some increasing marked inhibition probably helps hyperpolarization. There is no relevant change in conductors.

VAN GELDER: Certainly is!

PASANTES-MORALES: There is, and taurine has exactly the same effect as GABA in presynaptic conditions, but it is inhibited by both bicuculline and strychnine. So again it seems not to be specific.

MANDEL: I think that we are seeing some similarities of enzymes for GABA and for taurine. Now is the question, is it possible that there are also some similarities in their actions in the cell? At least, are they part of the same story?

KURIYAMA: Let's go to the next question. *(Laughter)* Is there a relationship between the functions of taurine and GABA? Are there interrelationships between CSAD and GAD?

HABER: Let me begin this by asking a question. Have the kind of criteria that have been used for the GABA receptor, in which there have been meticulous attempts to disassociate the receptor site from the binding or transport site, have these kind of criteria been applied to taurine? There's a sodium independent type of binding with a K_d in the nanomolar range, which has got nothing to do with the transport site which is sodium dependent.

KURIYAMA: Do you think sodium dependency is a big problem?

HABER: Yes it is.

KURIYAMA: How do you think that can be handled?

HABER: I'm not thinking, I'm not doing them, I'm asking you.

MANDEL: It is not sodium dependent, but sodium independent binding.

LOPEZ-COLOME: No, but this binding we found is sodium dependent.

MANDEL: Spinal cord?

LOPEZ-COLOME: No, in brain regions.

MANDEL: The trouble is, and I think that this is a basic assumption, that when we have to compare this situation in brain where you have a very high amount of taurine which is not a neurotransmitter with a small amount of binding of taurine which is a transmitter, it is extremely difficult to get an answer, so that you have to choose a situation where this relationship is in favor of the binding. Unless you can attempt this it is the same situation as if people would start to look for dopamine receptors on the cortex. They will say that dopamine receptors don't exist, simply because they cannot be found in the cortex.

BARBEAU: No, no, no, dopamine receptors do exist in the cortex! *(Several voices arguing)*

HUXTABLE: I'm worried that we are going to come to fist fights down at this end of the room, so I think we should maybe call it a close for the day.

BARBEAU: Oh come on, why is he stopping this, It's fun!

REFERENCES

1. Altura, B.M., Altura, B.T., Carella, A., and Turlapaty, P.D.M.V., 1980, Adverse effects of artificial buffers on contractile responses of arterial and venous smooth muscle, Brit. J. Pharmacol., 207-214.
2. Azari, J., Brumbaugh, P., and Huxtable, R., 1980, Prophylaxis by taurine in the hearts of cardiomyopathic hamsters, J. Mol. Cell. Cardiol., 12:1353-1366.
3. Azari, J., and Huxtable, R.J., 1980, The mechanism of the adrenergic stimulation of taurine influx in the heart, Eur. J. Pharmacol., 61:217-223.
4. Baskin, S.I., Zaydon, P.T., Kendrick, Z.V., Katz, T.C., and Orr, P.L., 1980, Pharmacokinetic studies of taurine in bovine Purkinje fibers, Circ. Res., 47:763-769.
5. Bigwood, E.J., 1962, Protein deficiency and aminoaciduria in Kwashiorkor, in Central Africa, Nutritio et Dieta, 4:17-50.
6. Chesney, R.W., Jax, D.K., Scriver, C.R., and Mohyuddin, F., Taurine transport in mammalian kidney, in: "Taurine in Neurological Disorders," R. Huxtable and A. Barbeau, Raven Press, New York (1978) pp.73-94.
7. Chovan, J.P., Kulakowski, E.C., Sheakowski, S., and Schaffer, S.W., 1980, Calcium regulation by the low-affinity taurine binding sites of cardiac sarcolemma, Mol. Pharmacol., 17: 295-300.
8. Collins, G.G.S., On the role of taurine in the mammalian nervous system, in: "Essays in Neurochemistry and Neuropharmacology," M.B.H. Youdim, W. Lovenberg, D.F. Sharman and J.R. Lagnado, eds., John Wiley, London (1977), pp.43-72.
9. Gruener, R., Bryant, H., Markovitz, D., Huxtable, R., and Bressler, R., Ionic actions of taurine on nerve and muscle

membranes: Electrophysiologic studies, in: "Taurine," R. Huxtable and A. Barbeau, eds., Raven Press, New York (1976), pp.225-242.

10. Hajós, F., 1975, An improved method for the preparation of synaptosomal fractions of high purity, Brain Res., 93:485-489.
11. Harogaya, S., and Schwartz, A., 1969, Rate of calcium binding and uptake in normal animal and failing human cardiac muscle, Card. Res., 25:781-794.
12. Hösli, L., and Hösli, E., 1978, Action and uptake of neurotransmitters in CNS tissue culture, Rev. Physiol. Biochem. Pharmacol., 81:135-188.
13. Jacobsen, J.G., and Smith, L.H., 1968, Biochemistry and Physiology of taurine and taurine derivatives, Physiol. Rev., 48: 424-511.
14. Kramer, J.H., Chovan, J.P., and Schaffer, S.W., The effect of taurine on calcium paradox and ischemic heart failure, Am. J. Physiol., in press.
15. Krnjevic, K., and Puil, E., Electrophysiological studies on actions of taurine, in: "Taurine," R. Huxtable and A. Barbeau, Raven Press, New York (1976), pp.179-189.
16. Lamb, A.J., Naewbanij, J., and Anzano, M.A., 1979, Altered taurine metabolism in vitamin A deficiency, Fed. Proc., 38:762 (Abstract).
17. Leach, M.J., 1979, Effect of taurine on release of ^{3}H-GABA by depolarizing stimuli from superfused slices of rat brain cerebral cortex in vitro, J. Pharmacy Pharmacol., 31:533-535.
18. Turlapaty, P.D.M.V., Altura, B.T., and Altura, B.M., 1979, Tris (hydroxymethyl) aminomethane inhibits calcium uptake in vascular smooth muscle, Biochim. Biophys. Acta, 551:459-462.
19. Williamson, M.B., and Passman, J.M., 1960, Excretion of taurine during healing of experimental wounds, Clin. Chem., 6:140-147.
20. Woelfel, C.G., Calhoun, M.C., Rousseau, J.E., and Eaton, H.D., 1963, Some biochemical constituents in urine of vitamin A deficient calves, J. Dairy Sci., 46:947-954.
21. Zile, M., Ahrens, H., and Deluca, H.F., 1972, Vitamin A deficiency and urinary calcium excretion in rats, J. Nutr., 102:1255-1258.

SESSION III

WHAT DOES TAURINE DO?

Moderator: Nico Van Gelder

VAN GELDER: We will use a different formula for this discussion. What I would like to do is to pose a problem, then designate a speaker to solve it, and then maybe afterwards we can get some opinions about whether the speaker had a lot of nonsense to say or was very wise. I would strongly request, although it is not an absolute prohibition, that no slides be shown. I think we all have an opportunity to show slides when our own turn comes and it might increase the discussion a little bit. So I'd like to pose the first problem: we are all, I think, in agreement that the pool of taurine in various tissues is very stable and the turnover is very slow. Nevertheless, as we have seen in the last two days, there are enzymes which regulate taurine and I would very much appreciate it if Dr. Fellman would give us an opinion, educated I'm sure, as to how he considers these enzymes. How may these enzymes regulate either the pool of taurine when it is steady or how can they replenish the pool when, for example, after stimulation of the tissue, this pool has gone down. We know that nutrition plays a part in replenishment of taurine, but certainly enzymes should also have some part in it. So I would like to ask Dr. Fellman to give us his opinion. Then we can see whether people agree with it or not.

FELLMAN: I can only fail to answer that challenge but I will try. I assume that I have 6 milliseconds to answer this and so I'll be very brief, but I would like to underscore some of the problems that, as I presently see it, exist in trying to sort out which enzymes are regulatory in the biosynthesis of taurine. I am persuaded using the model that we know from classical biochemistry, that the rate limiting enzyme for taurine biosynthesis is cysteine dioxygenase (Fig. 1). So I think that the first enzyme (Fig. 1, Step 5) in the sequence, which forms CSA, is regulatory for the overall biosynthesis of hypotaurine. Decarboxylation is important, despite the fact that there is strong evidence that the enzyme activity is not congruent with the appearance of the final product,

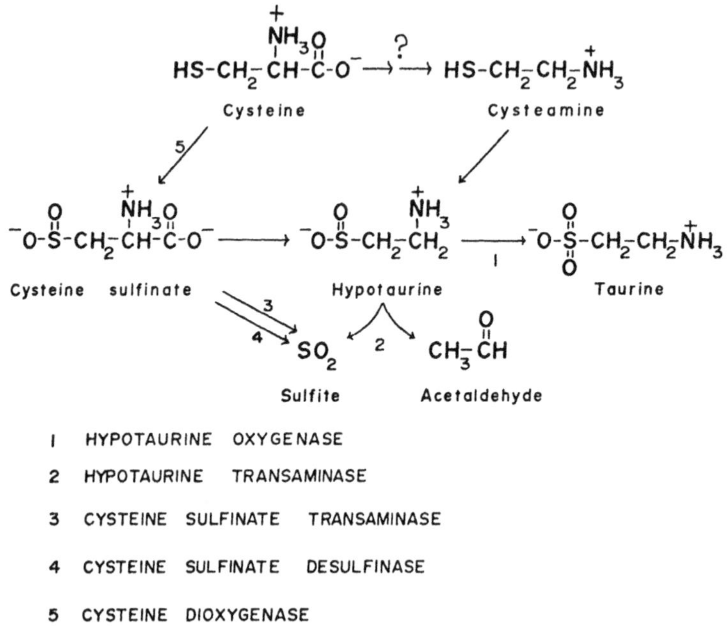

Fig. 1. FELLMAN

as was pointed out by Dave Rassin this morning. CSA is decarboxylated to hypotaurine and this intermediate, which one at least sees although in very small amount, may be transaminated to sulfite or must rapidly be converted to taurine. Now that process is clearly an oxidation. The nature of that oxidation, I submit, is still a mystery. It's certainly a mystery to me and I'd like to emphasize that until we know what the activity of this enzyme is, where it is deposited (I suspect the liver, but it may be in other tissues as well), I don't think we are going to understand clearly what regulates the biosynthesis of taurine.

I undertook the radiochemical synthesis of hypotaurine because I wanted to study this question. This leads me to the point of what the limitations of this undertaking are namely; when you synthesize hypotaurine, or when you buy it from Cal Biochem, what you in fact get is hypotaurine plus a certain amount of taurine, so when you undertake the study of hypotaurine oxidase you're adding a product as well as the substrate. Now the amount of that product that is there varies and one has to be very careful of the controls used in order to make certain that you are in fact seeing biosynthesis. At my best efforts I've obtained about 3% impurity. That is after I synthesize the hypotaurine and then put it through an ion exchange resin separation my best efforts give me a 3% contamination of hypotaurine by taurine. If I then conduct the enzymatic oxidation of hypotaurine goes to taurine using perfused liver, and rat liver homogenates and then with brain fractions, P_2 fractions from retina, and so forth, I have yet to see persuasive evidence for this enzyme activity until recently. I'll tell you a bit about that, but in any case for most of the undertakings that I've had, what I found was the major problem was sequestering or stabilizing the hypotaurine after the end of the enzymatic investigation for the separation of the product to demonstrate which product has been formed. So at least in my hands, trying to use retinal preparations, as has been similarly reported for a number of liver and brain preparations, I have not seen persuasive evidence that taurine is formed. Nevertheless if I take the radiolabeled hypotaurine, and inject it into say a mouse or a rat - I will get about 100% conversion to taurine in 8 to 12 hours. This is an endogenous event, because if I do it in germ-free mice I get just as brisk biosynthesis of taurine from the hypotaurine, and so I think that it is the nature of this enzyme which is the one at least I'm going to focus on to determine what regulates the biosynthesis of taurine.

VAN GELDER: I'd like to raise then the question if you get _in vivo_ conversion: can you get spontaneous oxidation of hypotaurine without an enzyme? I just wonder if Dr. Oja could comment on that.

OJA: Yes, we have been studying the oxidation of labeled hypotaurine for several years already (12), and in the beginning it was very difficult also for us to demonstrate any conversion of hypotaurine to taurine, and only after maybe 4 years ago, it was possible to do so definitely (7). Always you can get a little conversion which is probably not enzymatic. If you, for instance, incubate hypotaurine with boiled P_2 tissue preparations or if you add strong acids at the beginning and so on, there is a little nonenzymatic conversion but if you handle the situation carefully you can reduce this to such levels which can be tolerated. You can get a situation where, for instance, this, I'll say nonenzymatic or unspecific conversion, is less than 20% of the enzymatic conversion which is happening in the homogenates or subcellular fractions isolated from various tissues.

RASSIN: Jack, you mentioned you like the cysteine dioxygenase as a control unit and then you didn't say too much about it. I think it's worth pointing out that the work of Yamaguchi (19) may be quite compatible with that. He finds a lot of responses in activity of the enzyme to agents like glucocorticosteroids, and protein in the diet appears to regulate activity of the this enzyme. These responses make cysteine dioxygenase a very appealing control point in the pathway, which you don't see with the other enzymes.

HUXTABLE: As I said yesterday, I find it hard to accept cysteine dioxygenase as the regulating enzyme because this enzyme is also involved in the major pathway of sulfur metabolism through cysteine sulfinic acid and the transaminase. You're saying you have a regulatory step controlling the production of a compound which has three pathways of metabolism: the transaminase, the further oxidation to cysteic acid, and the conversion of hypotaurine.

RASSIN: Well, I think I like the controlling step at the bottle neck as opposed to a number of different steps in the pathway.

FELLMAN: But Ryan's point is that it is fair to say that 90% of cysteine is metabolized to sulfate through this pathway, and I think to underscore your point he is saying that this being the overall regulatory enzyme for cysteine metabolism, this could be a point of control for the synthesis.

RASSIN: Why do you think it's the regulatory step then?

FELLMAN: This enzyme here? Because it predetermines whether or not any of this oxidation will occur at all.

RASSIN: So which side of the scale are you coming down on here?

FELLMAN: Well, I think that the final regulation of synthesis of taurine is determined by the hypotaurine oxidase, or whatever that enzyme is, and that hypotaurine can be shunted off by transamination for example, as an alternate route, but that the final production of taurine must be also regulated by this enzyme.

VAN GELDER: This, I think, is the introduction to the next question: pools.

MANDEL: No! I have my question. *(Laughter)* No, I think that we have to be clear. Number one: That we are speculating. That is to say, we have no evidence until now for any hypothesis...

VAN GELDER: Oh! That's obvious.

MANDEL: And it's obvious, O.K., but it was not so obvious, because if you repeat 3 times a speculation people come to think that it's a hypothesis, and if you repeat a hypothesis 3 times they have a feeling that it is already proof. Which is not the case. Number one. Number two: Concerning regulation at the level of cysteine sulfinic transaminase, I think it is something which has to be taken into consideration, because when you block it, you have an increase of taurine and this is really a fact, it is no more speculation. And finally it's extremely rare that in a sequence of reactions the end reaction will be regulatory. Usually the regulation occurs before.

VAN GELDER: O.K. Now I would like to go on because we have a lot of things which were brought up yesterday which we never came to, and tomorrow we won't get to it again. I'd like to get to the problem of pools. Yesterday we had the discussion on how much of the taurine is staying behind when you homogenize tissue; how much is staying in organelles. Dr. Rassin is probably one of the best qualified to discuss this since he has worked with Dr. Whittaker, and so I would like to call on him to start this discussion. When you homogenize tissue, a certain percentage of taurine comes out, a certain percentage remains behind as bound (or free intraorganelle taurine, whatever we call it) and I would like to pose a very difficult question to start the discussion. Which of the pools does he think is the most stable, and could he speculate at all as to which of the pools would have a special physiological role to play, if any. Since he doesn't figure taurine has any physiological role, I would like to hear the other side for a change.

RASSIN: It's a lot to bite off, Nico! I think what we have to remember about subcellular fractionation is that taurine is not the only amino acid found in synaptosomes. The problem is rather what is special about that 20% or 25% of brain taurine that is found in synaptosomes, and in what ways is it different from all of the other amino acids that are found there in about the same amounts? When you listen to what went on yesterday and today, there's not a lot that is very special about that taurine and I find this disturbing. There are a few things, however, that you have to keep in mind. A couple of them include the findings of Bradford (4) a number of years ago that electrical stimulation or potassium stimulation could differentially release taurine compared with other amino acids like leucine, or alanine, examples of the nontransmitter amino acids. That's a suggestive fact and a satisfying difference. Also, when we go a little further down the scale of things and look at the organelles that we find inside the synaptosomes, the synaptic vesicles, there's a whole new question: Can a neurotransmitter exist not stored in a vesicle? I think we have progressed to the point where people will accept the fact that you can have transmitters in synaptosomes without them being in vesicles. But on the other hand Bradford (4), and ourselves (14,15), and Dr. Lähdesmäki (9) have found some taurine associated with synaptic vesicles, and that is a very difficult pool to deal with because, as Dr. Oja talked about earlier, this pool of taurine seems to be very tightly bound to synaptic vesicles. Certainly in our hands (14,15) and Bradford's (4), and I believe in Dr. Lähdesmäki's (9) if I am quoting his work correctly, you can't release that taurine the way you can release other transmitters. For instance, if you take isolated whole cholinergic vesicles and put them over a sephadex column you can separate vesicles that contain acetylcholine from all of the free compounds. You can then take those vesicles and osmotically shock them and release the acetylcholine. It's a very nice system. Roger Marchbanks (11) has done this, and it's an acceptable technique in the realm of neurotransmission. However, it doesn't work with taurine. You can shock those vesicles all you like; the taurine stays right inside. It's a little hard to understand that if these kind of gross treatments don't release the taurine, how, physiologically, is this transmitter released into a synaptic cleft to cause transmission? So just what is that very tightly bound taurine doing there? Is it physiological or is it some sort of artifact of these weird and wonderful separation techniques that we apply to the tissue? We don't really know yet. The final point I would like to make, and one that's resulted from the work that we did in our laboratory in New York over the years, is something that is positive too, and that is that during synaptosomal development one thing that you do see is an apparent protection of the pool of taurine that is isolated in synaptosomes. This finding you might expect if it's a transmitter pool of taurine. In other words, very early in development in fetal brain in rats there is a lesser percentage of taurine in the synaptosome compared to the total taurine and this percentage goes up during development, although the absolute amounts of taurine don't increase because you've got the decrease in total brain taurine concentration that I showed you earlier. So there is a tendency to maintain a more stable pool of taurine within the nerve endings, and perhaps that is the sort of thing you might expect from a neurotransmitter. You wouldn't want to have widely variant synaptosomal pools. A consistent pool to maintain neuronal function is a positive point. So I think I'll just stop here, Nico. Those are some things we're thinking of.

VAN GELDER: Sturman, would you like to comment?

STURMAN: Yes, Dave stopped one point short. We were questioned yesterday about how much of this is redistribution, and those very experiments were done simultaneously with the one Dave referred to, and you don't find labeled taurine added to the brain at homogenization in these other subcellular fractions (16). Secondly, it is worth noting that certainly after birth in the rat the injection of labeled taurine labels all of these pools, and as I mentioned yesterday, in rats that are labeled from conception I've recently come across a pool of taurine in brain which is not exchangeable by a subsequent postnatal injection of taurine (17).

GAULL: Going back to Dave's last point, the so called protection of the synaptosomal pool during development, I wonder, Dave, whether it necessarily implies neurotransmitter function. Harping back to Herminia's association of this with contractile protein, there is certainly contractile protein in the nerve endings and the presence of taurine would be consistent with taurine having this role in contractility. I think in fact we pointed this out in one of our reviews some time ago.

RASSIN: The fact is that you have two pools of taurine really, and redistribution is important here. Also I think maybe when we talk about redistribution we talk about two different things. When you grind up tissues you redistribute taurine in the sense that you break up all the cells and the taurine is released from the cells. You don't have redistribution in the sense that the taurine that is in nerve ending stays in the nerve endings as far as we can tell. It doesn't seem to be added to or subtracted from during the preparation of the subcellular fractions. The other thing is that, true, you've got 75% of your taurine in cell bodies, and 25% in the nerve ending. Is that really the relative amount of taurine that is present in brain? How important is the 25% in the nerve ending for contractile function versus neurotransmitter function, if it's neurotransmitter? I don't think we can measure that until we have some feeling for the relative amounts of contractile protein in these two areas. Maybe Herminia you could address yourself to that.

PASANTES-MORALES: I just want to say that even in the unlikely case that taurine is a neurotransmitter *(Laughter)* it is well proven, almost proven, that neurotransmitter amino acids are not located in synaptic vesicles. For GABA there is a good deal of evidence that it is not stored in synaptic vesicles and I would like to remind you that contractile proteins are involved in exocytosis. The mechanism for neurotransmitter release seems to be a contractile process and...

RASSIN: Do you know relatively how much of the contractile proteins are in nerve terminals versus the rest of the brain if you do some sort of fractionation?

PASANTES-MORALES: No, I don't remember figures. There should be a lot in the axons, of course, Anna Maria, do you know something about the figures?

LOPEZ-COLOME: No.

VAN GELDER: Dr. Fellman.

FELLMAN: I wonder if somebody could tell us something about the nature of the bonding of this unexspongeable taurine, this stable taurine, because hearing the description of the abrasive treatment of those membranes suggest that there's a kind of covalent bond that we may be overlooking.

HUXTABLE: Could I ask a way out question? Do you still see this tight association in delipidated synaptosomes?

OJA: I haven't tried to remove completely the lipids but we have treated this membrane, for instance, with several washings (8,10, 13).

HUXTABLE: Washing with what?

OJA: Triton X-100, weak acid, base and salt solutions.

HABER: But you didn't do chloroform-methanol?

OJA: Yes, we did this too.

HUXTABLE: So even after chloroform methanol extraction you still have the binding?

OJA: Yes, there is still some taurine left.

HUXTABLE: I just had a crazy idea, but I guess that's what we're meant to be putting out. Is it possible that you have aqueous inclusions in a lipid layer, which contain the taurine, and because it's impermeable to a lipid barrier, it cannot get out? So you are measuring apparently a tight binding whereas what in fact you are seeing is a permeability barrier preventing the loss of taurine. That's why perhaps you should look at membranes that have been very carefully delipidated to see if the apparently tightly bound pool of taurine remains.

SGARAGLI: I would like to make a comment about the procedure of trapping lipids with chloroform-methanol or other methods. As shown in Table 1, where the findings of an experiment carried out in my laboratory are reported, the two extraction procedures used in this case caused a reduction in the protein content of the treated preparation. So, my suggestion is that one involved in this kind of problems should analyze the content of lipids and proteins as well at the end of the extraction.

OJA: I agree.

BARBEAU: What could it be binding to? What's left?

VAN GELDER: Well, don't look at me. I never do these things. *(Laughter)*

IRVING: I think that one might want to look at the effect of pH because taurine might form very labile Schiff's base type complexes with an aldehyde group from which you might be able to release the taurine by lowering or raising the pH slightly without disruption of too much supra molecular structure.

VAN GELDER: Is it really an important problem when you do binding studies?

RASSIN: To put it a little in context, of course, that amount of taurine that is bound in these studies is a very, very small amount. You're talking about something less than 0.1% of the brain taurine, and again back to the question, how important is that amount of taurine physiologically?

Table 1

Effect of Different Lipid Extraction Procedures on Protein and Lipid Contents of Rat Liver Microsomal Fraction.

TREATMENT	PROTEIN	TOTAL LIPID	PHOSPHOLIPID
None	37.0	19.9	8.3
$CHCL_3/CH_3OH$[a]	22.2	1.5	0.1
$CH_3CH_2OH/$ $CH_3CH_2OCH_2CH_3$ [b]	2.7	7.2	1.5

Liver microsomal fraction depleted from lipids according to (a) J. Folch, M. Lees and G.H. Sloane Stanley (J. Biol. Chem., 226: 497, 1975); and (b) according to A.F. Horwitz, M.E. Hatten and M.M. Burger (Proc. Nat. Acad. Sci., 71:3115, 1974). Values (mg) represent a mean of three replicates. Total lipid content was calculated gravimetrically. Protein content was determined according to the method of H.H. Hess and E. Lewin (J. Neurochem., 12:205, 1965). Phospholipid content was determined as described by S. Ruggieri, A. Fallani and D. Tombaccini (J. Lipid Res., 17:456, 1976).

CRASS: Dr. Sgaragli, I just would like clarification of your comment on the incompleteness of 2:1 chloroform:methanol extraction: are you talking about neuronal tissues or are you generalizing and referring to muscle as well?

SGARAGLI: I have tried microsomal membranes from lung, liver, kidney and also from synaptic membranes of whole brain homogenates, and I got the same results from all preparations.

MANDEL: How much lipids remain on these microsomes?

SGARAGLI: Oh it was, I don't recall really the figures but...

VAN GELDER: This is becoming a little bit technical. Maybe we should leave that for afterwards.

HABER: I realize this is an important question that is being discussed, but seeing this is not a symposium on lipid methodology I'll ask another question so that we can digress away from that. In terms of stability of the binding, has anyone looked at the binding in another tissue? Let's forget about the brain for a moment. Let's look at an organ that has a lot of taurine and has a clear contractile function, such as in the heart. How tightly is taurine bound to subcellular particles prepared from the heart?

SCHAFFER: We've done studies on that. We use the same procedure as Anna Maria here. That is, all of our data is corrected for nonspecific binding. Thus, our binding data, by definition is reversible: it's not the really tight binding we've heard about today. I've never examined for really tight binding, but I agree with Jack Fellman, if you can't eliminate it by chloroform-methanol extraction, it's got to be covalent.

CHESNEY: In a slide I showed yesterday for the kidney where we were looking at sodium independent transport, which therefore would be binding because the uptake in the sodium independent system is going into vesicles, there is some binding but it's not particularly hard to get rid of. We certainly don't see this type of thing where you have to dynamite it off.

HABER: Just one quicky. It really doesn't have to be covalently bound. One good example is the history of the purification of γ-glutamyl transpeptidase where you have this halophobic polypeptide chain sitting on the enzyme, and if you don't pronase or pepsin-digest this thing off, you isolate an enzyme apparently with a molecular weight of millions whereas it's really only 60,000.

VAN GELDER: Yes, but this is not an enzyme. This is a single amino acid, with a molecular weight of 125.

HABER: I know, but lipids are wonderful things.

KLEIN: There is somewhat of a precedent with serotonin, which can be bound covalently to mitochondrial proteins. There is an intermediate aldehyde that's formed in the oxidation, and it forms some sort of a Schiff's base which is irreversible.

VAN GELDER: O.K. That brings up the next question. When we were discussing "receptors" yesterday, this came up with Herminia. If you look at the physiological effect of taurine, that is to say, the immediate effect on an external membrane which has ionophores, you'll find that the taurine effect seems to be either associated with the GABA receptor or chloride ionophore, because you can look at this with the same pharmacological tools as you use for the GABA receptors, or that affect GABA. Or when you go to spinal cord, the taurine effect seems to be associated with a glycine receptor because there you use pharmacological tools which block both taurine and glycine effects. As far as I know, and I stand corrected by anyone, there has not been found a tissue or a cell in, let's say, a central nervous system type of structure, where you can get a taurine-mediated chloride permeability which is not affected by either GABA or glycine at the same time. So the question then is, all the studies on taurine binding as are being done now seem to be very important because they should be compared to those performed for GABA, and also for glycine - what type of receptor are we looking at? The question came up, Dr. Mandel talked about that, he said that at zero degrees you can't get any transport. He is quite right, but let me propose then that at zero degrees, you still have a transporter and that transporter still has recognition sites. Could you be looking at a transporter which doesn't transport but still binds? How specific, and I think this may be transcending the taurine story, but I think it is well maybe to discuss it a little bit, how specific is the definition if you use a sodium associated binding or nonsodium specific binding to classify a postsynaptic or presynaptic, or physiological or transport receptor. I'd like somebody who has enough courage to discuss this.

MANDEL: Yes, I have courage. I think that now there is good evidence, for instance, GABA receptors, that they are still sodium dependent. That when you measure GABA receptors without sodium you have a part of the story, not the whole story. I should say also that proteins without sodium are artifacts. They do not exist, because a sodium-free water environment for these proteins does not exist, so that is an entirely artifactual situation.

VAN GELDER: We have a lot of our Japanese colleagues here who are very specialized in this field. I would love to hear from them. We would like to draw on their wisdom now that we have the opportunity. *(Pause)* Our Mexican colleague!

LOPEZ-COLOME: What I wanted to say is that Professor Mandel is right in saying that not all binding to postsynaptic receptors has necessarily to be sodium independent, for it has been seen first for the opiate receptor, by Snyder and his group, that the presence of sodium changes the configuration of the receptor and enhances the affinity for antagonists and lowers the affinity for agonists. This also seems to be true for GABA, because it has been seen that the bicuculline displacement curve shifts in the presence of sodium, so this could indicate that further studies should have been performed on this subject before one can really establish that this is a postsynaptic receptor.

VAN GELDER: Let me throw out an old idea, and the chemists probably can laugh me out of the room, but would it be possible to synthesize a taurine analogue which would bind covalently and specifically, like α-bungarotoxin, so that you could take the receptor out...

MANDEL: There are several pharmaceutical companies which are looking for that.

VAN GELDER: O.K. So we all seem to be in agreement that whether binding is nonspecific or specific is not a division. So what are we looking at?

HABER: I would like to throw out the comment that somehow along the way receptors have to have a function, and therefore somewhere along the way we have to have some coupling to some sort of function. I think we have perhaps an example today from one of our Japanese colleagues with this wonderful cysteine sulfinate which seems to activate adenylate cyclase. That's an example of coupling, like the ample evidence with alpha and beta receptors and H_1, H_2 receptors and with GABA and acetylcholine. They are all coupled to something, and I think what has been lacking in the taurine receptor field is what are we coupling it to? I mean, we are applying criteria which we developed for GABA and assuming that these are in terms of just binding, nothing else, and we are assuming that these are the criteria that operate across the board for all amino acids. I would submit as a point of argument that that is probably not so.

VAN GELDER: Could we come to a consensus and maybe think of the "physiological" effect of GABA - it's a misnomer I know - but could the direct membrane effect of taurine not be a nonspecific effect of taurine, one simply due to its coupling to either a GABA or glycine receptor? What I'm trying to drive at here is for a function of taurine which is removing itself further and further away from "the classical transmitter substance". One which is associated, as it is now starting to look from the evidence, to a regulation of either membrane stability or a calcium mechanism which is associated with membrane stability?

CRASS: Are you looking for an answer?

VAN GELDER: Yes, well, I would like to have an opinion. I'm not necessarily insisting on an answer!

SCHAFFER: We have tried to make these correlations in the heart that you are talking about by looking at binding of taurine and several of its analogues to component I (Schaffer et al., this book), and the effect of these analogues on calcium binding to the membrane. We also examined the effect of these same analogues on myocardial contractility using various calcium concentrations. We observe a good correlation between the interactions of these analogues with component I and their ability to alter both calcium binding and contractility in the heart. I think the same thing needs to be done in the central nervous system.

BASKIN: You say you have a sarcolemmal preparation when you don't have a pure sarcolemmal preparation (6). You have to correlate it with some activity other than contractility. Contractility doesn't occur to any extent in the sarcolemma. Perhaps you should correlate it with the ability of ions to cross the membrane to control rhythm. I think that might be a better correlation.

CRASS: I would like to expand on Dr. Haber's point by way of asking Dr. Huxtable to try to tell us about taurine uptake in cardiac muscle in which there's been speculated to be a carrier energy requirement and catecholamine stimulation. This seems to be a classic example of an apparently coupled process. Ryan could you kind of, you have a fairly clear view of how taurine is?

VAN GELDER: Could I place another mandate. Would it be possible to get into the mechanism in what you're telling us, that is, would you be able to make a unified...*(Laughter)* Well, we have to come somehow, you know, eventually to an understanding of what taurine does.

CRASS: An incredible number of physiological couplings, including cyclic AMP, are proposed in the literature. I think Dr. Huxtable knows more about such things. Could you summarize your findings on taurine transport in the heart?

HUXTABLE: I can briefly summarize what we know. There is a high concentration gradient of taurine between the myocyte and the serum, and this gradient is maintained by an active transport energy-requiring process. The turnover rate of taurine from the heart is very slow so the capacity of the uptake system is rather low. However, it is stimulated by cyclic AMP, or agents that give rise to cyclic AMP, and this is a very rapid response. At the shortest time period we can measure, which is within a few seconds, you can see a mild but significant stimulation of uptake of beta amino acids, without any effect on alpha amino acids, so it is specific (5). The kinetics of the stimulation are unusual, and as far as I know they are unique (Fig. 2). A Lineweaver-Burk plot reveals parallel lines for the hearts not exposed to isoproterenol and those that are, showing isoproterenol causes a decrease in both Km and Vmax. That is to say, the capacity of the system has been increased but the affinity of the system has been decreased. This can be rationalized on a simple sort of pseudo-Michaelis-Menten type of situation, where there is rapid and reversible association of ^{3}H-taurine with its receptor (R_o) on the outside of the heart, followed by a slower rate-limiting translocation to the inner membrane, followed by release of taurine (Fig. 3). The final step is shown as being essentially irreversible, because the ^{3}H-taurine is being released into a pool of unlabeled taurine. Now if translocation is the rate-limiting step, and this step is accelerated by cyclic AMP to increase the rate of breakdown of this complex, there will be an increase in Vmax, and you'll also see an increase in Km or a decrease in affinity because at a given taurine concentration the relative saturation of the receptor is decreased. Most of our studies have been done in the isolated perfused heart, where if conditions are changed and an effect is produced it's difficult to decide whether the effect is due to the biochemical change or the physiological change accompanying it. For example, if calcium concentration is changed, contractility also alters, and if an effect is observed is that due to the change in contractility or is it due

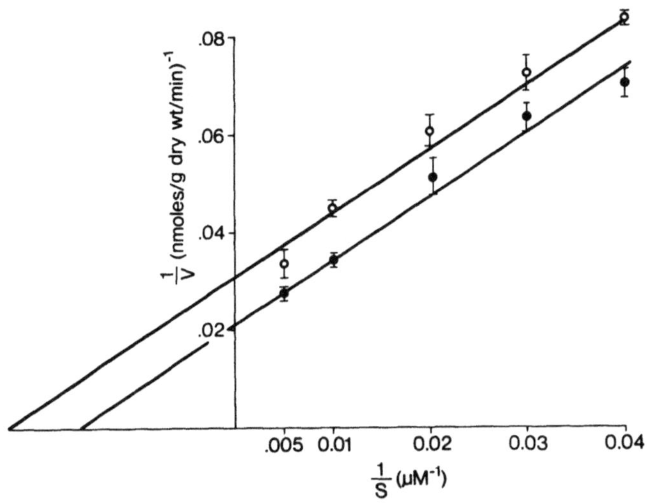

Fig. 2. HUXTABLE. Lineweaver-Burk plot for taurine uptake in Langendorff-perfused rat heart. Open symbols: 25 μM taurine in Krebs-Henseleit buffer. Closed symbols: with isoproterenol (4×10^{-7} M)

to the change in calcium *per se*? We have done a number of careful experiments to dissect out which of the effects produced by cyclic AMP is responsible for this stimulation, and ion movements *per se* do not affect it. Calcium, sodium or potassium movements do not cause this stimulation (1). The ultimate physiological responses to cyclic AMP, the increased contractility and increased heart rate, do not affect this. So it's an event produced rather directly by cyclic AMP.

There is one complication to this story in that another agent that stimulates taurine uptake but without affecting cyclic AMP, is a novel cardiotonic agent, amrinone (2). The company that's making this is touting it as being the replacement for digitalis. This compound will produce a not very marked increase in contractility, a maximum in the guinea pig heart of about 30%, and it will do this without affecting the sodium-potassium ATPase, so it's nothing like digitalis, and without affecting cyclic AMP levels. Yet it does increase taurine uptake. I should say that in the rat heart where it is not inotropic, in fact it is slightly negatively inotropic, it has no effect on taurine uptake.

VAN GELDER: But do you get the same kinetics?

HUXTABLE: We have not done the complete kinetic plot.

$$Tau_o + R_o \underset{}{\overset{k_1}{\rightleftharpoons}} Tau_oR_o \underset{k_{-2}}{\overset{k_2}{\rightleftharpoons}} Tau_iR_i \xrightarrow{k_3} Tau_i + R_i$$

$k_1>k_2<k_3$; $k_2>k_{-2}$ (for transport of ^3H-taurine)

Fig. 3. HUXTABLE. Transport of Taurine

VAN GELDER: Because if you get a change in receptor Km, what you're saying is that the protein has changed.

HUXTABLE: Yes. A phosphorylation, perhaps.

VAN GELDER: But the cyclic AMP response is surely calcium mediated isn't it? So it would still require an intermediary step where calcium would be involved. What I'm trying to get at, again, is to see if we can't come to an agreement somewhere along the line that between taurine and its effect there is a calcium mechanism underlying it.

CHESNEY: I have one question about this experiment. What was time course of the response to cyclic nucleotides? In other words, how long, you were using an isolated perfused heart, did you have to pre-perfuse it with cyclic AMP?

HUXTABLE: Anything that produces cyclic AMP; glucagon, adrenergic agonists, dibutyryl cyclic AMP, phosphodiesterase inhibitors, will produce this effect and you'll see it very rapidly (3). After the heart is exposed to the agent I think the first time point we could measure is within about 15 seconds. We already see the stimulated uptake. So it's a rapid response.

CHESNEY: What about verapamil then with this?

HUXTABLE: This is it. Verapamil by itself will have no effect and we use verapamil to dissect out the influence of inotropy and calcium movements from the effect of cyclic AMP itself (1). The experiment is this (Table 2): If we perfuse with isoproterenol, a beta adrenergic agonist, we see stimulation of uptake. There are also increases in contractility, in cyclic AMP levels, and in calcium influx. If we perfuse with the calcium blocker, verapamil, there is no effect on taurine transport, but a decrease in contractility, cyclic AMP content and calcium influx. Now if we perfuse with isoproterenol and verapamil simultaneously and adjust the concentrations such that the increased inotropy produced by isuprel is exactly antagonized by the negative inotropic effect of verapamil, under this conditions we have no change in contractility, no change in the calcium influx, we still have the increased cyclic AMP, and we still see the increase of taurine influx. This is one

Table 2

Effect of Verapamil and Isoproterenol on Taurine Influx

	Contractility	Cyclic AMP	Ca Uptake	Taurine Uptake
Control	-	-	-	-
Isoproterenol (iso)	↑	↑	↑	↑
Verapamil (ver)	↓	↓	↓	-
iso + ver	-	↑	-	↑

of several experiments whereby we dissected out which of the various consequences of adrenergic excitation were directly involved in the stimulation of taurine influx.

BASKIN: Did you control the heart rate?

HUXTABLE: Heart rate has no influence (5).

BASKIN: But did you control the heart rate?

HUXTABLE: Yes, yes. We looked at rates between 250 and 450 beats per minute.

VAN GELDER: If you balanced it out you're still stimulating cyclic AMP.

HUXTABLE: Agreed. That's what we're saying: there is still the increase in cyclic AMP.

VAN GELDER: But if cyclic AMP remains the same, the contractility, you have boosted it up.

HUXTABLE: No, there's no increase in contractility. Verapamil is acting at a separate site from isoproterenol. Isoproterenol is acting at the β1 receptor. Verapamil acts directly on calcium channels to antagonize influx so that calcium availability within the cell is not increased. Hence contractility is not increased. I should stress again that this stimulatory effect is only seen with β1 receptors not with β2.

MANDEL: I would like to make two comments. Number one: concerning the assumption that taurine is working through GABA receptors. We can maintain a hypothesis, if it fits with all experiments we have. If it doesn't fit with one single experiment, we have to remove it.

There is one single experiment, which is that strychnine blocks taurine effects and strychnine does nothing to GABA receptors. So the assumption that taurine is acting at GABA receptors cannot fit this story.

VAN GELDER: Strychnine is supposed to be much more of a glycine antagonist than a GABA antagonist according to Curtis.

MANDEL: It is what I say. It is what I say. Because of that I cannot accept that taurine is acting through a GABA receptor, because if it would be through a GABA receptor it will not work with strychnine. The second point concerns calcium. There is no doubt that calcium is involved in many many phenomena. Taurine interacts with calcium, at the high concentrations of taurine which are used. Now the question is, whether at physiological concentrations of taurine we will have the same effects on contractile proteins.

VAN GELDER: I think that probably we would have to go to an intact nervous tissue preparation, and I would suggest that the vision contingent is probably the people who are going to give that answer better than anyone else.

SCHAFFER: You can do this in the heart also. If you perfuse a heart under normal calcium conditions, taurine has essentially no effect. However, if the perfusate calcium concentration is decreased or if verapamil is introduced into the buffer, then taurine mediates a positive inotropic effect. We also mentioned yesterday that if one overloads the heart with calcium, one also observes an effect of taurine. This means, as stated by Ryan Huxtable, taurine acts to maintain calcium homeostasis. It tends to normalize the situation.

BASKIN: Most of you people use rat heart, maybe Ryan used another species, but if you were using overdrive suppression you'd have to run that heart with isoproterenol to about 600 or 700 beats a minute, and I maintain that there's a possibility from Bill Crass' work and other work, that you might have an ischemic situation. I'm not sure it's the best model although maybe not the worst model, but in the top ten, the problem is that there's negative staircase and there's a lot of problems with rat heart in terms of pharmacology, such as digitalis insensitivity.

VAN GELDER: I always, I'm being very simple, I always assume that the people who are working with the preparation know more about it than I do and probably have looked at those things. I may be naive.

BASKIN: I think that Ryan just said that he was using guinea pigs. If he can make that comparison I think that would be very good.

HUXTABLE: Yes, we have used guinea pig, hamster and rat heart and in terms of rate of contraction you can control this just with a pair of electrodes; you don't need a drug to do it.

VAN GELDER: You know, this homeostasis interests me also, because we have a homeostatic mechanism when the chloride permeability increases. We have a calcium homeostasis. I certainly can make a case for a glutamate homeostasis. Are there any other examples? Taurine, in effect, more than anything else, is trying to maintain something, or bring it back to normal. I think it is a fascinating concept.

CHESNEY: Well, we discussed the cell volume yesterday.

VAN GELDER: Yes, that's another one, so maybe osmotic regulation.

CRASS: I thank Dr. Huxtable for that explanation. One additional component that he didn't allude to is the evidence for carrier mediation of taurine uptake, such as saturation kinetics and counter transport criteria and this sort of thing. Could you comment, Ryan, on carrier mediation of uptake? I realize it's hard to put your finger on criteria there, but in the context of our discussion about uptake of taurine, some tissues actually release taurine. The difference here presumably is that we've got a large uphill gradient in one direction and a similarly large downhill gradient in the other direction. This is going to influence the factors that regulate uptake versus release. Could you comment on that in the context of possible carrier activity?

HUXTABLE: Well, the uptake is carrier mediated, it shows all the signs. It is saturable, whereas β-alanine is not, indicating diffusion across the membrane. If you plot uptake against perfusion concentration, taurine will appear as shown in Fig. 4, and β-alanine as in Fig. 5.

In terms of exchange, perhaps people try to make transport too complicated. Let's just look at the basic physical-chemical process of movement between two compartments, between an aqueous and an organic compartment (Fig 6). There is an uptake rate ($K_u \cdot C_{aq}$) and a release rate ($K_r \cdot C_{or}$). These rates are proportional to the substrate concentrations in the compartments from which movement is occurring, but are independent of the substrate concentrations in the compartments being supplied. The net movement of substrate, however, is the difference between uptake and release, and hence is affected by both processes. It is therefore dependent on the relative concentrations in the two compartments:

$$\text{Net Movement} = K_u \cdot C_{aq} - K_r \cdot C_{or}$$

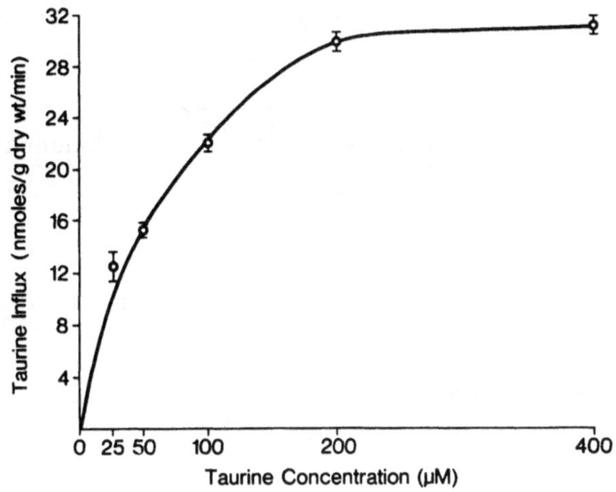

Fig. 4. HUXTABLE. Uptake of β-alanine in isolated perfused rat heart.

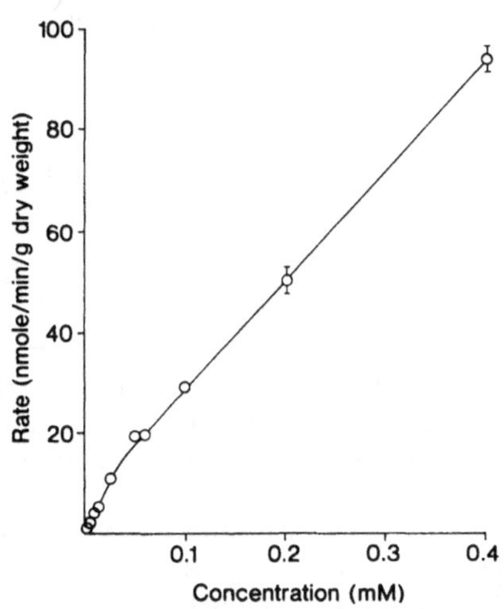

Fig. 5. HUXTABLE. Uptake of taurine in isolated perfused rat heart.

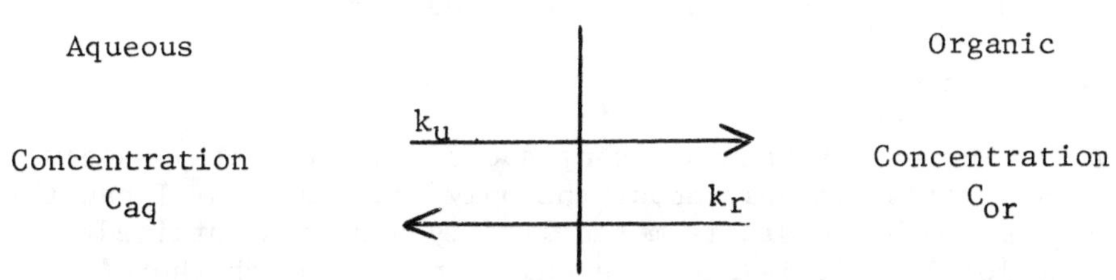

Fig. 6. HUXTABLE. Movement between two phases.

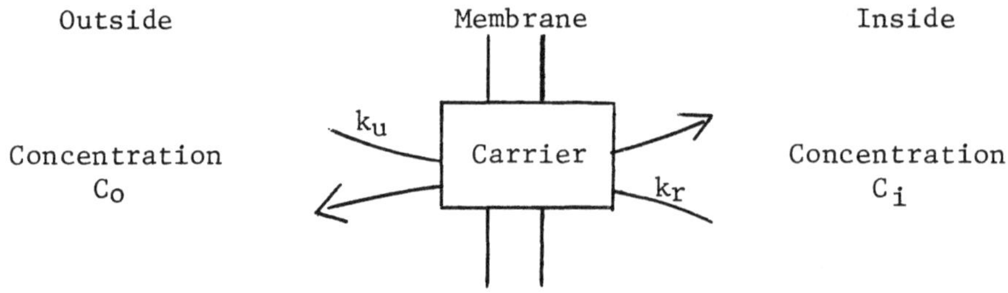

Fig. 7. HUXTABLE. Carrier-mediated transport.

At equilibrium, there is no net movement, and:

$$K_u \cdot C_{aq} = K_r \cdot C_{or}$$

Uptake into the heart can be considered as a 3 compartment system, the extracellular and intracellular compartments being aqueous, and the membrane being a lipid compartment. For carrier mediated influx (Fig. 7), at low substrate concentrations, influx is first order, and is proportional to extracellular substrate concentration:

$$\text{Influx Rate} = K_u \cdot C_o$$

As the carrier becomes saturated, the kinetics tend to zero order, and influx rate is given by:

$$\text{Influx Rate} = \text{Vmax}$$

However, under no condition is influx rate affected by the concentration of substrate within the heart. Furthermore, all the steps involved in active transport are reversible, so any factor that regulates influx via this system will also regulate release. Am I making the point clear?
In these studies we're doing with the heart we are specifically looking at an influx process and it's easy to do because you're looking at a process with a half life of three days, so over the time course of the experiment the rate of uptake of label is constant and will remain so for hours if you wanted to run the influx experiment that long. Did I answer your question?

CRASS: O.K. Thank you.

VAN GELDER: There is something funny about taurine which has always intrigued me. It's probably about the only compound that I can think of which seems to be so stable metabolically and is maintained at such stable levels. It is also the only compound which therefore could serve as a link between organs of the body and nutrition. It seems to be a perfect marker, against which you can measure concentrations. For example, among the many amino acids in all kinds of structures, this is the one which is maintained as a status quo. I just wondered what your opinion is if you look at taurine as a measuring level, against which the body is simply adjusting all the other metabolic parameters to take into consideration nutrition as well as internal processes such as stimulation, or calcium metabolism. That is to say, to explain this very peculiar homeostatic mechanism; we just mentioned four of them.

RASSIN: Nico, you wouldn't say that about sodium or calcium or any of the other ions. The only reason you're saying that taurine is stable is because it's an organic molecule and I'm not so sure that that is the best reason to say that. It's part of the amino acid pool and maybe we can use it as a measure of other amino acids, but there are a lot of molecules in our bodies that we don't metabolize and taurine is there only in a small concentration compared to all the rest of them. I don't know if you can say that it is the yardstick by which all the others are controlled.

VAN GELDER: Any other opinions?

HUXTABLE: When you have about 60-80 grams of taurine in you I wouldn't call that a small amount.

RASSIN: Compared to other unmetabolized components, there's a hell of a lot more sodium in you than there is taurine.

VAN GELDER: Sodium is metabolized and excreted.

GAULL: What is not metabolized? Everything is.

HUXTABLE: My mind's boggling at what you could possibly metabolize sodium to.

VAN GELDER: Now well, it depends again on a definition of metabolism.

GAULL: In any case your hypothesis could be tested. For instance, I know of a few studies which compare the plasma concentrations of amino acids in the fasting and postprandial states and they go up and down, up and down. John, did you ever do time courses in any of the studies that you did where you gave enormous amounts of taurine to rats?

STURMAN: We measured it in humans. Taurine levels go up extremely rapidly, after 9 grams a day, and come down relatively quickly. There are huge waves day by day (18).

KLEIN: An idea just popped into my head. The chemicals that we know that are constant - such as calcium and sodium - are regulated by hormones. So maybe someone should be searching for the taurine hormone. Or to turn it around does anybody have any data in their minds that demonstrate that any sort of ablation results in taurine levels going out of control?

VAN GELDER: There is certainly evidence that in hyperthyroidism and hypothyroidism, rheumatoid arthritis, there is a change of taurine excretion. The reverse now is also becoming apparent. I think the evidence is strong that taurine works on prolactin secretion, and, in general, that it works on the pituitary. So the interaction of taurine with hormones is starting to be important.

GAULL: Didn't Devivo do some studies on patients with seizures giving taurine, and didn't he show that there were changes in growth hormone levels?

VAN GELDER: The antiepileptic agents used are themselves now being shown to cause also, some of them at least, hormone changes and the question here is that these patients were all on medication. Was he measuring the effect of taurine, or was he measuring something else such as whether the action of the antiepileptics somehow was altered, or something like that.

STURMAN: Nico, if you want to test your idea of using taurine to measure other things against, I would think you would be better off in a species that normally has no dietary component, say the rabbit or the guinea pig, because then you could modify dietary intake. We know these species keep their taurine internally constant so you could see the effect of modifying external amino acids in the diet. Does that thought hold up?

VAN GELDER: But what intrigues me is exactly the fact that if your nutrition goes off, the internal organs respond by increasing taurine synthesis. If your nutrition is adequate then the internal clock tells you.

STURMAN: It's better to take the other side. Are you saying that a constant taurine is something that, say, glutamate is measured against and if that ratio changes it tells you do something about the glutamate?

VAN GELDER: Oh yes, it should be tested. As I say, our mandate was to talk about the possible functions of taurine and that's what I'm doing.

KLEIN: Actually putting together observations that Ryan has made and that I've made and then a little bit of theorizing, we could generate an interesting possibility that the catecholamines might serve as regulatory hormones. Any sort of stress that induces release of catecholamines may cause a release of taurine from tissue stores, which will make taurine more available for use in the heart.

BARBEAU: Stress was studied in the Jacobsen review. I remember that this was one of the major causes of increased taurine excretion in the urine.

KLEIN: Is that right? That's interesting.

BARBEAU: That was one of the first things found, if I remember.

KLEIN: That's very interesting then. Maybe our observations which on the surface don't seem to support each other, Ryan, are perhaps integral parts of reasonable hypothesis.

HUXTABLE: Why, I never thought that your data were in any way in conflict with mine.

VAN GELDER: My God I think on this sweet note we should immediately stop. We have reached a consensus.

REFERENCES

1. Azari, J., and Huxtable, R., 1980, The mechanism of the adrenergic stimulation of taurine influx in the heart, Eur. J. Pharmacol., 61:217-223.
2. Azari, J., and Huxtable, R.J., 1980, Differential effects of amrinone on contractility and taurine influx in rat and guinea pig hearts, Eur. J. Pharmacol., 67:347-353.
3. Chubb, J., and Huxtable, R., 1978, Isoproterenol-stimulated taurine influx in the perfused rat heart, Eur. J. Pharmacol., 48:369-376.
4. DeBellaroche, J.S., and Bradford, H.F., 1973, Amino acids in synaptic vesicles from mammalian cerebral cortex: A reappraisal, J. Neurochem., 21:441-451.
5. Huxtable, R., and Chubb, J., 1977, Adrenergic stimulation of taurine transport by the heart, Science, 198:409-411.
6. Jones, L.R., Besch, H.R., Jr., and Watanabe, A.M., 1978, Regulation of the calcium pump of cardiac sarcoplasmic reticulum, J. Biol. Chem., 253:1643-1653.
7. Kontro, P., and Oja, S.S., Hypotaurine oxidation by mouse tissues, in: "Natural Sulfur Compounds: Novel Biochemical and Structural Aspects," G.E. Gaull and V. Zappia, eds., Plenum Press, New York (1980), pp.201-212.
8. Kumpulainen, E., Olkinuora, M., and Lähdesmäki, P., 1979, Ability of calf brain synaptic membranes to bind ^{35}S taurine

to their Triton X-100 and chloroform extracts, Neurosci. Lett., 11:215-218.

9. Lähdesmäki, P., Karppinen, A., Saarni, H., and Winter, R., 1977, R. - Amino acids in the synaptic vesicle fraction from calf brain: content, uptake and metabolism, Brain Res., 138: 295-308.

10. Lähdesmäki, P., Kumpulainen, E., Raasakka, O., and Kyrki, P., 1977, Interaction of taurine, GABA and glutamic acid with synaptic membranes, J. Neurochem., 29:819-826.

11. Marchbanks, R.M., 1968, Exchangeability of radioactive acetylcholine with the bound acetylcholine of synaptosomes and synaptic vesicles, Biochem. J., 106:87-95.

12. Oja, S.S., Karvonen, M.-L., and Lähdesmäki, P., 1973, Biosynthesis of taurine and enhancement of decarboxylation of cysteine sulphinate and glutamate by the electrical stimulation of rat brain slices, Brain Res., 55:173-178.

13. Oja, S.S., Kontro, P., and Lähdesmäki, P., Transport of taurine in the central nervous system, in: "Transport Phenomena in the Nervous System. Physiological and Pathological Aspects, Advances in Experimental Medicine and Biology, G. Levi, L. Battistin and A. Lajtha, eds., Vol. 69, Plenum Press, New York (1976), pp.237-252.

14. Rassin, D.K., 1972, Amino acids as putative transmitter: Failure to bind to synaptic vesicles of guinea pig cerebral cortex, J. Neurochem., 19:139-148.

15. Rassin, D.K., Sturman, J.A., and Gaull, G.E., 1977, Taurine in developing rat brain: Subcellular distribution and association with synaptic vesicles of (^{35}S)-taurine in maternal, fetal and neonatal rat brain, J. Neurochem., 28:41-50.

16. Rassin, D.K., Sturman, J.A., Hayes, K.C., and Gaull, G.E., 1978, Taurine deficiency in the kitten: Subcellular distribution of taurine and (^{35}S)taurine in brain, Neurochem. Res., 3: 401-410.

17. Sturman, J.A., 1981, Origin of taurine in developing rat brain, Developmental Brain Res., in press (1981).

18. Sturman, J.A., Hepner, G.W., Hofmann, A.F., and Thomas, P.J., 1975, Metabolism of (^{35}S) taurine in man, J. Nutr., 105: 1206-1214.

19. Yamaguchi, K., Sakokibara, S., Asamizu, J., and Ueda, I., 1973, I. Induction and activation of cysteine oxidase of rat liver. II. The measurement of cysteine metabolism *in vivo* and the activation of *in vivo* activity of cysteine oxidase, Biochim. Biophys. Acta, 297:48-59.

SESSION IV

DO TAURINE AND ITS ANALOGS OR COGENERS HAVE ACTIONS IN COMMON?

Moderator: Stephen Schaffer

SCHAFFER: This table is taken from Grosso's study (6) on taurine uptake by fetal mouse heart (Table 1). I'm just using their work as an illustration because it's well recognized that beta amino acid uptake systems are extremely specific. For example, if you replace the sulfonic acid group by a phosphonate group, the new substance can no longer inhibit taurine uptake by that system. On the other hand, a carboxyl group in that position is permitted; beta alanine is clearly a potent inhibitor of taurine uptake. Another observation is that amino methane sulfonic acid, an analogue in which the carbon chain is decreased by one methylene, no longer interacts with the beta amino acid site. Likewise, if one lengthens the chain by one carbon to yield homotaurine, one also reduces the effectiveness of the inhibitor. If the amino group is removed or blocked, the compound ceases to act as an inhibitor indicating that the amino group is also an important factor in determining the effectiveness of the inhibitor. Perhaps one only requires a positive charge in that position, since the amino group can be replaced by a guanidinium group. It's these kinds of studies that suggest that the beta amino acid transport system is highly specific. I realize that uptake is one thing, and the actions of various analogues may be another, but there is some evidence to suggest - at least we saw some today - that the actions of taurine may also be highly specific, suggesting that there has to some sort of receptor. There are several things, of course, that one has to be concerned about. Are the actions of taurine really due to taurine or is there some other compound that is responsible for its actions? For example, Welty and Read (17) some time ago suggested that isethionic acid was responsible for the actions of taurine. Ryan, will you briefly comment on their hypothesis?

HUXTABLE: Right now?

SCHAFFER: Yes, right now.

TABLE 1

INHIBITION OF TAURINE UPTAKE BY TAURINE ANALOGUES AND

AMINO ACIDS IN FETAL MOUSE HEART

INHIBITOR[a]	UPTAKE AS PERCENT OF CONTROL[b]	
Hypotaurine	13.1 ± 1.1[c]	(4)
β-Alanine	13.4 ± 1.6[c]	(4)
γ-Aminobutyric acid	67.9 ± 4.4[d]	(4)
α-Alanine	94.9 ± 0.9	(7)
α-Aminoisobutyric acid	105.7 ± 4.1	(5)
Glycine	109.6 ± 10.8	(3)
Leucine	103.3 ± 2.6	(5)
Threonine	104.8 ± 9.9	(6)
Guanidinopropionic acid	20.7 ± 0.7[c]	(5)
Guanidinoethyl sulfonate	26.2 ± 2.8[c]	(3)
N-Acetyltaurine	110.6 ± 4.5	(4)
Isethionic acid	121.4 ± 5.1	(6)
2-Aminoethane phosphonic acid	83.2 ± 8.5	(4)
Aminomethane sulfonic acid	80.3 ± 5.5	(3)
3-Aminopropane sulfonic acid	89.2 ± 1.9	(4)

[a] Incubations were carried out for 1 hr with 0.2 mM (^{3}H)taurine and 20 mM inhibitor. Controls were incubated with 0.2 mM (^{3}H)taurine only.

[b] Controls were normalized to 100%. Values represent mean ± SEM. Numbers in parentheses indicate the number of hearts.

[c] Significantly different from control, P<0.001.

[d] Significantly different from control, P<0.01.

Data from Grosso et al. (6).

HUXTABLE: I think I can comment very quickly. In terms of considering what taurine does we can ignore isethionic acid in mammalian systems. The current state of knowledge can be summarized as follows. ISA occurs in rather small amounts in brain. It may or may not be formed from taurine, but if so this is not a quantitatively important route for the metabolism of taurine. Other routes by which ISA may be formed not involving the intermediacy of taurine have been proposed, and partially studied (5,15). Finally, the physiological actions of taurine are not predicated on its conversion to ISA. Although isethionic acid certainly is present in mammalian systems and it may have physiological actions of its own, I think the conversion of taurine to it is more myth than anything else, and so we can just ignore it. Details of this problem can be found in a recent review of mine (8).

SCHAFFER: Then we can go to homotaurine. Homotaurine has a lot of taurine-like activity (2,4,16). In fact there have been reports that it is more potent and effective than taurine itself. Steve, you've looked at homotaurine levels. Would you like to comment on that?

BASKIN: I would like to comment on your last slide. I agree with Ryan that 2-aminoethanephosphonic acid doesn't have the effect that you would expect by replacing sulfur with phosphorus, and part of the reason may be due to the fact that it's very rapidly inactivated and the Japanese (7) have shown that recently. If you use 3-aminopropanephosphonic acid you get a totally different picture than with 2-aminoethanephosphonic acid, at least in our hands. Our data on 2-aminoethanephosphonic acid is very much like your data, Ryan.

HUXTABLE: I know of no mammalian enzyme that can break the carbon-phosphorus bond; does anybody else?

SCHAFFER: That's purely unprecedented, Steve.

BASKIN: Its been published.

SCHAFFER: Well, getting back to the original question; you've looked at homotaurine levels in central nervous system.

BASKIN: Yes, it's a very small (12). There is some. We found it in the pineal and in the cerebellum. The levels are quite dwarfed by those of taurine, and in several of our chromatographic systems you don't find it, because it's under taurine and it's just buried.

SCHAFFER: Is its concentration too low to have any physiological effect?

BASKIN: That's a good question. According to Curtis and Watkins (3), it is more effective than taurine. If I were to pick an agent that was a potential neurotransmitter, I'd probably take homotaurine rather than taurine, but that's somewhat speculation at this time.

SCHAFFER: You've done some studies on homotaurine, Dr. Sgaragli, do you have a comment?

SGARAGLI: Homotaurine in the model I've used is more potent than taurine, and it shares most of the taurine effects. This is in contrast with the GABA binding studies on brain membranes which show that homotaurine antagonizes GABA binding.

HUXTABLE: In terms of the binding of taurine, the distance between the acidic and basic ends of the molecule is important. Aminopropane sulfonic acid has no affinity for the taurine transport system, but one molecule that we have all overlooked is aminobutane sulfonic acid, the one higher homolog. This compound is conformationally more versatile than the propane derivative, and can readily assume conformations that meet the same space requirements as for the acidic and basic ends of taurine.

BASKIN: How do you know that Ryan?

HUXTABLE: Molecular models.

BASKIN: How about acetylcholine and its congeners?

HUXTABLE: How about them?

BASKIN: I mean, there's a big difference between acetylcholine and the congeners in terms of effects at the neuromuscular junction.

HUXTABLE: What two compounds are you talking about, acetylcholine and what else?

BASKIN: Katherine Heber in Cambridge has looked at propinylcholine and butyrylcholine.

SCHAFFER: I don't quite follow what you're saying.

BASKIN: People have looked at classical models, and I think that all of us that are pharmacologists have looked at classical models and people looked at choline analogues, looking at acetylcholine, and the clam heart or other preparations and the propinyl derivative has different effects than the acetyl. Some people have done a great deal of these studies and I'm not sure why that would be any different than taurine analogues. You don't like that, Dave?

RASSIN: No, I think, as Ryan says, you've got to fit the receptor. Most of the acetylcholine congeners do that, the ones that have cholinergic activity, and Ryan's right, you have to have something that fits.

HUXTABLE: Even the spacing in curare alkaloids is the same as in acetylcholine...

SGARAGLI: Excuse me, Ryan, if I get your meaning, why doesn't taurine affect GABA binding to brain membranes? Why what does taurine do, this is my question.

HUXTABLE: Why does homotaurine affect GABA binding and...

SGARAGLI: Yes, why taurine does not?

HUXTABLE: I don't know, but I do know that the spacing between the acidic and basic ends of GABA can be duplicated by homotaurine. While we're talking about structure activity relationships, another interesting compound which we have only recently looked at is thiotaurine. Now with thiotaurine all you've done is to replace the SO_3H by S_2O_2H group, so it's still acidic, with similar properties to taurine. Thiotaurine has absolutely no affinity for the taurine transporting system in the heart, so clearly the acidic end of the molecule is important because the only difference between aminoethanephosphonic acid and taurine is that one is a dibasic acid and the other is monobasic. Even this may not mean too much, as the acid grouping is probably only mono-ionized at pH's around 7 (14). Otherwise the compounds are isosteric and isoelectronic. With thiotaurine the only difference is that you have a slightly larger grouping at the acid end, so there is extreme specificity there for some reason.

SCHAFFER: Let's move on to the third compound. I think the discussion today and yesterday indicated that hypotaurine might be a potential effector in the central nervous system. There's a lot of evidence supporting that view. Would you like to comment on that, Dr. Oja?

OJA: Yes, at least I think it is not excluded. We have little information besides that we have just shown in favor of a role for hypotaurine. At least in our studies the transport phenomena involving hypotaurine are very similar to those of amino acid transmitter candidates (10,11,13).

SCHAFFER: Jack, do you have any comment on possible functions for hypotaurine?

RASSIN: Steve, a comment on the amounts; there was a comment that homotaurine is only present in small amounts. Maybe this is more in accord with classical neurotransmitters. With taurine we usually suffer an embarrassment of riches, and things like homotaurine and hypotaurine are much closer to the amounts in the brain that we see with acetylcholine, serotonin, and catecholamines and I think that's something to keep in mind.

SCHAFFER: How about Kochakian's results (9) in male reproductive system?

RASSIN: Well, here we are not talking in terms of a neurotransmitter role.

FELLMAN: Presumably homotaurine comes from homocysteine. I wonder if there's any observed in homocystinuria. Normally the metabolism would be directed away from any accumulation of homocysteine, homocysteic acid and thus homotaurine.

GAULL: Is the pathway known?

FELLMAN: No, I'm just speculating. I'm going to assume that if there's homotaurine it must come from some such sequence, Dave, do you remember whether homotaurine comes from homocysteine?

RASSIN: The only thing you can really say about homocystinuria, is that taurine is never reduced in concentration in either the plasma or urine of affected individuals.

GAULL: I think probably that in man most of the taurine they found in the urine is derived from the diet.

STURMAN: How much homotaurine are you talking about?

BASKIN: I can't give you the exact numbers but it's in the area of acetylcholine. To get back to the point that Jack raised: I went looking at particularly Gerry Gaull's paper I think, his reviews. I'm looking to see if there was a high level in anybody's paper of homotaurine, because it seemed an obvious place to look in a homocystinuric patient or in a patient with homocystinuria.

GAULL: It obviously wasn't obvious to us.

BASKIN: My impression is that no one has really even looked for it, or they looked, and it was never found. Perhaps it's a labile compound _in vivo_. Perhaps it turns over very fast or the levels are very small or it's masked by taurine in conventional chromatographic systems.

RASSIN: There have been a variety of strange sulfur compounds shown to be excreted by homocystinurics but I don't recall homotaurine being one of them.

GAULL: I don't. If anybody would have looked for it they would have found it.

BASKIN: Maybe the urine is the one place to look for it.

GAULL: Well, I think we've pretty well accounted for everything in the plasma. The plasma is not that complicated. I was going to consult our renal expert but I guess he's not here. By and large amino acids that we don't have experience with usually are not very well reabsorbed by the kidneys so that one would expect that this would be found most readily in the urine if it was to be found. I have some urine for you if you'd like to...

SCHAFFER: Nobody has discussed the actions of taurine as an anti-epileptic agent. I was talking with Dr. Iwata earlier today and he has an interesting theory on the effects of cysteine sulfinic acid and other sulfur containing amino acids. Would you like to discuss your theory, Dr. Iwata?

IWATA: I don't like to say any speculation while in the hopeless stage you know, but I just wonder why the studies concerning the antiepileptic action of taurine so quickly disappeared. In the first meeting of this Symposium I heard so many beautiful Canadian, mostly Canadian, reports concerning this problem and I was so much stimulated about this...

BARBEAU: The answer is simple: they are now classics!!

IWATA: No, it is not classic now, and I dare to say I want to stay still in this position. That's my answer. I'll let you answer.

BARBEAU: Well, I think Nico probably will want to say a lot more, but I think the studies in animal have all been done at pharmacological doses and at these doses there's no question that in most models except the kindling and one other I forget, you know there are two models that don't work, but in all the other models it has an antiepileptic action. In man it hasn't turned out to be that good. Now whether it's because we don't know how much to give, or how to give it or what, at least in our experience, except in a few cases of temporal lobe epilepsy where we can suspect we have a blood brain barrier defect, we haven't really found clear cut evidence of improvement. I know there's a Japanese study that has better results than we have. I still think that it has clear cut action in animals.

SCHAFFER: Ryan, would you like to comment on that?

HUXTABLE: One of the reasons at least why we're not discussing it at this meeting is rather clear to me. As Dr. Barbeau has pointed out, the fact that taurine is a potent anticonvulsant agent is now well established. We have established the phenomenon beyond any doubt, the problem now is trying to work out the mechanism, and until we get some insights into mechanism continued reports of the anticonvulsant action of taurine in animals really do not have a great deal of interest.

GAULL: If it works in man it will have a great deal of interest.

HUXTABLE: Well, we don't know whether it works in man because it has never been tried. There have been...

BARBEAU: It has been tried in man but I don't think we know exactly how much to give or how to give it and so on. So studies in man are inconclusive still.

HUXTABLE: When I said it's never been tried in humans, I'm of course aware of all the publications that have appeared on taurine in humans (Table 2). None of these studies were controlled studies, and you cannot accept any evidence as binding that is not obtained from a controlled study. You're going to need a group of patients that are not basket cases, and yet how are you going to find these, how can you withhold from patients the treatment that is available to test them on taurine?

GAULL: The only ones you can really test them on are the ones that are unresponsive to standard anticonvulsants, and these...

HUXTABLE: And these are not controlled tests and they are the worst of all possible cases to try a new drug on.

VAN GELDER: Could I just interject a little thing that you forget. I think that the problems with the studies in man have been, first of all, that they have been done with the cases which are extremely drug-resistant. That is to say, everything else has been tried, and in final desperation taurine was tried. The impressive thing most people seem to ignore is that even in those patients, in 25% of these patients, there seems to be some effect. That's first. The second point, which has never been listened to in my opinion and which I find personally extremely frustrating, because each of these meetings has stressed this, is that every time people try taurine, that is when clinicians try it, they are either not prepared to wait a couple of months using low doses, or if they get an effect they keep on pumping the taurine in at a rate which in my opinion is toxic. There is such a thing, in my opinion, as a toxic dose of taurine and until the clinical trials are done in young children in moderate and small doses, in a well constructed clinical trial, well set up, you will never get much more information on this; it will never be accepted as a general "yes or no" antiepileptic. But to do that nowadays in North America, certainly, it requires an enormous amount of paperwork, permission, you need a hospital staff to observe and you need continuous monitoring of the EEG. Most clinicians are really not interested in it, and most studies of antiepileptic drugs are done by drug companies who want to make money. As Don Tower pointed out the other day, it costs, I believe, 64 million dollars before a drug comes to the market nowadays. So for something like taurine which is not patentable and which is not expensive, there's very little attraction.

IWATA: But I remember one compound which was introduced by Dr. Baskin at the second meeting in Tucson I think.

BASKIN: We made a compound which was a cross between diphenylhydantoin and taurine. We made taurine-phenytoin (1). One of the problems I found with that compound was that it had a high amount of plasma protein binding, in fact higher than that of phenytoin. If someone could remove the two carbonyl groups I think we would

TABLE 2

CLINICAL TRIALS OF TAURINE AS AN ANTIEPILEPTIC AGENT

Barbeau and Donaldson, The Lancet II, 387, 1973.
Bergamini and Mutani, Rev. N. Sper. Freniat., 97 783-787, 1973.
Barbeau and Donaldson, Arch. Neurol., 30 52-58, 1974.
Bergamini, Mutani, Delsedime and Durelli, Europ. Neurol. 11 261, 1974.
Sbarbaro, Acta Neurol. (Napoli), 29 33-37, 1974.
Striano and Buscaino, Acta Neurol. (Napoli), 29 537-542, 1974.
Bergamini and Mutani, Le Concours Medical (Paris), 35 25-28, 1975[c].
Borromei, Sacquegna and Coccagna, Phronesis, 17 401-415, 1975[b].
Marchesi, Angeleri and Quattrini, Riv. Neurol., 45 123-127, 1975.
Mutani, Monaco, Durelli and Delsedime, Epilepsia, 16 765-769, 1975.
Van Gelder, Sherwin, Sacks and Andermann, Brain Res., 94 297-306, 1975[a].
Barbeau, Tsukada and Inoue, in "Taurine," Raven Press, 253-263, 1976.
Mutani, Bergamini and Durelli, in "Taurine and Neurological Disorders," Raven Press, 359-374, 1978.
Takahashi and Nakane, in "Taurine and Neurological Disorders," Raven Press, 375-386, 1978.
Mantovani and DeVivo, Arch. Neurol., 36 672-674, 1979[d].

[a] No studies on clinical effects were performed.
[b] Quoted in Barbeau, Can. J. Neurol. Sci., 343-346, 1975.
[c] Quoted in Mutani, Bergamini and Durelli, in "Taurine and Neurological Disorders," Raven Press, 359-374, 1978.
[d] No anticonvulsant effect observed.

probably have a good compound. Maybe even a cheaper way is put taurine in a liposome and try to introduce taurine as a liposome so you can get it past the blood brain barrier. Someone should do that.

VAN GELDER: The blood brain barrier is not a problem here.

GAULL: The blood brain barrier is no problem. And furthermore, in a sense we speak about epilepsy as if it is an entity, and in fact it would be a little bit like if we went back 40 years and somebody came up with penicillin and decided that we ought to try this drug on fever. In a certain number of cases it would work, but it would work because it affected the bacteria and not because it affected the fever. I think that, I have to say, until we understand the mechanism so that we can select the patient in some way other than the fact that the patient doesn't respond to the standard anticonvulsants, you have to have some sort of marker or understanding of how to select the patients.

SCHAFFER: Steve you've always emphasized the fact that one should be able to modify taurine to produce a new substance which will more easily penetrate the blood brain barrier. Based on our discussion thus far, how much can we alter the structure of taurine and still observe its effects? Or how can we change taurine to potentiate its activity?

BASKIN: I'm not sure for all the actions of taurine, only the ones we have measured.

BARBEAU: The first question we should ask is whether it's true that taurine does not cross the blood brain barrier. Are we absolutely sure of that, I'm not convinced. I think some of it crosses.

MANDEL: At least in the rat, we increased taurine in plasma about 20 times and there was not any increase of taurine in brain.

HUXTABLE: How long did you run the experiments?

MANDEL: Two weeks.

HUXTABLE: You're too impatient.

GAULL: But that doesn't mean it doesn't cross the blood brain barrier.

BARBEAU: That only means that you're maintaining a homeostatic mechanism in the brain and it may be just turning over faster. It doesn't mean that it doesn't get there.

GAULL: It doesn't really mean that it doesn't get in. On the contrary.

VAN GELDER: From my experience, it's very simple. If you lower the taurine levels by certain artifical means, for example by inducing epilepsy in an animal which has very high brain taurine, and you subsequently give taurine, you see an immediate restoration, practically in a couple of days, of the taurine levels. If you have an animal where the taurine levels have not gone down, such as the cat, you will find that you can pump in as much as you want, and the taurine levels remain completely stable. The simple answer is, if you perfuse a cortex - open cortex - with taurine, you will find you can go up to 60 mM, which is ridiculous in terms of concentration, and you find that no more gets in than what you can account for by entrapment in the extracellular space. If you lower your taurine levels first you can put in as little as 10 μM and immediately see an uptake. The same is true for brain slices as a matter of fact. Anybody can try it.

STURMAN: So you're saying that taurine is effective only in those seizures or epilepsies where taurine concentrations are reduced?

VAN GELDER: No, because that is contradicted by people working with the epileptic baboon; it is also contradicted by the work of Fariello. There are a large number of seizure models - such as metrazole convulsions - where there is no decrease in taurine, but taurine still seems to be effective. It is not from my laboratory. What I'm saying is that the taurine levels in the brain can be restored very easily by intravenous, intramuscular, or even feeding experiments. Look at Emson's data, look at my data, look at many others. But once the taurine levels are back to normal you will not see any further increases.

STURMAN: Has anybody ever looked at treating such a model where there's no taurine change with radiolabeled taurine to see what its disposition is in the brain?

VAN GELDER: No, I have not, because I am always confused by labels. No, I'm just not capable, I'm not an expert in tracer techniques and there I never know what to do.

BASKIN: Does the blood brain barrier to taurine breakdown during seizure?

VAN GELDER: There is absolutely no evidence for that. With these animals, by the way, you're talking about spontaneous epilepsy, and with spontaneous epilepsy you don't have an animal which is continuously in seizures. An epileptic state may mean that once every two weeks you observe a seizure but it doesn't always mean that you have status epilepticus which is a big difference. You have to distinguish between seizures and epilepsy.

BASKIN: When they are having convulsions, does the blood brain barrier to taurine or any other substance breakdown?

VAN GELDER: Well, there are some people working with the CAT Scanner, and work done at Montreal Neurological Institute. There may be a breakdown of the barrier if you have anoxia or hypoxia, but that again does not necessarily occur during seizures.

BASKIN: Couldn't that explain part of your data.

VAN GELDER: What?

BASKIN: Couldn't that explain part of your data? That when the animal seizures there more taurine goes in the brain because there's a breakdown of the barrier?

VAN GELDER: From my experience with spontaneous epilepsies, that is to say epilepsies where I haven't touched the cortex, and from what I've seen from studies by others, you get an abnormal EEG, you give taurine for a month and the EEG stays abnormal: you come back the next morning and all of a sudden you see an animal which is practically normal. This effect is dramatic to the extent that my colleague - not my colleague but my superior, Dr. Jasper, who is really one of the greatest experts in epilepsy - was so incredulous, that I went back and took the animal off the taurine. When we did that we got the epilepsy back. We gave exactly the same treatment again, took again a month but then after having gotton rid of the abnormal EEG, you put a gram of taurine in there, you watch out. You see an enormous EEG change which is absolutely abnormal.

GAULL: Nico, has anybody in any of these models that respond to taurine, harping back to our hostess's universal hypothesis, has anybody looked at calcium - phosphorous fluxes, because they are known to be involved in seizures, and has anybody looked at the effect of pH, especially the administration of bicarbonate which seems to be critical for the action of taurine?

VAN GELDER: That is the classical work of Dixon Woodbury, of course, with diamox.

GAULL: Has it been done with taurine?

VAN GELDER: No, I mean modifying the pH and a ketogenic diet.

GAULL: Yes, this is well known, but what I'm trying to do is to work Herminia's ideas into your observations. Herminia pointed out the importance of calcium, phosphorus and bicarbonate to the action of taurine and so I'm asking, has anybody studied the other factors in the taurine responsive epileptic models?

VAN GELDER: I don't think so.

PASANTES-MORALES: Barbeau and Izumi demonstrated that when EDTA is injected with taurine, taurine has no protective effect. This means, in that case at least, the effect of taurine is related to free calcium present in the extracellular space. It is just a hypothesis of course, but probably taurine is maintaining calcium in some way chelated or in the form of a compound, a complex, close to the membrane, thus decreasing excitability in a general way. This could explain the decrease of seizures or convulsions coming from a variety of experimental models.

VAN GELDER: There have been a large number of studies done on the extracellular - intracellular calcium in spontaneous epilepsy proving that (and I would say on the average that people have never been convinced, that is to say, one electrophysiologist has never been able to convince another electrophysiologist) there is a clearcut change in calcium extracellular - intracellular, with the onset, the duration or offset of a seizure. On the other hand, I personally have always expressed the opinion that the effluxes of glutamic acid which I see and which you see in practically any hyperirritable tissue including just ordinary provoked stimulation, I have always been convinced that what I was looking at was the tail end of some other mechanism and I've always, because of this, guessed that glutamate binding to some extent was connected somehow with this calcium mechanism.

GAULL: It may well be the tail end, and if it is, then we have to have some other probe for the system in order to examine the taurine effect, I think.

VAN GELDER: I couldn't agree more.

GAULL: In order to select the patients.

VAN GELDER: Sure, there is one other way of selecting patients. We are doing it now, and it seems to be holding up. But Dr. Huxtable will probably tell us more about it in a few years also as he and Dr. Laird are now studing amino acid changes in epileptic patients. You can pick up glutamate changes practically in the blood of many epileptic patients and their relatives, you seem to see a change in the taurine/glutamate pattern. That seems to be true in Finland, it seems to be true in Italy, and seems to be true in Montreal.

SCHAFFER: I..excuse me.

VAN GELDER: I'm sorry, I'm sorry, I'm running off at the mouth.

SCHAFFER: It's getting kind of late, and I'd like to have Ryan sum things up. I think one of the things that we have to think about is that the effects of taurine are generally not potent effects. So one has to ask, is there a substance that can be synthesized that would be much more potent than taurine itself and I'm going to pose this question to Ryan. Or is that a nasty question?

HUXTABLE: It's not a very fair question at this time of day. I'm not really sure that we need taurine derivatives to increase delivery to the brain. I take Nico's point, that if there are disturbances, if taurine levels in the brain are lowered, you can raise them by giving exogenous taurine. We have to remember, however, that the exchange of taurine from the brain is rather slow. The uptake capacity of the brain for taurine is rather low, so it takes time, and the experiments that were being done only a few years ago where patients were being given 10 grams a day, or 15 grams a day, it was overkill! Nico was the first who propounded the advantage of giving small repeated doses of taurine, and when you do this you can increase the retention of taurine. To give you an example, we treated a patient in Tucson, a child with intractable epilepsy, with taurine; we just gave him 500 mg of taurine per day and we followed his urine taurine excretion. His baseline excretion was about 30 mg per day at home and 12-13 mg per day in hospital (Fig. 1). Each day he's taking this taurine he's excreting rather more, so the first day he takes 500 plus 30 in his diet a total of 530 mg, and he excreted 160 mg (Fig. 1). After a period of about 3 or 4 weeks he's excreting approximately 400 mg per day, the same amount he is taking in (Fig. 2), and if you calculate the difference between what he's taking and what he has excreted, and it adds up to a considerable amount of taurine. In the first 10 days he has retained 1.8 g taurine. Now I don't know whether it's in the brain or in the muscle, but it's certainly somewhere in the body. The serum levels, by the way, remained rather constant throughout (Fig. 3). So maybe we don't need to be in too much of a hurry to rush into trying to make derivatives which will only complicate the picture in terms of toxicity testing and so on so forth.

SCHAFFER: I guess we better terminate now.

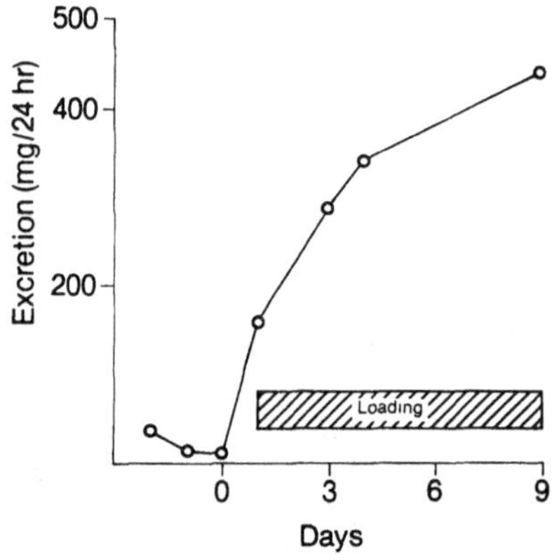

Fig. 1. HUXTABLE. 24 hr Urinary excretion of taurine in epileptic patient given 500 mg taurine per day by mouth starting at day one.

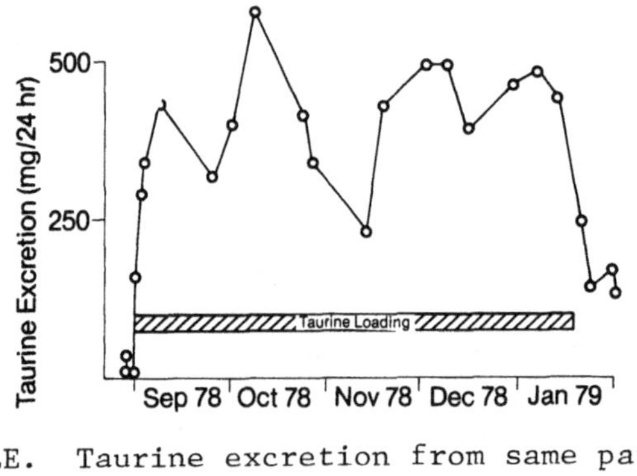

Fig. 2. HUXTABLE. Taurine excretion from same patient over 4½ months of oral taurine treatment.

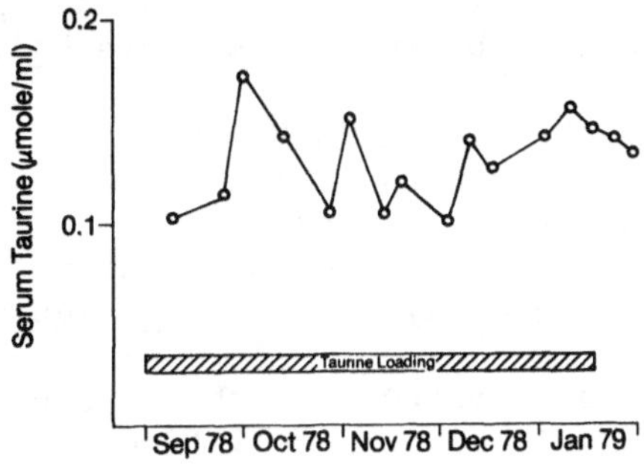

Fig. 3. HUXTABLE. Serum taurine levels in same patient.

REFERENCES

1. Baskin, S.I., Leibman, A.J., and Cohn, E.M., in: "First and Second Messengers, New Vistas, E. Costa, E. Giacobini, and R. Pasletti, eds., Raven Press (1976).
2. Crawford, J.M., and Curtis, D.R., 1964, The excitation and depression of mammalian cortical neurones by amino acids, British J. Pharmacol., 23:313-329.
3. Curtis, D.R., and Watkins, J.C., 1960, The excitation and depression of spinal neurones by structurally related amino acids, J. Neurochem., 6:117-141.
4. Fariello, R.G., 1979, Action of inhibitory amino acids on acute epileptic foci: An electrographic study, Exp. Neurology, 66:55-63.
5. Federici, G., Dupre, S., Rosei, M.A., Granata, F., and Orlando, M., 1974, The metabolism of taurine in the living rat, Physiol. Chem. Phys., 6(5):411-416.
6. Grosso, D.S., Roeske, W.R., and Bressler, R., 1978, Characterization of a carrier-mediated transport system for taurine in the fetal mouse heart in vitro, J. Clin. Invest., 61:944-952.
7. Horigane, A., Horiguchi, M., and Matsumoto, T., 1979, Metabolism of 2-amino-3-phosphono(3-^{14}C)proprionic acid in cell-free preparations of rat liver, Biochim. Biophys. Acta, 584G: 254-260.
8. Huxtable, R.J., Insights on function: metabolism and pharmacology of taurine in the brain, in: "The Role of Peptides and Amino Acids as Neurotransmitters," J. B. Lombardini and A. Kenny, eds., J.R. Liss, New York (1981) in press.

9. Kochakian, C.D., 1975, Free amino acids of sex organs of the mouse: regulation by androgen, Am. J. Physiol., 228:1231-1235.
10. Kontro, P., 1980, Transport and metabolism of taurine and hypotaurine in the brain, Acta Univ. Oul. A 88, Biochem., 27:1-52.
11. Kontro, P., 1980, Inhibition of hypotaurine uptake by structural analogues and drugs in brain slices, Acta Univ. Oul. A 97, Biochem., 29:111-116.
12. Kruz, J.C., Kendrick, Z.V., and Baskin, S.I., Pineal taurine content in aging in nonhuman primates, D. Bowden, ed., Van Nostrand Reinhold, New York (1979).
13. Oja, S.S., and Kontro, P., 1980, Hypotaurine uptake by brain slices from adult and 8-day-old mice, J. Neurochem., 35: 1303-1308.
14. Okaya, 1966, Crystal structure of the stable modification of 2-aminoethylphosphonic acid, β-ciliatine, Acta Cryst., 20: 712-715.
15. Scandurra, R., Federici, G., Dupre, S., and Cavallini, D., 1978, Taurine and isethionic acid production in mammals, Bull. Mol. Biol. Med., 3:141-147.
16. Sgaragli, G.P., Carla, V., Magnani, M., and Giotti, A., 1978, Homotaurine and muscimol mimic taurine and GABA effects on muscle tone and temperature regulation, Naunyn-Schmiedeberg's Arch. Pharmacol., 305:155-158.
17. Welty, J.D., and Read, W.O., 1964, Studies on some cardiac effects of taurine, J. Pharmacol. Expt. Therap., 144:110-115.

SESSION V

IS TAURINE ESSENTIAL IN DEVELOPMENT?

Moderator: John Sturman

STURMAN: I think what we should do is take the most difficult part of this first. Is taurine essential in developing brain? And then consider developing animals. Then later we'll go on to the most important animal; man. I just thought I'd show you a few often quoted figures just to show some facts but not functions (Table 1). Put the slide on please. This is well known to most people. It shows the concentration of taurine in brain; the amounts found in the mature brain and in the newborn animal. If I want to take an optimistic note you can say, well, every species from birth to maturity has a nice substantial decrease in taurine. If you want to take the opposite side, you can say well what the hell can you make of that? Look how different all these values are! One difficulty is trying to find a universal function for taurine, especially when one looks at the species differences and one looks at what state they're born. Animals like the rabbit or the rat or the mouse are born very immature. Animals like the guinea pig and monkey are born fairly well formed. Guinea pig has already undergone more than 80% of brain development. So I think these are some of the problems that we have tried to tackle in the past, but never came up with anything. The next slide shows one other fact: that taurine is axonally transported, at least in the visual system (Fig. 1). Much more taurine is transported in developing animals (Table 2). These are the ratios in a 10 day old animal compared to the adult, and you can see much more was transported in the 10 day old animal than the 20 day old animal. These are facts not functions. We've speculated about these numbers but we haven't proved a function. And then with the last slide that I want to show (Table 3), you've seen most of this at this meeting anyway, it's hard for me to conceive that taurine isn't involved in development, probably brain development, in some way when we have these quite high concentrations of taurine in mammals. Why would a mother bother to supply her progeny with milk if it wasn't needed?

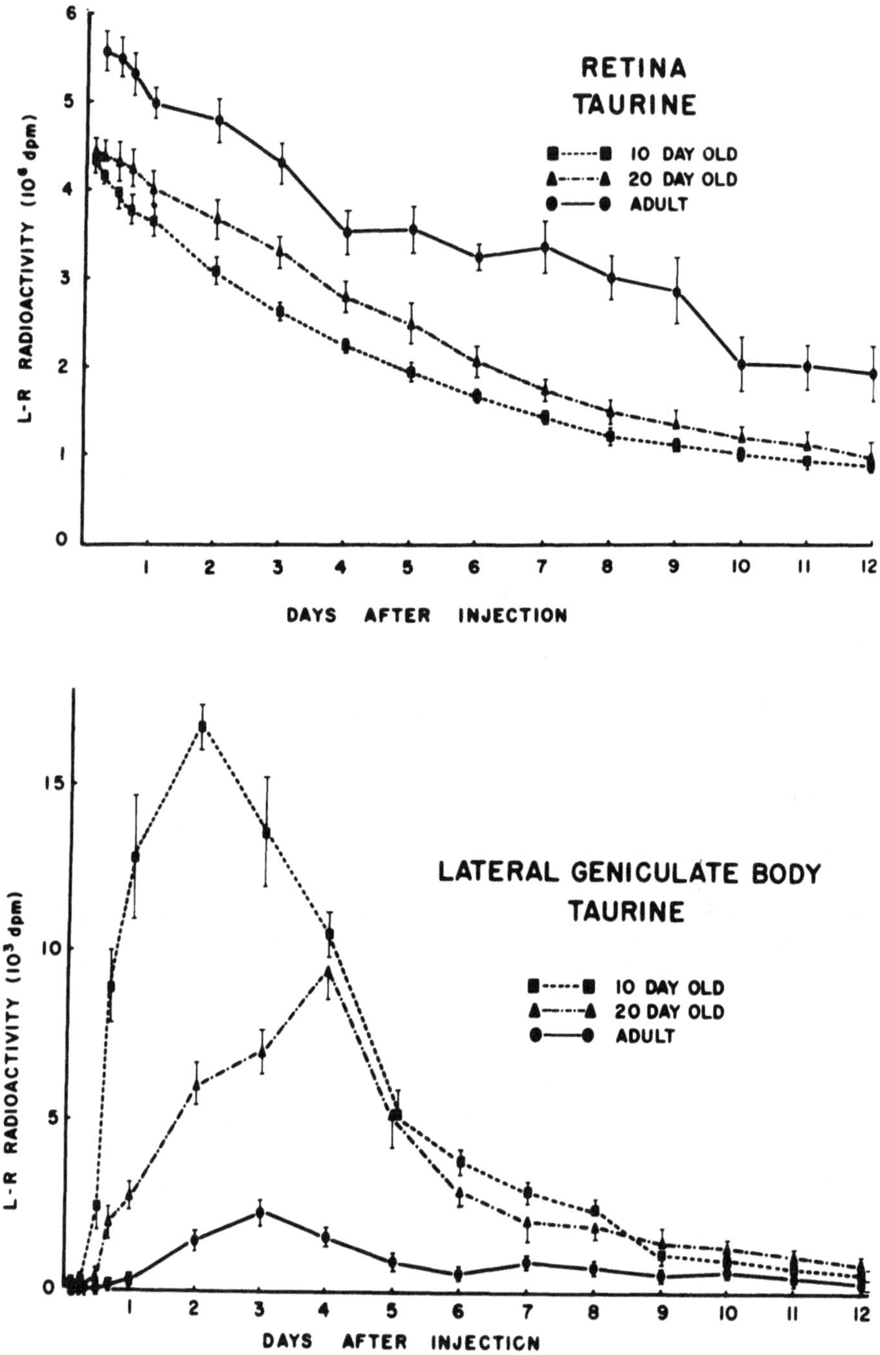

Fig. 1. STURMAN

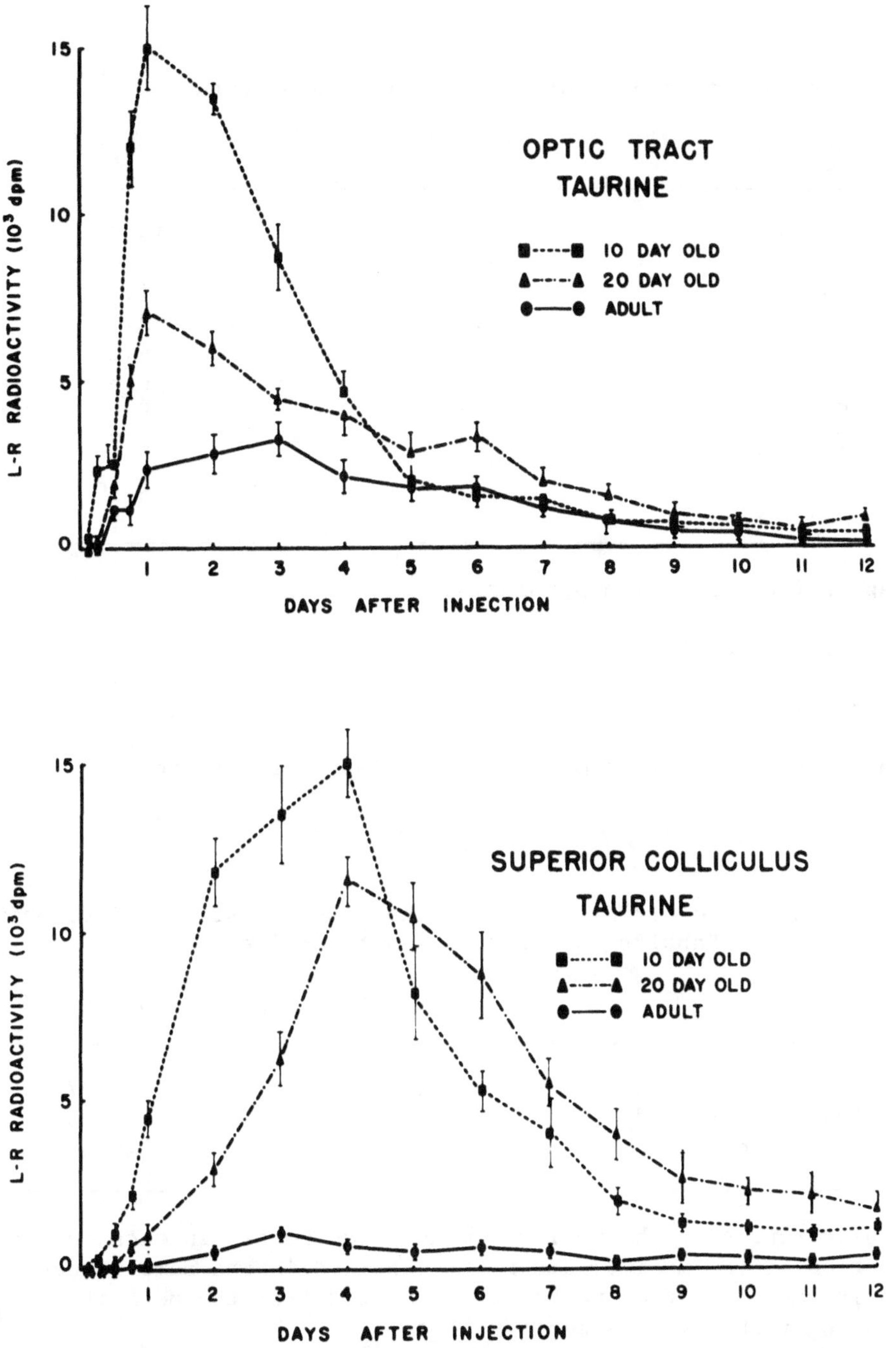

Fig. 1. STURMAN. (^{35}S)Taurine in right retina and left minus right (^{35}S)taurine in optic tract, lateral geniculate body, or superior colliculus of rabbits at various times after the intravitreal injection of 5 μCi (^{35}S)taurine into the right eye. Each point represents the mean ± SEM from 6 neonatal or 4 adult rabbits.

Table 1

Concentration of Taurine in Adult and Newborn Brain[a]

Species	Adult[b]	Newborn[b]
Mouse	8.6	15.3
Rat	4.4	16.6
Gerbil	6.5	21.2
Guinea Pig	1.0	2.0
Rabbit	1.2	5.1
Dog	1.3	6.8
Cat	2.3	9.2
Chick	2.3	8.9
Monkey[c]	2.1	6.9
Man	1.4	3.3[d]

[a] Adapted from (15) with permission.

[b] μmole/g wet weight.

[c] Occipital cortex.

[d] Gray matter only, mean of 5 children 1-5 years of age.

Table 2

Axonal Transport of (^{35}S)Taurine in the Visual System of Developing Rabbits compared to Adult Rabbits[a]

	10-day old	20-day old
Optic tract	4.26	1.97
Lateral geniculate body	7.26	4.13
Superior colliculus	13.73	10.45

[a] Results represent the maximum amount transported in each case as a fraction of the maximum amount transported in the same structure in adult rabbits. Each value is calculated from the mean of 6 developing rabbits and 4 adult rabbits.

Table 2 (continued)

Rates of Axonal Transport of Taurine and Proteins in the Visual System of Developing and Adult Rabbits

Age	Taurine[a]	Protein[a]	
		Fast	Slow
10-day old	32	128	4
20-day old	20	152	3
Adult	30	208	2

[a] mm/day.

Table 3

Taurine in Milk

Species	Less than 5 days after birth[a]	More than 5 days after birth[a]
Gerbil		595
Cat		287
Dog	264	191
Mouse		75
Rhesus monkey	61	56
Baboon		38
Man	41	34
Chimp	71	26
Guinea pig		17
Rat	63	15
Java monkey		14
Rabbit		14
Sheep	68	14
Horse		3
Cow	31	1
Pig	56	

2 μmole/100 ml

LOMBARDINI: But John, couldn't you order it the other way around: that the cow has such a little amount in its milk that maybe taurine is not necessary?

GAULL: The milk that is fed to the calf has about 30 times the concentration of dairy cow milk.

STURMAN: During the first 5 days after birth, but the cow doesn't ever get any taurine in its diet.

VAN GELDER: May I show something about taurine in the milk (Fig. 2)? This is just a series of experiments we did with 8% protein diets and we looked at the survival rate of neonatal litters which were matched either 7 per litter, 4 per litter, 2 per litter, sex matched, and weight matched. White bars are pups whose mothers received 8% protein and no other supplements. The other group of mothers received the same diet but had 0.02% taurine in the water (black bars). You can see that in each of these groups the survival rate in the neonates whose mothers received taurine was higher. Though very interesting it was not until a colleague of mine said well, what about the milk itself that I eventually found an explanation. In Table 4, this now is milk, milligrams milk, in the neonates' stomachs, so we were absolutely sure that this is what the neonates were getting. You see here that in the taurine neonates, the milk production increases (Table 4, column I versus II). This is normal, I have another slide for it, but generally in the second week, milk quantities increase in normal neonates (Table 4, columns III,V).

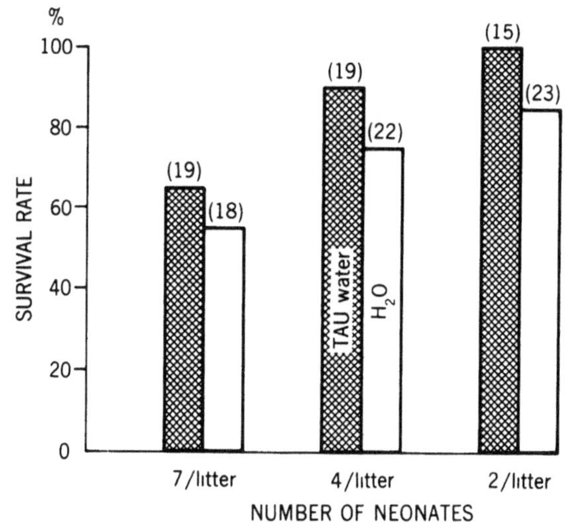

Fig. 2. VAN GELDER. Neonatal survival during nursing at 16 days. Mothers on 8% protein diet.

Table 4

Milk Content (mg) of Mouse Neonate Stomach[1]; Mothers Fed One of Three Diets for 16 Days ± Taurine

Age of Neonates	Low Protein[2]		Normal Protein		High Protein	
	I	II	III	IV	V	VI
	H_2O	Tau Water[3]	H_2O	Tau Water	H_2O	Tau Water
8 days	64±23 (4)	64±28 (4)	-	-	-	-
12 days	78±30 (8)	75±32 (16)	229±30 (14)	202±77 (14)	184±19 (14)	156±27 (14)*
16 days	59±33 (16)	93±35 (20)*	239±52 (21)	259±38 (14)	208±43 (14)	198±43 (14)
Milk Tau Content (μmole/100 gm)	3.66 (2)	21.72 (2)	11.82 (2)	35.73 (2)	27.78 (2)	32.32 (2)

1. Milk content: Neonates removed from mother for 4 hr, then allowed to nurse for exactly 30 min; average content ± SD; number of neonates/nursing mother - Low prot.:4; Normal and High prot.:7.
2. All diets were identical except for casein/starch ratio; Low protein: 8% casein; Normal protein: 27% casein; High protein: 64% casein.
3. 0.02% taurine (w/v) in drinking water; ( ) number of neonate stomach contents.

* $p < 0.05$ vs "water" group; both Low and High protein: $p < 0.01$ versus Normal protein group ± taurine at 12 days.

Note increase in TAU content of milk when mothers fed taurine (II.IV,VI); protein content of diet also influences TAU content of milk (I,III,V); with High protein diet, effect of taurine in diet on milk TAU content is minimal, suggesting that a maximum taurine concentration has been reached, which may be attained by increasing either protein <u>or</u> taurine content of a maternal diet. Both taurine or protein "in excess" may therefore be detrimental to providing optimum milk supply while "in moderation" both dietary factors seem beneficial (summating effect).

Now you see that around 16 days the 8% neonates received less milk than at 12 days, but the mothers which had taurine, they still increased their milk production, or the young could still take in more milk (Table 4, columns I,II). It is quite significant. Now if you look at the bottom line (Table 1), you'll see the milk taurine content - now remember this is at 16 days - this is not initial milk, as you can see, you find very low taurine content in milk of the 8% mothers. If you give these mothers taurine, then the taurine content in the milk goes up quite considerably.

STURMAN: This is just simply putting taurine in the mother's drinking water?

VAN GELDER: That's right. The next slide and that is the last I promise, shows you what the protein content of milk is, and here you see 8% protein diet, normal protein diet, and high protein diet (Fig. 3). The protein content of the milk generally tends to go up as the mothers receive more protein. Now if you give taurine to the 8% mothers, you'll find that the protein content is going towards the normal level. So taurine has an influence on protein content of the milk, and they come together at around 12-20% protein after which either increasing your taurine or increasing your protein has no more effect on taurine and protein content in milk. So a lot of the developmental funny things we saw were simply due to the quality of the milk. My conclusion here is that in marginally sufficient diets, taurine seems to play a very important role in that it can mobilize the last protein reserves in the mother and maintain both the quantity and the quality of the milk. Commensurate with this is of course that the mothers lose more weight when they're on taurine because they're mobilizing more protein than the 8% nonsupplemented mothers.

STURMAN: I take it with this type of experiment you're using partially purified diets with no taurine in them anyway?

VAN GELDER: There's no taurine or free amino acids in these diets. These are casein starch diets from Nutritional Biochem.

STURMAN: So what your experiments would say then is that you've now found the tissue in the rat which can be taurine-depleted.

VAN GELDER: That could be.

STURMAN: Nutritionally, taurine depleted?

VAN GELDER: Well, I showed you yesterday or the day before, what happens to the cerebellum of these animals (See Van Gelder, this book). Of course, when maintained, for the rest of their lives they are taurine deficient in fact.

STURMAN: Would you like to comment, Dave? You're on the negative side of taurine in brain development these days.

RASSIN: Nico hasn't said anything about brain development here; he said something about survival. I think the big question we've been wondering about in the discussions here is the question of where or what that taurine is doing in early development especially in terms of what cells it is associated with in the brain. Here we go back to the old observation that you made, John, that the total brain content goes up, what does that mean? Does that imply that as neuronal and glial development occur taurine is redistributing, and you've got new structures diluting the total taurine pool?

STURMAN: It also is occurring at different rates in different species, which doesn't fit generally.

HAYES: Before we leave Nico's data, I would like to pursue one observation there. Were the taurine-supplemented pups, in other words, those that received milk with taurine in it, larger than the others?

VAN GELDER: No, now I'll give you really the last slide. Just put it on (Fig. 4). It just shows you the growth rate, in effect, of these pups, and you'll find that these pups don't grow better, just more of them survive because they get more milk. But these are the three groups. Now you can see in the high protein diets that taurine-water is starting to have a detrimental effect and you can also see that, from the milk output in the mothers on the high protein (Table 4), the milk output is about half of what it is with a normal protein diet.

RASSIN: Nico, what happened to the ones that didn't survive? Have you only included the survivors or have you included all the animals?

VAN GELDER: These are only intact litters, the same number of neonates in the litters which survived. That is to say, we only compared the litters when there were, let's say, 4 per litter or 7 per litter and these litters grew up to 14 days without mortality.

RASSIN: Yes, but what you've thrown out there though are the ones that haven't lived.

VAN GELDER: Yes, well, then I threw the whole litter out. This comparison is per litter, this is not per neonate.

GAULL: There is only so much protein to go around.

VAN GELDER: I don't understand.

RASSIN: Some animals may be more susceptible than others, so you

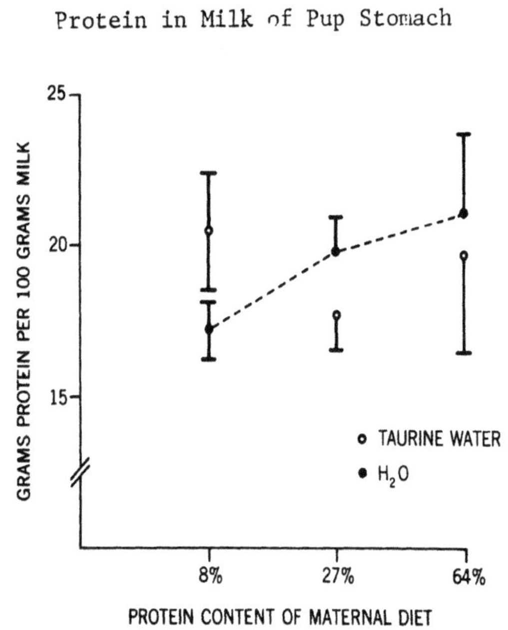

Fig. 3. VAN GELDER. Data with permission, Neurochemical Research Vol. 6, 1981, Plenum Press, New York.

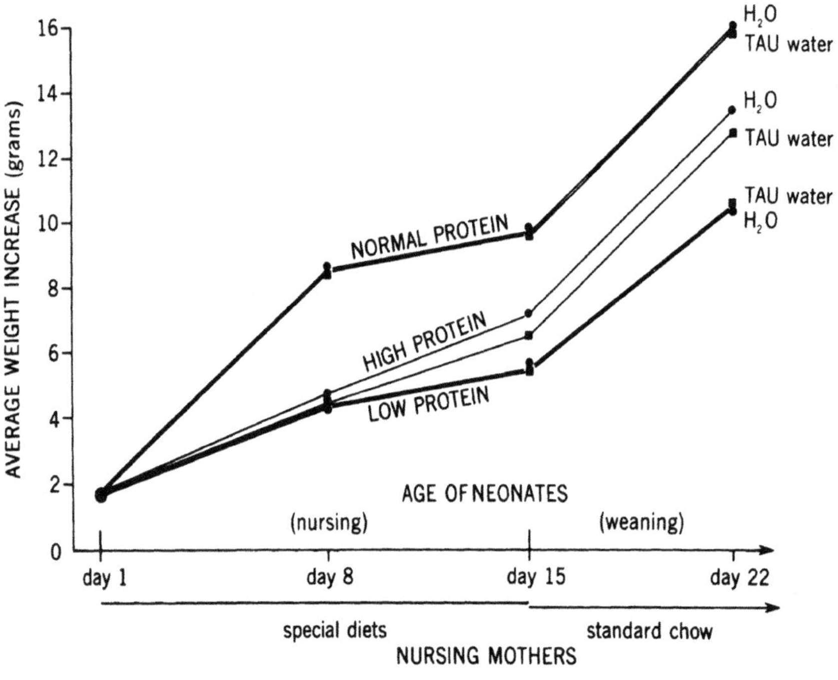

Fig. 4. VAN GELDER. Average weight gain of neonates nursed by mothers on diets having different protein contents ± taurine supplement.

have thrown out the susceptible ones and we're only looking at ones that are not dramatically affected by this diet.

VAN GELDER: No, they're all affected.

HAYES: They're all influenced but some are affected more than others; and the ones that were affected the most are the ones that did not survive. Why did they not survive? What was the pathogenesis of their demise?

VAN GELDER: Because they were starving to death, and as soon as the litter became incomplete, I threw them out because you can't compare litters of 3 and litters of 4 neonates: the 3 will get more milk proportionally than the 4 so these are only the average of intact litters...

GAULL: All of which survived.

VAN GELDER: ...all survived. They were always matched in number and in sex, that is all females litters, all surviving.

HAYES: When did the survival rate drop off?

VAN GELDER: Between the 12th and 16th day of course. That's where the mortality rate is increasing dramatically.

HAYES: Well, it will be very interesting to look at day 12 and compare weights of all your animals for instance, or kill them all at that point, and weigh organs and try to find out where that protein has gone.

STURMAN: Between 16 to 18 days is the time of the highest taurine content in developing brain. Just before we go to Russ, have you ever run into this sort of thing, Gerry, in humans, that the milk content is in any way related to their development?

GAULL: This is an area of controversy in human nutrition. The current dogma is that undernutrition affects the volume of milk but does not affect, at least in man, the quality of the milk except for water soluble vitamins, for vitamins in general, and for fatty acids. Those are relatively easily manipulated. As a matter of fact, David and I looked in a preliminary way at upper class versus lower class mothers in a vegetarian area of India. The preliminary results indicate that although there was a tendency for the lower class mothers who had poorer diets to have lower taurine concentrations, this was not statistically significant in milk.

HAYES: What about plasma?

GAULL: We didn't look at the plasma, we didn't get plasmas, but that experiment is going to be rerun exactly that way and we may have an answer. Nico told me about these experiments the night before the Symposium, and these were the ones I was trying to get him to show two days ago. I think they're very exciting, and I think that, harping back to the Pasantes hypothesis again, this is another tissue in which contractile proteins are very important and which you're getting an effect of taurine. I'm not certain that this will be directly applicable to man, but we don't really know.

STURMAN: Russell, do you have a comment?

CHESNEY: My only point is that in looking at this now for about two years on these different diets that we've been using, there's no question that the animals that are on the low taurine diet have a much higher mortality rate. The only problem with mine is that my diet is minimally sufficient in methionine so I don't want to make any great conclusions from this, and again they have a weight reduction, as I showed in the slides, but again I don't want to make too much of it. But it certainly is the case that there is a much higher mortality rate in these pups.

STURMAN: Do you have a comment, Bernie, on this?

HABER: I was wondering if I might have your permission to switch this a little bit, and I was wondering if the question of whether taurine is involved in brain development might not perhaps be a little bit too global and perhaps we could ask if there are areas of the brain in which taurine may have a role in development for which the evidence might may be a little more compelling than for other areas.

STURMAN: I think that's what will happen in the end. That we'll maybe find one or two areas that are important. I have feelings for olfactory bulb and hippocampus but I think other people have their favorite places.

HABER: In relation to that, might I ask if anyone has looked at taurine-deficient diets on development of zinc containing enzymes in the brain?

STURMAN: I don't know. Do you know Casey?

HAYES: Interactions with zinc?

STURMAN: You were talking about taurine and zinc, or protein and zinc?

HABER: No, I'm talking about enzymes which have a requirement for zinc, and their developmental appearance, let us say, in the taurine-deficient diets.

HAYES: I think that information is available but...

STURMAN: Not on a taurine-deficient diet.

HAYES: No, not on a taurine-deficient diet, but the basic question is, has anybody detailed a pattern for the zinc influence on enzyme development and whether we can superimpose our taurine data on that to see how they fit? I think the data may be available.

STURMAN: Probably it is available in the rat, but then that's not an animal affected much by taurine.

GAULL: I'd like to go back to Nico's experiment. I'm not clear about your size, your growth data, and the fact that all the litters were of the same size and same sex. What about the effect of taurine-deficiency in the mother on survival of the pups? That ought to be an easy number to derive.

VAN GELDER: We started out with six groups of animals. The 8% groups had 25 mothers in both groups, that is twice 25 X 4 or X 7 or X 2 litters. Don't forget that. Then we had 7 per litter for the normal groups and those were 15 mothers each, and we did that three times, so that there were a total of 45 litters and then we did the high protein and again there were about 35 litters in each group. Of the mothers themselves very few died. It was our experience that the mother who could not provide enough milk showed an almost standard pattern of behavior: at the 8% level you would find one pup had died and we put the litter aside. They're no longer in the statistics. But the next day very often you would find that the three others had disappeared (eaten) and the mother then was perfectly able to survive. The mothers lost weight, considerable weight. They lost over a period of 16 days about 15 grams, which is a lot for a 30 gram mouse which is starving. *(Laughter)* Well, I just want to emphasize that. But we never took it past the 16 days, because then we put all litters and mothers on a normal diet; because we are interested in temporary malnutrition in relation to what happens to these animals later on in life.

RASSIN: Just a minute. Let's pursue maybe the mechanism of this a tad, and maybe Casey could comment. What's the possibility that what you're doing is affecting taurocholate in these animals, that you are making them less able to absorb the fats from mothers milk, so in a sense making them fatty acid deficient?

HAYES: That's a question I originally addressed in a sense when I asked him whether the pups grew faster. In other words, was low intake affecting the pups directly, or their mothers. I think that's an important question; was taurine depletion affecting the mothers' mobilization of protein and nutrients, or was it that the milk from supplemented dams provided the pups with a better bile acid pool and more muscle taurine for better growth?

RASSIN: We don't know what happened to the dead pups.

HAYES: If he knows when they died and how much they weigh I think...

VAN GELDER: Oh! We have them all.

HAYES: You can evaluate some very interesting possibilities.

STURMAN: One interesting sidelight. It's just come to light from a little collaboration with Jack Fellman. I've been working with rats with milk labeled with various compounds. There is always another radioactive compound produced, even when the rats have only been given labeled taurine, (^{35}S)taurine (Fig. 5). It's a frontal elution on an amino acid analyzer. When we first reported this (14) we thought it was, could be, isethionic acid because it co-chromatographed on electrophoresis. However we've now done a different analysis of this. It is not inorganic sulfate, and that's usually the only other compound that we see in different places, plasma and urine, but it turns out to be a sulfate ester, very possibly, I suspect, linked to cholesterol or a similar compound.

VAN GELDER: All I can say is that the body weight of the mothers was commensurate with the amount of milk the pups were receiving. That normal mothers gained weight as a matter of fact. The weight of the high protein mothers stayed stable, and they produced less milk, while 8% mothers lost considerable weight while trying to put out milk for the young. So for the rest I can't tell anything about cholesterol or bile acids because my feeling is that all these effects on growth in the first 16 days could be directly attributed to the amount of milk these pups got. It is true that in the 8% group the milk protein content was higher in the pups with a taurine supplement than in the other. But whether that had much of an influence? It certainly didn't have an influence on the rate of growth of those litters which survived the full experimental period. That is obvious.

STURMAN: That's fascinating. There's one other thing that leads me to suspect that taurine must have something to do in brain. From birth to about the time you're talking about, 16 or 18 days, there is a net transfer of taurine from the rest of the body to the brain (Fig. 6). I mean the brain is accumulating some of its taurine from the rest of its body, and this doesn't occur thereafter. Brain taurine is then decreasing. So it needs a little leading.

HUXTABLE: Yes, we also see that (Huxtable, this book).

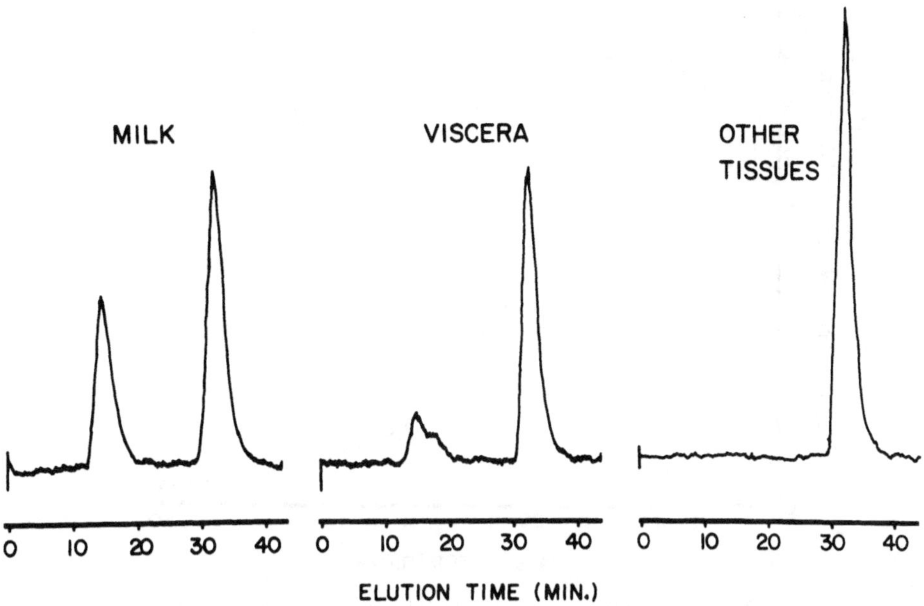

Fig. 5. STURMAN. Flow-cell chromatograms of extracts of milk, viscera and other tissues from lactating female rats receiving daily injections of (^{35}S)taurine, and from pups suckling on such females.

STURMAN: I think perhaps Casey if you could address overall growth with different species to see where we touched on growth here.

HAYES: Before I pick up on that point, can we assume this sulfated compound is a steroid?

STURMAN: That is totally speculation.

FELLMAN: Let me just simply say the material which co-chromatographed with isethionate is hydrolyzeable, and that all of the radioactivity turned out to be sulfate. The R group is as yet not clear, and we're looking at it now. I don't want to make any comments so what that R grcup is. It's an ester of sulfate.

STURMAN: That was a big jump.

CHESNEY: Could it be vitamin D sulfate?

FELLMAN: Sure, sure.

HAYES: Or a bile acid? How much is there?

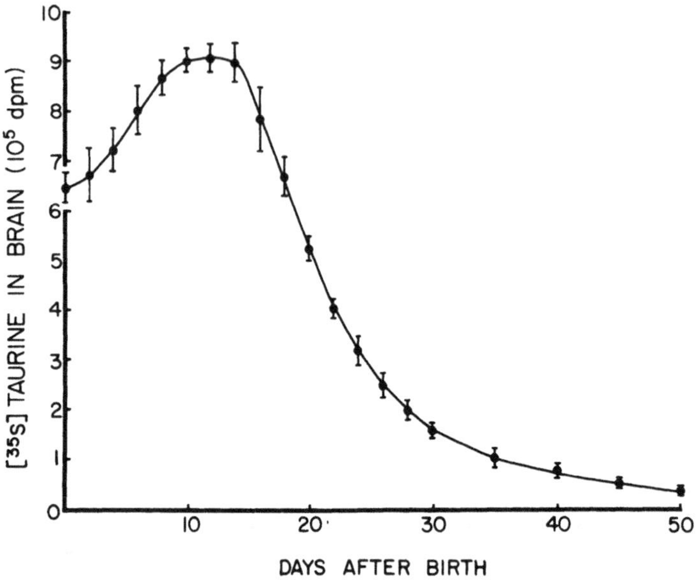

Fig. 6. STURMAN. (^{35}S)Taurine in brain of pups born to female rats receiving daily injections of (^{35}S)taurine and cross-fostered at birth to lactating female rats receiving daily injections of unlabeled taurine. Sturman, unpublished.

STURMAN: It's always constant. It's approximately 40% of total radioactivity in milk, just as we reported in that one short experiment.

FELLMAN: Maybe when the book goes to press we'll have the data and we'll include it in it, but I don't have it with me now.

HUXTABLE: You better work fast.

STURMAN: Casey, you want to give us a little rundown on growths and taurine growths and animals?

HAYES: Well, I suppose that goes back to my Fig. 4 (Hayes, this book) concerning species differences in growth rate. If you remember, the rat increased its birth weight 100% whereas other species dropped off rather dramatically with the cat, appreciably higher than man or monkey. The point that I was trying to make, which would integrate our thinking with concentrations of taurine in milk, various tissues' supply and demand, and overall requirement for taurine in a given species, is that Mother Nature has put together a combination of factors influencing our supply, such as synthesis and diet (or milk in the case of the neonate) which results in different requirements. The most important pool of taurine determining that requirement appears to be the bile pool and whether taurine is needed to conjugate bile acids. So the cat is the number one sufferer, if you will, in terms of having the major requirement to meet. The rat and the dog must also meet this requirement, but they do so readily and early in life from synthesis and so they get off the hook. New World monkeys apparently have an obligate requirement for taurine conjugation like the cat, but because they grow slower they may avoid taurine depletion and degeneration of the retina.

STURMAN: Our ideas of biosynthesis though are going to be stretched out a little bit probably, from what Ryan reported. The rabbit really is in there close to the cat knowing the activities of CSAD in the rabbit tissue which are not very large at all.

HAYES: John, there's also a ringer in there in that both of these species of primates have essentially nothing for CSAD activity in the liver, so they cannot synthesize taurine very well at all and we didn't give them any dietarily. Again I get back to the difference in growth rate and/or a different pathway of synthesis not involving CSAD.

GAULL: I'd like to show one slide that I forgot to show when I gave my talk. I think it's very pertinent to what Casey is saying. These are the glycine:taurine ratios in duodenal bile from the latest series of experiments that Dave Rassin and I were doing. These are preterm infants and this is the glycine:taurine in infants fed pooled human milk. This is the glycine:taurine ratio which you can see going up in the infants fed a whey-predominant formulae unsupplemented with taurine, and you can see here is the glycine: taurine ratio on the infants fed the same formulae supplemented with the amount of taurine in human milk.

HAYES: Which is what, Gerry?

GAULL: Which is about 30 μmole/100 ml. Now the other thing that is interesting is that Warden in Birmingham, England, has shown in similar types of studies that if you supplement the formulae with cysteine instead of with taurine, they behave as if they were unsupplemented. So that the human infant, as we predicted, as the rest of our evidence I think shows, is not able to convert cysteine to taurine very well. As a matter of fact there are some old experiments in the literature that even adult man doesn't convert cysteine to taurine in anything like the rate of the conversion that John Sturman found in his experiments with ^{14}C labeled cysteine, in the rat.

STURMAN: But Gerry just to go on from there; would you conclude that taurine is an essential amino acid to humans?

GAULL: Well, this depends a little bit about how you define "essential". I was recently at a symposium and one of my fellow speakers, a classical nutritionist, said to me afterward, he put his arm around me and said, 'Now really, heart to heart, now do you really think that taurine is essential?' And I said, 'Well, heart to heart, I think that we can't synthesize it, and I think we're having a hard time demonstrating that it's essential.' 'But, ah! come on,' he said, 'even your data shows it has no effect on growth and no effect on nitrogen balance'! I think that we discussed this before; I think that in one sense "essential" means that you are unable to synthesize it. You have no ability to synthesize it. From the classical nutritional sense it also means that you can demonstrate a deleterious effect structurally, functionally or morphologically in the animal. The problem is, we have now had all of these symposia on taurine and we still haven't settled what the central function of it is, and I think it makes it very difficult to pin down what we should be looking for. Furthermore, I think that classical nutritionists have always looked for short term changes, and I think that we have to start thinking over a longer time scale. Not what happens in the next day, two weeks or month but what happens 20 years from now when you get your myocardial infarction, your dissecting aneurysm and what have you.

STURMAN: Yes, I think it's almost certainly that way. Otherwise something would have shown up.

GAULL: The other thing I'd like to say apropos, and in support of, what Casey said: I don't think that the cat is going to turn out to be all that different from man. I think there are a number of reasons that we have not seen epidemics of blindness. One is the growth rate that Casey brought out. The other is the fact that we have this safety factor in converting to glycine conjugations. But also in societies where milk substitutes have been fed there is also a tendency to give these infants early on what would have been weaning foods, that they would ordinarily have received much later. You can hear mothers in the United States bragging about how their child at the age of a month and a half is already eating meat. Well, if that kid is already eating meat then he doesn't need the taurine in his mother's milk.

STURMAN: Could it be another effect, more in your area, André. We know from monkey studies and, if man is at all similar to monkey, we get decreased brain concentrations of taurine during development, not as dramatic as the cat, but nevertheless a decrease, I can't remember off the top of my head - 20 to 30% maybe after some months. In the normal situation, you might pick up any obvious syndrome associated with this, but supposing such a child were predisposed to one of these diseases which you have apparently linked with taurine. Could this little shift be the thing that triggers it over the top?

BARBEAU: I think that's very important, and that's what we clinicians would like to call a relative deficiency, in that in normal situations probably these patients have sufficient amount of whatever substance we're talking about, whether it be taurine or something else, but under stress conditions, metabolic stresses, or I guess the stress of life, then these subliminal types of deficiencies would become clinically apparent. I think that is a general mechanism that's becoming more and more possible for degenerative disorders, and this may even be genetically determined.

HABER: I wonder if I might make a comment, just to make sure that I'm understanding things right. What I'm getting from all this discussion right now is that in terms of the essentiality of taurine and its role in development, the essentiality of taurine is probably going to turn out to be a diet-dependent variable, and it's probably going to be very hard to come out with a blind statement as to its essentiality. Am I reading you right Gerry?

GAULL: You have understood completely.

HAYES: Yes, but that's not the definition of dietary essentiality.

RASSIN: Yes, it's not really correct. If it's a diet-dependent variable then it's not essential.

GAULL: What I'm saying is that you have to change your time frame and you have to get a better understanding of what taurine does. My analogy would be vitamin E. Vitamin E is now known to be a potent antioxidant, I don't think anybody would argue about it's being a vitamin, being essential, and yet it's very difficult to demonstrate the requirement for it. The way you can do it most easily was done by accident by the formulae companies in feeding infants when they changed the fatty acid composition of the formulae and they increased the polyunsaturated fatty acids. This increased the requirement for vitamin E and preterm infants, which are born with decreased stores of vitamin E, started getting hemolytic anemia. Then when they added iron to the formulae in order to prevent the decrease in hemoglobin that is seen in every mammal postnatally, they increased the hemolysis even more because it turns out that iron is a cofactor in the peroxidation of the long chain unsaturated fatty acids. And so the way you could demonstrate the essentiality of vitamin E best was only when you had distorted the other dietary components.

STURMAN: Well, there is one other point. I think perhaps, that looking through some of Ryan's data, it may not even be the tissue content or concentration of taurine which finally may be important. It may be where that taurine came from, the proportion synthesized to that obtained from the diet, isn't that right? From some of your results you showed different effects later on?

HAYES: You mean different pools of taurine exist, an endogenous different from the exogenous?

STURMAN: I think I drew that from your data, is that correct?

HUXTABLE: I think you're drawing more from my data than I drew.

STURMAN: I thought you said there was something that you could stress or manipulate and the effect was different.

HUXTABLE: Yes, you're referring to some work that we've never published, which is rather interesting. This was at the time we were trying to lower taurine levels in the rat by putting them on taurine-deficient diets, and we found, and others have found, that we couldn't lower tissue taurine concentrations. However, we did find that the animals on these diets became extremely sensitive to isoproterenol. The LD50 of isoproterenol was sharply decreased for animals on the taurine-deficient diet.

GAULL: For instance a lot of the infants and children are given theophylline. Now, it is well known that the metabolism of theophylline is changed considerably by manipulating the carbohydrate versus the protein content of the diet. Is it possible that some problems that were seen, that we are seeing now, that are apparently pharmacological problems, in fact, are really not pharmacological problems. They're really nutritional problems.

IRVING: This is pure speculation now, to carry on the idea of imposing some kind of stress, based on the fat delivery enterohepatic role of taurine. If you put animals on high lipid or fat diets and decrease the percent of protein and carbohydrate so that lipids or fatty acids will be the primary source of calories, perhaps you can induce the kind of stress which would amplify species differences.

STURMAN: Have you done that Nico?

VAN GELDER: Could you reformulate the diet for lipids?

STURMAN: If one is talking about stressing, with altering lipid proportions in the diet.

VAN GELDER: I haven't really done that.

RASSIN: Just to maybe finish off this essentiality business, if I may try to clarify my thoughts on that. I don't think man can make much taurine, and if taurine has a function in man it is essential in the sense it must be supplied in the diet, and I wonder to what extent anybody disagrees with that?

FELLMAN: Can I raise a question, a point of information that some of you may be able to answer. To what extent does taurine occur in vegetable sources?

GAULL: Very low.

HAYES: Almost none.

GAULL: Almost non-existent.

FELLMAN: There are reports in literature that fungi and some plants can make taurine.

GAULL: But the higher plants have very little, virtually none.

FELLMAN: So a vegetarian diet is clearly taurine deficient. But how do they obtain it?

HAYES: Unless they are strict vegetarians, some of them eat cheese, a little milk, or eggs. But vegetarians are not taurine deficient, so I would suggest they are synthesizing enough taurine, switching to glycine conjugation of bile acids so that even a small amount is sufficient to meet demands.

RASSIN: The question is are they synthesizing any taurine? Do they have to?

HAYES: Have you ever injected radiolabeled cysteine in humans?

STURMAN: I'd kinda like to.

GAULL: One of the things that might be done further to examine the Pasantes hypothesis is, and as soon as I go home, I'm going to ask my wife, if there are any plants that she knows about, she's a botanist, that have contractile proteins. I think it would be fascinating to see whether the plants with contractile proteins are plants that have taurine.

BASKIN: How about Mimosa pudica?

HAYES: I think the seaweeds are, and that's where the most of the taurine...

GAULL: Is that right?

BASKIN: Asparagus has taurine in it, I believe.

BARBEAU: And contractile proteins?

BASKIN: I'm not sure about contractile proteins.

HABER: In terms of trying to find a unifying concept for what taurine might or might not do, it's clearly calcium that seems to be emerging as a candidate. I wonder if it would be fair at this time to ask if tissues which have a clear contractile or secretory or excitability function are either more dependent on taurine in their development, or if they tend to conserve the taurine pool more than others.

STURMAN: Well, I reported, probably inadvertently, at the first taurine symposium some information that was given to me, I don't even know the fellow's name now, in Denver. He was just submitting it to Science. He showed that taurine can replace cyclic GMP in microtubule assembly and he had assured me of this and sent me a copy of the preprint and then he never published it. He withdrew it. So it's mentioned somewhere. Yes, this is a possibility, although I don't know of anybody actively looking at effects of taurine with these sort of structural proteins, actin, myosin, microtubules, or nerve filaments. I don't know of anybody that even has the slightest passing interest in taurine who works in this area.

CRASS: Ryan, did you ever assess tritiated taurine uptake in hearts from rats that have been placed on a taurine-deficient diet for any extended period of time?

HUXTABLE: Yes, we've done many studies on uptake of tritiated taurine.

CRASS: In rats placed on taurine-enriched diets or taurine-deficient diets?

HUXTABLE: Both. The deprived animals had a higher rate of influx, and responded more sharply to isoproterenol stimulation (3).

CRASS: Is it conceivable that a major deficiency in a certain intracellular pool could be refilled by a dietary source of taurine as opposed to an endogenous biosynthesized taurine?

HAYES: To that point, I would go back to our cat studies where we injected radiolabeled taurine in very deficient cats having 1/10 to 1/20 of their normal pools, and the question was which pools sequester the taurine? Supposedly, they would be the ones that require the most taurine. The bile has first dibs, the retina second and the olfactory lobe had third, and then nothing else came close...

STURMAN: But the heart was undistinguished.

HAYES: The cat represents a really depleted model that can be used to answer the question: which tissue really needs taurine, and then you ask, well why is it needed?

HUXTABLE: The cat is an unfair animal to quote, because it can only use the taurine conjugate in the bile, as you pointed out here the other day.

HAYES: I realize this, but still that brings the bile pool into the total requirement concept, which is why I suspect man can avoid severe depletion.

GAULL: I'd like to throw another thought in here. It has come to my attention recently that the disposition of xenobiotics in some animals are as taurine conjugates, and I wonder whether this is something that we ought to think about in terms of the draws on the taurine pool. In addition, quantitatively how significant is it?

BASKIN: Williams in England (4) has shown that the ferret particularly conjugates phenylacetic acids. A lot of the insecticides, such as arylacetic acids, that we see in our environment now are conjugated with taurine primarily in the ferret but he's looked at other animals.

HUXTABLE: A whole slew of other animals also conjugate with taurine; fish (6), shark, mouse (2), dog (7) and so on.

STURMAN: To go away completely from mammalian species, I've picked up a very interesting tidbit when I was visiting in England. I saw a film of Dictyostelium discoideum which is a slime mold which grows....just like slime mold, *(Laughter)* but it grows rapidly and eventually the cells slowly aggregate and form into a slug or a grex, which is a fruiting body. Now this thing grows up out of this slime mold. If you put glucose into the medium, it prevents the development, so to speak, of this slime mold. But taurine added at the same time as glucose will overcome the glucose inhibition. So taurine prevents the glucose effect on this system.

HUXTABLE: Even more interestingly, in slime molds taurine stimulates the release of cyclic AMP extracellularly (12).

CHESNEY: In Dictyostelium discoideum there's this substance acrasin which causes them to aggregate which turns out to be cyclic AMP (9,11), so if taurine's role is release of cyclic AMP, then it would make sense that that would be, and again goes with some of the stuff we heard yesterday.

BARBEAU: I feel more at ease with man. Just a wild idea, but thinking of all the areas that have high taurine in the brain of man, all of them are related to sensory input pathways rather than output pathways, and that is true even in the spinal cord distribution. So for whatever it's worth, taurine may be somehow related to integration or modulation of sensory inputs.

GAULL: Well, to go back to sea beasts, there's a spiny lobster that has a receptor for taurine, for beta amino acids, to pick out food sources from the background of α-amino acids present in seawater. Now that seems to be a very interesting thought that it's a sensory input.

IRVING: I'd like to direct this question to Herminia. We have seen that taurine is associated with rods. Does one also find taurine in usually large concentrations in halobacteria which use bacterial rhodopsin?

PASANTES-MORALES: I think that there is no taurine in that bacteria.

HUXTABLE: There's certainly a lot of bacteria that metabolize taurine (see e.g. Refs. 1,5,8,13,17). There is also one - Gluconobacter cerinus - that has a taurine-containing lipid (16).

BASKIN: You made a comment about neurofilaments. There's been a couple of studies, I think the first one was by Jim Kocsis (10), in 1961, Steve can correct me, but he showed that addition of colchicine would cause massive loss of taurine from the muscle cells.

STURMAN: Isn't that just simple disruption of the membrane.

BASKIN: I'm not sure that colchicine really does that at pharmacological doses. Certainly we use colchicine as a therapeutic agent. I'm not sure that people have shown that it causes a massive destruction of muscle membrane.

PASANTES-MORALES: Well, cytochalasin produces a very large increase in spontaneous release of GABA, glycine, taurine and serotonin...

BASKIN: Isn't cytochalasin the agent that pops nuclei out of cells. Or are you using it for some other reason?

BARBEAU: Pursuing my thoughts about the sensory input, and thinking about Friedreich's ataxia where I'll show rapidly that there may be a taurine deficiency: there is of course, sensory nerve damage and no involvement of motor nerve damage, and if you add a glycine deficiency, which in animals can give you spasticity, in man you get another which is spastic ataxia. You have the same ataxia, you have the same sensory nerve findings, but then you get added onto this the pyramidal tract increase of reflexes and spasticity, and that's when the glycine is low. Maybe compensation can occur through changes in the taurine: glycine ratio. For whatever it's worth, I think it has to be checked out and we're going to look at this in bile.

STURMAN: Can you make a human glycine deficient?

BARBEAU: There are some experiments in animals, of course, and we are doing some studies now with feeding glycine to spastic people and are getting some very interesting results.

GAULL: You can make man glycine deficient?

BARBEAU: No, no, no, we have some preliminary evidence that in spastic ataxia there is a glycine deficiency in the spinal cord. Now I don't know how they became glycine deficient, that I don't know.

STURMAN: I think we covered about everything we can cover in this session unless anybody has any final points. Let's go to lunch.

REFERENCES

1. Braun, R. and Fromageot, P., 1962, Désamination de la taurine par Aspergillus niger, Biochim, Biophys. Acta, 62:548-555.
2. Hutson, D.H., and Casida, J.E., 1978, Taurine conjugation in metabolism of 3-phenoxybenzoil acid and the pyrethroid insecticide cypermethrin in mouse, Xenobiotica, 8:565-571.
3. Huxtable, R.J., The regulation and function of taurine in the heart and other organs, in: "Natural Sulfur Compounds: Novel Biochemical and Structural Aspects," D. Cavallini, G.E. Gaull and V. Zappia, eds., Plenum Press (1980), pp.277-293.
4. Idle, J.R., Millburn, P., and Williams, R.T., 1978, Taurine conjugates as metabolites of arylacetic acids in the ferret, Xenobiotica, 8:253-264.
5. Ikeda, K, Yamada, H., and Tanaka, S., 1963, Bacterial degradation of taurine, J. Biochem., 54:312-316.
6. James, M.O., and Bend, J.R., 1976, Taurine conjugation of 2,4-dichlorophenoxy acetic acid and phenyl acetic acid in two marine species, Xenobiotica, 6:393-398.
7. Jordan, B.J., and Rance, M.J., 1974, Taurine conjugation of fenclofenac in the dog, Pharmacol., 26:359-361.
8. Kondo, H., and Ishimoto, M., 1975, Purification and properties of sulfoacetaldehyde sulfo-lyase, a thiamine pyrophosphate-dependent enzyme forming sulfite and acetate, J. Biochem., (Tokyo), 78:317-325.
9. Konign, T.M., Van de Meene, J.G., and Bonner, J.T., 1967, The acrasin activity of adenosine-3',5'-cyclic phosphate, Proc. Natl. Acad. Sci., (USA), 58:1152-1154.

10. Kastos, V., and Kocsis, J., 1961, Effect of colchicine on taurine excretion, Proc. Soc. Exptl. Biol. Med., 106:659-660.
11. Mato, J.M., Jastorff, B., Morr, M., and Konijn, T.M., 1978, A model for cyclic-AMP chemoreceptor interactions in Dictyostelium discoideum, Biochim, Biophys. Acta, 544: 309-314.
12. Raizuda, M.K., and Murti, C.R.K., 1973, Binding of taurine to Hartmannella culbertsoni membranes and the synthesis of cyclic AMP, Curr. Sci., 42:202-204.
13. Shimamoto, G., and Berk, R.S., 1980, Taurine catabolism, III evidence for the participation of the glyoxylate cycle (BBA 29400), Biochim. Biophys. Acta, 632:399.
14. Sturman, J.A., Rassin, D.K., and Gaull, G.E., 1977, Taurine in developing rat brain: Transfer of (^{35}S)taurine to pups via the milk, Pediat. Res., 11:28-33.
15. Sturman, J.A., Rassin, D.K., and Gaull, G.E., Taurine in the development of the central nervous system, in: "Taurine and Neurological Disorders," A. Barbeau and R.J. Huxtable, eds., Raven Press, New York (1978), pp.49-71.
16. Tahara, Y., Kameda, M., Yamada, Y., and Kondo, K., 1976, A new lipid: The ornithine and taurine-containing "cerilipin", Agrie Biol. Chem., 40:243-244.
17. Toyama, S., Misono, H., and Soda, K., 1978, Properties of taurine: α-ketoglutarate aminotransferase of Achromobacter Superficialis, Biochim. Biophys. Acta, 523:75-81.

SESSION VI

DOES TAURINE HAVE CLINICAL SIGNIFICANCE?

Moderator: André Barbeau

BARBEAU: We can attempt to start the last session of this most important meeting. I've been given a number of questions here, but we've covered some of these over the last three days, and I think what we should do is really have a free ranging discussion on what people around the table think taurine does, could do, will do and doesn't do. There are many more things that it doesn't do than things that it does, as I learned from this meeting. I will put no restriction on the time, except I don't want any slides. The only restriction is that in exactly 1 hour, 32 seconds this will end, even in the middle of a sentence. That's the only restriction we have. Otherwise I think we should have a totally free discussion. The first question I think we should address, and we have heard pros and cons for this and around this, I would like people to address the question; is taurine a neurotransmitter, and if so where? Will anybody try to summarize data that was presented? The evidence as I read it is against it, but are there some people who feel strongly in favor of taurine being a neurotransmitter somewhere in the world? Nico.

VAN GELDER: No, I was just getting my slides back! *(Laughter)*

BARBEAU: If somebody raises his hand he has had it. You have to talk now Nico! Anna Maria, did I see you raise your hand?

LOPEZ-COLOME: I didn't raise my hand, but I do not favor a role as a transmitter for taurine because the calcium dependent release of taurine has not been shown. Even if this calcium dependent release is not a specific condition for all transmitters, it has been seen for other amino acid neurotransmitters and for taurine this is very doubtful. I've seen the results from Sieghart and Heckl and from Placheta et al., and they just can't find this in spite of the slide we saw from Dr. Kuriyama. So this doesn't point to taurine being a neurotransmitter. Furthermore, receptors for taurine as they have been described for other amino acids apparently are not present, so either taurine has a role as a neurotransmitter in a very restricted subpopulation of cells somewhere, and the sensitivity of the methods we have available is not enough to detect it, or I should say the interaction with membranes that taurine shows is not one of a transmitter in the "classical" sense, although it could be a modulator.

BARBEAU: In the classical sense, I should say binding as for noradrenaline, adrenaline and acetylcholine. Not, as compared to some of the other so-called neurotransmitters even including the peptides where we're getting more and more confused.

LOPEZ-COLOME: Yes, one could think that it could be acting through glycine or GABA receptors in certain conditions.

BARBEAU: Could we just wait for the receptors. If we list sites of possible taurine function: the retina; is there anybody who still believes taurine is a neurotransmitter in the rétina? *(Negative head shakes all around)* The other place that was postulated in man is in the cerebellum, in the stellate cells. Now nobody mentioned the evidence for or against that here.

SGARAGLI: I would like to bring to the discussion the data of Davidson from Aberdeen (1), I guess, because he's the first author of evidence that taurine <u>in vivo</u> is taken up by Purkinje cell body and dendrites but not by stellate or basket cells and it is not associated with nerve endings at least in those experimental conditions.

BARBEAU: Could it not be released by the stellate cells and taken up in the Purkinje cells?

SGARAGLI: I'm talking about the histological data.

BARBEAU: Yes, I know what the evidence is.

KURIYAMA: Dr. Sgaragli, I was puzzled by some of your data you presented yesterday. You said that autoradiographically you found many grains on cerebellar Purkinje cells. But many investigators, including myself, carefully determined the endogenous level of taurine. It is very, very low; so we may have a discrepancy between your autographic data and the endogenous levels.

RASSIN: Could I comment on this a bit. I said once before that because a compound is low in concentration it doesn't mean it cannot be a neurotransmitter. As a matter of fact, personally, that is more satisfying than the larger amounts of taurine that are present, and I think you should reemphasize whether you believe in it or not.

VAN GELDER: I found it rather striking to see these inverse correlation in the cerebellum between a GABA pool and a taurine pool.

I'm not quite sure that this is in favor of taurine being a transmitter or not. It seems to indicate somehow that there is an indirect regulation between these two pools, so if we suppose that GABA and aspartic acid are transmitter pools mostly, although that might not be true either, I don't know, then one would suppose that there is interrelated mutually exclusive regulation.

BARBEAU: In conclusion it would seem that taurine in not a classical neurotransmitter. Steve you wanted to say something on that.

BASKIN: I think some of the evidence, it's just that it still could be. I don't think it can be ruled out.

BARBEAU: Well, what is the evidence in favor?

BASKIN: Some of the evidence in favor I'm not sure has ever been published as yet, so I'm not sure I can comment on it. There are papers in press or papers that are submitted.

BARBEAU: If they are accepted you can talk!

BASKIN: I don't know if they have been accepted. But I don't think we can rule it out. That's all I'd like to say at this time.

BARBEAU: Alright. With the evidence we heard on the pineal which can in some way be considered a sensory relay along the line, that could serve as a possibility that it may modulate sensory input, but not necessarily be a neurotransmitter there.

VAN GELDER: In there the evidence is fairly strong that there is a sort of diurnal cycle going on. There seems to be a light adaption and dark adaption and I think it seems to indicate that this has a strong influence on the physiological environment, that taurine therefore may in this way mediate a response through the rest of the body. It doesn't mean very much but as a general statement it seems to me again a link between an environmental condition and the internal state.

BARBEAU: O.K. Now we have discussed in man. Is there any evidence that people are aware of in lower species, where taurine would definitely be considered a neurotransmitter. I believe you have some?

FRANCONI: We have none. However there is a paper that has demonstrated that taurine acts at 10^{-10} M in a chemosensory system. Giampietro do you remember the name of that mollusk? Oh! I remember: Panuliris argus. In this mollusk taurine acts at 10^{-10} M in a chemosensory system (14). Perhaps in this case, since taurine exerts its action at very low concentrations, it is a neurotransmitter.

VAN GELDER: What about the large amount of taurine in squid axon? That is a motor axon surely?

BARBEAU: Yes.

VAN GELDER: What is the transmitter at the motor axon in squid?

HUXTABLE: Who said there's a large amount of taurine in squid axon?

BASKIN: The fellow from Chicago.

GAULL: Hoskins. But even there the transmitter is acetylcholine.

FELLMAN: I think so, yes, it's acetylcholine.

BARBEAU: To call taurine a neurotransmitter we need a great number of criteria which taurine obviously does not meet, as we classically understand them. One of them is, do we have a specific receptor? What do you feel around the table on the validity of the studies that are being done on the taurine receptors? Is there such a thing as a specific taurine receptor?

LOPEZ-COLOME: Yes, but not postsynaptic.

BARBEAU: Where do you put it?

LOPEZ-COLOME: Possibly as a modulating a receptor which...

BARBEAU: But where physically do you put it, if it's not postsynaptic. Is it presynaptic or is it along the axon?

LOPEZ-COLOME: It is difficult to say, because when you measure these receptors in synaptosomal membranes, you have both presynaptic and postsynaptic membranes.

BARBEAU: How certain can we be that this is really specific for the taurine receptor? The evidence is the displacement of radioactive taurine binding, that's how I understand it, and that is more or less specific.

VAN GELDER: Well, there's the work of Sieghart a long time ago, who showed that in the synaptic vesicle population there seems to be a specific population binding taurine. Although when you look at the data, the density profile difference between that population and the ones which contain, I think, glycine and GABA is very, very small.

BARBEAU: When you prepare a synaptosomal preparation for your studies, you include all the vesicles, you include the membranes of these vesicles, and what evidence do you have that really taurine is not just part of that membrane or part of the tubulin system or the exocytosis system?

VAN GELDER: None.

BARBEAU: The fact is that some of that taurine in membranes may be bound very tightly, we heard that we can't seem to get rid of it, so could it be that we don't really have a postsynaptic receptor in the sense we normally understand it for a neurotransmitter, but that we have some area within these membranes that somehow uses taurine. That doesn't mean it's a receptor.

RASSIN: I just want to make it clear that, although I don't necessarily believe that taurine in the vesicles is transmitter taurine, I think both Bradford's (10) work and our work (28,29) shows that taurine is associated with vesicular structures based on the EM fractions. It's not just any old membrane, and what that taurine is doing and why it's held there is another question, but I think there is at least some of it is associated with synaptic vesicles.

BARBEAU: But to take up the Herminia hypothesis, maybe we should call it that, at least it sounds nice, could it be that these vesicles which necessarily must have tubulin and contractile elements to be able to exocytose, could it be that the taurine is tightly bound to these parts of the vesicle? We know it is in there, but we don't know what it is attached to, or if it's free.

RASSIN: Acetylcholine has been suggested to be attached to molecules within the synaptic vesicles, including protein structures, but it certainly is much more labile than taurine. You can release acetylcholine, but you can't release the taurine. I guess it's just hard to imagine how you could release that strong binding physiologically, or how it's functional.

BARBEAU: Maybe you don't release it at all. Some of it you do, but not necessarily all. No, the reason I was bringing that up is that I have not been personally convinced that the receptor studies indicate really that this is a receptor in the sense we normally see with the adrenergic or the cholinergic receptor. That it means it's just attached to some component of the whole synaptic thing that we're looking at and that's all we can say. Ryan?

HUXTABLE: I was wondering whether you think it would be appropriate at this time if I were to categorize the actions of taurine into three groups and see if we can use this as basis for trying to simplify discussion on how it's acting.

BARBEAU: I don't know what your three groups are, so you're on!

HUXTABLE: These are some things I jotted down yesterday at the end of Anna Maria's talk (Table 1). She gave me the impetus for it

TABLE 1

CLASSIFICATION OF THE ACTIONS OF TAURINE

CLASS OF ACTION	EXAMPLES
Homeostasis	
- maintenance of normal physiological state, or physiological levels of metabolites	Certain amino acids (38) Calcium/Inotropy (11,12,13) Osmolarity (26,36) Membrane potential (16) Membrane function (23)
Inhibition	
- depression of behavior	Eating (37) Drinking (21) Epilepsy - See table, this volume Psychomotor activity (4) Temperature regulation (22, 33)
- depression of physiological processes	Neuronal firing (9,17) Excitation-secretion (25)
Effects of Hormones	Prolactin secretion (30) Growth hormone (31)

Neither the list nor the references are intended to be exhaustive.

in the questioning. You may find it an incomplete list, but it seems to me that one group of actions of taurine are essentially homeostatic actions. That is, if a response or a concentration moves too far beyond or below a certain point, taurine will tend to normalize it. Nico in his work on glutamate feels that taurine is normalizing levels there, and Herminia yesterday was indicating that she felt taurine was inhibiting the release of certain amino acids in the central nervous system. There is certainly a homeostatic action on the effects of calcium in the heart and possibly the central nervous system. Taurine affects osmolarity in primitive marine creatures all the way up to the brain in mammals in that taurine levels will alter to maintain a given osmolarity. You see the same with membrane potential. This has been shown in lobsters and we showed in muscle fibers some years ago that you get a reversal potential and taurine will maintain the potassium potential. Another group of effects seems to me to be depression of behavior or physiological processes. The ones I thought of are eating, drinking, barbiturate sleeping, epilepsy, psychomotor activity and inhibition of or depression of excitation-secretion couplings, which are, of course, calcium dependent processes. You see this inhibition in many places affecting release of acetylcholine and norepinephrine from the adrenals.

GAULL: You put eating and drinking as a behavior in the same column with seizures.

HUXTABLE: Is epilepsy not a behavior?

GAULL: I wouldn't have thought so.

BARBEAU: It's an abnormal one. It's the extreme of a normal response.

GAULL: No, it's quite involuntary.

HUXTABLE: In the same way that overeating by the genetically obese mouse is an extreme of a normal response, and is quite involuntary (37).

VAN GELDER: My suggestion to that is that neuronal depression can explain all the others in the list, if it occurs in the right brain region.

HUXTABLE: O.K. and the third group would be hormonal interactions, and I could only think offhand, I wasn't thinking very hard, of effects on growth hormone.

VAN GELDER: Thyroid?

HUXTABLE: So you can just stare at that list, as you're talking about these.

BARBEAU: Where would you fit in, for example, sensory modulation, which I think is one possibility one has to consider here? Well, we can put it as number four, sensory modulation and then put an arrow when we know where to put it.

GAULL: It's in hormonal homeostasis.

BARBEAU: O.K. getting back to the neurotransmitter role, you can see it is not clearly indicated here at all. Let's go now to the effect of taurine on membranes and forget the neurotransmitter role, which I think we have taken care of to the limit of our present knowledge. What is the summary of our feeling on the role of taurine on membranes, and the possible relationship between that and zinc and fatty acid metabolism and so on?

PASANTES-MORALES: Well, I think that the main problem is considering if taurine is acting via specific receptors or if it is acting in a general nonspecific way on membranes. For the first point, until we are able to have specific antagonists of taurine actions, we can not advance more in this matter, as Ryan has pointed out. Concerning this second point, that taurine has a general action not mediated by specific receptors I see it is very difficult how this could be realized.

BARBEAU: I wish somebody would address the membrane stabilization role or a role upon phosphorylation for taurine. I don't think that necessarily implies we must have receptors, it could be a metabolic role. It could be there are other ways it could stabilize membranes than acting necessarily through a receptor. Would anybody care to look at that?

VAN GELDER: This is very speculative and probably shows a great deal of ignorance on my part, but could taurine in some way act, let's say, like a local anesthetic does? That is to say, bind to a calcium site within a membrane and stick to it perhaps with a greater affinity constant than calcium itself. In a sense like the local anesthetics do, which will stabilize the membrane. It can be displaced by calcium if you use enough calcium and it can be displaced when you stimulate but it always needs more stimulation or more calcium than in the absence of taurine.

SCHAFFER: We've done some studies with the calcium paradox heart failure model that indirectly deals with this problem. I think my comments also pertain to the cardiomyopathic hamsters. There definitely is membrane damage in these hearts and, I'd have to say, taurine prevents that damage. Now I'm not sure what the mechanism is, but in our model, it prevents the loss of sarcolemmal ATPase activity, it normalizes sarcolemmal calcium binding and it prevents CPK release from the heart (24). These results suggest that taurine prevents the membrane from becoming abnormally altered. I'm not sure of the mechanism. Perhaps we could invoke Herminia's hypothesis. It could account for what we call membrane stabilizing actions of taurine.

BASKIN: I was going to comment on Nico's observation about a local anesthetic. A number of people have looked at, tried to reproduce the work of Japanese of 1936 (35) without complete success. It has been examined as an analgesic agent. One of the models is to put mice on a hot plate and if there was analgesic, or local anesthetic effect, you might forestall the response. I know a few studies that were never published because they were negative. I think that the evidence in general is that it doesn't have any local anesthetic effect.

VAN GELDER: I would object to that because we are not dealing with an abnormal binding here, we are dealing with a normal function so these nerves have to respond.

BASKIN: You can't object to them because they are not my studies. It is just a way that people tried to see if there was any analgesic effect, any local anesthetic effect.

BARBEAU: I think an analgesic effect can be central and peripheral and all you have ruled out is the peripheral effect.

GAULL: I'd like to ask a question. We were having some very interesting discussion just before we broke for refreshments about the relationship of the structure of taurine to its possible actions. I'm wondering, I don't know much about membrane structure, I am wondering whether people like Charles Irving and Steve Schaffer could tell us how they think that the chemical configuration of the molecule could interact with a membrane and stabilize it.

SCHAFFER: My feeling is that taurine has to interact with proteins, since its actions are too specific to be due to some nonspecific interaction. Why can taurine cause some effects, but beta alanine can't? Maybe I'm prejudiced because of my previous training in this regard, but it seems that the most highly specific effects in biology are mediated through proteins: enzymes or other macromolecules. I don't see how an interaction with just phospholipids or the carbohydrate moieties can account for taurine's action. I think it has to be with specific binding sites on the membrane.

BARBEAU: Dr. Franconi, you had raised your hand. Is that on the same topic? I think Herminia wanted to continue on that same one so...

GAULL: I think we've entirely covered it.

PASANTES-MORALES: I disagree that the action of taurine is related to a specific protein or enzyme. We don't know about the other molecules present at the enormous concentrations that we know for taurine. Concentrations in the order of millimolar! I cannot imagine how it could be related to a modification of an enzyme of a protein. I'd rather think in terms of a general action on membranes, but not related to specific enzymes or proteins. Could it be possible to have the amino group of taurine interacting with a negative part of membranes and having the sulfonate free for interactions with metals and calcium for instance, is that possible?

IRVING: It's possible, but I think, if one looks at the structure of taurine there is nothing unique about the amino end of it. I think that receptors would probably interact with the sulfonic acid end. To go back to the idea of taurine interacting with macromolecules, one might think of taurine as a glue or string that holds certain arrays of macromolecules together, in a way that could result in cooperative interactions, as occurs with hemoglobin subunits.

SCHAFFER: I'll agree with that.

IRVING: And it can serve as a bridge to bring two plus's together or two subunits together. When it's released, then one initiates all kinds of cooperativity that one might want to look at if one can demonstrate cooperativity in calcium binding. How does it change in the absence and presence of taurine? I'm not sure that sulfonic acid is a very good group for interactions with metal ions. It certainly doesn't bind the alkaline earths. I don't think we have enough physical evidence now to know which kind of protein groups the sulfonic acids will have a much greater avidity for coordinating with. If we can find out that, then we might have a good clue about what it's doing. I think as far as the interactions with membranes, didn't you do some electron spin resonance measurements, Steve?

SCHAFFER: We did, yes, in our studies.

IRVING: Didn't you show that adding taurine to vesicles and so forth doesn't change mobility of fluidity? I think this argues against a global effect of taurine that somehow stabilizes head group interactions.

SCHAFFER: There's one point I'd like to make regarding our studies with verapamil (7), and Flavia's with D600 (Franconi et al., this book). Both of these substances interact very strongly with the membrane, and both exhibit some type of competition with taurine.

GAULL: Isn't it true, Steve, that wherever you find taurine you find it in tissues that have very intricate and large membrane structures, and isn't it also correlated with cholesterol which is also a membrane constituent?

BARBEAU: I'm not sure I'm reading all the biophysics, but am I understanding that the possibility could be that the interaction of taurine with a membrane would be through something else, another substance, either a phospholipid, or zinc or calcium? Is that a very small summary of the last issue?

SCHAFFER: I'd prefer to say that taurine interacts directly with the membrane, more specifically with a component of the membrane.

BARBEAU: You could take a specific component of the membrane here, *(pointing)* which then in turn would have other effects.

IRVING: What we are trying to say is that it's interacting with protein or glycoprotein or something which is embedded within the membrane, rather than just with the phospholipids themselves.

SCHAFFER: That's right.

VAN GELDER: I have one suggestion how you could test that. Could you look at denervation sensitivity and see that it doesn't occur in the presence of taurine?

BASKIN: It does.

VAN GELDER: It does?

BARBEAU: How does it affect it, Steve?

BASKIN: People looked at ganglion in the 50's and they showed that there is the denervation sensitivity. They were looking at IPSP's, the ganglia, and they showed that after denervation there's a supersensitivity or desensitivity, I think, to taurine. It changed the sensitivity to the taurine (34).

BARBEAU: That's not denervation supersensitivity, though! That's not quite the same thing.

STURMAN: I'm just trying to see how I can fit zinc binding into this, because I think from what Charles said one would guess, not knowing as much as he does, that two molecules of taurine would be involved with one molecule of zinc. Now my guess is that the binding is quite tight, because zinc is not removed by perfusion whereas we know that when zinc is present in a very ionic state, say like in the dog tapetum where it exists as zinc cysteinate, it's all washed out. In a normal perfusion, there is something different about the zinc in the membrane, that is different from those zinc ions which are relatively free. It appears that if you take either the zinc away or the taurine away from the tapetum, membrane surrounding the tapetal rods crumbles.

BASKIN: I'm not sure it is a simple salt. There's an old observation in the New York Academy of Science by Green and Day (15) who saw it "complex" with histamine. They were measuring histamine by measuring conductivity. They were looking at heparin-histamine salt formation and when they used a number of anions to form salts with histamine they obtained a conductivity of 1, but when they substituted taurine they obtained a conductivity greater than 1. They were a bit perplexed by that. What it would say is that there is not exactly a 1:1 salt formation, it was something greater. I don't think they ever took it any further, probably because it was not well understood, but I'm not sure it would be a straight salt formation; it might be something else.

STURMAN: In the membrane I don't think it can be straight-salt formation. Within the tapetal rod of the dog it surely is.

SCHAFFER: John, I think your data suggest that zinc is associated with the membrane.

STURMAN: If you take it away the membrane collapses. We know that.

BARBEAU: Do other membranes also collapse when you take away zinc?

I'm sorry, I cut you off.

SCHAFFER: That's O.K. Zinc is often associated with proteins: carboxypeptidase, carbonic anhydrase, and alcohol dehydrogenase to name a few. Perhaps zinc is bound to a protein in the membrane. In that case the binding of taurine could influence the binding of zinc. They could somehow be related.

STURMAN: I think it has to be related, because if you take the taurine away then you lose your zinc from the membrane structure.

HUXTABLE: But the membrane significance of zinc is that it protects against lipid peroxidation (8), which would be an apparent contradiction to your theories of the proteins.

SCHAFFER: Not necessarily, because it can be a lipoprotein. When you're talking about proteins embedded in membrane, you're talking usually about glycoproteins that also contain lipids, and if you take that lipid away, as with sodium potassium ATPase, you lose enzyme activity.

BARBEAU: There is something else I'd like to ask here, it may be a very long shot but I thought it would be of interest that Dr. Davignon in our institute has shown recently that in zinc deficiency you also get membrane deficiency in an essential unsaturated fatty acid, linoleic acid and arachidonic acid respectively (unpublished), and since some of these membranes, particularly the mitochondrial membrane, but others also, contain phospholipids with very high concentration of linoleic acid. Cardiolipin, for example, is a well known ionophore, and you may completely change the membrane properties just by changing the zinc availability in there. I don't know how this ties in with taurine as protecting against that loss of zinc, for example, but I think it ought to be mentioned that unsaturated fatty acids may come in somewhere.

PASANTES-MORALES: Well, you partially answered my question. Zinc binding is a more or less general property of membranes, and this interaction occurs in intracellular membranes because I think that we too forget that taurine actions are occurring inside the cell mainly, not outside the cell, so we should remember that any physiological action of taurine should occur in inner membranes, or probably at the inner side of the cell membrane, or to a lesser extent at the outside of the membrane, but only by exchange of taurine. So the question is zinc binding is a general property of membranes, and does it occur in intracellular membranes?

BARBEAU: Can anybody answer that. Is zinc binding a general property of membranes? Particularly does it bind to tubulin?

PASANTES-MORALES: If it is binding to phospholipids, it should be.

GAULL: One thing about it is that the form in which it is supplied in milks is as a ligand with a carrier protein, and it doesn't do any good just to throw in zinc chloride. You have to get availability. It must be bound to proteins if it's going to be biologically available.

BARBEAU: It has, in a way, to be internalized before it does anything.

CHESNEY: I think as long as we are being this speculative. I think it's interesting, and perhaps the next meeting, we'll all come back with better knowledge of zinc, but if zinc does have a property as preventing the lipid peroxidation of unsaturated fatty acids in the membrane, certainly selenium serves the same role. Selenium is also a metal, and I wonder about the interactions of taurine with selenium, because selenium is clearly involved in both the intracellular and the extracellular site of the membrane as being very important in stabilizing this element. I'm going to throw that out.

BARBEAU: I think I'll allow just two more questions on membranes and then we'll go on to something else.

HAYES: I would like to pursue the lipid peroxidation-zinc relationship. I think Ryan mentioned it. Do you know the mechanism for that association? Is it through an enzyme?

HUXTABLE: Maybe because it acts as scavenger, I'm not sure, but it markedly decreases malondialdehyde production from membranes. It also decreases release of β-glucuronidase from lysosomes (27).

HAYES: When it's added _in vitro_, you mean, or _in vivo_?

HUXTABLE: Both. If you have zinc deficient rats you'll see an increase in lipid oxidation. So it seems to be a normal physiological process.

HAYES: Another nutritional factor associated with lipid peroxidation, that we mentioned, is alpha-tocopherol which is itself in the membrane. It can be linked to zinc and polyunsaturated fat because it protects against lipid peroxidation in that system (6). So I think we are homing in on an area of interrelationships. It brings our focus back to the membrane.

BARBEAU: O.K. one last question, Dr. Franconi.

FRANCONI: Yes, I should like to mention there is a paper entitled "Zinc, Taurine and Epilepsy" (3).

BARBEAU: Yes, I know that paper. *(Laughter)* I remember writing it one day. That was then speculation. It remains speculative, but I think we are getting closer to this becoming a hypothesis. Can we leave membranes now? I think it's pretty obvious that, if I may make a conclusion, that membranes will be one of the major sites of focusing of our attention, including the zinc, including the taurine, including the phospholipids and including your proteins, and I think this is where there's a lot of research that should be done. Until we understand the molecular types of interaction we are going to be still speculating. So let's hope we get more information. I'd like to spend the remaining time on trying to elicit if taurine has a role in any disease state, because, as I said when I gave my own lecture, I think this is what will eventually launch the research on taurine, and grants and so on. Not that we're interested in grants at all, but I think the experience of Parkinson's disease and dopamine is one that I remember vividly. I think it has certainly been an impetus for various types of research, and I would hope that we could delineate some type of disease state where taurine may be of importance also. I'd like for us to focus on a few of these: on retinitis pigmentosa. We haven't mentioned this, and of course it is not a tapetal defect, it is not quite the same thing, but I would like for a few comments to be made on our present knowledge of taurine in retinitis pigmentosa and I would like to go a little bit further to dark adaptation in humans. Whether taurine or not is involved, I would like to know if anybody around here has done studies on taurine in retinitis pigmentosa? I know of two studies; our own and those of Campanella in Italy, in Napoli. Campanella and Filla (unpublished) have just done a study on 14 patients, and they measured visual fields and vision as measured by the standard table, and after treatment with taurine in all these 14 patients they have absolutely no objective evidence of improvement. I can confirm this; we have given taurine in 5 patients with retinitis pigmentosa. We gave 250 mg a day following Nico's advice not to give too much, but we have not modified at all the aspect of the fundi, the distribution of the pigments, or the visual field, but all of the patients, without exception, as well as all of their spouses, confirmed that they move much better in the dark than they did before. Now how does one measure that, and I know this is not an objective study, but I think it should not be pushed away and forgotten. If somebody has a way of measuring adaptation in the dark and ability to see in the dark when sudden darkness comes on or when you wake up in the night and have to go somewhere *(Laughter)*, now how do you measure that? When all of the patients and all the spouses say these people stumbled all the time before, and now they don't see any better during the daytime, their objective findings have not changed, their fields are the same, pigments are the same, but they don't stumble in the dark anymore. How does one measure this?

HAYES: I think you should be able to pick that up with an ERG, looking for differences in amplitude and perhaps in cone timing sequences. The cat was of interest to Berson (32) as a model for retinitis pigmentosa, because the cat was the only experimental model that responded with an inducible cone-timing delay characteristic of retinitis pigmentosa in man. However, at the electron microscopic level, there are differences in the lesions. Apparently these humans have a problem with phagocytosis of the photoreceptor outer segments, leading to their accumulation and blindness. In the cat model, the photoreceptors were disrupted, but phagocytosis was normal, so the primary defect was a breakdown of unstable outer segments, similar to the lesion in vitamin A deficiency (19,20). Thus, the fine points of the lesions in retinitis pigmentosa and the taurine-depleted cat are different. Berson also measured plasma taurine levels in his patients and, as I recall, found no differences attributable to taurine (5).

BARBEAU: I just have to complete because I didn't express myself clearly. Campanella's study was an open one, ours was a crossover with placebos, and there's no question that during the placebo period they came back to their previous level.

VAN GELDER: How long was the treatment with taurine?

BARBEAU: Nine months on the average, between 9-12 months. I don't try to make any claim at all on that. I think if somebody can help me on how to quantify that better, then we could do a better study.

RASSIN: What stage in the disease was this? You expect maybe it would be more appropriate to look for effects in the early stages and to prevent the progression of the disease as opposed to trying to reconstruct structures which may be long gone.

BARBEAU: I agree, and they were just cases that we happen to have in our collection, if you want, but I think your points on the difference between retinitis pigmentosa and what we actually are seeing are very important. We may not even have a good reason to do this study. Kanji?

IZUMI: I'd like to ask whether taurine deficiency can produce retinitis pigmentosa or retinal degeneration only?

BARBEAU: I think what you are asking is whether it's possible to produce retinitis pigmentosa with taurine deficiency?

IZUMI: There seems to be some confusion.

BARBEAU: I think that's what we answered essentially a few minutes ago, that what is reported in animals is different from retinitis pigmentosa. I think the original idea to do our study was based on a wrong interpretation of the pictures, in that we thought it was retinal and not just tapetal involvement and I don't know why retinitis pigmentosa came to mind immediately but it did apparently with a number of people. O.K. Let's move on, we only have a few minutes left. I think we have properly discussed the problem of taurine in epilepsy the day before yesterday. Is there anything further to add on this? It was obvious at the conclusion of the discussion, first that human studies were very difficult to do now with the Food and Drug Regulations, because the only studies they would permit are on people that are very far advanced, therefore the worst ones to choose from, and secondly, the patients you'd like to get are epileptic children, and they are the ones where drug studies will be refused unless we're much more advanced. So we have a problem there, but are there any further thoughts? I don't think that the story of taurine in epilepsy is ended by far, and I don't think it should be ended. I know there are studies going on in Japan. Would somebody from that country like to tell us about the present studies going on in epilepsy? Not in experimental but in human epilepsy.

KURIYAMA: In Japan, several groups are working on clinical trials concerning taurine and epilepsy. As yet we have no firm conclusions as to whether or not taurine is really effective on epilepsy. By the way, in Japan a couple of groups are also working on the effects of taurine on alcoholic dementia or maybe on alcohol withdrawal syndromes. Again, their results seem confusing.

BARBEAU: I'm happy to see that confusion is not only in this part of the world *(Laughter)*, but Herminia?

PASANTES-MORALES: I just want to say that probably taurine is not very useful as a treatment for human epilepsy. However, I think that people working on the protective effects of taurine on experimental models of seizures could give us an answer as to the mechanism by which taurine is acting in nervous tissue. I would like, for instance, to know if the injection of EDTA with taurine has the same action on all the wide variety of convulsions against which taurine is effective, or is it only on metrazole seizures? I think that people working on this shouldn't drop out, because it can give us many answers on the effects of taurine on nervous tissue.

BARBEAU: That's quite true. Ryan?

HUXTABLE: One thing the clinicians might do would be to check the incidence of pediatric epilepsy in babies who were breast fed as against those who were bottle fed to see if the deprivation of maternal taurine during the early periods had any subsequent effect on seizure susceptibility.

BARBEAU: Yes, that could be checked. Metracos certainly would have the population to look at if he so desired. But I'm not quite as pessimistic as you are, Herminia, saying that it probably will not do in human epilepsy. I think we just don't know and even though I've been personally involved in these studies, I think they were not done under sufficiently careful conditions. With the available knowledge at the time we had this stupid idea that everything would work like DOPA; that the more you gave the better results you got. But it doesn't work like DOPA and that's very obvious and now we know that the studies were done with doses that would not be acceptable anymore. Now maybe they ought to be started all over again, if one can find an area where we can do that with permission. I think they ought to be done, I don't think we can rule them out. The last item I'd like for us to say a few words about is the cardiomyopathy and of course, Ryan has been involved in this. I think that we should have sort of a short state of the art thing on the cardiomyopathy which was referred to many times during this meeting, but not really discussed. Cardiomyopathy is of course a cardinal sign in Friedreich's ataxia, a disease with a membrane defect, and loses in taurine. The cardiomyopathy is presence in other entities also. Ryan, would you like to say a few words? I don't think we should forget taurine and cardiomyopathy.

HUXTABLE: What we know is purely a phenomenon: that taurine will protect against calcium-overload cardiomyopathy, or cardiomyopathies in which there is a defect in the cell membrane (2). How it's doing this we have no idea. I don't know whether it has any clinical applicability because there are other drugs that are much more effective in preventing calcium overload, such as verapamil or D600. I'll answer any questions anybody has, but I don't think I should just go on talking. It's getting late.

BARBEAU: Anybody has any questions on cardiomyopathy?

VAN GELDER: Would you suggest a clinical trial to see if recuperation or...

BARBEAU: These are now being done in Friedreich's ataxia. There is a triple trial of verapamil on one side, taurine alone and taurine plus verapamil.

VAN GELDER: But what about ordinary cardiac infarctions?

BARBEAU: Infarctions? That's a different matter. We're talking about hypertrophic cardiomyopathy. Before we close, is there any other disease entity that should be mentioned? Yes, Dr. Kuriyama.

KURIYAMA: I want to bring up one point. In this session, for example, we did not have any discussion on the hepatic problems. So if we have time I want to discuss this topic a little bit.

BARBEAU: Would you say a few words?

KURIYAMA: The reason I'm saying this, in Japan, recently many clinical people have been interested in the possible therapeutic efficacy of taurine on hepatic disorders including hepatitis. I want to know whether or not similar types of clinical trials are going on in other countries.

BARBEAU: Anybody outside of Japan that has done any work on the hepatic, I'm not talking about biliary, but hepatic functions?

VAN GELDER: Well, there is some; I think Andres Hamberger (18). But I'd like to pitch into hepatic coma, but in a way that is to some extent related to glutamic acid metabolism. Just the same rationale for that type of hyperexcitability as there is for epilepsy. But the origin may be different. Dr. Furukawa, would you like to say something?

FURUKAWA: I mentioned yesterday that taurine seems to have some adaptative effect but that's experiment from animal when taurine is the most effective as a hypertension agent, so we make some teams that consist of pharmacologists, internists and so on. We wanted to observe homotaurine as an antihypertensive agent but even homotaurine affect this weak inhibition if we gave homotaurine repeatedly it developed some tolerance so we gave up such kind of trial so at least at present I suspected that taurine analogs is effective agent as an antihypertensive drug agent.

BARBEAU: I think it is exactly 10 seconds short of the time I had picked on my watch. I think it is my privilege as the Chairman of the last session to express on your behalf and on behalf of everyone who has participated here our most sincere thanks to the two organizers, Herminia and Ryan, for this magnificent Congress, for the magnificent food which will totally destroy my own diet, and for the very pleasant exchange and atmosphere we've had, so I'm sure I express your wish for all of us.

REFERENCES

1. Assumpcao, J.A., Bernardi, N., Dacke, G.C., Davidson, N., and Eichelberger, H.G., 1979, Purkinje cell uptake in vivo of (^{3}H) taurine in the rat cerebellum, J. Physiol., 298:35-36.
2. Azari, J., Brumbaugh, P., and Huxtable, R., 1980, Prophylaxis by taurine in the hearts of cardiomyopathic hamsters, J. Mol. Cell. Cardiol., 12:1353-1366.
3. Barbeau, A., and Donaldson, J., 1974, Zinc, taurine, and epilepsy, Arch. Neurol., 30(1):52-58.
4. Baskin, S.I., Hinkamp, D.L., Marqus, W.J., and Tilson, H.A., 1974, Effects of taurine on psychomotor activity in the rat, Neuropharmacology, 13(7):591-594.
5. Berson, E.L., Schmidt, S.Y., and Rabin, A.R., 1976, Plasma amino acids in hereditary retinal disease: ornithine, lysine and taurine, Brit. J. Ophthalmol., 60:142-147.
6. Bettger, W.J., Reeves, P.G., Savage, J.E., and O'Dell, B.L., 1980, Interaction of zinc and vitamin E in the chick, Proc. Soc. Exp. Biol. Med., 163:432-436.
7. Chovan, J.P. Kulakowski, E.C., Sheakowski, S. and Schaffer, S., 1980, Calcium regulation by the low-affinity taurine binding

sites of cardiac sarcolemma, Mol. Pharm., 17:295-300.
8. Chvapil, M., Peng, Y.M., Aronson, A.L., and Zukoski, C., 1974, Effect of zinc on lipid peroxidation and metal content in some tissues of rats, J. Nutr., 104:434-443.
9. Curtis, D.R., and Watkins, J.C., 1965, The pharmacology of amino acids related to gamma-aminobutyric acid, Pharmacol. Rev., 17: 347-391.
10. Debellaroche, J.S., and Bradford, H.F., 1973, Amino acids in synaptic vesicles from mammalian cerebral cortex: A reappraisal, J. Neurochem., 21:441-451.
11. Dietrich, J. and Diacono, J., 1971, Comparison between ouabain and taurine effects on isolated rat and guinea pig hearts in low-calcium medium, Life Sciences, 10;499.
12. Dolara, P., Agresti, A., Giotti, A., and Pasquini, G., 1973, Effect of taurine on calcium kinetics of guinea pig heart, Eur. J. Pharmacol., 24:352-358.
13. Fujimoto, S., 1977, Inhibitory effect of taurine on decrease in the inotropic action of ouabain at high concentrations in isolated atria, Experientia, 33:1350-1352.
14. Fuzessery, Z.M., Carr, W.E.S., and Barry, W.A., 1978, Antennular chemosensitivity in the spiny lobster, Panuliris Argus: studies of taurine-sensitive receptor, Biol. Bull., 154:226-240.
15. Green, J.P., and Day, M., 1963, Biosynthetic pathways in mastocytoma cells in culture and in vivo, Ann. N.Y. Acad. Sci., 103: 334-350.
16. Gruener, R., Markovitz, D., Huxtable, R., and Bressler, R., 1975, Excitability modulation by taurine. Transmembrane measurements of neuromuscular transmission, J. Neurol. Sci., 24:351-360.
17. Haas, H.L., and Hösli, L., 1973, The depression of brain stem neurones by taurine and its interaction with strychnine and bicuculline, Brain Res., 52:399-402.
18. Hamberger, A., Jacobson, I., Molin, S.-O, Nystrom, B., and Sandberg, M., Glutamate as a CNS neurotransmitter: the regulation of its biosynthesis by ammonium ions. A model for hepatic encephalopathy, in: "Synaptic Constituents in Health and Disease," M. Brzin, D. Sket, and H. Bachelard, eds., Pergamon Press, Ljubljana (1980) pp.494-511.
19. Hayes, K.C., 1974, Retinal degeneration in monkeys induced by deficiencies of vitamin E or A, Invest. Ophthalmol., 13:499-510.
20. Hayes, K.C., Rabin, A.R., and Berson, E.L., 1975, An ultrastructural study of nutritionally induced and reversed retinal degeneration in cats, Am. J. Pathol., 78:505-524.
21. Hruska, R.E., Thut, P.D., Huxtable, R.J., and Bressler, R., 1975, Suppression of conditioned drinking by taurine and related compounds, Pharmacol. Biochem. Behav., 3:593-599.
22. Hruska, R.E., Thut, P.D., Huxtable, R., and Bressler, R., Hypothermia produced by taurine, in: "Taurine," R. Huxtable and A. Barbeau, eds., Raven Press, New York (1976), pp.347-356.
23. Huxtable, R., and Bressler, R., 1973, Effect of taurine on a muscle intracellular membrane, Biochim. Biophys. Acta, 323:

573-583.

24. Kramer, J.H., Chovan, J.P., and Schaffer, S.W., 1981, Effect of taurine on calcium paradox and ischemic heart failure, Am. J. Physiol., 240:H238-H246.
25. Kuriyama, K., and Nakagawa, K., Role of taurine in adrenal gland: A preventive effect on stress-induced release of catecholamine from chromaffin granules, in: "Taurine," R. Huxtable and A. Barbeau, eds., Raven Press, New York (1976), pp.335-345.
26. Lange, R., 1963, The osmotic function of amino acids and taurine in the mussel, mytilus edulis, Comp. Biochem. Physiol., 10: 173-179.
27. Ludwig, J.C., and Chvapil, M., 1980, Reversible stabilization of liver lysosomes by zinc ions, J. Nutr., 110:945-953.
28. Rassin, D.K., 1972, Amino acids as putative transmitter: Failure to bind to synaptic vesicles of guinea pig cerebral cortex, J. Neurochem., 19:139-148.
29. Rassin, D.K., Sturman, J.A., and Gaull, G.E., 1977, Taurine in developing rat brain: Subcellular distribution and association with synaptic vesicles of (^{35}S)-taurine in maternal, fetal and neonatal rat brain, J. Neurochem., 28:41-50.
30. Scheibel, J., Elsasser, T., and Ondo, J.G., 1980, Stimulation of prolactin secretion by taurine, a neurally depressant amino acid, Neuroendocrinology, 30:350-354.
31. Scheibel, J., Elsasser, T., and Ondo, J.G., 1980b, Stimulation of LH and FSH secretion following intraventricular injection of cysteic acid but not taurine, Brain Res., 201:99.
32. Schmidt, S.Y., Berson, E.L., and Hayes, K.C., 1976, Retinal degeneration in the taurine-deficient cat, Trans. Am. Acad. Ophthalmol. Otolaryngol., 81:687-693.
33. Sgaragli, G., and Pavan, F., 1972, Effect of amino acid compounds injected into cerebrospinal fluid spaces on colonic temperature, arterial blood pressure and behavior of the rat, Neuropharmacology, 11:45-56.
34. Stanton, H.C., and Woodhouse, F.H., 1960, The effect of gamma-amino-N-butyric acid and some related compounds on the cardiovascular system of anesthetized dogs, J. Pharmacol. Exptl. Therap., 128:233-242.
35. Sugohara, H., Nagasawas, S., and Okaba, H., 1936, Experumtelle und klinasche untarsuchungen uber taurin, Klin. Wschr., 15: 546-551.
36. Thurston, J.H., Hauhart, R.E., and Dirgo, J.A., 1980, Taurine: A role in osmotic regulation of mammalian brain and possible clinical significance, Life Sci., 26:1561-1568.
37. Thut, P.D., Hruska, R.E., Huxtable, R., and Bressler, R., Effect of taurine on eating and drinking behavior, in: "Taurine," R. Huxtable and A. Barbeau, eds., Raven Press, New York (1976), pp.357-364.
38. Van Gelder, N.M., Rectification of abnormal glutamate levels by taurine, in: "Taurine," R. Huxtable and A. Barbeau, eds., Raven Press (1976), pp.293-302.

PARTICIPANTS

Akemichi Baba
Department of Pharmacology
Faculty of Pharmaceutical Sciences
133-1 Yamada-kami
Suita, Osaka, Japan

A. Barbeau
Department of Neurobiology
Clinical Research Institute of Montreal
110 Pine Ave West
Montreal, Quebec
H2W 1R7

S. I. Baskin
Departments of Pharmacology and Medicine
The Medical College of Pennsylvania
3300 Henry Avenue
Philadelphia, Pennsylvania

Russell W. Chesney
Pediatric Renal Disease Laboratory
Department of Pediatrics
University of Wisconsin Center for the Health Sciences
600 North Highland Avenue
Madison, Wisconsin 53792

M. F. Crass, III
Department of Physiology
School of Medicine
Texas Tech University
Health Sciences Center
Lubbock, Texas 79430

J. H. Fellman
Department of Biochemistry
School of Medicine
University of Oregon Health Sciences Center
Portland, Oregon 97201

F. Franconi
Istituto Interfacolta' di Farmacologia e Tossicologia
dell' Universita' di Firenze
Viale Morgagni 65
Firenze, Italy

T. Furukawa
Department of Pharmacology
Fukuoka University
Fukuoka 814, Japan

Gerald E. Gaull
Department of Human Development and Nutrition
New York State Institute for Basic Research in Mental Retardation
Staten Island, New York
Department of Pediatrics
Mount Sinai School of Medicine of the City University of New York
New York, New York

Bernard Haber
Neurochemistry Section
Marine Biomedical Institute
University of Texas Medical Branch
Galveston, Texas 77550

K.C. Hayes
Department of Nutrition
Harvard School of Public Health and Nutrition Division
Harvard Medical School
New England Regional Primate Research Center
Southborough, Massachusetts

Ryan J. Huxtable
Department of Pharmacology
Health Sciences Center
University of Arizona
Tucson, Arizona 85724

Charles S. Irving
Children's Nutrition Research Center
Department of Pediatrics
Baylor College of Medicine
Houston, Texas 77030

Heitaroh Iwata
Department of Pharmacology
Faculty of Pharmaceutical Sciences
133-1 Yamada-kami
Suita, Osaka, Japan

Kanji Izumi
Department of Neurology
Institute of Clinical Medicine
University of Tsukuba
Ibaraki 305, Japan

D. C. Klein
Section on Neuroendocrinology
Laboratory of Developmental Neurobiology
National Institute of Child Health and Human Development
National Institutes of Health
Bethesda, Maryland 20205

Pirjo Kontro
Department of Biomedical Sciences
University of Tampere
Box 607
SF-33101, Tampere 10
Finland

Kinya Kuriyama
Department of Pharmacology
Kyoto Prefectural University of Medicine
Kawaramachi-Hirokoji
Kamikyo-ku, Kyoto 602
Japan

J. M. Lipton
Departments of Physiology and Neurology
University of Texas Health Science Center at Dallas
Dallas, Texas 75235

J. Barry Lombardini
Departments of Internal Medicine and Pharmacology and Therapeutics
Texas Tech University
Health Sciences Center
Lubbock, Texas 79430

A. M. López-Colomé
Departamento de Neurociencias
Centro de Investigaciones en Fisiologia Celular
Universidad Nacional Autónoma de México
Apartado Postal 70-600
México 20, D.F., México

P. Mandel
Centre National de la Recherche Scientifique
Centre de Neurochimie
11, Rue Humman
67085 Strasbourg, Cedex
France

S. S. Oja
Department of Biomedical Sciences
University of Tampere
Box 607
SF-33101, Tampere 10
Finland

H. Pasantes-Morales
Departamento de Neurociencias
Centro de Investigaciones
en Fisiología Celular
Universidad Nacional Autónoma de México
Apartado Postal 70-600
México 20, D.F. México

David K. Rassin
Division of Developmental Nutrition and Metabolism
Department of Pediatrics
University of Texas Medical Branch
Galveston, Texas 77550

S. W. Schaffer
University of South Alabama
Department of Pharmacology
Mobile, Alabama 36688

Tomio Segawa
Department of Pharmacology
Institute of Pharmaceutical Sciences
Hiroshima University
School of Medicine
Kasumi 1-2-3
Hiroshima 734, Japan

Nico M. Van Gelder
Centre de Recherche en Sciences Neurologiques
Départment de Physiologie
Faculté de Médecine
Université de Montréal
C.P. 6128, Succursale A
Montréal, Québec
H3C 3J7 (Canada)

Giampietro Sgaragli
Istituto Interfacolta' di Farmacologia e Tossicologia
Della Universita' di Firenze
Viale G.B. Morgagni 65
50134 Firenze, Italy

J. A. Sturman
Department of Pathological Neurobiology
Institute for Basic Research in Mental Retardation
Staten Island, New York 10314

INDEX

GPSR Compliance
The European Union's (EU) General Product Safety Regulation (GPSR) is a set of rules that requires consumer products to be safe and our obligations to ensure this.

If you have any concerns about our products, you can contact us on

ProductSafety@springernature.com

In case Publisher is established outside the EU, the EU authorized representative is:

Springer Nature Customer Service Center GmbH
Europaplatz 3
69115 Heidelberg, Germany

www.ingramcontent.com/pod-product-compliance
Lightning Source LLC
LaVergne TN
LVHW082015200926
848481LV00010B/2866
* 9 7 8 1 4 7 5 7 0 4 0 4 4 *